Engineering
Economy

Engineering
Economy

Third Edition

Leland T. Blank, P. E.
Department of Industrial Engineering
Assistant Dean of Engineering
Texas A & M University

Anthony J. Tarquin, P. E.
Department of Civil Engineering
Assistant Dean of Engineering
The University of Texas at El Paso

McGraw-Hill, Inc.
New York St. Louis San Francisco Auckland Bogotá
Caracas Lisbon London Madrid Mexico City Milan
Montreal New Delhi San Juan Singapore
Sydney Tokyo Toronto

ENGINEERING ECONOMY

Copyright © 1989, 1983, 1976 by McGraw-Hill, Inc. All rights reserved. Printed in the United States of America. Except as permitted under the United States Copyright Act of 1976, no part of this publication may be reproduced or distributed in any form or by any means, or stored in a data base or retrieval system, without the prior written permission of the publisher.

90 AGMAGM 998765

ISBN 0-07-062982-X

This book was set in Times Roman by Syntax International.
The editors were Anne T. Brown and Scott Amerman;
the designer was Merrill Haber;
the production supervisor was Diane Renda.
Arcata Graphics/Martinsburg was printer and binder

Library of Congress Cataloging-in-Publication Data

Blank, Leland T.
    Engineering economy/Leland T. Blank, Anthony J.
    Tarquin. — 3rd ed.
        p.   cm.
    Bibliography: p.
    Includes index.
    ISBN 0-07-062982-X
    1. Engineering economy.   I. Tarquin, Anthony J.
    II. Title.
TA177.4.B58 1989
658. 1'52—dc19   88-9051

This book is printed on acid-free paper.

Contents

Preface   xiii

Introduction   xvii

Level One   **1**

1   Terminology and Cash-Flow Diagrams   **3**
   1.1   Basic Terminology   4
   1.2   Interest Calculations   5
   1.3   Equivalence   7
   1.4   Simple and Compound Interest   9
   1.5   Symbols and Their Meaning   12
   1.6   Cash-Flow Diagrams   14

2   Factors and Their Use   **24**
   2.1   Derivation of Single-Payment Formulas   25
   2.2   Derivation of the Uniform-Series Present-Worth Factor and the Capital-Recovery Factor   27
   2.3   Derivation of the Uniform-Series Compound-Amount Factor and the Sinking-Fund Factor   28
   2.4   Standard Factor Notation and Use of Interest Tables   29
   2.5   Definition and Derivation of Gradient Formulas   30
   2.6   Derivation of Present Worth of Escalating-Series Equation   36
   2.7   Interpolation in Interest Tables   37
   2.8   Present-Worth, Future-Worth, and Equivalent Uniform Annual-Series Calculations   39
   2.9   Present Worth and Equivalent Uniform Annual Series of Conventional Gradients   42

2.10   Calculations Involving an Escalating Series   45
2.11   Calculation of Unknown Interest Rates   46
2.12   Calculation of Unknown Years   48

3   Nominal and Effective Interest Rates and Continuous Compounding   59
3.1   Nominal and Effective Rates   60
3.2   Effective Interest-Rate Formulation   62
3.3   Calculation of Effective Interest Rates   63
3.4   Effective Interest Rates for Continuous Compounding   66
3.5   Calculations for Payment Periods Equal to or Longer than Compounding Periods   67
3.6   Calculations for Payment Periods Shorter than Compounding Periods   71

4   Use of Multiple Factors   80
4.1   Location of Present Worth and Future Worth   81
4.2   Calculations for a Uniform Series That Begins after Period 1   83
4.3   Calculations Involving Uniform-Series and Randomly Distributed Amounts   85
4.4   Equivalent Uniform Worth of Both Series and Single Payment Amounts   87
4.5   Present Worth and Equivalent Uniform Series of Shifted Gradients   88
4.6   Decreasing Gradients   94

Level Two   113

5   Present-Worth and Capitalized-Cost Evaluation   115
5.1   Present-Worth Comparison of Equal-Lived Alternatives   116
5.2   Present-Worth Comparison of Different-Lived Alternatives   117
5.3   Capitalized-Cost Calculations   120
5.4   Capitalized-Cost Comparison of Two Alternatives   122

6   Equivalent-Uniform-Annual-Worth Evaluation   134
6.1   Study Period for Alternatives Having Different Lives   135
6.2   Salvage Sinking-Fund Method   136
6.3   Salvage Present-Worth Method   137
6.4   Capital-Recovery-Plus-Interest Method   138
6.5   Comparing Alternatives by EUAW   139
6.6   EUAW of a Perpetual Investment   141

7   Rate-of-Return Computations for a Single Project   153
7.1   Overview of Rate-of-Return Computation   154
7.2   Rate-of-Return Calculations by the Present-Worth Method   156
7.3   Rate-of-Return Calculations by the Equivalent-Uniform-Annual-Worth Method   159

7.4   Multiple Rate-of-Return Values   160
7.5   Internal and Composite Rates of Return   162

8   Rate-of-Return Evaluation for Multiple Alternatives   172
8.1   The Need for Incremental Analysis   173
8.2   Tabulation of Net Cash Flow   174
8.3   Interpretation of Rate of Return on Extra Investment   176
8.4   Incremental-Rate-of-Return Evaluation Using a Present-Worth
      Equation   177
8.5   Incremental-Rate-of-Return Evaluation Using an EUAW
      Equation   181
8.6   Selection from Mutually Exclusive Alternatives Using Rate-of-Return
      Analysis   182

Level Three   195

9   Benefit/Cost Ratio Evaluation   197
9.1   Classification of Benefits, Costs, and Disbenefits   198
9.2   Benefits, Disbenefits, and Cost Calculations of a Single Project   198
9.3   Alternative Comparison by Benefit/Cost Analysis   202
9.4   Benefit/Cost Analysis for Multiple Alternatives   203
9.5   Selection from Mutually Exclusive Alternatives Using Incremental
      Benefit/Cost Ratio Analysis   204

10   Replacement Analysis   212
10.1   The Defender and Challenger Concepts in Replacement
       Analysis   214
10.2   Replacement Analysis Using a Specified Planning Horizon   215
10.3   Conventional and Cash-Flow Approach to Replacement
       Analysis   218
10.4   Replacement Analysis for One-Additional-Year Retention   220
10.5   Minimum-Cost Life Analysis   222

11   Bonds   233
11.1   Bond Classifications   234
11.2   Bond Terminology and Interest   235
11.3   Bond Present-Worth Calculations   236
11.4   Rate of Return on Bond Investment   239

12   Inflation and Cost Estimation   246
12.1   Present-Worth Calculations with Inflation Considered   247
12.2   Future-Worth Calculations with Inflation Considered   251
12.3   Capital-Recovery and Sinking-Fund Calculations with Inflation
       Considered   254
12.4   Cost Indexes   255
12.5   Cost Estimating   258

Level Four   267

13   Capital Recovery and Depletion Models   269
  13.1   Terminology   270
  13.2   Straight-Line (SL) Depreciation   271
  13.3   Declining-Balance (DB) and Double-Declining-Balance (DDB) Depreciation   274
  13.4   Types of Property and Their Recovery Periods   276
  13.5   Switching between Depreciation Models   277
  13.6   Modified Accelerated Cost Recovery System (MACRS)   280
  13.7   Sum-of-Year-Digits (SYD) Depreciation   283
  13.8   Capital-Expensing Alternative and Investment Tax Credit   285
  13.9   Depletion Methods   286

14   Basics of Taxation for Corporations   291
  14.1   Definitions Useful in Tax Computations   293
  14.2   Basic Tax Formulas and Computations   294
  14.3   Tax Treatment of Capital Gains and Losses   296
  14.4   Tax Laws on Operating Losses   298
  14.5   Definition of Tax Cash-Flow Terms   298
  14.6   Effect of Depreciation Models on Taxes   299
  14.7   Tax Effects of Different Recovery Periods   301
  14.8   Before-Tax and After-Tax Rates of Return   302

15   After-Tax Economic Analysis   308
  15.1   Effect of Income Taxes on Cash Flow   309
  15.2   PW and EUAW Computations on Cash Flow after Taxes   312
  15.3   Rate-of-Return Computations for Cash Flow after Taxes   313
  15.4   After-Tax Replacement Analysis   316
  15.5   After-Tax Evaluation Using Revenue Requirements   318

Level Five   327

16   Determination of Breakeven Values   329
  16.1   Breakeven Value of a Variable   330
  16.2   Computation of Breakeven Points between Two or More Alternatives   335
  16.3   Determination and Use of Payback Period   338
  16.4   Life-Cycle Costing (LCC)   341

17   Capital Rationing under Budget Constraints   350
  17.1   The Capital-Budgeting Problem   351
  17.2   Capital Budgeting Using Present-Worth Analysis   351
  17.3   Capital Budgeting Using Mathematical Programming   356

18   Establishing the Minimum Attractive Rate of Return   360
  18.1   Types of Financing and the Cost of Capital   361
  18.2   Variations in MARR   362

18.3   Weighted Average Cost of Capital and the Debt-Equity Mix   363
18.4   Cost of Debt Capital   365
18.5   Cost of Equity Capital   366
18.6   Setting the MARR Relative to Cost of Capital   369
18.7   Effect of Debt-Equity Mix on Investment Risk   369

19   Sensitivity Analysis and Decision Trees   378
19.1   The Approach of Sensitivity Analysis   379
19.2   Determination of the Sensitivity of Estimates   379
19.3   Sensitivity Analysis Using Estimates of Pertinent Factors   382
19.4   Economic Uncertainty and the Expected Value   384
19.5   Expected Value of Alternatives   385
19.6   Alternative Evaluation Using Decision Trees   386

20   Decision Making for Large Capital Investments   405
20.1   Role of Economic Evaluation in Capital-Investment Studies   406
20.2   Distinguishing Elements in Investment Evaluation Studies   408
20.3   Cost and Benefit Components in Strategic Investment Studies   410
20.4   Multiple-Criteria Evaluation of Investment Alternatives   413
20.5   Use of Sensitivity Analysis in Large Investment Decisions   419

Appendixes   427

A Interest Factors for Discrete Compounding, Discrete Cash Flow   429

B The Modified Accelerated Cost Recovery System (MACRS) for Capital Recovery   469
B.1 Accelerated Cost Recovery System (ACRS)   470
B.2 Modified ACRS Depreciation Method   471
B.3 Derivation of the MACRS Depreciation Rates   474

C Basics of Accounting and Cost Allocation   477
C.1   The Balance Sheet   478
C.2   The Income Statement and the Cost-of-Goods-Sold Statement   479
C.3   Business Ratios   480
C.4   Allocation of Factory Expense   483
C.5   Factory-Expense Computation and Variance   485

D Sensitivity Analysis for Business Planning   492

E Computer Applications   504

F Answers to Selected Problems   507

Bibliography   521

Index   523

Preface

The first two editions of this text presented, in a clearly written fashion, the basic principles of economic analysis for application in the decision-making process. Our objective was to present the material in the clearest, most concise method possible without sacrificing coverage or true understanding on the part of the reader. In this third edition, we have attempted to do the same thing while retaining the basic structure of the text developed in the previous editions.

Material and organizational changes

In addition to the rewording and sentence restructing that always take place in new editions of textbooks, material has been added and other material updated to render this edition an even more valuable resource than the previous ones. Specifically, a new chapter has been added (Chapter 20) to provide a wide ranging overview of the decision-making process, including economic as well as less-quantifiable factors, especially as applicable to large investments. A new section on decision trees has been added to Chapter 19 to enhance the treatment of risk in economic analysis. Chapter 16 now includes expanded breakeven analysis sections. New sections on cost indexes and cost estimating, useful for estimating the cost of large-scale engineering design projects, have been added to Chapter 12. Finally, the latest tax law information has been incorporated into Chapters 13 and 14, with detailed information provided in Appendix B. In making any changes, the overriding consideration has been the preservation of the free flowing, easy to understand format which characterized editions one and two. We are confident that this text represents an up-to-date, well-balanced presentation on economic analysis with coverage that is particularly relevant to engineers and other decision makers.

Use of text

This text has been prepared in any easy-to-read fashion for use in teaching and as a reference book for the basic computations used in an engineering economic analysis.

It is best suited for use in a one-semester or a one-quarter undergraduate course in engineering economic analysis, project analysis, or engineering cost analysis. The students should have at least a junior standing. A background in calculus is not necessary to understand the material contained herein, but a basic understanding of economics and accounting (especially from the cost viewpoint) is helpful. However, the building block approach used in the text's design allows a practitioner unacquainted with economics and engineering principles to use the text to learn, understand, and correctly applying the techniques in the process of decision making.

Computer programs for microprocessors which assist in economic analysis are discussed briefly and are detailed in the Instructor's Guide. The solution to all odd-numbered problems at the end of each chapter are included in an appendix.

Composition of text

Each chapter contains an overall objective and several specific behavioral objectives followed by the study material with section headings that correspond to the specific objectives. Section 5.1, for example, contains the material which discusses objective 5.1. All sections contain one or more illustrative solved examples which are separated from the textual material and include comments about the solution and pertinent relations to other topics in the book. Many sections have reference to additional examples at the end of the chapter and unsolved problems which the reader should now be able to understand and solve. The final answers to the odd-numbered problems are contained in Appendix F. This approach allows the opportunity to apply material on a section-by-section basis or wait until a chapter is completed.

Text overview

The text is composed of 20 chapters in five levels as shown on the flowchart on the facing page. Coverage of the material should approximate the flow in this chart to ensure understanding of the material. The material in Level One emphasizes basic computational skills. Level Two discusses the three most widely used techniques to evaluate alternatives and Level Three extends these analysis techniques to other methods commonly used by the engineering economist. Level Four presents the important topics of depreciation (capital recovery) and corporation taxation while Level Five presents supplementary and advanced analysis procedures, especially risk and sensitivity analysis. Much of this material may be considered optional if a shortened version of the text is of interest. Finally, the appendices contain the interest tables and supplementary information. The computer programs described in the Instructor's Guide are constantly being revised and supplemented so that even between new editions of this book, improvements and additions to this software will be available to those who prefer computerized analyses.

Leland T. Blank
Anthony J. Tarquin

Composition by Level

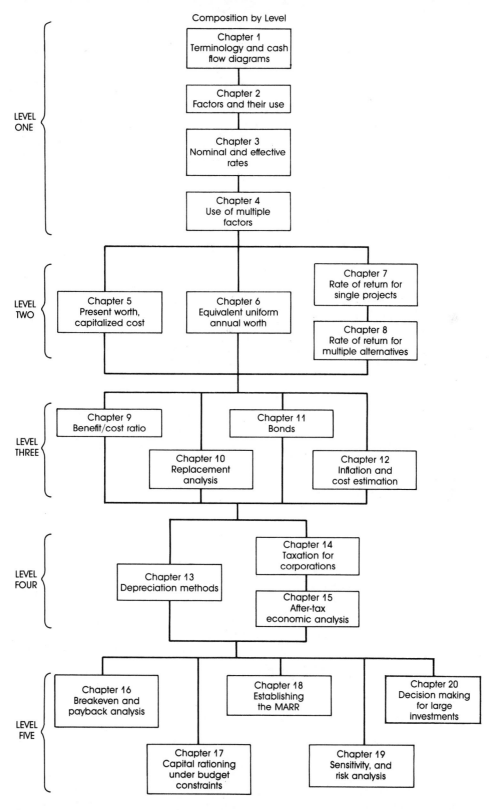

LEVEL ONE

Chapter 1
Terminology and cash flow diagrams

Chapter 2
Factors and their use

Chapter 3
Nominal and effective rates

Chapter 4
Use of multiple factors

LEVEL TWO

Chapter 5
Present worth, capitalized cost

Chapter 6
Equivalent uniform annual worth

Chapter 7
Rate of return for single projects

Chapter 8
Rate of return for multiple alternatives

LEVEL THREE

Chapter 9
Benefit/cost ratio

Chapter 10
Replacement analysis

Chapter 11
Bonds

Chapter 12
Inflation and cost estimation

LEVEL FOUR

Chapter 13
Depreciation methods

Chapter 14
Taxation for corporations

Chapter 15
After-tax economic analysis

LEVEL FIVE

Chapter 16
Breakeven and payback analysis

Chapter 17
Capital rationing under budget constraints

Chapter 18
Establishing the MARR

Chapter 19
Sensitivity, and risk analysis

Chapter 20
Decision making for large investments

Introduction

In most cases when a specific task is to be performed there are alternative ways to accomplish the task. In a business setting and in personal life much of the information for each alternative can be expressed quantitatively in terms of dollar incomes and disbursements. When engineering alternatives involve capital investments for equipment, materials, and labor, the techniques of economic analysis may be used to assist in determining which is the best alternative. Usually the monetary values are *future estimates* that would result if each alternative were selected for implementation. These estimates are based on facts, experience, judgment, and comparison with similar projects.

In many cases an engineer, rather than an economist, accountant, financier, banker, or tax expert, performs the analysis, because the technical details are already known by the engineer. It is usually easier for the engineer to learn and perform the analysis procedures than it is for people in other fields to learn the technical details. Therefore, economic analysis and engineering economy are extremely important for the engineer in his or her professional, as well as in personal life, for evaluating alternatives with respect to investments, automobile purchases, and the like.

Some typical industrially based questions addressed by the techniques covered in this text may be stated in general terms as follows:

Given a buy or lease plan for a computer, which one should be selected?

Should a particular processing line be fully automated or should automation be developed at each station?

How long should a presently owned asset remain in service before it pays for itself and returns 20% per year on the investment?

Which of several separate investment projects should be selected given that only a fixed amount of investment capital is available?

What rate of return can be realized on an investment in a particular project if the yearly revenues and disbursements are reliably estimated?

Can tax dollars be saved for additional investment if a different method of depreciation (capital recovery) or recovery period is considered?

In short, engineering economy will allow you to take into account the fact that money makes money. The engineering design to accomplish a specific task may be the best possible, but if it is not economically competitive, the design should not be implemented.

The life cycle of any project, product, or service starts with the analysis of needs; it continues with the requirements and specifications; it goes through design and implementation; and it finishes with support and maintainence throughout the use phase. Usage usually gives rise to the start of the life cycle for a new project, product, or service as innovation takes place. It is important to introduce economic analysis early in the life cycle as an integral component of the decision-making process. Economic analysis should be used in the needs-analysis phase to assist in scoping the project. Design decisions should include economic analysis to ensure that the product can be manufactured—economically and with high quality. Detailed implementation and integration plans must be economically evaluated.

The figure on page xix shows how economic analysis fits into the life cycle of a project. The application of economic analysis techniques requires a unique approach at different phases of the life cycle; however, the integration of economic-based decision making into the development of a project helps accomplish a smooth transition from one life-cycle step to the next. This blending of economic analysis techniques with other decision criteria assist in answering early in the design process the two important questions: "Is it cost effective?" and "Does it appear to be a profitable venture?"

Some of the features you will understand and learn to include in your own economic analyses are:

Time value of money
Return on investment
Breakeven
Bonds
Cash flow
Effective interest rate
Minimum cost life
Inflation
Cost estimation
Depreciation
Income taxes
Cost of capital
Replacement analysis
Sensitivity analysis and risk
Decision analysis

It is important to remember throughout your work in engineering economic analysis that the numerical data you use in the evaluations are only estimates of what is expected to take place. Therefore, the more accurate the estimates at the time the analysis is made, the better the decision when the alternative is chosen.

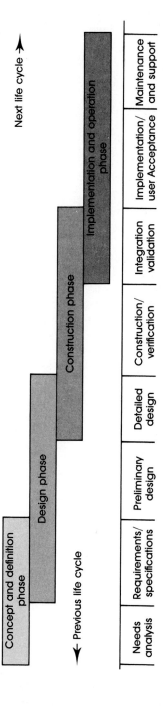

Engineering Economy

LEVEL ONE

In the first four chapters you will learn how to correctly account for the *time value of money* and how to construct and use a cash-flow diagram. The concept that money has different values over different time intervals is called *equivalence*. The movement of incomes and disbursements through time (with equivalence) will require the understanding and use of economy factors which greatly simplify otherwise complicated computations.

There are different ways of stating and using interest rates in economic computations. The explanation of nominal and effective rates is included here so you can correctly use the economy factors tabulated in the appendix section of this text. The effective and nominal rates, and the factors are directly applicable to individual, as well as industrial and government, economic and investment evaluations.

Chapter 1 Terminology and cash-flow diagrams
Chapter 2 Factors and their use
Chapter 3 Nominal and effective interest rates
Chapter 4 Use of multiple factors

Terminology and Cash-flow Diagrams

1

The purpose of this chapter is to provide you with the basic terminology of engineering economy and the fundamental concepts that form the basis for economic analysis. This chapter will also teach you the meaning of the symbols used in engineering economy and how to construct a cash-flow diagram. The material you learn here will be used throughout the remainder of this book. In particular, you will find the cash-flow diagram exceptionally useful in simplifying complicated descriptive problems.

Section Objectives

After completing this chapter, you should be able to do the following:

1.1 Define the terms *engineering economy, alternative, evaluation criterion, intangible factors, time value of money, interest,* and *principal.*

1.2 Define *interest rate* and *interest period* and calculate the interest that has been accrued in one interest period, given the principal and the interest rate or total accrued amount.

1.3 Define *equivalence*; calculate the future amount of money equivalent to a present sum, given the present sum, the future or past date of equivalence, and the interest rate; calculate the interest rate per year at which different sums separated by 1 year are equivalent; and state how different loan-repayment schemes can be equivalent for a given principal, interest rate, and time period.

1.4 Define *simple interest* and *compound interest* and calculate the total amount of money accrued after one or more years using simple- and compound-interest methods, given the annual interest rate and the original principal.

3

1.5 Define and recognize in a problem statement the economy symbols *P*, *F*, *A*, *n*, and *i*.

1.6 Define *cash flow*, state what is meant by *end-of-period convention*, and construct a cash-flow diagram, given a statement describing the amount and times of the cash flows.

Study Guide

1.1 Basic Terminology

Before we begin to develop the terminology and fundamental concepts upon which engineering economy is based, it would be appropriate to define what is meant by *engineering economy*. In the simplest terms, engineering economy is a collection of mathematical techniques which simplify economic comparisons. With these techniques, a rational, meaningful approach to evaluating the economic aspects of different methods of accomplishing a given objective can be developed. Engineering economy is, therefore, a decision assistance tool by which one method will be chosen as the most economical one.

In order for you to be able to apply the techniques, however, it is necessary for you to understand the basic terminology and fundamental concepts that form the foundation for engineering-economy studies. Some of these terms and concepts are described below.

An *alternative* is a stand-alone solution for a give situation. We are faced with alternatives in virtually everything we do, from selecting the method of transportation we use to get to work every day to deciding between buying a house or renting one. Similarly, in engineering practice, there are always several ways of accomplishing a given task, and it is necessary to be able to compare them in a rational manner so that the most economical alternative can be selected. The alternatives in engineering considerations usually involve such items as purchase cost (first cost), the anticipated life of the asset, the yearly costs of maintaining the asset (annual maintenance and operating cost), the anticipated resale value (salvage value), and the interest rate (rate of return). After the facts and all the relevant estimates have been collected, an engineering-economy analysis can be conducted to determine which is best from an economic point of view. However, it should be pointed out that the procedures developed in this book will enable you to make accurate economic decisions *only about those alternatives which have been recognized as alternatives*; these procedures will not help you identify what the alternatives are. That is, if alternatives *A*, *B*, *C*, *D*, and *E* have been identified as the only possible methods to solve a particular problem when method *F*, which was never recognized as an alternative, is really the most attractive method, the wrong decision is certain to be made because alternative *F* could never be chosen, no matter what analytical techniques are used. Thus, the importance of alternative identification in the decision-making process cannot be overemphasized, because it is only when this aspect of the process has been thoroughly completed that the analysis techniques presented in this book can be of greatest value.

In order to be able to compare different methods for accomplishing a given objective, it is necessary to have an *evaluation criterion* that can be used as a basis for judging the alternatives. That is, the evaluation criterion is that which is used to answer the question "How will I know which one is best?" Whether we are aware of it or not, this question is asked of us many times each day. For example, when we drive to work, we subconsciously think that we are taking the "best" route. But how did we define *best*? Was the best route the safest, shortest, fastest, cheapest, most scenic, or what? Obviously, depending upon which criterion is used to identify the best, a different route might be selected each time! (Many arguments could have been avoided if the decision makers had simply stated the criteria they were using in determining the best). In economic analysis, *dollars* are generally used as the basis for comparison. Thus, when there are several ways of accomplishing a given objective, the method that has the lowest overall cost is *usually* selected. However, in most cases the alternatives involve *intangible factors*, such as the effect of a process change on employee morale, which cannot readily be expressed in terms of dollars. When the alternatives available have approximately the same equivalent cost, the nonquantifiable, or intangible, factors may be used as the basis for selecting the best alternative.

For items of an alternative which can be quantified in terms of dollars, it is important to recognize the concept of the time value of money. It is often said that money makes money. The statement is indeed true, for if we elect to invest money today (for example, in a bank or savings and loan association), by tomorrow we will have accumulated more money than we had originally invested. This change in the amount of money over a given time period is called the *time value of money*; it is the most important concept in engineering economy. You should also realize that if a person or company finds it necessary to borrow money today, by tomorrow more money than the original loan will be owed. This fact is also explained by the time value of money.

The manifestation of the time value of money is termed *interest*, which is a measure of the increase between the original sum borrowed or invested and the final amount owed or accrued. Thus, if you invested money at some time in the past, the interest would be

$$\text{Interest} = \text{total amount accumulated} - \text{original investment} \qquad (1.1)$$

On the other hand, if you borrowed money at some time in the past, the interest would be

$$\text{Interest} = \text{present amount owed} - \text{original loan} \qquad (1.2)$$

In either case, there is an increase in the amount of money that was originally invested or borrowed, and the increase over the original amount is the interest. The original investment or loan is referred to as *principal*.

Probs. 1.1 to 1.4

1.2 Interest Calculations

When interest is expressed as a percentage of the original amount per unit time, the result is an *interest rate*. This rate is calculated as follows:

$$\text{Percent interest rate} = \frac{\text{interest accrued per unit time}}{\text{original amount}} \times 100\% \qquad (1.3)$$

By far the most common time period used for expressing interest rates is 1 year. However, since interest rates are often expressed over periods of time shorter than 1 year (i.e., 1% per month), the time unit used in expressing an interest rate must also be identified and is termed an *interest period*. The following two examples illustrate the computation of interest rate.

Example 1.1

The Get-Rich-Quick (GRQ) Company invested $100,000 on May 1 and withdrew a total of $106,000 exactly one year later. Compute (*a*) the interest gained from the original investment and (*b*) the interest rate from the investment.

Solution
(*a*) Using Eq. (1.1),

   Interest = 106,000 − 100,000 = $6000

(*b*) Equation (1.3) is used to obtain

$$\text{Percent interest rate} = \frac{6000 \text{ per year}}{100,000} \times 100\% = 6\% \text{ per year}$$

Comment   For borrowed money, computations are similar to those shown above except that interest is computed by Eq. (1.2). For example, if GRQ borrowed $100,000 now and repaid $110,000 in 1 year, using Eq. (1.2) we find that interest is $10,000, and the interest rate from Eq. (1.3) is 10% per year.

Example 1.2

Joe Bilder plans to borrow $20,000 for 1 year at 15% interest. Compute (*a*) the interest and (*b*) the total amount due after 1 year.

Solution
(*a*) Equation (1.3) may be solved for the interest accrued to obtain

   Interest = 20,000(0.15) = $3000

(*b*) Total amount due is the sum of principal and interest or

   Total due = 20,000 + 3000 = $23,000

Comment   Note that in part (*b*) above, the total amount due may also be computed as

   Total due = principal(1 + interest rate) = 20,000(1.15) = $23,000

In each example the interest period was 1 year and the interest was calculated at the end of one period. When more than one yearly interest period is involved (for example, if we had wanted to know the amount of interest Joe Bilder would owe on the above loan after 3 years), it becomes necessary to determine whether the interest is payable on a *simple* or *compound* basis. The concepts of simple and compound interest are discussed in Sec. 1.4.

Additional Examples 1.12 and 1.13
Probs. 1.5 to 1.7

1.3 Equivalence

The time value of money and interest rate utilized together generate the concept of *equivalence*, which means that different sums of money at different times can be equal in economic value. For example, if the interest rate is 12% per year, $100 today (i.e., at present) would be equivalent to $112 one year from today, since

$$\text{Amount accrued} = 100 + 100(0.12) = 100(1 + 0.12) = 100(1.12)$$
$$= \$112$$

Thus, if someone offered you a gift of $100 today or $112 one year from today, it would make no difference which offer you accepted, since in either case you would have $112 one year from today. The two sums of money are therefore equivalent to each other when the interest rate is 12% per year. At either a higher or a lower interest rate, however, $100 today is not equivalent to $112 one year from today. In addition to considering future equivalence, one can apply the same concepts for determining equivalence in previous years. Thus, $100 now would be equivalent to $100/1.12 = \$89.29$ one year ago if the interest rate is 12% per year. From these examples, it should be clear that $89.29 last year, $100 now, and $112 one year from now are equivalent when the interest rate is 12% per year. The fact that these sums are equivalent can be established by computing the interest rate as follows:

$$\frac{112}{100} = 1.12, \text{ or } 12\% \text{ per year}$$

and

$$\frac{100}{89.29} = 1.12, \text{ or } 12\% \text{ per year}$$

The concept of equivalence can be further illustrated by considering different loan-repayment schemes. Each scheme represents repayment of a $5000 loan in 5 years at 15%-per-year interest. Table 1.1 presents the details for the four repayment methods described below. (The methods for determining the amount of the payments are presented in Chaps. 2 and 3.)

- *Plan 1* No interest or principal is recovered until the fifth year. Interest accumulates each year on the total of principal and all accumulated interest.
- *Plan 2* The accrued interest is paid each year and the principal is recovered at the end of 5 years.
- *Plan 3* The accrued interest and 20% of the principal, that is, $1000, is paid each year. Since the remaining loan balance decreases each year, the accrued interest decreases each year.

Table 1.1 Different repayment schedules of $5,000 at 15% for 5 years

(1) End of year	(2) = 0.15(5) Interest for year	(3) = (2) + (5) Total owed at end of year	(4) Payment per plan	(5) = (3) − (4) Balance after payment
Plan 1				
0				$5,000.00
1	$ 750.00	5,750.00	$ 0	5,750.00
2	862.50	6,612.50	0	6,612.50
3	991.88	7,604.38	0	7,604.38
4	1,140.66	8,745.04	0	8,745.04
5	1,311.76	10,056.80	10,056.80	0
			$10,056.80	
Plan 2				
0				$5,000.00
1	$750.00	$5,750.00	$ 750.00	5,000.00
2	750.00	5,750.00	750.00	5,000.00
3	750.00	5,750.00	750.00	5,000.00
4	750.00	5,750.00	750.00	5,000.00
5	750.00	5,750.00	5,750.00	0
			$8,750.00	
Plan 3				
0				$5,000.00
1	$750.00	$5,750.00	$1,750.00	4,000.00
2	600.00	4,600.00	1,600.00	3,000.00
3	450.00	3,450.00	1,450.00	2,000.00
4	300.00	2,300.00	1,300.00	1,000.00
5	150.00	1,150.00	1,150.00	0
			$7,250.00	
Plan 4				
0				$5,000.00
1	$750.00	$5,750.00	$1,491.58	4,258.42
2	638.76	4,897.18	1,491.58	3,405.60
3	510.84	3,916.44	1,491.58	2,424.86
4	363.73	2,788.59	1,491.58	1,297.01
5	194.57	1,491.58	1,491.58	0
			$7,457.90	

- *Plan 4* Equal payments are made each year with a portion going toward princi-pal recovery and the remainder covering the accrued interest. Since the loan balance decreases at a rate which is slower than in plan 3 because of the equal end-of-year payments, the interest decreases, but at a rate slower than in plan 3.

Note that the total amount repaid in each case would be different, even though each repayment scheme would require exactly 5 years to repay the loan. The difference in the total amounts repaid can of course be explained by the time value of money, since the amount of the payments is different for each plan. With respect to equiva-lence, the table shows that when the interest rate is 15% per year, $5000 at time 0 is equivalent to $10,056.80 at the end of year 5 (plan 1), or $750 per year for 4 years and $5750 at the end of year 5 (plan 2), or the decreasing amounts shown in years 1 through 5 (plan 3), or $1,491.58 per year for 5 years (plan 4). Using the formulas developed in Chaps. 2 and 3, we could easily show that if the payments in each plan (column 4) were reinvested at 15% per year when received, the total amount of money available at the end of year 5 would be $10,056.80 from each repayment plan.

Additional Examples 1.14 and 1.15
Probs. 1.8 and 1.9

1.4 Simple and Compound Interest

The concepts of interest and interest rate were introduced in Secs. 1.1 and 1.2 and used in Sec. 1.3 to calculate for one interest period past and future sums of money equivalent to a present sum (principal). When more than one interest period is involved, the terms *simple* and *compound* interest must be considered.

Simple interest is calculated using the principal only, ignoring any interest that was accrued in preceding interest periods. The total interest can be computed using the relation

$$\text{Interest} = (\text{principal})(\text{number of periods})(\text{interest rate}) = Pni \qquad (1.4)$$

Example 1.3

If you borrow $1000 for 3 years at 14%-per-year simple interest, how much money will you owe at the end of 3 years?

Solution The interest for each of the 3 years is

Interest per year = 1000(0.14) = $140

Total interest for 3 years from Eq. (1.4) is

Total interest = 1000(3)(0.14) = $420

Finally, the amount due after 3 years is

1000 + 420 = $1420

Comment The $140 interest accrued in the first year and the $140 accrued in the second year did not earn interest. The interest due was calculated on the principal only. The results of this loan are tabulated in Table 1.2. The end-of-year figure of zero represents the present, that is, when the money is borrowed. Note that no payment is made by the borrower until the end of year 3. Thus, the amount owed each year increases uniformly by $140, since interest is figured only on the principal of $1000.

Table 1.2 Simple-interest computation

(1) End of year	(2) Amount borrowed	(3) Interest	(4) = (2) + (3) Amount owed	(5) Amount paid
0	$1,000			
1	...	$140	$1,140	$ 0
2	...	140	1,280	0
3	...	140	1,420	1,420

In calculations of *compound interest*, the interest for an interest period is calculated on the principal *plus the total amount of interest accumulated in previous periods*. Thus, compound interest means "interest on top of interest" (i.e., it reflects the effect of the time value of money on the interest too).

Example 1.4

If you borrow $1000 at 14%-per-year *compound* interest, instead of simple interest as in the preceding example, compute the total amount due after a 3-year period.

Solution   The interest and total amount due for each year is computed as follows:

$$\text{Interest, year } 1 = 1000(0.14) = \$140$$

$$\text{Total amount due after year } 1 = 1000 + 140 = \$1140$$

$$\text{Interest, year } 2 = 1140(0.14) = \$159.60$$

$$\text{Total amount due after year } 2 = 1140 + 159.60 = \$1299.60$$

$$\text{Interest, year } 3 = 1299.60(0.14) = \$181.94$$

$$\text{Total amount due after year } 3 = 1299.60 + 181.94 = \$1481.54$$

Comment   The details are shown in Table 1.3. The repayment scheme is the same as that for the simple-interest example; that is, no amount is repaid until the principal plus all interest is due at the end of year 3. The time value of money is especially recognized in compound interest. Thus, with compound interest, the original $1000 would accumulate an extra $1481.54 − $1420 = $61.54 compared with simple interest in the 3-year period. If $61.54 does not seem like a significant difference, remember that the beginning amount here was only $1000. Make these same calculations for an initial amount of $10 million, and then look at the size of the difference!

The power of compounding can further be illustrated through another interesting exercise called "Pay Now, Play Later". It can be shown (by using the equations that will be developed in Chap. 2) that at an interest rate of 12% per year, approximately $1,000,000 will be accumulated at the end of a 40-year time period by either of the

Table 1.3 Compound-interest computation

(1) End of year	(2) Amount borrowed	(3) Interest	(4) = (2) + (3) Amount owed	(5) Amount paid
0	$1,000			
1	...	$140.00	$1,140.00	$ 0
2	...	159.60	1,299.60	0
3	...	181.94	1,481.54	1,481.54

following investment schemes:

- *Plan 1* Invest $2610 each year for the first 6 years and then nothing for the next 34 years, or
- *Plan 2* Invest nothing for the first 6 years, and then $2600 each year for the next 34 years!!

Note that the total investment in plan 1 is $15,660 while the total required in plan 2 to accumulate the same amount of money is nearly six times greater at $88,400. Both the power of compounding and the wisdom of planning for your retirement at the earliest possible time should be quite evident from this example.

An interesting observation pertaining to compound-interest calculations involves the estimation of the length of time required for a single initial investment to double in value. The so-called rule of 72 can be used to estimate this time. The rule is based on the fact that the time required for an initial lump-sum investment to double in value when interest is compounded is approximately equal to *72 divided by the interest rate that applies.* For example, at an interest rate of 5% per year, it would take approximately 14.4 years (i.e., 72/5 = 14.4) for an initial sum of money to double in value. (The actual time required is 14.3 years, as will be shown in Chap. 2.) In Table 1.4, the times estimated from the rule of 72 are compared to the actual times required for doubling at various interest rates and, as you can see, very good estimates are obtained.

Conversely, the interest rate that would be required in order for money to double in a specified period of time could be estimated by dividing 72 by the specified time period. Thus, in order for money to double in a time period of 12 years, an interest rate of approximately 6% per year would be required (i.e., 72/12 = 6). It should be obvious that for simple-interest situations, the "rule of 100" would apply, except that the answers obtained will always be exact.

In Chap. 2, formulas are developed which simplify compound-interest calculations. The same concepts are involved when the interest period is less than a year. A discussion of this case is deferred until Chap. 3, however. Since real-world calculations almost always involve compound interest, the interest rates specified herein refer to compound interest rates unless specified otherwise.

Additional Example 1.16
Probs. 1.10 to 1.26

Table 1.4 Doubling time estimated from rule of 72 versus actual time

Interest rate, % per period	Doubling time, no. of periods	
	Estimated from rule	Actual
1	72	70
2	36	35.3
5	14.4	14.3
10	7.2	7.5
20	3.6	3.9
40	1.8	2.0

1.5 Symbols and Their Meaning

The mathematical relations used in engineering economy employ the following symbols:

P = value or sum of money at a time denoted as the present; dollars, pesos, etc.

F = value or sum of money at some future time; dollars, pesos, etc.

A = a series of consecutive, equal, end-of-period amounts of money; dollars per month, dollars per year, etc.

n = number of interest periods; months, years, etc.

i = interest rate per interest period; percent per month, percent per year, etc.

The symbols P and F represent single-time occurrence values: A occurs at *each* interest period for a specified number of periods with the same value. It should be understood that a present sum P represents a single sum of money at some time prior to a future sum or uniform series amount and therefore does not necessarily have to be located at time $t = 0$. Example 1.11 shows a P value at a time other than $t = 0$. The units of the symbols aid in clarifying their meaning. The present sum P and future sum F are expressed in dollars; A is referred to in dollars per interest period. It is important to note here that in order for a series to be represented by the symbol A, it must be uniform (i.e., the dollar value must be the same for each period) and the uniform dollar amounts must extend through *consecutive* interest periods. Both conditions must exist before the dollar value can be represented by A. Since n is commonly expressed in years or months, A is usually expressed in units of dollars per year or dollars per month, respectively. The compound-interest rate i is expressed in percent per interest period, for example, 5% per year. Except where noted otherwise, this rate applies throughout the entire n years or n interest periods. The i value is often the minimum attractive rate of return (MARR).

All engineering-economy problems must involve at least four of the symbols listed above, with at least three of the values known. The following four examples illustrate the use of the symbols.

Example 1.5

If you borrow $2000 now and must repay the loan plus interest at a rate of 12% per year in 5 years, what is the total amount you must pay? List the values of P, F, n, and i.

Solution   In this situation P and F, but not A, are involved, since all transactions are single payments. The values are as follows:

$P = \$2000$ $F = ?$ $i = 12\%$ per year $n = 5$ years

Example 1.6

If you borrow $2000 now at 17% per year for 5 years and must repay the loan in equal yearly payments, what will you be required to pay? Determine the value of the symbols involved.

Solution

$P = \$2000$

$A = ?$ per year for 5 years

$i = 17\%$ per year

$n = 5$ years

There is no F value involved.

In both examples, the P value of $2000 is a receipt and F or A is a disbursement. It is equally correct to use these symbols in reverse roles, as in the examples below.

Example 1.7

If you deposit $500 into an account on May 1, 1988, which pays interest at 17% per year, what annual amount can you withdraw for the following 10 years? List the symbol values.

Solution

$P = \$500$

$A = ?$ per year

$i = 17\%$ per year

$n = 10$ years

Comment The value for the $500 disbursement P and receipt A are given the same symbol names as before, but they are considered in a different context. Thus, a P value may be a receipt (Examples 1.5 and 1.6) or a disbursement (this example).

Example 1.8

If you deposit $100 into an account each year for 7 years at an interest rate of 16% per year, what single amount will you be able to withdraw after 7 years? Define the symbols and their roles.

Solution In this example, the equal annual deposits are in a series A and the withdrawal is a future sum, or F value. There is no P value here.

$A = \$100$ per year for 7 years

$F = ?$

$i = 16\%$ per year

$n = 7$ years

Additional Example 1.17
Probs. 1.27 to 1.29

1.6 Cash-Flow Diagrams

Every person or company has cash receipts (income) and cash disbursements (costs) which occur over a particular time span. These receipts and disbursements in a given time interval are referred to as *cash flow*, with positive cash flows usually representing receipts and negative cash flows representing disbursements. At any point in time, the net cash flow would be represented as

$$\text{Net cash flow} = \text{receipts} - \text{disbursements} \qquad (1.5)$$

Since cash flow normally takes place at frequent and varying time intervals within an interest period, a simplifying assumption is made that all cash flow occurs at the end of the interest period. This is known as the *end-of-period convention*. Thus, when several receipts and disbursements occur within a given interest period, the net cash flow is assumed to occur at the end of the interest period. However, it should be understood that although the dollar amounts of F or A are always considered to occur at the *end of the interest period*, this does not mean that the end of the period is December 31. In the situation of Example 1.7, since investment took place on May 1, 1988, the withdrawals will take place on May 1, 1989 and each succeeding May 1 for 10 years (the last withdrawal will be on May 1, 1998, not 1999). Thus, *end of the period* means one time period from the date of the transaction (whether it be receipt or disbursement). In the next chapter you will learn how to determine the equivalent relations between P, F, and A values at different times.

A *cash-flow diagram* is simply a graphical representation of cash flows drawn on a time scale. The diagram should represent the statement of the problem and should include what is given and what is to be found. That is, after the cash-flow diagram has been drawn, an outside observer should be able to work the problem by looking at only the diagram. Time 0 is considered to be the present and time 1 is the end of time period 1. (We will assume that the periods are in years until Chap. 3.) The time scale of Fig. 1.1 is set up for 5 years. Since it is assumed that cash flows occur only at the end of the year, we will be concerned only with the times marked 0, 1, 2, ..., 5.

The direction of the arrows on the cash-flow diagram is important to problem solution. Therefore, in this text, a vertical arrow pointing up will indicate a positive cash flow. Conversely, an arrow pointing down will indicate a negative cash flow. The cash-flow diagram in Fig. 1.2 illustrates a receipt (income) at the end of year 1 and a disbursement at the end of year 2.

It is important that you thoroughly understand the meaning and construction of the cash-flow diagram, since it is a valuable tool in problem solution. The three examples below illustrate the construction of cash-flow diagrams.

Figure 1.1 A typical cash-flow time scale.

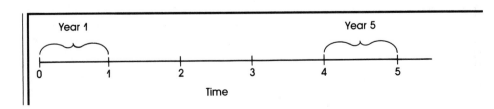

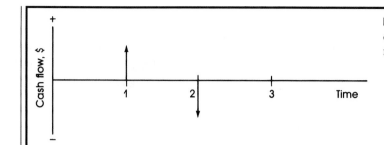

Figure 1.2 Example of positive and negative cash flows.

Example 1.9

Consider the situation presented in Example 1.5, where $P = \$2000$ is borrowed and F is to be found after 5 years. Construct the cash-flow diagram for this case, assuming an interest rate of 12% per year.

Solution   Figure 1.3 presents the cash-flow diagram.

Comment   While it is not necessary to use an exact scale on the cash-flow axes, you will probably avoid errors later on if you make a neat diagram. Note also that the present sum P is a *receipt* at year 0 and the future sum F is a *disbursement* at the end of year 5.

Example 1.10

If you start now and make five deposits of $1000 per year ($A$) in a 17%-per-year account, how much money will be accumulated (and can be withdrawn) immediately after you have made the last deposit? Construct the cash-flow diagram.

Solution   The cash flows are shown in Fig. 1.4.

Comment   Since you have decided to start now, the first deposit is at year 0 and the fifth deposit and withdrawal occur at the end of year 4. Note that in this example, the amount accumulated after the fifth deposit is to be computed; thus, the future amount is represented by a question mark (i.e., $F = ?$)

Figure 1.3   Cash-flow diagram for Example 1.9.

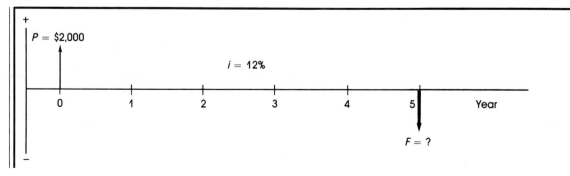

Figure 1.4 Cash-flow diagram for Example 1.10.

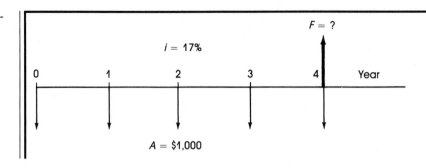

Additional Examples 1.18 to 1.20
Probs. 1.30 to 1.46

Example 1.11

Assume that you want to deposit an amount P into an account 2 years from now in order to be able to withdraw $400 per year for 5 years starting 3 years from now. Assume that the interest rate is $15\frac{1}{2}\%$ per year. Construct the cash-flow diagram.

Solution   Figure 1.5 presents the cash flows, where P is to be found. Note that the diagram shows what was given and what is to be found and that a P value is not necessarily located at time $t = 0$.

Additional Examples

Example 1.12

Calculate the interest and total amount accrued after 1 year if $2000 is invested at an interest rate of 15% per year.

Solution

$$\text{Interest earned} = 2000(0.15) = \$300$$

$$\text{Total amount accrued} = 2000 + 2000(0.15) = 2000(1 + 0.15)$$
$$= \$2300$$

Figure 1.5 Cash-flow diagram for Example 1.11.

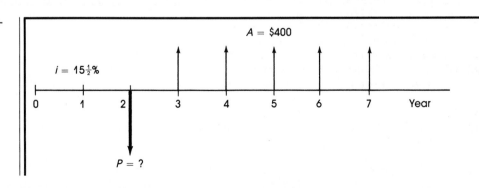

Example 1.13

(a) Calculate the amount of money that must have been deposited 1 year ago for you to have $1000 now at an interest rate of 5% per year.
(b) Calculate the interest that was earned in the same time period.

Solution
(a) Total amount accrued = original deposit + (original deposit)(interest rate). If X = original deposit, then

$$1000 = X + X(0.05) = X(1 + 0.05)$$

$$1000 = 1.05X$$

$$X = \frac{1000}{1.05} = 952.38$$

Original deposit = $952.38

(b) By using Eq. (1.1), we have

Interest = $1000 - 952.38 = \$47.62$

Example 1.14

Calculate the amount of money that must have been deposited 1 year ago for the investment to earn $100 in interest in 1 year, if the interest rate is 6% per year.

Solution   Let a = total amount accrued and b = original deposit.

Interest = $a - b$

Since $a = b + b$ (interest rate), interest can be expressed as

Interest = $b + b$ (interest rate) $- b$

Interest = b (interest rate)

$$\$100 = b(0.06)$$

$$b = \frac{100}{0.06} = \$1666.67$$

Example 1.15

Make the calculations necessary to show which of the statements below are true and which are false, if the interest rate is 5% per year:
(a) $98 now is equivalent to $105.60 one year from now.
(b) $200 one year past is equivalent to $205 now.
(c) $3000 now is equivalent to $3150 one year from now.
(d) $3000 now is equivalent to $2887.14 one year ago.
(e) Interest accumulated in 1 year on an investment of $2000 is $100.

Solution
(a) Total amount accrued = 98(1.05) = $102.90 ≠ $105.60; therefore false. Another way to solve this is as follows: Required investment = 105.60/1.05 = $100.57 ≠ $98. Therefore false.
(b) Required investment = 205.00/1.05 = $195.24 ≠ $200; therefore false.

(c) Total amount accrued $\doteq 3000(1.05) = \$3150$; therefore true.
(d) Total amount accrued $= 2887.14(1.05) = \$3031.50 \neq \$3000$; therefore false.
(e) Interest $= 2000(0.05) = \$100$; therefore true.

Example 1.16

Calculate the total amount due after 2 years if $2500 is borrowed now and the compound-interest rate is 8% per year.

Solution The results are presented in the table to obtain a total amount due of $2916.

(1) End of year	(2) Amount borrowed	(3) Interest	(4) $\doteq$ (2) + (3) Amount owed	(5) Amount paid
0	$2,500			
1	...	$200	$2,700	$0
2	...	216	2,916	2,916

Example 1.17

Assume that you plan to make a lump-sum deposit of $5000 now into an account that pays 6% per year, and you plan to withdraw an equal end-of-year amount of $1000 for 5 years starting next year. At the end of the sixth year, you plan to close your account by withdrawing the remaining money. Define the engineering-economy symbols involved.

Solution

$P = \$5000$

$A = \$1000$ per year for 5 years

$F = ?$ at end of year 6

$i = 6\%$ per year

$n = 5$ years for A

Figure 1.6 Cash-flow diagram for Example 1.18.

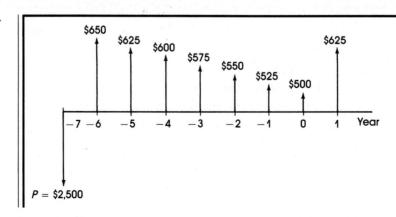

$P = \$2,500$

Example 1.18

The Hot-Air Company invested $2500 in a new air compressor 7 years ago. Annual income from the compressor was $750. During the first year, $100 was spent on maintenance, a cost that increased each year by $25. The company plans to sell the compressor for salvage at the end of next year for $150. Construct the cash-flow diagram for the piece of equipment.

Solution The income and cost for years −7 through 1 (next year) are tabulated below with net cash flow computed using Eq. (1.5). The cash flows are diagrammed in Fig. 1.6.

End of year	Income	Cost	Net cash flow
−7	$0	$2,500	$−2,500
−6	750	100	650
−5	750	125	625
−4	750	150	600
−3	750	175	575
−2	750	200	550
−1	750	225	525
0	750	250	500
1	750 + 150	275	625

Example 1.19

Suppose that you want to make a deposit into your account now such that you can withdraw an equal annual amount of $A_1 = \$200$ per year for the first 5 years starting 1 year after your deposit and a different annual amount of $A_2 = \$300$ per year for the following 3 years. How would the cash-flow diagram appear if i is $14\frac{1}{2}\%$ per year?

Solution The cash flows would appear as shown in Fig. 1.7.

Comment The first withdrawal (positive cash flow) occurs at the end of year 1, exactly one year after P is deposited.

Figure 1.7   Cash-flow diagram for two different A values, Example 1.19.

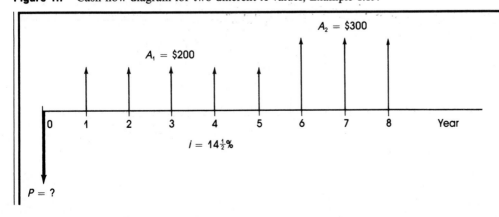

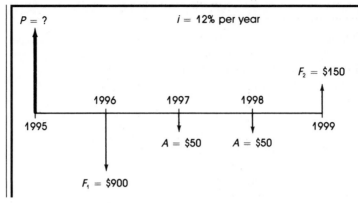

Figure 1.8
Cash-flow diagram for Example 1.20.

Example 1.20

If you buy a new television set in 1996 for $900, maintain it for 3 years at a cost of $50 per year, and then sell it for $200, diagram your cash flows and label each arrow as P, F, or A with its respective dollar value so that you can find the single amount in 1995 that would be equivalent to all of the cash flows shown. Assume an interest rate of 12% per year.

Solution Figure 1.8 presents the cash-flow diagram.

Comment The two $50 negative cash flows form a series of two equal end-of-year values. As long as the dollar values are equal and in two or more consecutive periods, they can be represented by A, regardless of where they begin or end. However, the $150 positive cash flow in 1999 is a single-occurrence value in the future and is therefore labeled an F value. It is possible, however, to view all of the individual cash flows as F values. The diagram could be drawn as shown in Fig. 1.9. In general, however, if two or more equal end-of-period amounts occur consecutively, by the definition in Sec. 1.5 they should be labeled A values because, as is described in Chap. 2, the use of A values when possible simplifies calculations considerably. Thus, the interpretation pictured by the diagram of Fig. 1.9 is discouraged and will not generally be used further in this text.

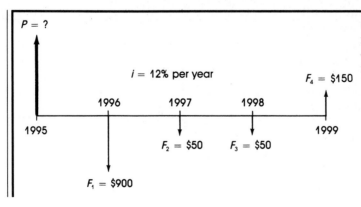

Figure 1.9 A cash flow for Example 1.20 considering all values as future sums.

1.1 What is meant by the *time value of money*?

1.2 What is the difference between the terms *principal* and *present amount owed*?

1.3 List at least three criteria which might be used besides dollars for evaluating each of the following: (*a*) restaurants; (*b*) airplanes; (*c*) apartments; (*d*) engineering-economy books.

1.4 List three items that might be regarded as intangible factors.

1.5 If a bank advertises 14%-per-year interest compounded semiannually, what is the interest period?

1.6 Find the interest due after 1 year on a loan of $5000 if interest is 12% per year.

1.7 Find the original amount of a loan if interest is $1\frac{1}{2}$% per month payable monthly and the borrower just made the first monthly payment of $25 in interest.

1.8 At what interest rate are $450 a year ago and $550 one year from now equivalent?

1.9 How do you explain the fact that two different amounts of money can be equivalent to each other?

1.10 If the Get-Rich-Quick Company invested $50,000 in a new process 1 year ago and has just now realized a profit of $7500, what was its rate of return based on this investment?

1.11 Why is the minimum attractive rate of return of a business organization greater than the interest rate obtainable from a bank or savings and loan association?

1.12 Assume that you have been offered an investment opportunity in which you may invest $1000 at 7%-per-year simple interest for 3 years or you may invest the same $1000 at 6%-per-year compound interest for 3 years. Which investment offer would you accept?

1.13 (*a*) How much total interest would you pay if you borrowed $600 at $1\frac{1}{2}$% per month compounded monthly for 3 months?
(*b*) What percent of the principal is this interest amount?

1.14 Work the two parts of Prob. 1.13 for $1\frac{1}{2}$%-per-month simple interest.

1.15 How much money will your friend owe after 4 years if she borrows $1000 now at 7%-per-year simple interest?

1.16 How much money will be owed after 2 years if a person borrows $500 at 1%-per-month simple interest?

1.17 How much money can you borrow now if you repay the lender $850 in 2 years and the interest rate is 6% per year compounded yearly?

1.18 If you borrow $1500 now and must repay $1800 two years from now, what is the interest rate on your loan? Assume that interest is compounded yearly.

1.19 If you invest $10,000 now in a business venture that promises to return $14,641, how soon must you receive the $14,641 in order to make at least 10% per year compounded yearly on your investment?

1.20 Your friend tells you that he has just repaid a loan he got 3 years ago at 10%-per-year simple interest. If, upon questioning, you learn that his payment was $195, how much did he borrow?

1.21 If you invest $3500 now in return for a guaranteed $5000 income at a later date, when must you receive your money in order to earn at least 8%-per-year simple interest?

1.22 Section 1.4 shows that $1000 at 6%-per-year simple interest is equivalent to $1180 in 3 years. Find the compound-interest rate per year for which the equivalence is also correct.

1.23 At an interest rate of 8% per year, estimate the time required for money to double at (*a*) simple, and (*b*) compound interest.

1.24 Repeat Prob. 1.23, except use an interest rate of 12% per year.

1.25 Estimate the interest rate at which money would double in approximately 12 years if the interest is (*a*) simple, and (*b*) compound.

1.26 Repeat Prob. 1.25, except use a doubling time of 8 years.

1.27 Five equal deposits of $1000 will be made every 2 years starting next year at 10% per year and the total accrued amount withdrawn when the last deposit is made. List the economy symbols and values involved in this problem.

1.28 The GRQ Company plans to deposit $709.90 now at 6% per year and withdraw $100 per year for the next 5 years and $200 per year for the following 2 years. What are the economy symbols and their respective values?

1.29 How many years would it take for $1400 to triple in value at an interest rate of 10% per year? Define the economy symbols.

1.30 What is meant by *end-of-period convention*?

1.31 Diagram the yearly net cash flows for Example 1.5.

1.32 Diagram the yearly net cash flows for Example 1.7.

1.33 Assume that you have developed an investment plan that is carried out as follows: Invest $500 now and every other year through year 10 and withdraw $300 every year starting 5 years from now and continuing for 8 more years. Diagram the yearly net cash flows.

1.34 Construct cash-flow diagrams for the loan-repayment schedules in Table 1.1 (col. 4).

1.35 If you plan to make a deposit now such that you will have $3000 in your account 5 years from now, how much must you deposit if the interest rate is 8% per year? Draw the cash-flow diagram.

1.36 Your uncle has agreed to make five $700-per-year deposits into a savings account for you starting now. You have in turn agreed not to withdraw any money until the end of year 9, at which time you plan to remove $3000 from the account. Further, you plan to withdraw the remaining amount in three equal year-end installments after the initial withdrawal. Diagram the cash flows for your uncle and yourself.

1.37 The president of a company wants to make two equal lump-sum deposits, one 2 years and the second 4 years from now, so he can make five $100-per-year withdrawals starting when the second deposit is made. Further, he plans to withdraw an additional $500 the year after the withdrawal series ends. Draw his cash-flow diagram.

1.38 You want to invest money at 8% per year so that 6 years from now you can withdraw an amount F in a lump sum. The investment consultant at the bank has developed the following two plans for you: (*1*) Deposit $351.80 now and $351.80 three years from now. (*2*) Deposit $136.32 per year starting next year and ending in year 6. Draw the cash-flow diagram for each plan if F is to be found.

1.39 How much could you spend now in order to avoid spending $580 eight years from now if the interest rate is 6% per year? Draw the cash-flow diagram.

1.40 If you deposit $100 per year for 5 years starting one year from now, how much will you have in your account 15 years from now if the interest rate is 10% per year? Draw the cash-flow diagram.

1.41 What is the present worth of an expenditure of $1200 five years from now and $2200 eight years from now if the interest rate is 10% per year? Construct the cash-flow diagram.

1.42 Calculate the present worth of an expenditure of $85 per year for 6 years that starts 3 years from now if the interest rate is 20% per year. Construct the cash-flow diagram.

1.43 If you invest $10,000 now in a real estate venture, how much must you sell your property for 10 years from now if you want to make a 12% per year rate of return on your investment? Define the economy symbols and draw the cash-flow diagram.

1.44 If you invest $4100 now and receive $7500 five years from now, what is the rate of return on your investment? Define the economy symbols and construct the cash-flow diagram.

1.45 How much money would be accumulated in 6 years if a person deposited $500 now and the deposits increase by $50 per year for the next 6 years? Assume i is 16% per year and draw the cash-flow diagram.

1.46 What uniform payment for 8 years beginning 1 year from now would be equivalent to spending $4500 now, $3300 three years from now, and $6800 five years from now if the interest rate is 8% per year? Define the economy symbols and draw the cash-flow diagram.

2 Factors and Their Use

The objective of this chapter is to teach you the derivation of the engineering-economy factors and the use of these basic factors in economy computations. This chapter is one of the most important in the book, since the concepts presented here will be used throughout the remainder of the text.

Section Objectives

After completing this chapter, you should be able to define and derive formulas for the following:

2.1 Single-payment compound-amount factor and single-payment present-worth factor
2.2 Uniform-series present-worth factor and capital-recovery factor using the single-payment present-worth factor
2.3 Uniform-series compound-amount factor and sinking-fund factor using the single-payment compound-amount factor and the capital-recovery factor

You should also be able to do the following:

2.4 Find the correct numerical value of a factor in a table, given the standard factor notation.
2.5 Define and develop the uniform-gradient present-worth and annual-series factors using the single-payment present-worth factor.

24

2.6 Define *escalating series* and derive the formula for calculating the present worth of such a series.

2.7 Linearly interpolate to find a correct factor value, given an interest rate and/or a number of periods not listed in the tables.

2.8 Calculate the present worth P, future worth F, or equivalent uniform annual series A of an investment, given the interest rate i, the number of periods n, and the monetary value of one of the terms P, F, or A.

2.9 Calculate the present worth P and equivalent uniform annual series A of alternatives involving a *conventional* uniform gradient, given the interest rate and statement of the problem.

2.10 Calculate the present worth, equivalent uniform annual series, or future worth of an escalating series, given the cash-flow series, escalation rate, interest rate, and length of the series.

2.11 Calculate the interest rate (rate of return) of a sequence of cash flows, given the number of periods and two of the following: present worth P, future worth F, or uniform series A starting at the end of period 1 and ending at the end of period n.

2.12 Determine the number of years n required for equivalence for a sequence of cash flows, given the interest rate i and two of the following: present worth P, future worth F, uniform series A starting at the end of period 1 and ending at the end of period n, or uniform gradient G starting at the end of period 2 and continuing through period n.

Study Guide

2.1 Derivation of Single-Payment Formulas

In this section, a formula is developed which allows determination of the amount of money that is accumulated (F) after n years (or periods) from a *single* investment (P) when interest is compounded one time per year (or period). In this chapter, an interest period of 1 year will be assumed. However, you should recognize that the i and n symbols in the formulas developed here apply to *interest periods*, not just years, as will be discussed in Chapter 3.

You will recall from Chap. 1 that compound interest refers to interest paid on top of interest. Therefore, if an amount of money P is invested at some time $t = 0$, the amount of money F_1 that will be accumulated 1 year hence at an interest rate of i per year will be

$$F_1 = P + Pi$$

$$F_1 = P(1 + i)$$

At the end of the second year, the amount of money accumulated (F_2) will be equal to the amount that accumulated after year 1 plus the interest from the end of year 1 to the end of year 2. Thus,

$$F_2 = F_1 + F_1 i \tag{2.1}$$
$$= P(1 + i) + P(1 + i)i$$

or

$$F_2 = P(1 + i + i + i^2)$$
$$= P(1 + 2i + i^2)$$
$$= P(1 + i)^2$$

Similarly, the amount of money accumulated at the end of year 3, using Eq. (2.1), will be

$$F_3 = F_2 + F_2 i$$
$$= [P(1 + i) + P(1 + i)i] + [P(1 + i) + P(1 + i)i]i$$
$$= P(1 + i) + 2P(1 + i)i + P(1 + i)i^2$$

Factoring out $P(1 + i)$, we have

$$F_3 = P(1 + i)(1 + 2i + i^2)$$
$$= P(1 + i)(1 + i)^2$$
$$= P(1 + i)^3$$

From the preceding values, it is evident by mathematical induction that the formula can be generalized for n years as

$$F = P(1 + i)^n \tag{2.2}$$

The expression $(1 + i)^n$, called the *single-payment compound-amount factor* (SPCAF), will yield the future amount F of an initial investment P after n years at interest rate i.

Expressing P from Eq. (2.2) in terms of F results in the expression

$$P = F\left[\frac{1}{(1 + i)^n}\right] \tag{2.3}$$

The expression in brackets is known as the *single-payment present-worth factor* (SPPWF). This expression will allow determination of the present worth P of a given future amount F after n years at interest rate i. The cash-flow diagram for this formula is shown in Fig. 2.1. Conversely, if you used the SPCAF for the diagram in Fig. 2.1, you could find F, given P.

It is important to note that the two formulas derived here are *single-payment* formulas; that is, they are used to find the present or future amount when only one

Figure 2.1
Cash-flow diagram
to find P given F
using the SPPWF.

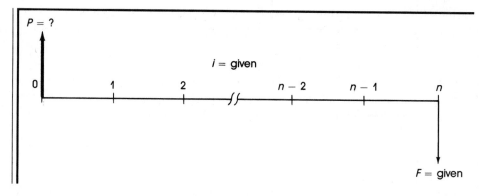

payment or receipt is involved. In the next two sections, formulas are developed for calculating the present or future worth when several uniform payments or receipts must be considered.

2.2 Derivation of the Uniform-Series Present-Worth Factor and the Capital-Recovery Factor

The present worth of the uniform series shown in Fig. 2.2 can be determined by considering each A value as a future worth F in the single-payment present-worth formula in Eq. (2.3) and then summing the present-worth values. The general formula is

$$P = A\left[\frac{1}{(1 + i)^1}\right] + A\left[\frac{1}{(1 + i)^2}\right] + A\left[\frac{1}{(1 + i)^3}\right] + \cdots + A\left[\frac{1}{(1 + i)^{n-1}}\right]$$
$$+ A\left[\frac{1}{(1 + i)^n}\right]$$

where the terms in brackets represent the SPPWF for years 1 through n, respectively. Factoring out A,

$$P = A\left[\frac{1}{(1 + i)^1} + \frac{1}{(1 + i)^2} + \frac{1}{(1 + i)^3} + \cdots + \frac{1}{(1 + i)^{n-1}} + \frac{1}{(1 + i)^n}\right] \qquad (2.4)$$

Equation (2.4) may be simplified by multiplying both sides by $1/(1 + i)$ to yield

$$\frac{P}{1 + i} = A\left[\frac{1}{(1 + i)^2} + \frac{1}{(1 + i)^3} + \frac{1}{(1 + i)^4} + \cdots + \frac{1}{(1 + i)^n} + \frac{1}{(1 + i)^{n+1}}\right] \qquad (2.5)$$

Subtracting Eq. (2.4) from Eq. (2.5) yields

$$\frac{P}{1 + i} - P = A\left[-\frac{1}{(1 + i)^1} + \frac{1}{(1 + i)^{n+1}}\right]$$

Factoring out P and rearranging, we have

$$P\left(\frac{1}{1 + i} - 1\right) = A\left[\frac{1}{(1 + i)^{n+1}} - \frac{1}{1 + i}\right]$$

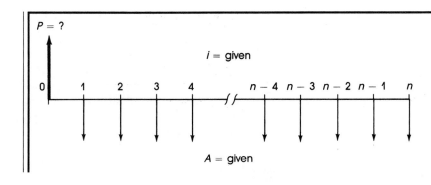

P = ?

i = given

A = given

Figure 2.2
Diagram used to determine the present worth of a uniform series.

Simplifying both sides of the equation yields

$$P\left(\frac{-i}{1+i}\right) = A\left(\frac{1}{1+i}\right)\left[\frac{1}{(1+i)^n} - 1\right]$$

Dividing by $-i/(1+i)$ yields the following for $i \neq 0$.

$$P = A\left(\frac{1}{1+i}\right)\frac{[1/(1+i)^n - 1]}{-i/(1+i)}$$

$$= A\left(\frac{1}{-i}\right)\left[\frac{1-(1+i)^n}{(1+i)^n}\right]$$

$$= A\left[\frac{(1+i)^n - 1}{i(1+i)^n}\right] \qquad i \neq 0 \tag{2.6}$$

The term in brackets is called the *uniform-series present-worth factor* (USPWF). This equation will give the present worth P of an equivalent uniform annual series A which begins *at the end of year 1* and extends for n years at an interest rate i.

By rearranging Eq. (2.6), we can express A in terms of P:

$$A = P\left[\frac{i(1+i)^n}{(1+i)^n - 1}\right] \tag{2.7}$$

The term in brackets, called the *capital-recovery factor* (CRF), yields the equivalent uniform annual worth A over n years of a given investment P when the interest rate is i.

It is very important to commit to memory the fact that these formulas were derived with the present worth P and the first uniform annual-worth value A *one year* (*period*) *apart*. That is, the present sum P *must always* be located one period *prior* to the first A. The correct use of these factors is illustrated in Sec. 2.7.

2.3 Derivation of the Uniform-Series Compound-Amount Factor and the Sinking-Fund Factor

While the *sinking-fund factor* (SFF) and the *uniform-series compound-amount factor* (USCAF) could be derived using the SPCAF, the simplest way to derive the formulas is to substitute into those already developed. Thus, if P from Eq. (2.3), which uses the SPPWF, is substituted into Eq. (2.7), the following formula results:

$$A = F\left[\frac{1}{(1+i)^n}\right]\left[\frac{i(1+i)^n}{(1+i)^n - 1}\right] = F\left[\frac{i}{(1+i)^n - 1}\right] \tag{2.8}$$

The expression in brackets in Eq. (2.8) is the sinking-fund factor. Equation (2.8) is used to determine the uniform annual worth series that would be equivalent to a given future worth F. This is shown graphically in Fig. 2.3. Note that the uniform series A begins at the end of period 1 and continues through the period of the given F. This is unlike the uniform-series present-worth formulas in the preceding section, where the P and the first A are always one period apart.

Equation (2.8) can be rearranged to express F in terms of A:

$$F = A\left[\frac{(1+i)^n - 1}{i}\right] \tag{2.9}$$

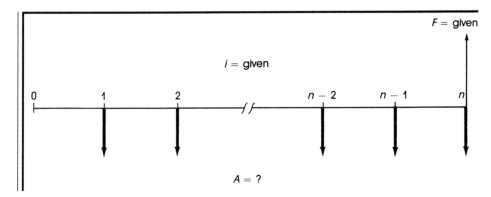

Figure 2.3
Transformation of
a given F value
into an equivalent
A series.

The term in brackets is called the *uniform-series compound-amount factor* (USCAF) which, when multiplied by the given uniform annual amount A, yields the future worth of the uniform series. The cash-flow diagram for this case would be just the opposite of that shown in Fig. 2.3. Again, it is important to remember that the future amount F occurs in the same period as the last A. As an exercise, the reader should show that the USCAF could be obtained by multiplying the respective formulas from the SPCAF in Eq. (2.2) and the USPWF in Eq. (2.6) for a given i and n.

2.4 Standard Factor Notation and Use of Interest Tables

To avoid the cumbersome task of writing out the formulas each time one of the factors is used, a standard notation has been adopted which represents the various factors. This standard notation, which also includes the interest rate and the number of periods, will always be of the general form $(X/Y, i\%, n)$. The first letter inside the parentheses represents what you want to find, while the second letter Y represents what is given. For example, F/P means "find F when given P." The i is the interest rate in percent and the n represents the number of periods involved. Thus, $(F/P, 6\%, 20)$ means obtain the factor which when multiplied by a given P allows you to find the future amount of money F that will be accumulated in 20 periods if the interest rate is 6% per period.

The standard notation is simpler than factor names for identifying factors and will be used exclusively hereafter. Table 2.1 shows the standard notation for the formulas derived thus far in this chapter. For ready reference the formulas used in computations are collected in Table 2.2. It is also easy to use the standard notation to remember how the factors may be derived. For example, the A/F factor may be

Table 2.1 Standard factor notations

Factor name	Standard notation
Single-payment present-worth (SPPWF)	$(P/F, i\%, n)$
Single-payment compound-amount (SPCAF)	$(F/P, i\%, n)$
Uniform-series present-worth (USPWF)	$(P/A, i\%, n)$
Capital-recovery (CRF)	$(A/P, i\%, n)$
Sinking-fund (SFF)	$(A/F, i\%, n)$
Uniform-series compound-amount (USCAF)	$(F/A, i\%, n)$

Table 2.2 Computations using standard notation

To find	Given	Factor	Equation	Formula
P	F	$(P/F, i\%, n)$	$P = F(P/F, i\%, n)$	$P = F[1/(1 + i)^n]$
F	P	$(F/P, i\%, n)$	$F = P(F/P, i\%, n)$	$F = P(1 + i)^n$
P	A	$(P/A, i\%, n)$	$P = A(P/A, i\%, n)$	$P = A\left[\dfrac{(1 + i)^n - 1}{i(1 + i)^n}\right]$
A	P	$(A/P, i\%, n)$	$A = P(A/P, i\%, n)$	$A = P\left[\dfrac{i(1 + i)^n}{(1 + i)^n - 1}\right]$
A	F	$(A/F, i\%, n)$	$A = F(A/F, i\%, n)$	$A = F\left[\dfrac{i}{(1 + i)^n - 1}\right]$
F	A	$(F/A, i\%, n)$	$F = A(F/A, i\%, n)$	$F = A\left[\dfrac{(1 + i)^n - 1}{i}\right]$

derived by multiplying the P/F and A/P factor formulas. In equation form, this is,

$$A = F(P/F, i\%, n)(A/P, i\%, n)$$
$$= F(A/F, i\%, n)$$

The *equivalent* of algebraic cancellation of the P makes this relation easier to remember.

In order to simplify the routine engineering-economy calculations involving the factors above, tables of factor values have been prepared for interest rates from 0.5 to 50% and time periods from 1 to 100. These tables, found in Appendix A and identified as Tables A-1 through A-30, are arranged with various factors across the top and the number of periods n down the left and right column. The word "discrete" in the title of each table is printed to emphasize that these tables are for factors which utilize the end-of-period convention (Sec. 1.6) and that interest is compounded once each interest period. For a given factor, interest rate, and time, the correct factor value would be found in the respective interest-rate table at the intersection of the given factor and n. For example, the value of the factor $(P/A, 5\%, 10)$ is found in the P/A column of Table A-7 at period 10 as 7.7217. The value 7.7217 could, of course, have been computed using the mathematical expression for the USPWF in Eq. (2.6).

$$(P/A, 5\%, 10) = \frac{(1 + i)^n - 1}{i(1 + i)^n} = \frac{1.05^{10} - 1}{0.05(1.05)^{10}} = 7.7217$$

Table 2.3 presents several examples of the use of the interest tables in Appendix A.

Prob. 2.2

2.5 Definition and Derivation of Gradient Formulas

A *uniform gradient* is a cash-flow *series* which either increases or decreases *uniformly*. That is, the cash flow, whether income or disbursements, changes by the same amount

Table 2.3 Use of interest tables

Standard notation	$i,\%$	n	Table	Factor value
$(F/A, 10\%, 3)$	10	3	A-12	3.310
$(A/P, 7\%, 20)$	7	20	A-9	0.09439
$(P/F, 25\%, 35)$	25	35	A-25	0.0004

in each interest period. The *amount* of the increase or decrease is the *gradient*. For example, if a clothing manufacturer predicts that the cost of maintaining a robot will increase by $500 per year until the machine is retired, a gradient series is involved and the amount of the gradient is $500. Similarly, if the company expects income to decrease by $3000 per year for the next 5 years, the decreasing income represents a negative gradient in the amount of $3000 per year.

The formulas previously developed for uniform-series cash flows were generated on the basis of year-end payments of equal value. In the case of a gradient, each year-end cash flow is different, so a new formula must be derived. In developing a formula which can be used for uniform gradients, it is convenient to assume that the payment that occurs at the end of year 1 is not a part of the gradient series but is rather a *base payment*. This is convenient because in actual applications, the base payment is usually larger or smaller than the gradient increase or decrease. For example, if you purchase a new car with a 12,000-mile complete guarantee, you might reasonably expect to have to pay for only the gasoline during the first year of operation. Let us assume that this cost is $900; that is, $900 is the base amount. After the first year, however, you would have to absorb the cost of repair or replacement yourself, and these costs could reasonably be expected to increase each year that you own the car. So if you estimate your operation costs to increase by $50 each year, the amount you would pay after the second year would be $950, after the third $1000, and so on to year n, when the total cost would be $900 + (n - 1)50$. The cash-flow diagram for this is shown in Fig. 2.4. Note that the gradient is first observed between year 1 and year 2, and the first (base) payment ($900) is not equal to the gradient ($50). We will now define a new symbol for gradients:

G = uniform arithmetic change in the magnitude of receipts or disbursements from one time period to the next.

The value of G may be positive or negative. If we ignore the base payment, we can construct a generalized uniformly increasing-gradient cash-flow diagram as shown in Fig. 2.5. Note that the gradient begins between years 1 and 2. This is called a *conventional gradient*.

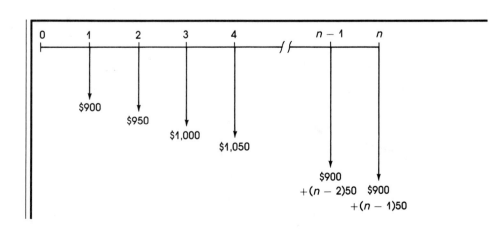

Figure 2.4
Diagram of a uniform-gradient series with a gradient of $50.

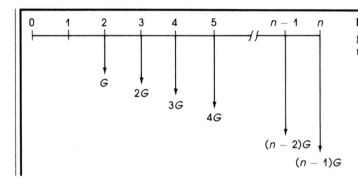

Figure 2.5   Uniform-gradient series ignoring the base amount.

Example 2.1

The Free Spirit Liquor Company expects to realize a revenue of $47,500 next year from the sale of its product. However, sales are expected to increase uniformly with the introduction of a new commodity (moonshine) to a level of $100,000 in 8 years. Determine the gradient and construct the cash-flow diagram.

Solution

$$\text{Base amount} = \$100,000$$

$$\text{Revenue gain by year 8} = 100,000 - 47,500 = \$52,500$$

$$\text{Gradient} = \frac{\text{gain}}{n - 1}$$

$$= \frac{52,500}{8 - 1} = \$7500 \text{ per year}$$

The cash-flow diagram is shown in Fig. 2.6.

Figure 2.6
Diagram for gradient series, Example 2.1.

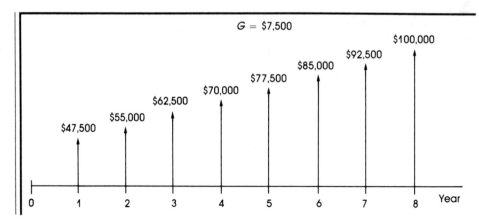

There are several ways by which the uniform-gradient factors can be derived. We will use the single-payment present-worth factor $(P/F, i\%, n)$ but the same result can be obtained using single-payment compound-amount factors, uniform-series compound-amount factors, or uniform-series present-worth factors.

Referring to Fig. 2.5, we find that the present worth at year 0 of the gradient payments would be equal to the sum of the present worths of the individual payments. Thus,

$$P = G(P/F, i\%, 2) + 2G(P/F, i\%, 3) + 3G(P/F, i\%, 4)$$
$$+ \cdots + [(n - 2)G](P/F, i\%, n - 1) + [(n - 1)G](P/F, i\%, n)$$

Factoring out G yields

$$P = G[(P/F, i\%, 2) + 2(P/F, i\%, 3) + 3(P/F, i\%, 4)$$
$$+ \cdots + (n - 2)(P/F, i\%, n - 1) + (n - 1)(P/F, i\%, n)]$$

Replacing the symbols with the appropriate single-payment present-worth factor expression of Eq. (2.3) yields

$$P = G\left[\frac{1}{(1 + i)^2} + \frac{2}{(1 + i)^3} + \frac{3}{(1 + i)^4} + \cdots + \frac{n - 2}{(1 + i)^{n-1}} + \frac{n - 1}{(1 + i)^n}\right] \qquad (2.10)$$

Multiplying both sides of Eq. 2.10 by $(1 + i)^1$ to simplify yields

$$P(1 + i)^1 = G\left[\frac{1}{(1 + i)^1} + \frac{2}{(1 + i)^2} + \frac{3}{(1 + i)^3} + \cdots + \frac{n - 2}{(1 + i)^{n-2}} + \frac{n - 1}{(1 + i)^{n-1}}\right] \quad (2.11)$$

Subtracting Eq. (2.10) from Eq. (2.11), noting that the first term of Eq. (2.11) and the last term of Eq. (2.10) have no matching terms, yields the following relations.

$$P(1 + i)^1 - P = G\left[\frac{1}{(1 + i)^1} + \frac{(2 - 1)}{(1 + i)^2} + \frac{(3 - 2)}{(1 + i)^3}\right.$$
$$\left. + \cdots + \frac{(n - 1) - (n - 2)}{(1 + i)^{n-1}} - \frac{n - 1}{(1 + i)^n}\right]$$

$$= G\left[\frac{1}{(1 + i)^1} + \frac{1}{(1 + i)^2} + \frac{1}{(1 + i)^3} + \cdots + \frac{1}{(1 + i)^{n-1}} + \frac{1 - n}{(1 + i)^n}\right]$$

If we write the left side of this equation as $P + Pi - P$, factor out the n in the last term, and divide by i, we have

$$P = \frac{G}{i}\left[\frac{1}{(1 + i)^1} + \frac{1}{(1 + i)^2} + \frac{1}{(1 + i)^3} + \cdots + \frac{1}{(1 + i)^{n-1}} + \frac{1}{(1 + i)^n}\right] - \frac{Gn}{i(1 + i)^n}$$

Since the expression in the brackets is the present worth of a uniform series of 1 for n years, we can substitute the expression for the P/A factor from Eq. (2.6).

$$P = \frac{G}{i}\left[\frac{(1 + i)^n - 1}{i(1 + i)^n}\right] - \frac{Gn}{i(1 + i)^n}$$

$$= \frac{G}{i}\left[\frac{(1 + i)^n - 1}{i(1 + i)^n} - \frac{n}{(1 + i)^n}\right] \qquad (2.12)$$

34   Level One

Figure 2.7   Conversion diagram from a uniform gradient to a present worth.

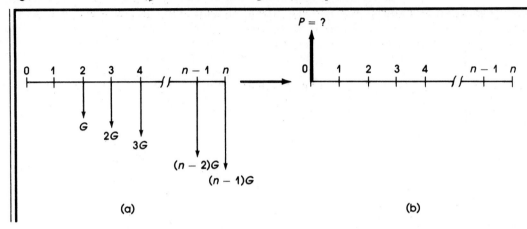

(a)   (b)

Equation (2.12) is the general relation to convert a uniform gradient G for n years into a present worth at year 0; that is, Fig. 2.7a is converted into the equivalent cash flow shown in Fig. 2.7b. The *uniform-gradient present-worth factor* in standard factor notation is

$$(P/G, i\%, n) = \frac{1}{i}\left[\frac{(1+i)^n - 1}{i(1+i)^n} - \frac{n}{(1+i)^n}\right]$$

Note that the gradient starts in year 2 in Fig. 2.7a and P is found in year 0. Equation (2.12) is represented in standard factor notation as

$$P = G(P/G, i\%, n)$$

The equivalent uniform annual worth of the gradient G is found by multiplying the present worth in Eq. (2.12) by the $(A/P, i\%, n)$ factor expression from Eq. (2.7). Using standard factor notation,

$$A = G(P/G, i\%, n)(A/P, i\%, n)$$
$$= G(A/G, i\%, n)$$

Note that in standard form, the equivalent of algebraic cancellation of P can be used to obtain the $(A/G, i\%, n)$ factor. In equation form,

$$A = \frac{G}{i}\left[\frac{(1+i)^n - 1}{i(1+i)^n} - \frac{n}{(1+i)^n}\right]\left[\frac{i(1+i)^n}{(1+i)^n - 1}\right]$$
$$= G\left[\frac{1}{i} - \frac{n}{(1+i)^n - 1}\right] \tag{2.13}$$

The expression in brackets in Eq. (2.13) is called the *uniform-gradient annual-worth factor* and is identified by $(A/G, i\%, n)$. This factor converts Fig. 2.8a into Fig. 2.8b. You should realize that the annual worth is nothing but an A value equivalent to the gradient. Note from Fig. 2.8 that the gradient starts in year 2 and the A values occur from year 1 to year n inclusive.

An F/G factor (*uniform-gradient future-worth factor*) could readily be generated by multiplying the P/G and F/P factors for the same interest rate and n values as

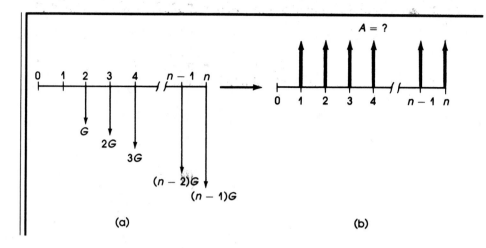

Figure 2.8
Conversion
diagram of a
uniform gradient
to an equivalent
uniform annual
series.

follows:

$$(P/G, i\%, n)(F/P, i\%, n) = (F/G, i\%, n)$$

Such a factor would yield an F value in the same year as the last gradient amount. As an exercise, carry out the multiplication suggested above to obtain the following F/G equation:

$$F = \frac{G}{i}\left[\frac{(1+i)^n - 1}{i} - n\right]$$

In standard factor notation, the formulas used to compute P and A from gradient cash flows are

$$P = G(P/G, i\%, n) \tag{2.14}$$

$$A = G(A/G, i\%, n) \tag{2.15}$$

Even though P in Fig. 2.7b is drawn as in income, it is actually the present worth of the disbursements shown in the gradient of Fig. 2.7a. This convention is used to avoid drawing equivalent cash flows over actual cash flows in more complicated situations. In fact, this confusion would be present in Fig. 2.8 if the annual series were superimposed on the gradient.

Tables A-31 through A-38 present the P/G and A/G factors for interest rates up to 50%. Both factors are arranged and can be used in the same manner as the compound-interest tables. Table 2.4 lists several examples of gradient factors taken

Table 2.4 Examples of gradient factors

Value to be computed	Standard notation	i, %	n	Table	Factor
P	$(P/G, 5\%, 10)$	5	10	A-31	31.652
P	$(P/G, 30\%, 24)$	30	24	A-34	10.943
A	$(A/G, 6\%, 19)$	6	19	A-36	7.287
A	$(A/G, 35\%, 8)$	35	8	A-38	2.060

from the tables. Throughout this chapter we will assume the n value is given in years. In Chap. 3 you will learn how to use the tables in Appendix A for interest periods other than years.

Probs. 2.3 to 2.5

2.6 Derivation of Present Worth of Escalating-Series Equation

In Sec. 2.5 uniform-gradient factors were introduced which could be used for calculating the present worth or equivalent uniform annual amount of a series of payments which increase or decrease by a *constant amount* in consecutive payment periods. Oftentimes, cash flows change by a *constant percentage* in consecutive payment periods instead of by a constant dollar amount. This type of cash flow, called an *escalating series*, is shown in general form in Fig. 2.9, where D represents the dollar amount in year 1 and E represents the escalation rate. The derivation of the equation for calculating the present worth P_E of an escalating series is found by computing the present worth of the Fig. 2.9 cash flow using the P/F formula, $1/(1 + i)^n$.

$$P_E = \frac{D}{(1 + i)^1} + \frac{D(1 + E)}{(1 + i)^2} + \frac{D(1 + E)^2}{(1 + i)^3} + \cdots + \frac{D(1 + E)^{n-1}}{(1 + i)^n}$$

$$= D\left[\frac{1}{1 + i} + \frac{1 + E}{(1 + i)^2} + \frac{(1 + E)^2}{(1 + i)^3} + \cdots + \frac{(1 + E)^{n-1}}{(1 + i)^n}\right] \qquad (2.16)$$

Multiply both sides by $(1 + E)/(1 + i)$:

$$P_E\left(\frac{1 + E}{1 + i}\right) = D\left[\frac{1 + E}{(1 + i)^2} + \frac{(1 + E)^2}{(1 + i)^3} + \frac{(1 + E)^3}{(1 + i)^4} + \cdots + \frac{(1 + E)^n}{(1 + i)^{n+1}}\right] \qquad (2.17)$$

Subtract Eq. (2.16) from Eq. (2.17) and factor out P_E to obtain

$$P_E\left(\frac{1 + E}{1 + i} - 1\right) = D\left[\frac{(1 + E)^n}{(1 + i)^{n+1}} - \frac{1}{1 + i}\right]$$

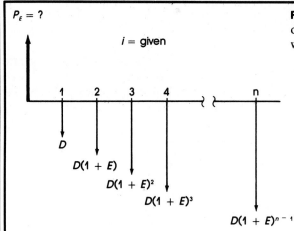

$P_E = ?$

$i = $ given

Figure 2.9 Cash-flow diagram of escalating series and its present worth P_E.

Solve for P_E and simplify to obtain

$$P_E = \frac{D[(1 + E)^n/(1 + i)^{n+1} - 1/(1 + i)]}{[(1 + E)/(1 + i) - 1]}$$

$$= \frac{D[(1 + E)^n/(1 + i)^n - 1]}{E - i} \qquad E \neq i \qquad (2.18)$$

where P_E indicates the present worth of an escalating series starting in year 1 at D dollars. For the condition $E = i$, use L'Hospital's rule to modify Eq. (2.18).

$$\frac{dP_E}{dE} = \frac{D[n(1 + E)^{n-1}(1 + i)^{-n}]}{1} = D\left[\frac{n}{(1 + E)^{1-n}(1 + i)^n}\right] \qquad (2.19)$$

Since $E = i$, P_E may be written from Eq. (2.19) as

$$P_E = D\left(\frac{n}{1 + E}\right) \qquad E = i \qquad (2.20)$$

The equivalent value P_E occurs in the year prior to the cash flow D, as shown in Fig. 2.9. Calculations for escalating series are discussed in Sec. 2.10.

Prob. 2.6

2.7 Interpolation in Interest Tables

Sometimes it is necessary to locate a factor value for an interest rate i or number of periods n that is not in the interest tables. When this occurs, the desired factor value can be obtained in one of three ways: (1) by using the formulas that were derived in Secs. 2.1 to 2.3 and Sec. 2.5, (2) by interpolating between the tabulated values on both sides of the unlisted desired value, or (3) by using a calculator which has the economy factors programmed into it. It is generally easier and faster to use the formulas rather than to interpolate for determining the factor value which corresponds to the unlisted i or n value. However, *linear* interpolation is acceptable and is considered sufficient as long as the values of i or n are not too distant from each other.

The first step in linear interpolation is to set up the known and the unknown values as shown in Table 2.5. A ratio equation is then set up and solved for c, as follows

$$\frac{a}{b} = \frac{c}{d} \qquad \text{or} \qquad c = \frac{a}{b}d \qquad (2.21)$$

where a, b, c, and d represent the differences between the numbers shown in the interest tables. The value of c from Eq. (2.21) is added to or subtracted from value 1, depending

Table 2.5 Linear interpolation setup

i or n	Factor
tabulated	value 1
desired	unlisted
tabulated	value 2

on whether the factor is increasing or decreasing in value, respectively. The following examples illustrate the procedure just described.

Example 2.2

Determine the value of the A/P factor for an interest rate of 7.3% and n of 10 years; that is, $(A/P, 7.3\%, 10)$.

Solution The values of the A/P factor for interest rates of 7 and 8% are listed in Tables A-9 and A-10, respectively. Thus we have the following situation:

$$b \begin{bmatrix} a \begin{bmatrix} \rightarrow 7\% & 0.14238 \\ \rightarrow 7.3\% & X \leftarrow \end{bmatrix} c \\ \rightarrow 8\% & 0.14903 \leftarrow \end{bmatrix} d$$

The unknown X is the desired factor value. From Eq. (2.21),

$$c = \frac{7.3 - 7}{8 - 7}(0.14903 - 0.14238)$$

$$= \frac{0.3}{1} 0.00665$$

$$= 0.00199$$

Since the factor is increasing in value as the interest rate increases from 7 to 8%, the value of c must be *added* to the value of the 7% factor. Thus,

$$X = 0.14238 + 0.00199 = 0.14437$$

Comment It is good practice to check the "reasonableness" of your final answer by verifying that X lies *between* the values of the known factors used in the interpolation in approximately the correct proportions. In this case, since 0.14437 is less than 0.5 of the distance between 0.14238 and 0.14903, the answer seems reasonable. Rather than interpolating, a simpler procedure in some cases may be to use the formula to compute the factor value directly.

Example 2.3

Find the value of the $(P/F, 4\%, 48)$ factor.

Solution From Table A-6 for 4% interest, the values of the P/F factor for 45 and 50 years can be found as follows:

$$b \begin{bmatrix} a \begin{bmatrix} \rightarrow 45 & 0.1712 \\ \rightarrow 48 & X \leftarrow \end{bmatrix} c \\ \rightarrow 50 & 0.1407 \leftarrow \end{bmatrix} d$$

Again, from Eq. (2.21),

$$c = \frac{a}{b}d = \frac{48 - 45}{50 - 45}(0.1712 - 0.1407) = 0.0183$$

Since the value of the factor decreases as n increases, the value of c must be *subtracted* from the value for $n = 45$. Thus,

$X = 0.1712 - 0.0183 = 0.1529$

Additional Example 2.15
Probs. 2.7 to 2.10

2.8 Present-Worth, Future-Worth, and Equivalent Uniform Annual-Series Calculations

The first and probably most important step in solving engineering-economy problems is construction of a cash-flow diagram. In addition to more clearly illustrating "the problem," the cash-flow diagram immediately shows which formulas should be used and whether the conditions of cash flow presented allow straightforward application of the formulas as derived in the preceding sections. Obviously, the formulas can be used only when the cash flow of the problem conforms exactly to the cash-flow diagram for the formulas. For example, the uniform-series factors could not be used if payments or receipts occurred *every other year* instead of every year. It is very important, therefore, to remember the conditions for which the formulas apply. The correct use of the formulas for finding P, F, or A is illustrated in the examples that follow. The equations used are shown in Table 2.2. See the Additional Examples for cases in which some of these formulas cannot be applied.

Example 2.4

If a woman deposits $600 now, $300 two years from now, and $400 five years from now, how much will she have in her account 10 years from now if the interest rate is 5% per year?

Solution The first step is to draw the cash-flow diagram (Fig. 2.10), which indicates that an F value is to be computed. Since each value is different and they do not take place each year, the future worth F is equal to the sum of the individual single payments at

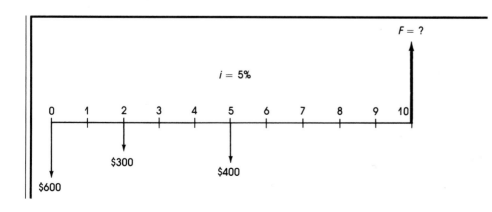

Figure 2.10
Diagram for a future value, Example 2.4.

year 10. Thus,

$$F = 600(F/P, 5\%, 10) + 300(F/P, 5\%, 8) + 400(F/P, 5\%, 5)$$
$$= 600(1.6289) + 300(1.4775) + 400(1.2763)$$
$$= \$1931.11$$

Comment The problem could also be solved by finding the present worth in year 0 of the $300 and $400 deposits using the P/F factors and then finding the future worth of the total.

$$P = 600 + 300(P/F, 5\%, 2) + 400(P/F, 5\%, 5)$$
$$= 600 + 300(0.9070) + 400(0.7835)$$
$$= \$1185.50$$

Then,

$$F = 1185.50(F/P, 5\%, 10)$$
$$= 1185.50(1.6289)$$
$$= \$1931.06$$

It should be obvious that there are a number of ways the problem could be worked, since any year could be used to find the equivalent total of the deposits before finding the value at year 10. As an exercise, you should work the problem using year 5 for finding the equivalent total before determining the final amount in year 10. All answers should be the same, except for round-off error.

Example 2.5

How much money would a man have in his account after 8 years if he deposited $1000 per year for 8 years at 14% per year starting 1 year from now?

Solution The cash-flow diagram is shown in Fig. 2.11. Since the payments start at the end of year 1 and end in the year the future worth is desired, the F/A formula can be used. Thus,

$$F = 1000(F/A, 14\%, 8) = 1000(13.23) = \$13,230$$

Example 2.6

How much money should you be willing to spend now in order to avoid spending $500 seven years from now if the interest rate is 18% per year?

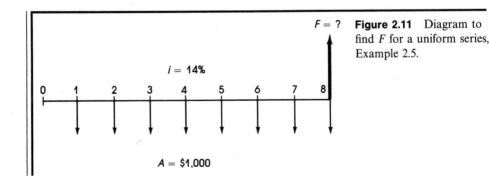

i = 14%

0 1 2 3 4 5 6 7 8

A = $1,000

F = ? **Figure 2.11** Diagram to find *F* for a uniform series, Example 2.5.

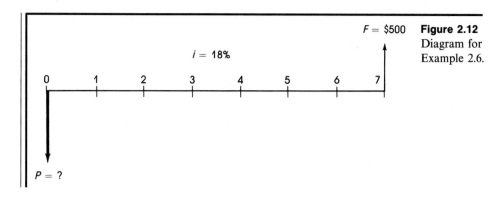

$F = \$500$ **Figure 2.12**
Diagram for
Example 2.6.

Solution   The cash-flow diagram appears in Fig. 2.12. The problem might be easier if it were stated in another manner, such as, what is the present worth of $500 seven years from now if the interest rate is 18% per year; or, what present amount would be equivalent to $500 seven years hence if the interest is 18% per year; or, what initial investment is equivalent to spending $500 seven years from now at an interest rate of 18% per year? In all cases F is given and P is to be computed.

$P = 500(P/F, 18\%, 7) = 500(0.3139) = \$156.95$

Comment   Although there are several ways to state the same problem, the cash-flow diagram remains the same in each case.

Example 2.7

How much money should you be willing to pay now for a note that will yield $600 per year for 9 years starting next year if the interest rate is 17% per year?

Solution   The cash-flow diagram is shown in Fig. 2.13. Since the cash-flow diagram fits the P/A uniform-series formula, the problem can be solved directly.

$P = 600(P/A, 17\%, 9) = 600(4.4506) = \$2670.36$

Comment   You should recognize that P/F factors could be used for each of the 9 receipts and the resulting present worths added to get the correct answer. Another way would be to find the future worth F of the $600 payments and then find the present worth of the

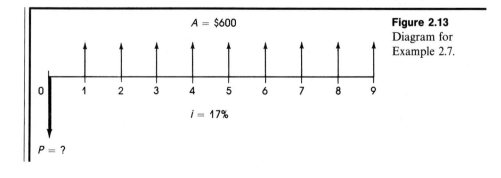

$A = \$600$ **Figure 2.13**
Diagram for
Example 2.7.

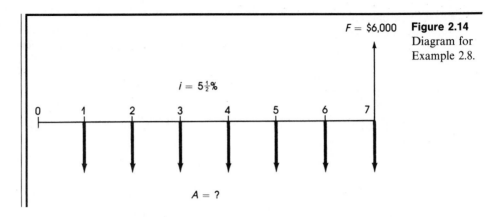

$F = \$6,000$ **Figure 2.14**
Diagram for
Example 2.8.

$i = 5\frac{1}{2}\%$

$A = ?$

F value. There are many ways to solve an engineering-economy problem. Usually only the most direct method will be presented here, but you should work the problems in at least one other way to become more familiar with the use of the formulas.

Example 2.8

How much money must a person deposit every year starting one year from now at $5\frac{1}{2}\%$ per year in order to accumulate $6000 seven years from now?

Solution   The cash-flow diagram is shown in Fig. 2.14. The cash-flow diagram fits the A/F formula as derived. Thus,

$A = 6000(A/F, 5.5\%, 7) = 6000(0.12096) = \$725.76$ per year

Comment   The A/F factor value of 0.12096 was computed using Eq. (2.8).

Additional Example 2.16
Probs. 2.11 to 2.39

2.9   Present Worth and Equivalent Uniform Annual Worth of Conventional Gradients

If the gradient begins between years 1 and 2, year 0 for the gradient and year 0 of the entire cash-flow diagram coincide, and the gradient is referred to as *conventional*. In this case the present worth P or equivalent uniform annual worth A of the *gradient only* can be determined by using Eq. (2.14) or (2.15), respectively. The cash flow that forms the base amount of the gradient must be considered separately. Thus, for cash-flow situations involving conventional gradients:

1. The base amount is the uniform series amount A that begins in year 1 and extends through year n.
2. For an increasing gradient, the gradient amount must be added to the uniform series amount.

3. For a decreasing gradient, the gradient amount must be subtracted from the uniform series amount.

The general equations for calculating the present worth of conventional gradients, therefore, are

$$P_T = P_A + P_G \quad \text{and} \quad P_T = P_A - P_G$$

The present-worth calculation for an increasing gradient is illustrated in the next example.

Example 2.9

A couple plan to start saving money by depositing $500 into their savings account 1 year from now. They estimate that the deposits will increase by $100 each year for 9 years thereafter. What would be the present worth of the investments if the interest rate is 5% per year?

Solution   The cash-flow diagram is shown in Fig. 2.15. Two computations must be made: the first to compute the present worth of the base amount (P_A), and a second to compute the present worth of the gradient (P_G). Then the total present worth P_T would be equal to P_A plus P_G, since P_A and P_G occur in year 0. This is clearly illustrated by the partitioned cash-flow diagram in Fig. 2.16. The present worth would be calculated as follows:

$$
\begin{aligned}
P_T &= P_A + P_G \\
&= 500(P/A, 5\%, 10) + 100(P/G, 5\%, 10) \\
&= 500(7.7217) + 100(31.652) \\
&= \$7026.05
\end{aligned}
$$

Comment   It is important to emphasize again that the gradient factor represents the present worth of the *gradient only*. Any other cash flow involved must be considered separately.

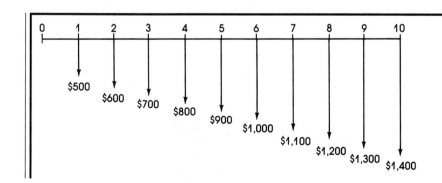

Figure 2.15
Cash flow,
Example 2.9.

Figure 2.16
Partitioned
diagram for
Example 2.9,
$(a) = (b) + (c)$.

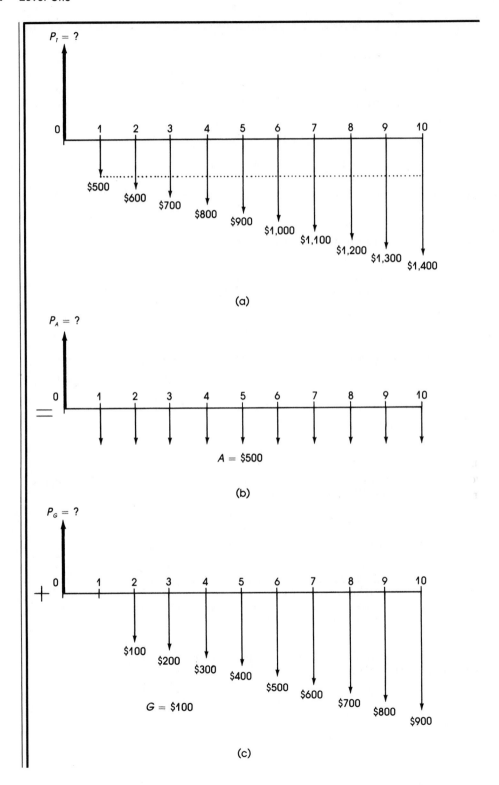

Example 2.10

Work Example 2.9 solving for the equivalent uniform annual worth series.

Solution   Here, too, it is necessary to consider the gradient and the other costs involved in the cash flow separately. From the diagrams shown in Fig. 2.16b and c, the annual worth would be

$$A = A_1 + A_G$$

where A_1 = equivalent annual worth of the base amount \$500
A_G = equivalent annual worth of the gradient = $100(A/G, 5\%, 10)$

Then,

$$A = 500 + 100(A/G, 5\%, 10) = 500 + 100(4.099)$$
$$= \$909.90 \text{ per year for years 1 through 10}$$

Comment   It is often helpful to remember that the present worth of the base amount and gradient can simply be multiplied by the appropriate A/P factor to get A. Here,

$$A = P_T(A/P, 5\%, 10) = 7026.05(0.12950)$$
$$= \$909.87$$

Probs. 2.40 to 2.51

2.10   Calculations Involving an Escalating Series

As shown in Sec. 2.6, the present worth of an escalating series can be determined through the use of Eq. (2.18) or (2.20). The equivalent uniform annual worth or future worth of the series can be calculated by converting the present worth with the appropriate interest factor. The use of Eq. (2.18) is illustrated in Example 2.11.

Example 2.11

A new pickup truck has a first cost of \$8000 and is expected to last 6 years with a \$1300 salvage value. The operating cost of the vehicle is expected to be \$1700 the first year, increasing by 11% per year thereafter. Determine the equivalent present cost of the truck if the interest rate is 8% per year.

Solution   The cash-flow diagram is shown in Fig. 2.17. Since $E \neq i$, Eq. (2.18) is used to calculate P_E. The total P_T is

$$P_T = 8000 + P_E - 1300(P/F, 8\%, 6)$$

$$= 8000 + 1700 \frac{\{[(1 + 0.11)/(1 + 0.08)]^6 - 1\}}{0.11 - 0.08} - 1300(P/F, 8\%, 6)$$

$$= 8000 + 1700(5.9559) - 819.26 = \$17,305.85$$

Figure 2.17
Cash-flow diagram
for Example 2.11.

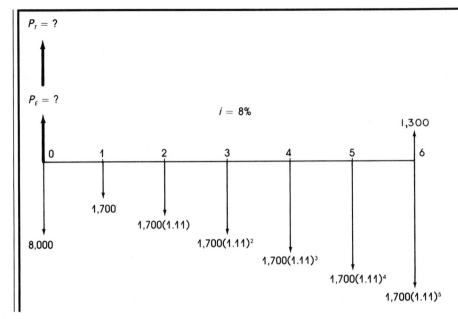

$P_I = ?$

$P_E = ?$

$i = 8\%$

1,300

0   1   2   3   4   5   6

1,700

1,700(1.11)

8,000

1,700(1.11)²

1,700(1.11)³

1,700(1.11)⁴

1,700(1.11)⁵

Comment The equivalent uniform annual worth of the truck could be determined by multiplying the $17,305.85 by $(A/P, 8\%, 6)$.

Probs. 2.52 to 2.57

2.11 Calculation of Unknown Interest Rates

In some cases, the amount of money invested and the amount of money received after a specified number of years are known, and it is desired to determine the interest rate or rate of return. When only a single payment and single receipt, or a uniform series of payments or receipts, or a conventional gradient of payments or receipts are involved, the unknown interest rate can be determined by direct solution of the economy equation. When nonuniform payments or several factors are involved, however, the problem must be solved by the trial-and-error method. In this section, only single-payment, uniform-series, or conventional gradient-series cash-flow problems are considered. The more complicated trial-and-error problems are deferred until Chap. 7, which deals with rate-of-return analysis.

The single-payment formulas can be rather easily rearranged and expressed in terms of i, but for the uniform series and gradient equations, it is necessary to *solve for the value of the factor* and to look up the interest rate in the interest tables. This method is illustrated in the examples that follow.

Example 2.12

(*a*) If a person can make a business investment requiring an expenditure of $3000 now in order to receive $5000 five years from now, what would be the rate of return on the investment?

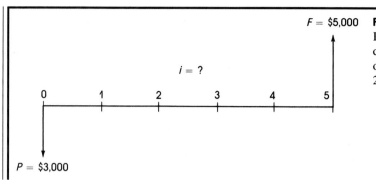

Figure 2.18
Diagram used to determine the rate of return, Example 2.12a.

(b) If the same person can receive 9% per year interest from certificates of deposit, which investment should be made?

Solution

(a) The cash-flow diagram is shown in Fig. 2.18. The interest rate can be found by setting up the P/F or F/P equations and solving directly for the factor value. Using P/F,

$$P = F(P/F, i\%, n)$$

$$3000 = 5000(P/F, i\%, 5)$$

$$(P/F, i\%, 5) = \frac{3000}{5000} = 0.6000$$

From the interest tables, a P/F factor of 0.6000 for $n = 5$ lies between 10 and 11%. Interpolating between these two values using Eq. (2.21), we have

$$c = \left(\frac{0.6209 - 0.6000}{0.6209 - 0.5935}\right)(11 - 10) = \frac{0.0209}{0.0274}(1) = 0.7628$$

Therefore,

$$i = 10 + 0.76 = 10.76\%$$

It is good practice to insert the calculated value back into the equation to verify the correctness of the answer. Thus,

$$3000 = 5000(P/F, 10.76\%, 5)$$

$$= 5000\left[\frac{1}{(1 + 0.1076)^5}\right]$$

$$= 5000(0.5999)$$

$$= 3000 \quad (\text{O.K.})$$

(b) Since 10.76% is greater than the 9% available in certificates of deposit, the person should make the business investment.

Figure 2.19
Diagram to determine the rate of return, Example 2.13.

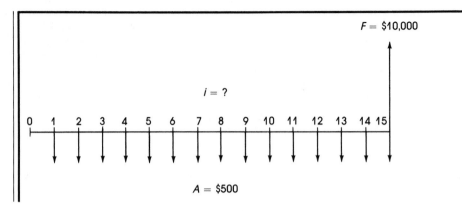

Comment Since the higher rate of return would be received on the business investment, the investor would probably select this option instead of the certificates of deposit. However, the degree of risk associated with the business investment was not specified. Obviously the amount of risk associated with a particular investment is an important parameter and oftentimes causes selection of the lower-rate-of-return investment. Unless specified to the contrary, the problems in this text will assume equal risks for all alternatives.

Example 2.13

Parents wishing to save money for their child's education purchased an insurance policy that will yield $10,000 fifteen years from now. The parents must pay $500 per year for the 15 years starting 1 year from now. What will be the rate of return on their investments?

Solution The cash-flow diagram is shown in Fig. 2.19. Either the A/F or F/A factor could be used. Using A/F,

$$A = F(A/F, i\%, n)$$

$$500 = 10,000(A/F, i\%, 15)$$

$$(A/F, i\%, 15) = 0.0500$$

From the interest tables under the A/F column for 15 years, the value 0.0500 is found to lie between 3 and 4%. By interpolation, $i = 3.98\%$.

Comment As a confirming exercise, insert the value $i = 3.98\%$ into the A/F formula to determine that 0.0500 is obtained.

Probs. 2.58 to 2.65

2.12 Calculation of Unknown Years

In breakeven economic analysis, it is sometimes necessary to determine the number of years (periods) required before an investment pays off. Other times it is desirable to be able to determine when given amounts of money will be available from a proposed investment. In these cases, the unknown value is n, and techniques similar to those of the preceding section on unknown interest rates can be used to find n.

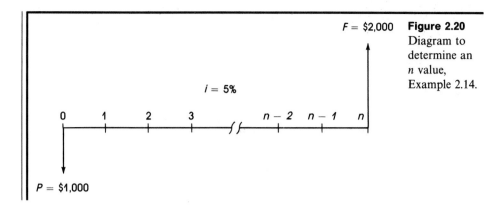

Figure 2.20
Diagram to
determine an
n value,
Example 2.14.

Though some of these problems can be solved directly for n by proper manipulation of the single-payment and uniform-series formulas, it is generally easier to solve for the factor value and interpolate in the interest tables, as illustrated below.

Example 2.14

How long will it take for $1000 to double if the interest rate is 5% per year?

Solution The cash-flow diagram is shown in Fig. 2.20. The problem can be solved using either the F/P or P/F factor. Using the P/F factor,

$$P = F(P/F, i\%, n)$$

$$1000 = 2000(P/F, 5\%, n)$$

$$(P/F, 5\%, n) = 0.500$$

From the 5% interest table, the value 0.500 under the P/F column lies between 14 and 15 years. By interpolation, $n = 14.2$ years.

Comment Problems of this type become more complicated when two or more non-uniform payments are involved. See the Additional Examples for an illustration using trial and error.

Additional Example 2.17
Probs. 2.66 to 2.74

Additional Examples

Example 2.15

A new building has been purchased by Waldorf Concession Stands, Inc. The present worth of future maintenance costs is to be calculated with a P/A factor. If $i = 13\%$ per year and the life is expected to be 42 years, estimate what factor value is correct by using two-way interpolation in the tables at $i = 12$ and 15% per year.

Solution The P/A factor requires two-way interpolation for i and n. First, we will find the P/A factor for $i = 13\%$ at $n = 40$ and $n = 45$ using 12 and 15%.

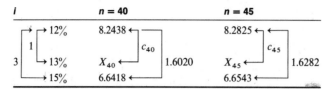

The subscripts correspond to the n value for which the factor is computed.

$c_{40} = \frac{1}{3}(1.6020) = 0.5340$ $\quad$ $X_{40} = 8.2438 - 0.5340 = 7.7098$

$c_{45} = \frac{1}{3}(1.6282) = 0.5427$ $\quad$ $X_{45} = 8.2825 - 0.5427 = 7.7398$

Now estimate the P/A factor for $n = 42$.

$X_{42} = 7.7098 + \frac{2}{5}(0.0300) = 7.7218$

Thus, we have

$(P/A, 13\%, 42) = 7.7218$

The correct value using Eq. (2.6) is 7.6469.

Example 2.16

Explain why the uniform-series factors *cannot* be used to compute P or F *directly* for the cash flows of Fig. 2.21.

Solution
(a) The P/A factor cannot be used to compute P since the $100-per-year receipt does not occur each year from year 1 through year 5.
(b) Since there is no $A = \$550$ in year 5, the F/A factor cannot be used. The relation $F = 550(F/A, i\%, 4)$ would furnish the future worth in year 4, not year 5 as desired.
(c) The first $A = \$1000$ value occurs in year 2. Use of the relation $P = 1000(P/A, i\%, 4)$ will compute P in year 1, not year 0.
(d) The receipt values are unequal; thus the relation $F = A(F/A, i\%, 3)$ cannot be used to compute F.

Comment Naturally, there are ways to compute P or F without resorting to only P/F and F/P factors; these methods are discussed in Chap. 4.

Example 2.17

If an investor deposits $2000 now, $500 three years from now, and $1000 five years from now, how many years will it take from now for his total investment to amount to $10,000 if the interest rate is 6% per year?

Solution The cash-flow diagram (Fig. 2.22) requires that the following equation be satisfied:

$F = P_1(F/P, i\%, n) + P_2(F/P, i\%, n - 3) + P_3(F/P, i\%, n - 5)$

$10,000 = 2000(F/P, 6\%, n) + 500(F/P, 6\%, n - 3) + 1000(F/P, 6\%, n - 5)$

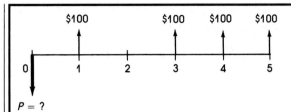

Figure 2.21 Cash-flow diagrams, Example 2.16.

(a)

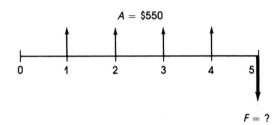

(b)

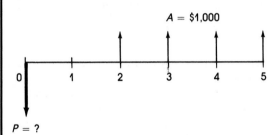

(c)

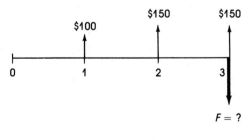

(d)

Figure 2.22
Diagram to determine *n* for a nonuniform series, Example 2.17.

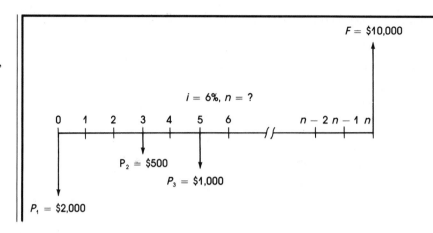

Table 2.6 Trial-and-error solution for *n*

n	2,000(F/P, 6%, n)	500(F/P, 6%, n − 3)	1,000(F/P, 6%, n − 5)	F	Comment
5	$2,676.40	$ 561.80	$1,000.00	$ 4,238.20	Too small
15	4,793.20	1,006.10	1,790.80	7,590.10	Too small
20	6,414.20	1,346.40	2,396.60	10,157.20	Too large

This relation must be solved by selecting various values of *n* and solving until the equation is satisfied. Interpolation for *n* will be necessary to obtain an exact equality. The procedure shown in Table 2.6 indicates that 20 years is too long and 15 years is too short. Therefore, we interpolate between 15 and 20 years.

$$c = \frac{10{,}000 - 7590.10}{10{,}157.20 - 7590.10}(20 - 15) = 4.69$$

$$n = 15 + c$$
$$= 19.69 \text{ years}$$

Comment   The final answer is close to 20 years, as it should be, since the *F* value calculated for *n* = 20 in Table 2.6 is close to the desired value of $10,000.

Problems   **2.1**   Construct the cash-flow diagrams and derive the SPPWF, USPWF, and USCAF formulas for beginning-of-year amounts rather than the end-of-year convention. The *P* value should take place at the same time as for the end-of-year convention.

2.2   Find the correct numerical value for the following factors from the interest tables:

1. $(A/F, 6\%, 20)$
2. $(F/P, 8\%, 5)$
3. $(F/A, 20\%, 8)$
4. $(A/P, 10\%, 25)$
5. $(F/A, 15\%, 20)$

2.3 Construct the cash-flow diagram for the following deposits (year $= k$):

Year, k	1	2	3	4–7
Deposit, $	60	60	60	$100 + 10(k - 4)$

2.4 Use the factor formulas to show that the following expressions are correct for the uniform-gradient factors:
(a) $(A/G, i\%, n) = (P/G, i\%, n)(A/P, i\%, n)$
(b) $(P/G, i\%, n) = (A/G, i\%, n)(P/A, i\%, n)$

2.5 Find the value of the factor to convert a 12-year gradient to an equivalent uniform annual series if interest is at 20% per year.

2.6 What is the difference between a conventional uniform gradient and an escalating series?

2.7 Find the numerical value of the following factors by (a) interpolation and (b) use of the appropriate formula:

1. $(P/A, 8.5\%, 13)$
2. $(F/A, 37\%, 24)$
3. $(P/F, 7.7\%, 9)$
4. $(A/F, 49\%, 28)$

2.8 Find the numerical value of the following factors by (a) interpolation and (b) use of the appropriate formula:

1. $(F/P, 3\%, 39)$
2. $(A/P, 10\%, 9.8)$
3. $(A/F, 6\%, 52)$
4. $(P/F, 18\%, 37)$

2.9 Find the numerical value of the following factors by (a) interpolation and (b) use of the appropriate formula:

1. $(P/F, 3.8\%, 7.7)$
2. $(P/A, 9.7\%, 68)$
3. $(F/A, 23\%, 11.6)$
4. $(A/F, 17\%, 23)$

2.10 Find the correct factor value for the following from the interest tables and, when necessary, interpolation:

1. $(P/G, 10\%, 8)$
2. $(A/G, 15\%, 5)$
3. $(A/G, 17\%, 23)$
4. $(P/G, 28\%, 41)$

2.11 How much money will Mr. Jones have in his bank account in 12 years if he deposits $3500 now at an interest rate of 12% per year?

2.12 If Ms. James wants to have $8000 in her account 8 years from now to buy a new sports car, how much money will she have to deposit every year starting 1 year from now if the interest rate is 9% per year?

2.13 What is the present worth of $700 now, $1500 four years from now, and $900 six years from now at an interest rate of 8% per year?

2.14 How much money would be accumulated in 14 years if $1290 were deposited each year starting 1 year from now at an interest rate of 15% per year?

2.15 If Mr. Savum borrowed $4500 with a promise to make 10 equal annual payments starting 1 year from now, how much would his payments be if the interest rate were 20% per year?

2.16 How much money must be deposited in a lump sum 4 years from now in order to accumulate $20,000 eighteen years from now if the interest rate is 8% per year?

2.17 Ms. Lendup would like to know the present worth of a 35-year $600-per-year annuity beginning 1 year from now at an interest rate of $6\frac{1}{2}\%$ per year?

2.18 How much money can you borrow now if you promise to pay $600 per year beginning 1 year from now for 7 years at an interest rate of 17% per year?

2.19 How much money now will be equivalent to $5000 six years from now at an interest rate of 7% per year?

2.20 What uniform annual amount must you deposit for 5 years to have an equivalent present-investment sum of $9000 at an interest rate of 10% per year?

2.21 How much money would be accumulated in 25 years if $800 were deposited 1 year from now, $2400 six years from now, and $3300 eight years from now, all at an interest rate of 18% per year?

2.22 What is the future worth of a uniform annual series of $1000 for 10 years at an interest rate of $8\frac{3}{4}\%$ per year?

2.23 What is the present worth of $600 per year for 52 years beginning 1 year from now at an interest rate of 10% per year?

2.24 How much money would be accumulated in 43 years from an annual deposit of $1200 per year starting 1 year from now if the interest rate were $19\frac{1}{4}\%$ per year?

2.25 I plan to buy some property which my uncle has generously offered to me. The payment scheme is $700 every other year through year 8 starting 2 years from now. What is the present worth of this generous offer if the interest rate is 17% per year?

2.26 If a college student can save $600 per year from her part-time job, how long will it take her to save enough money to purchase a $2500 dune buggy if she can get 10% per year interest on her money?

2.27 What single amount of money will have to be deposited 4 years from now if you want to have $8000 in your account 11 years from now? Use an interest rate of 10% per year.

2.28 For the cash-flow diagram shown below, calculate the amount of money in year 3 that would be equivalent to all of the cash flows shown, using an interest rate of 11% per year.

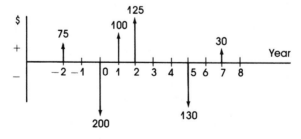

2.29 What uniform year-end payment will be required to pay off a $25,000 debt in 9 years if the first payment is to be made 1 year from now and the interest rate is 15% per year?

2.30 How much money must be deposited into a savings account each year, starting in 1985, if you want to have $150,000 when you retire in the year 2020? The interest rate is 16% per year.

2.31 If annual deposits of $1000 are made into a savings account for 30 years beginning 1 year from now, how much will be in the fund immediately after the last deposit if the fund pays interest at a rate of 10% per year?

2.32 What single payment 12 years from now would be equivalent to a payment of $6200 five years from now at an interest rate of 13% per year?

2.33 How much money would have to be invested at the end of each year for 25 years starting next year into a fund which is to amount to $180,000 at the end of the 25 years? Assume that the interest rate is 14% per year.

2.34 The 4-Sight Steel Fabricating Co. is planning to make two equal deposits such that 10 years from now the company will have $49,000 to replace a small machine. If the first deposit is to be made 2 years from now and the second is to be made 8 years from now, how much must be deposited each time if the interest rate is 15% per year?

2.35 Rework Prob. 2.34 for deposits made in years 1 and 9.

2.36 The GRQ Company is planning to borrow $58,000 at 15% per year. The company expects to repay the loan with six equal annual payments at the end of each year, beginning 1 year after the loan is received. Determine the amount of interest that would be charged in the first and second year's payment.

2.37 The Lee-Key Roof Company has offered a small company two methods by which to pay for some needed roof repairs. Method 1 involves a payment of $2500 as soon as the job is done, i.e., now. Method 2 allows the company to defer payment for 5 years, at which time a payment of $5000 would be required. If the interest rate is 16% per year, compute the P value for each method and select the one with the smaller P value.

2.38 If a company has an opportunity to invest $33,000 now for 14 years at 15% per year simple interest or 13% per year compound interest, which investment should be made?

2.39 A plant manager is trying to decide whether to buy a new machine now or wait and purchase a similar one 3 years from now. The machine at the present time would cost $25,000, but 3 years from now it is expected to cost $39,000. If the interest rate the company uses is 20% per year, should the plant manager buy now or should she buy 3 years from now?

2.40 A cash-flow sequence starts in year 1 at $200 and increases to $354 in year 8. Do the following: (*a*) construct the cash-flow diagram; (*b*) determine the amount of the annual gradient; (*c*) locate the gradient present worth on the diagram; and (*d*) determine the value of n for the gradient factor.

2.41 For the cash-flow shown below, calculate (*a*) the equivalent uniform annual cost in years 1 through 5 and (*b*) the present worth of the cash flow. Assume that the interest rate is 12% per year.

Year	1	2	3	4	5
Cash flow, $	5000	5400	5800	6200	6600

2.42 For the diagram shown below, find the value of x that will make the negative cash flows equal to the positive cash flow of $800 at time 0. Assume that $i = 15\%$ per year.

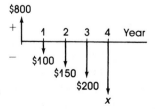

2.43 Find the present worth of an income series wherein the cash flow in year 1 is $1200 and it increases by $300 per year through year 11. Use an interest rate of 15% per year.

2.44 Determine the equivalent uniform annual worth value for a machine which has costs of $8500 at the end of year 1 and costs increasing by $500 per year through year 8. Assume that the interest rate is 25% per year.

2.45 Determine the present worth of a machine which has an initial cost of $10,000 and operating costs of $1200 the first year, $1350 the second year, and amounts increasing by $150 per year through year 10. Use an interest rate of 18% per year.

2.46 Determine the equivalent uniform annual worth of a process which will involve an initial outlay of $70,000 followed by costs of $8000 in year 1, $9000 in year 2, and amounts increasing by $1000 per year through its 13-year life. Assume that the interest rate is 14% per year.

2.47 A company borrows $15,000 at an interest rate of 15% per year with the agreement that the loan will be repaid over an 8-year period. The repayment scheme will be such that each payment will be $250 larger than the preceding one, with the first payment to be made 1 year after the loan is negotiated. Determine the amount of the third payment.

2.48 Assume that the GRQ Company wants to have $500,000 available for investment 10 years from now. The company plans to invest $4000 the first year and amounts increasing by a uniform gradient thereafter. If the company's interest rate is 20% per year, what must be the size of the gradient in order for GRQ to meet its objective?

2.49 Find the present worth of the cash flow shown below, using an interest rate of 18% per year.

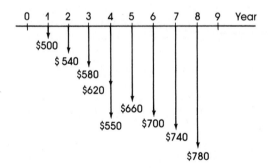

2.50 Calculate the value of G in the cash-flow series shown below such that the present worth of the cash flow will be $28,500 when the interest rate is 15% per year.

Year	1	2	3	4	5	6	7
Cash flow, $	4000	4000 + G	4000 + 2G	4000 + 3G	4000 + 4G	4000 + 5G	4000 + 6G

2.51 The Cimeron Cinnamon Co. is undertaking a program to reduce operating costs. The vice president in charge of operations has established a goal of saving an equivalent total present amount of $90,000 in the next 4 years through reduced finished-product losses. He estimates that the company will be able to save $40,000 the first year, but cost reductions will become more difficult each year. If the reductions are expected to follow a uniformly decreasing gradient, what must the reductions be in years 2, 3, and

4 in order for the company to meet its established goal? The company's interest rate is 15% per year.

2.52 Calculate the present worth of $2500 in year 1, $2750 in year 2, and amounts increasing by 10% per year thereafter until year 11, if the interest rate is 16% per year.

2.53 Calculate the present worth of a machine which has an initial cost of $17,000, a salvage value of $2000, and an operating cost of $9000 in year 1, $9180 in year 2, and amounts increasing by 2% each year for 12 years. Use an interest rate of 17% per year.

2.54 If you deposit $300 now into a savings account and increase your deposit by 10% each month, how much will you have after 3 years if the account earns interest at a rate of 1% per month?

2.55 A company is planning to make deposits such that each one is 12% larger than the preceding one. How large must the first deposit be (at the end of year 1) if the company wants to accumulate $21,000 by the end of year 16? Assume that the interest rate the fund earns is 12% per year.

2.56 What amount would the deposit in Prob. 2.55 have to be if the fund earns interest at a rate of 14% per year?

2.57 How long will it take for a savings fund to accumulate to $15,000 if $1000 is deposited at the end of year 1 and the amount of the deposit is increased by 10% each year? Assume that the interest rate is 10% per year.

2.58 If a company invests $3000 now and will receive $5000 twelve years from now, what is the rate of return on the investment?

2.59 The Playmore Company, a group of investors, is considering the attractiveness of purchasing a piece of property for $18,000. The group anticipates that the value of the property will increase to $21,500 in 5 years. What is the rate of return on this investment?

2.60 The Juicer Utility Company has a retirement program in which employees invest $1200 every year for 25 years, starting 1 year after their initial employment. If the company guarantees at least $50,000 at the time of retirement, what is the rate of return on the investment?

2.61 If a person borrows $6000 for a new car with an agreement to pay the loan company $2500 per year for 3 years, what is the interest rate on the loan?

2.62 At what interest rate per year would $1500 accumulate to $2500 in 5 years?

2.63 If a person purchased stock for $8000 twelve years ago and received dividends of $1000 per year, what is the rate of return on the investment?

2.64 At what rate of return would $900 per year for 9 years accumulate to (a) $8100 and (b) $10,000?

2.65 The Gotcha Loan Company offers qualified borrowers the opportunity to borrow $1000 now and repay the loan in 10 "easy" yearly installments of $155, the first installment due 1 year from now. At what interest rate are their customers borrowing money?

2.66 How many years would it take for $1750 to triple in value if the interest rate were 12% per year?

2.67 If a person deposits $5000 now at 8% per year interest and plans to withdraw $500 per year every year starting 1 year from now, how long can the full withdrawals be made?

2.68 What is the minimum number of years a person must deposit $400 per year in order to have at least $10,000 on the date of the last deposit? Use an interest rate of 8% per year and round off to the higher integer year.

2.69 What is the minimum number of year-end deposits that have to be made before the total value of the deposits is at least ten times greater than the value of a single year-end deposit if the interest rate is $12\frac{1}{2}$% per year?

2.70 How many years would it take for a $800 deposit now and a $1600 deposit 3 years from now to accumulate to $3500 at an interest rate of 8% per year?

2.71 If you start saving money by depositing $1000 per year into a bank account which pays 11%-per-year interest, how many years will it take to accumulate $10,000 if the first deposit is made 1 year from now?

2.72 How long would it take to recover an investment of $10,000 which pays 15% per year interest if $1000 were saved the first year and $1100 the second year and the amounts continued to increase by $100 per year?

2.73 If a company invests $15,000 in energy-saving devices, how long will it take to recover the investment if the annual savings are $2000 the first year and they increase by $500 per year thereafter? Use an interest rate of 25% per year.

2.74 The cash flow associated with a particular project is expected to be $2500 in year 1 and $2800 in year 2 and amounts increasing by $300 per year. How many years will it take to recover an initial investment of $20,000 if the interest rate is 18% per year?

Nominal and Effective Interest Rates and Continuous Compounding

3

This chapter teaches you how to make engineering-economy computations using interest periods and compounding frequencies which are other than 1 year. The material of this chapter is helpful for handling personal financial matters which often involve monthly, daily, or continuous time periods.

Section Objectives

After completing this chapter, you should be able to do the following:

3.1 Define *compounding period*, *nominal interest rate*, *effective interest rate*, and *payment period*.

3.2 Write the formula for computing the effective interest rate and define each term in the formula.

3.3 Compute the *effective interest rate* and find the numerical value of any specific engineering-economy factor for that rate, given the nominal interest rate and number of compounding periods.

3.4 Derive and use the effective interest-rate formula for *continuous compounding*.

3.5 Calculate the present worth or future worth of a specified cash flow when the payment period is equal to or *longer* than the compounding period, given the amount and times of the payments, the compounding period, and the nominal or effective interest rate.

3.6 Calculate the present worth or future worth of a specified cash flow when the payment period is *shorter* than the compounding period, given the amount and times of the payments, the compounding period, and the nominal or effective interest rate.

Study Guide

3.1 Nominal and Effective Rates

In Chap. 1, the concepts of simple- and compound-interest rates were introduced. The basic difference between the two is that compound interest includes interest on the interest earned in the previous period while simple interest does not. In essence, nominal and effective interest rates have the same relationship to each other as do simple and compound interest. The difference is that nominal and effective interest rates are used when the *compounding period* (or interest period) is less than 1 year. Thus, when an interest rate is expressed over a period of time shorter than a year, such as 1% per month, the terms *nominal* and *effective* interest rates must be considered.

A dictionary definition of the word "nominal" is purported, so-called, ostensible, or professed. These synonyms imply that a nominal interest rate is not a correct, actual, genuine, or effective rate. As discussed below, nominal interest rates must be converted into effective rates in order to accurately reflect time–value considerations. Before discussing effective rates, however, let us define a *nominal interest rate r* as the period interest rate times the number of periods. In equation form,

$$r = \text{interest rate per period} \times \text{number of periods}$$

A nominal interest rate can be found for any time period which is longer than the originally stated period. For example, a period interest rate listed as 1.5% per month could also be expressed as a *nominal* 4.5% per quarter (that is, 1.5% per month × 3 months per quarter), or 9.0% per semiannual period, or 18% per year, or 36% per 2 years, etc. The calculation of a nominal interest rate obviously ignores the time value of money, similar to the calculation of simple interest. When the time value of money *is* taken into consideration in calculating interest rates from period interest rates, the rate is called an *effective interest rate*. Just as was true for nominal interest rates, effective rates can be determined for any time period longer than the originally stated period as shown in the next two sections of this chapter. It is important to recognize that all of the formulas derived in Chap. 2 were based on compound interest and, therefore, *only* effective interest rates can be used in the equations.

A discussion of nominal and effective interest rates would not be complete without commenting about the various ways that interest rates can be expressed. There are three general ways of expressing interest rates as shown by the three groups of statements in Table 3.1. The three statements in the top part of the table show that an interest rate can be stated over some designated time period *without specifying the compounding period.* Such interest rates are assumed to be *effective rates* with the compounding period (CP) assumed to be that over which the interest is stated.

For the interest statements presented in the middle of Table 3.1, three conditions prevail: (1) The compounding period is identified, (2) this compounding period is shorter than the time period over which the interest is stated, and (3) the interest rate is neither designated as nominal nor effective. In such cases, the interest rate is

Table 3.1 Various interest statements and their interpretation

Interest rate statement	Interpretation	Comment
$i = 12\%$ per year	$i = $ *effective* 12% per year compounded yearly	When no compounding period is given, interest rate is an effective rate, with compounding period assumed to be equal to stated time period.
$i = 1\%$ per month	$i = $ *effective* 1% per month compounded monthly	
$i = 3\frac{1}{2}\%$ per quarter	$i = $ *effective* $3\frac{1}{2}\%$ per quarter compounded quarterly	
$i = 8\%$ per year, compounded monthly	$i = $ *nominal* 8% per year compounded monthly	When compounding period is given without stating whether the interest rate is nominal or effective, it is assumed to be nominal. Compounding period is as stated.
$i = 4\%$ per quarter compounded monthly	$i = $ *nominal* 4% per quarter compounded monthly	
$i = 14\%$ per year compounded semiannually	$i = $ *nominal* 14% per year compounded semiannually	
$i = $ effective 10% per year compounded monthly	$i = $ *effective* 10% per year compounded monthly	If interest rate is stated as an effective rate, then it *is* an effective rate. If compounding period is not given, compounding period is assumed to coincide with stated time period.
$i = $ effective 6% per quarter	$i = $ *effective* 6% per quarter compounded quarterly	
$i = $ effective 1% per month compounded daily	$i = $ *effective* 1% per month compounded daily	

assumed to be *nominal* and the *compounding period is equal to that which is stated.*

For the third group of interest rate statements, the word "effective" precedes or follows the specified interest rate and the compounding period is also given. These interest rates are obviously effective rates over the respective time periods stated. Likewise, the compounding periods are equal to those which are stated. Similarly, if the word "nominal" had preceded any of the interest statements, the interest rate would be a nominal rate.

The importance of being able to recognize whether a given interest rate is nominal or effective cannot be overstated with respect to the reader's understanding the remainder of the material in this chapter and indeed the rest of the book. Table 3.2 is a listing of several interest statements (col. 1) along with their interpretations (cols. 2 and 3).

Table 3.2 Specific examples of interest statements and interpretations

Interest statement (1)	Type of interest (2)	Compounding period (3)
15% per year compounded monthly	Nominal	Monthly
15% per year	Effective	Yearly
Effective 15% per year compounded monthly	Effective	Monthly
20% per year compounded quarterly	Nominal	Quarterly
Nominal 2% per month compounded weekly	Nominal	Weekly
2% per month	Effective	Monthly
2% per month compounded monthly	Effective	Monthly
Effective 6% per quarter	Effective	Quarterly
Effective 2% per month compounded daily	Effective	Daily
1% per week compounded continuously	Nominal	Continuously
0.1% per day compounded continuously	Nominal	Continuously

Now that the concept of nominal and effective interest rates has been introduced, in addition to considering the compounding period (which is also known as the interest period), it will also be necessary to consider the frequency of the payments or receipts within the cash-flow time interval. For simplicity, the frequency of the payments or receipts is known as the *payment period* (PP). It is important to distinguish between the compounding period and the payment period because in many instances the two do not coincide. For example, if a company deposited money each month into an account that pays a nominal interest rate of 14% per year compounded semiannually, the payment period would be 1 month while the compounding period would be 6 months. Similarly, if a person deposits money each year into a savings account which compounds interest quarterly, the payment period is 1 year, while the compounding period is 3 months. Hereafter, for problems which involve either uniform-series or uniform-gradient cash-flow amounts, it will be necessary to determine the relationship between the compounding period and the payment period as a first step in the solution of the problem (Sec. 3.5).

Probs. 3.1 to 3.6

3.2 Effective Interest-Rate Formulation

To illustrate the difference between nominal and effective interest rates, the future worth of $100 after 1 year is determined using both rates. If a bank pays 12% interest compounded annually, the future worth of $100 using an interest rate of 12% per year is

$$F = P(1 + i)^n = 100(1.12)^1 = \$112.00 \tag{3.1}$$

On the other hand, if the bank pays interest that is compounded semiannually, the future worth must include the *interest on the interest earned in the first period*. An interest rate of 12% per year compounded semiannually means that the bank will pay 6% interest two times per year (i.e., every 6 months). Figure 3.1 is the cash-flow diagram for semiannual compounding for a nominal interest rate of 12% per year. Equation (3.1) obviously ignores the interest earned in the first period. Taking interest into consideration, therefore, the future worth of the $100 would actually be

Figure 3.1 Cash-flow diagram for semiannual compounding periods.

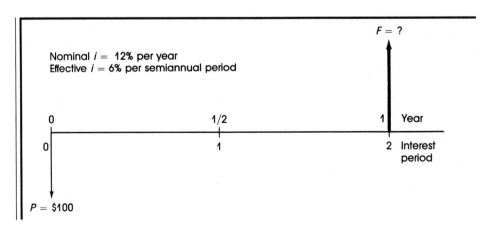

$$F = 100(1 + 0.06)^2 = 100(1.06)^2$$
$$= 100(1.1236)$$
$$= \$112.36 \tag{3.2}$$

where 6% is the *effective semiannual interest rate*. The effective annual interest rate, therefore, would be 12.36%, instead of 12%, since $12.36 instead of $12.00 interest would be earned in 1 year.

The equation for acquiring the effective interest rate from the nominal interest rate may be generalized as follows

$$i = \left(1 + \frac{r}{m}\right)^m - 1 \tag{3.3}$$

where i = effective interest rate per period
r = nominal interest rate per period
m = number of compounding periods

Equation (3.3) is referred to as the effective interest-rate equation. As the number of compounding periods increases, m approaches infinity, in which case the equation represents the interest rate for *continuous compounding*. A detailed discussion of this subject is presented in Sec. 3.4.

Prob. 3.7

3.3   Calculation of Effective Interest Rates

Effective interest rates can be calculated for any time period longer than the actual compounding period through the use of Eq. (3.3). That is, an effective interest rate of 1% per month, for example can be converted into effective rates per quarter, per semiannual period, per 1 year, per 2 years, or per any other period longer than 1 month (the compounding period). It is important to remember that in Eq. (3.3) the time units on i and r must always be the same. Thus, if an effective interest rate per semiannual period is desired, then r must be the nominal rate per semiannual period. The m in Eq. (3.3) is always equal to the number of times that interest would be compounded in the period of time over which i is sought. The next example illustrates these relationships.

Example 3.1

A national credit card carries an interest rate of 2% per month on the unpaid balance. (*a*) Calculate the effective rate per semiannual period. (*b*) If the interest rate is stated as 5% per quarter, find the effective rates per semiannual and annual time periods.

Solution
(*a*) In this part of the example, the compounding period is monthly. Since the effective interest rate per semiannual period is what is desired, the r in Eq. (3.3) must be the nominal rate per 6 months, or

r = 2% per month × 6 months per semiannual period
  = 12% per year

The m in Eq. (3.3) is equal to 6, since interest would be compounded 6 times in a 6 month time period. Thus,

$$i \text{ per 6 months} = \left(1 + \frac{0.12}{6}\right)^6 - 1$$

$$= 12.62\%$$

(b) For an interest rate of 5% per quarter, the compounding period is quarterly. Therefore, in a semiannual period, $m = 2$ and $r = 10\%$. Thus,

$$i \text{ per 6 months} = \left(1 + \frac{0.10}{2}\right)^2 - 1$$

$$= 10.25\%$$

The effective interest rate per year can be determined using $r = 20\%$ and $m = 4$, as follows:

$$i \text{ per year} = \left(1 + \frac{0.20}{4}\right)^4 - 1$$

$$= 21.55\%$$

Comment   Note that the term r/m in Eq. (3.3) is always equal to the interest rate (effective) per compounding period. In part (a), this was 2% per month while in part (b), it was 5% per quarter.

Table 3.3 presents the effective interest rate i for various nominal interest rates r using Eq. (3.3) and compounding periods of 6 months, 3 months, 1 month, 1 week, and 1 day. The continuous-compounding column is discussed in the next section. Note that as the interest rate increases, the effect of more frequent compounding becomes more pronounced. When Eq. (3.3) is used to find an effective interest rate, the answer is usually an interest rate which is not an integer number, as illustrated in Example 3.1 and Table 3.3. When this occurs, the factor values desired must be obtained either through interpolation in the interest tables or through direct use of the equations developed in Chap. 2. Example 3.2 shows these calculations.

Example 3.2

A university credit union advertises that its interest rate on loans is 1% per month. Calculate the annual effective interest rate and use the tables in Appendix A to find the corresponding P/F factor for $n = 8$.

Solution   Substituting $r/m = 0.01$ and $m = 12$ into Eq. (3.3) yields

$$i = (1 + 0.01)^{12} - 1$$

$$= 1.1268 - 1$$

$$= 0.1268 \quad (12.68\%)$$

Table 3.3 Tabulation of effective interest rates for nominal period rates

Nominal rate, r%	Semiannually (m = 2)	Quarterly (m = 4)	Monthly (m = 12)	Weekly (m = 52)	Daily (m = 365)	Continuously (m = ∞; $e^r - 1$)
0.25	0.250	0.250	0.250	0.250	0.250	0.250
0.50	0.501	0.501	0.501	0.501	0.501	0.501
0.75	0.751	0.752	0.753	0.753	0.753	0.753
1.00	1.003	1.004	1.005	1.005	1.005	1.005
1.50	1.506	1.508	1.510	1.511	1.511	1.511
2	2.010	2.015	2.018	2.020	2.020	2.020
3	3.023	3.034	3.042	3.044	3.045	3.046
4	4.040	4.060	4.074	4.079	4.081	4.081
5	5.063	5.095	5.116	5.124	5.126	5.127
6	6.090	6.136	6.168	6.180	6.180	6.184
7	7.123	7.186	7.229	7.246	7.247	7.251
8	8.160	8.243	8.300	8.322	8.328	8.329
9	9.203	9.308	9.381	9.409	9.417	9.417
10	10.250	10.381	10.471	10.506	10.516	10.517
11	11.303	11.462	11.572	11.614	11.623	11.628
12	12.360	12.551	12.683	12.734	12.745	12.750
13	13.423	13.648	13.803	13.864	13.878	13.883
14	14.490	14.752	14.934	15.006	15.022	15.027
15	15.563	15.865	16.076	16.158	16.177	16.183
16	16.640	16.986	17.227	17.322	17.345	17.351
17	17.723	18.115	18.389	18.497	18.524	18.530
18	18.810	19.252	19.562	19.684	19.714	19.722
19	19.903	20.397	20.745	20.883	20.917	20.925
20	21.000	21.551	21.939	22.093	22.132	22.140
21	22.103	22.712	23.144	23.315	23.358	23.368
22	23.210	23.883	24.359	24.549	24.598	24.608
23	24.323	25.061	25.586	25.796	25.849	25.860
24	25.440	26.248	26.824	27.054	27.113	27.125
25	26.563	27.443	28.073	28.325	28.390	28.403
26	27.690	28.646	29.333	29.609	29.680	29.693
27	28.823	29.859	30.605	30.905	30.982	30.996
28	29.960	31.079	31.888	32.213	32.298	32.313
29	31.103	32.309	33.183	33.535	33.626	33.643
30	32.250	33.547	34.489	34.869	34.968	34.986
31	33.403	34.794	35.807	36.217	36.327	36.343
32	34.560	36.049	37.137	37.578	37.693	37.713
33	35.723	37.313	38.478	38.952	39.076	39.097
34	36.890	38.586	39.832	40.339	40.472	40.495
35	38.063	39.868	41.198	41.740	41.883	41.907
40	44.000	46.410	48.213	48.954	49.150	49.182
45	50.063	53.179	55.545	56.528	56.788	56.831
50	56.250	60.181	63.209	64.479	64.816	64.872

In order to find the P/F factor, it is necessary to interpolate between $i = 12\%$ and $i = 13\%$.

$$b \begin{bmatrix} a \begin{bmatrix} \rightarrow 12\% \\ \hookrightarrow 12.68\% \end{bmatrix} \\ \rightarrow 13\% \end{bmatrix} \qquad \begin{matrix} 0.4039 \\ P/F \\ 0.3762 \end{matrix} \begin{bmatrix} c \\ \end{bmatrix} d$$

$$c = \frac{0.68}{1}(0.0277) = 0.0188$$

Then,

$$(P/F, 12.68\%, 8) = 0.4039 - 0.0188 = 0.3851$$

If the values $i = 12.68\%$ and $n = 8$ are substituted into the P/F factor relation, Eq. (2.3), the result is

$$(P/F, 12.68\%, 8) = \frac{1}{(1 + 0.1268)^8} = 0.3848$$

Comment The difference in answers between the two methods occurs because the interpolation method assumes a linear functional relationship when in fact all of the equations are nonlinear. The interpolation method, therefore, yields approximate answers while the equations produce exact results.

Additional Example 3.7
Probs. 3.8 to 3.13

3.4 Effective Interest Rates for Continuous Compounding

As the compounding period becomes shorter and shorter, the value of m, number of compounding periods per interest period, increases. In the situation where interest is compounded continuously, m approaches infinity and the effective interest-rate formula in Eq. (3.3) may be written in a new form. First recall the definition of the natural logarithm base e.

$$\lim_{h \to \infty} \left(1 + \frac{1}{h}\right)^h = e = 2.71828+ \qquad (3.4)$$

The limit of Eq. (3.3) as m approaches infinity is found using $r/m = 1/h$, which makes $m = hr$.

$$\lim_{m \to \infty} i = \lim_{m \to \infty} \left(1 + \frac{r}{m}\right)^m - 1$$

$$= \lim_{h \to \infty} \left(1 + \frac{1}{h}\right)^{hr} - 1 = \lim_{h \to \infty} \left[\left(1 + \frac{1}{h}\right)^h\right]^r - 1$$

$$i = e^r - 1 \qquad (3.5)$$

Equation (3.5) is used to compute the effective continuous interest rate. For example, if $r = 15\%$ per year, the effective continuous rate is

$$i = e^{0.15} - 1 = 0.16183 \qquad (16.183\%)$$

For convenience, Table 3.3 includes the effective continuous rate for many nominal rates as computed by Eq. (3.5).

Example 3.3

(a) For an interest rate of 18% per year compounded continuously, calculate the effective monthly and annual interest rates.
(b) If an investor requires at least an effective return of 15% on his money, what is the minimum annual nominal rate that is acceptable if continuous compounding takes place?

Solution
(a) The nominal monthly rate is $\frac{18}{12} = 1\frac{1}{2}\%$. From Eq. (3.5), the effective monthly rate is:

$$i \text{ per month} = e^{0.015} - 1$$
$$= 1.511\%$$

Similarly, the effective annual rate is

$$i \text{ per year} = e^{0.18} - 1$$
$$= 19.72\%$$

(b) Use Eq. (3.5) with $i = 15\%$ to solve for r by taking the natural logarithm.

$$e^r - 1 = 0.15$$
$$e^r = 1.15$$
$$\ln e^r = \ln 1.15$$
$$r = 0.13976 \quad (13.976\%)$$

Therefore, a rate of 13.976% per year compounded continuously will generate an effective 15% per year return.

Comments The general formula to find the nominal rate given the effective continuous rate i is

$$r = \ln(1 + i)$$

Additional Examples 3.8 and 3.9
Probs. 3.14 to 3.27

3.5 Calculations for Payment Periods Equal to or Longer than Compounding Periods

When the compounding period of an investment (or loan) does not coincide with the payment period, it becomes necessary to manipulate the interest rate and/or payment period in order to determine the correct amount of money accumulated or paid at various times. Remember that if the payment and compound periods do not agree, the interest tables cannot be used until appropriate corrections are made. In this section, we consider the situation where the payment period (for example, year) is

Table 3.4 Various *i* and *n* values for single-payment equations
(*r* = 12% per year, compounded monthly)

Effective interest rate, *n*	Units for *n*
1% per month	Months
3.03% per quarter	Quarters
6.15% per 6 months	Semiannual periods
12.68% per year	Years
26.97% per 2 years	2-year periods

equal to or longer than the compounding period (for example, month). The two conditions that occur are as follows:

1. The cash flows require the use of the single-payment factors (*P/F*, *F/P*).
2. The cash flows require the use of uniform-series or gradient factors.

3.5.1 Single-Payment Factors   There are essentially an infinite number of correct procedures that can be used when only single factors are involved. This is because there are only two requirements which must be satisfied: (1) An effective rate must be used for *i*, and (2) the units on *n* must be the same as those on *i*. In standard factor notation, then, the single-payment equations can be generalized as follows:

$P = F(P/F, \text{effective } i \text{ per period, no. of periods})$

$F = P(F/P, \text{effective } i \text{ per period, no. of periods})$

Thus, for a nominal interest rate of 12% per year compounded monthly, any of the *i* and corresponding *n* values shown in Table 3.4 could be used (as well as many others not shown) in the single-payment formulas. For example, if an effective rate per month is used for *i*, then the *n* term must be in months. If an effective quarterly interest rate is used for *i*, then the *n* term must be in quarters. Example 3.4 demonstrates these procedures.

Example 3.4

If a woman deposits $1000 now, $3000 four years from now, and $1500 six years from now at an interest rate of 12% per year compounded semiannually, how much money will she have in her account 10 years from now?

Solution   The cash-flow diagram is shown in Fig. 3.2. Let us assume that we have decided to use an annual interest rate in solving the problem. Since only effective interest rates can be used in the equations, the first step is to find the effective annual rate. From Table 3.3, for *r* = 12% and semiannual compounding, effective *i* = 12.36%; or by Eq. (3.3),

$$i \text{ per year} = \left(1 + \frac{0.12}{2}\right)^2 - 1$$

$$= 0.1236 \quad (12.36\%)$$

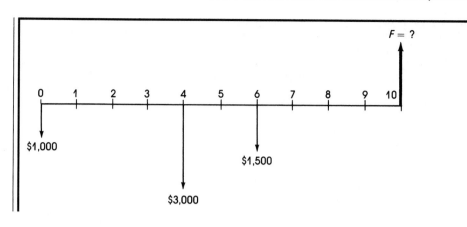

Figure 3.2 Cash-flow diagram, Example 3.4.

Since i has units of "per year," n must be expressed in years. Thus,

$F = 1000(F/P, 12.36\%, 10) + 3000(F/P, 12.36\%, 6) + 1500(F/P, 12.36\%, 4)$
$\quad = \$11,634.50$

Alternatively, we could have decided to use the effective rate of 6% per semiannual period and then use semiannual periods for n. In this case, the future worth would be:

$F = 1000(F/P, 6\%, 20) + 3000(F/P, 6\%, 12) + 1500(F/P, 6\%, 8)$
$\quad = \$11,634.50$

Comment   The second method is the easier of the two because the interest tables can be used directly without interpolation.

3.5.2 Uniform-Series Factors

3.5.2 Uniform-Series Factors   When the cash flow of the problem dictates the use of one or more of the uniform-series or gradient factors, the first step in solving the problem is to determine the relationship between the compounding period, CP, and payment period, PP. The relationship between them must be one of the following three cases:

Case 1 The payment period is equal to the compounding period, PP = CP.
Case 2 The payment period is greater than the compounding period, PP > CP.
Case 3 The payment period is shorter than the compounding period, PP < CP.

In this section, the procedure for solving problems which belong to one of the first two categories will be presented. Case 3 problems are discussed in the following section.

After a problem has been determined to involve either a uniform series or a gradient, the *first step* is to identify which of the above three cases it represents. If it is either case 1 or case 2, where PP = CP or PP > CP, the following procedure *always* applies:

1. Count the number of payments and use that number as n. (For example, if payments are made quarterly for 5 years, n is 20 quarters.)

Table 3.5 Examples of *n* and *i* values where PP = CP or PP > CP

Payment scheme (1)	Interest rate (2)	Factor involved (3)	Standard notation (4)
$500 semiannually for 5 years	16% per year compounded semiannual	Given A, find P	$P = 500(P/A, 8\%, 10)$
$75 monthly for 3 years	24% per year compounded monthly	Given A, find F	$F = 75(F/A, 2\%, 36)$
$180 quarterly for 15 years	5% per quarter	Given A, find F	$F = 180(F/A, 5\%, 60)$
$25 per month increase for 4 years	1% per month	Given G, find P	$P = 25(P/G, 1\%, 48)$
$5,000 per quarter for 6 years	1% per month	Given P, find A	$A = 5,000(A/P, 3.03\%, 24)$
$12,000 in 8 years with semiannual payments	18% per year compounded monthly	Given F, find A	$A = 12,000(A/F, 9.34\%, 16)$

2. Find the *effective* interest rate over the same time period as *n*. (For example, if *n* from step 1 is expressed in quarters, then the effective interest rate per quarter must be found.)
3. Use these values of *n* and *i* (and only these!) in the standard factor notation equations or formulas.

Table 3.5 shows the correct formulation (col. 4) for several hypothetical cash-flow schemes (col. 1) and interest rates (col. 2). Note from col. 4 that *n* is always equal to the number of payments and *i* is an effective rate *expressed over the same time period as n.*

Example 3.5

If a woman deposits $500 every 6 months for 7 years, how much money will she have in her account after she makes her last deposit if the interest rate is 20% per year compounded quarterly?

Solution　The cash-flow diagram is shown in Fig. 3.3. Since the compounding period (quarterly) is less than the payment period (semiannually), this is a case 2 problem wherein

Figure 3.3
Diagram of semi-annual deposits used to determine *F*, Example 3.5.

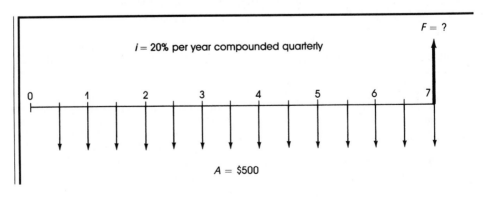

the first step is to record n equal to the number of payments (14):

$$F = 500(F/A, i\%, 14)$$

Now, since n is *semiannual* periods. an effective *semiannual* rate must be used. In order to get an effective semiannual rate from the interest rate that was given, it will be necessary to use Eq. (3.3) as follows:

$$i \text{ per 6 months} = \left(1 + \frac{0.10}{2}\right)^2 - 1$$

$$= 0.1025 \quad (10.25\%)$$

Alternatively, the effective semiannual interest rate could have been obtained from Table 3.3 by using the r value of 10% with "semiannual" compounding to get $i = 10.25\%$.

The value $i = 10.25\%$ seems reasonable, since we expect the effective rate to be slightly higher than the nominal rate of 10% per 6-month period. The effective rate can now be used in the F/A formula to find the future worth of the semiannual deposits, where $n = 2(7) = 14$ periods. Thus,

$$F = A(F/A, 10.25\%, 14)$$

$$= 500(28.4891)$$

$$= \$14,244.50$$

Comment It is important to note that the effective interest *per payment period* (6 months) was used for i and that the *number of payment periods* was used for n.

Additional Example 3.10
Probs. 3.28 to 3.42

3.6 Calculations for Payment Periods Shorter than Compounding Periods

This is the case 3 situation previously mentioned in Sec. 3.5.2. When the compounding period occurs less frequently than the payment period, the procedure that must be followed to calculate the future amount or present worth depends on the conditions specified (or assumed) regarding the interperiod compounding. *Interperiod compounding,* as used here, refers to the handling of the payments made *between* compounding periods. The two cases considered here are as follows:

1. There is no interest paid on the money deposited (or withdrawn) between compounding periods.
2. The money deposited (or withdrawn) between compounding periods earns simple interest. That is, interest is not paid on the interest earned in the preceding interperiod.

In the first case any amount of money that is deposited or withdrawn between compounding periods is regarded as having been *deposited* at the *beginning of the next compounding period* or *withdrawn* at the *end of the previous compounding period.* This is the usual mode of operation of banks and other lending institutions. Thus, if the compounding period were a *quarter,* the actual transactions shown in Fig. 3.4*a* would be treated as shown in Fig. 3.4*b.* To find the present worth of the cash flow represented by Fig. 3.4b, the nominal yearly interest rate is divided by 4 (since interest is compounded quarterly) and the appropriate P/F or F/P factor is used.

For the second case, any amount of money that is deposited between compounding periods earns simple interest; in order to obtain the interest earned in the interperiod, *each interperiod deposit* must be *multiplied by*

$$\left(\frac{M}{N}\right)i \tag{3.6}$$

where N = number of periods in a compounding period
M = number of periods prior to the end of a compounding period
i = interest rate per compounding period

Note that the value obtained by Eq. (3.6) yields only the interest that has accumulated and is not the total end-of-period amount. Example 3.6 illustrates the calculations described here.

Figure 3.4 Diagram of cash flows for quarterly compounding periods using no interperiod interest.

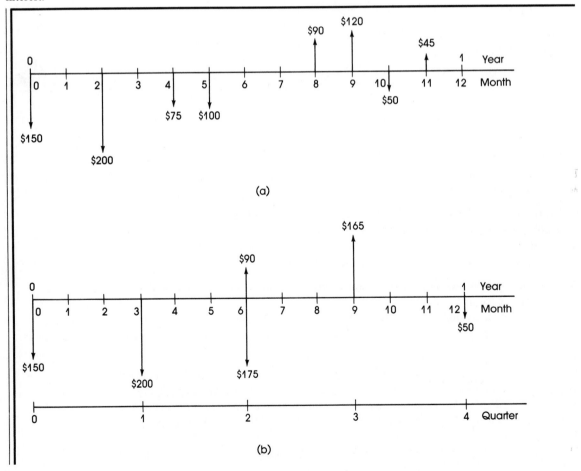

(a)

(b)

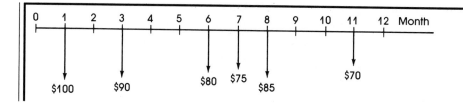

Figure 3.5
Actual deposits made with simple interperiod interest paid. Example 3.6.

Example 3.6

Calculate the amount of money that would be in a person's savings account after 12 months if deposits were made as shown in Fig. 3.5. Assume that the bank pays 6% per year compounded semiannually and pays simple interest on the interperiod deposits.

Solution   The first step is to find the amount of money that will be accumulated at each compounding period (i.e., every 6 months) using the effective rate of 3% semiannually. The future or present worth of such deposits can then be calculated with the regular interest formulas. Thus, for the deposits made within the first compounding period, the total value F_6 after 6 months is

$$F_6 = [100 + 100(\tfrac{5}{6})0.03] + [90 + 90(\tfrac{3}{6})0.03] + 80$$
$$= (100 + 2.50) + (90 + 1.35) + 80$$
$$= \$273.85$$

Similarly, the amount F_{12} accumulated in the second compounding period is

$$F_{12} = [75 + 75(\tfrac{5}{6})0.03] + [85 + 85(\tfrac{4}{6})0.03] + [70 + 70(\tfrac{1}{6})0.03]$$
$$= (75 + 1.88) + (85 + 1.70) + (70 + 0.35)$$
$$= \$233.93$$

The cash-flow diagram has now been reduced to that shown in Fig. 3.6. Thus, the future worth F at the end of the year is

$$F = 273.85(F/P, 3\%, 1) + 233.93 = \$516$$

A lot of trouble, right?

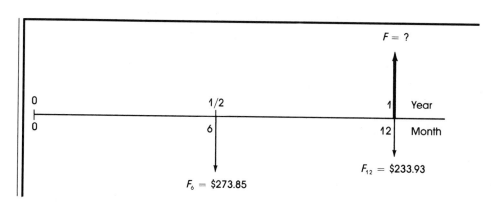

Figure 3.6
Equivalent deposits after simple interest is computed for interperiods, Example 3.6.

Comment   In calculating the amounts of money accumulated after 6 months (F_6) and 12 months (F_{12}), note that the amount of the deposit was added to each interest term, since Eq. (3.6) represents only the *interest accumulated*, not the total amount.

Determination of the compound-period amount (i.e., end-of-period amount) is simplified when the interperiod payments are regular and uniform. That is, if the same amount of money is deposited in *each* interperiod, the equivalent compound-period amount can be calculated from the following equation.

$$A = ND + \frac{N-1}{2}(iD) \tag{3.7}$$

where A = equivalent uniform compound-period payment
D = value of the deposit made in each interperiod
N = number of interperiods per compounding period

See the Additional Examples for an illustration.

Additional Example 3.11
Probs 3.43 to 3.51

Additional Examples

Example 3.7

Ms. Jones plans to place money in a JUMBO certificate of deposit that pays 18% per year compounded daily. What effective rate will she receive (*a*) yearly and (*b*) semiannually?

Solution
(*a*) Using Eq. (3.3), with $m = 365$,

$$i \text{ per year} = \left(1 + \frac{0.18}{365}\right)^{365} - 1$$

$$= 0.19716 \quad (19.716\%)$$

That is, Ms. Jones will get an effective 19.716% per year on her deposit.
(*b*) Here $r = 0.09$ per 6 months and $m = 182$ days:

$$i \text{ per 6 months} = \left(1 + \frac{0.09}{182}\right)^{182} - 1$$

$$= 0.09415 \quad (9.415\%)$$

Example 3.8

Mr. Blunder and Mr. Smart both plan to invest $5000 for 10 years at 10% per year. Compute the future worth for both men if Mr. Blunder figures interest compounded annually and Mr. Smart assumes continuous compounding.

Solution

Mr. Blunder For annual compounding the future worth is

$$F = P(F/P, 10\%, 10) = 5000(2.5937) = \$12,969$$

Mr. Smart Using the continuous-compounding relation, Eq. (3.5), find the effective i per year.

$$i = e^{0.10} - 1 = 0.10517 \qquad (10.517\%)$$

Then, the future worth is

$$F = P(F/P, 10.517\%, 10) = 5000(2.7183) = \$13,591$$

Comment Continuous compounding represents a $622, or 4.8%, increase in earnings. Just for comparison, note that a savings and loan association might compound daily, which yields an effective rate of 10.516% ($F = \$13,590$), whereas 10% continuous compounding offers only a very slight increase to 10.517%.

Example 3.9

Compare the present worth of $2000 a year for 10 years at 10% per year (*a*) compounded annually and (*b*) compounded continuously.

Solution

(*a*) For annual compounding,

$$P = 2000(P/A, 10\%, 10) = 2000(6.1446) = \$12,289$$

(*b*) For continuous compounding,

$$i \text{ per year} = e^{0.10} - 1 = 10.517$$

Then,

$$P = 2000(P/A, 10.517\%, 10)$$
$$= 2000(6.0104)$$
$$= \$12,021$$

As expected, the present worth for continuous compounding is less than that for annual compounding because higher interest rates require greater discounts of future cash flows.

Example 3.10

Ms. Warren wants to purchase a new compact car for $8500. She plans to borrow the money from her credit union and to repay it monthly over a period of 4 years. If the nominal interest rate is 12% per year compounded monthly, what will her monthly installments be?

Solution The compounding period equals the payment period (case 1), with an effective monthly rate of $i = 1\%$ per month and $n = 12(4) = 48$ payments. The monthly payments are

$$A = 8500(A/P, 1\%, 48) = 8500(0.02633) = \$223.81$$

════════════ *Example 3.11* ════════════

Calculate the equivalent compound-period amount if $200 is deposited every month at an interest rate of 6% per year compounded semiannually with simple interest paid on inter-period deposits.

Solution From Eq. (3.7) and a semiannual interest rate of 3%,

$$A = 6(200) + \tfrac{5}{2}(0.03)(200) = \$1215$$

Thus, $200 per month is equivalent to $1215 every 6 months, since the compounding period is semiannual.

Comment The present worth or future worth could be calculated using the P/A or F/A factors for $i = 3\%$ and $n =$ number of compounding periods (for example, 3 years is six compounding periods).

Problems **3.1** What is the nominal interest rate per year if interest is (a) 0.50% every 2 weeks and (b) 2% every semiannual period?

3.2 What time period is usually used in expressing a nominal interest rate?

3.3 What is meant by payment period?

3.4 Identify the following interest rates as either nominal or effective. (a) $i = 2\%$ per month, (b) $i = 14\%$ per year compounded quarterly, (c) $i = 15\%$ per year compounded annually, (d) $i =$ effective 13% per year compounded monthly, (e) $i = 1\%$ per month compounded weekly.

3.5 For each of the interest statements in Prob. 3.4, identify the compounding period.

3.6 Complete the table shown below.

Interest statement	Type of interest	Compounding period
$i = 3\%$ per semiannual period		
$i =$ effective 3% per semiannual period		
$i = 2\%$ per month compounded monthly		
$i =$ effective 10% per year compounded continuously		
$i = 0.5\%$ per week compounded continuously		
$i = 0.03\%$ per day compounded daily		
$i = 0.03\%$ per day compounded continuously		
$i = 0.03\%$ per day		

3.7 What is the dimensional unit of the term r/m in the effective interest-rate equation?

3.8 Calculate the nominal and effective interest rates per year for a finance charge of $1\tfrac{1}{2}\%$ per month.

3.9 What are the nominal and effective interest rates per year for an interest charge of 4% every 6 months?

3.10 What effective interest rate per year is equivalent to a nominal rate of 12% per year compounded semiannually?

3.11 What effective interest rate per year is equivalent to a nominal rate of 16% per year compounded quarterly?

3.12 Calculate the nominal and effective interest rates per year for a finance charge of $1\frac{3}{4}$% per month and find the value of the P/A factor for 10 years for each rate.

3.13 What quarterly interest rate is equivalent to an effective annual rate of 6%?

3.14 Justin had just invested money at 14.5% per year compounded continuously. What is the effective annual rate to be expected?

3.15 Is it true that the effective rate yielded by a nominal rate of 15% per year compounded quarterly is just about the same as the nominal rate of 14.726% per year compounded continuously? Why or why not?

3.16 What is the annual nominal rate necessary to yield the following annual effective rates if continuous compounding is in effect: (a) 22%, (b) 13.75%.

3.17 For an interest rate of 12% per year compounded continuously, determine (a) the effective yearly rate, (b) the nominal monthly rate, and (c) the effective monthly rate.

3.18 For an interest rate of 4% per quarter compounded continuously, determine (a) the nominal quarterly rate, (b) the effective monthly rate, (c) the nominal annual rate, and (d) the effective semiannual rate.

3.19 What uniform annual amount would you have to deposit for 5 years to have an equivalent present-investment sum of $10,000 at an interest rate of 10% per year compounded continuously?

3.20 Determine the difference in future worth of the following annual cash flows if interest is 12% per year compounded annually and 12% per year compounded continuously.

Year	0	1	2	3	4
Discrete cash flow, $	$-15,000$	5,000	5,000	5,000	12,000

3.21 If Ms. Watson has agreed to repay a loan of $4500 in 10 equal annual payments starting 1 year from now, determine the size of each payment if interest is 15% per year compounded continuously.

3.22 Deposits of $900 per year for 9 years are to be made. If $10,000 is to be accumulated, determine (a) the continuously compounded nominal annual rate (b) the continuously compounded effective annual rate.

3.23 Mr. Adams plans to borrow $5000 now and repay the loan in eight equal payments of $875 per year starting 1 year from now. Compute the interest rate if interest is (a) compounded annually and (b) compounded continuously.

3.24 J. Smith plans to purchase a new computer for $200,000 and sell data-processing services for the next 10 years. What is the total annual revenue requirement to yield a nominal 20% per year on the investment if revenues are assumed to be received at the end of each year and interest is compounded continuously?

3.25 Find the present worth of $5000 per month cash flow for 10 years if the effective continuous rate is 20% per year.

3.26 S. Johnson wants to accumulate $10,000 in 5 years. If the nominal annual rate is 10% per year, determine the total annual deposit under the following conditions and compare the amounts for each plan to determine which requires the smallest annual deposit.
(a) End-of-year deposits, semiannual compounding
(b) 6-month deposits, semiannual compounding
(c) End-of-year deposits, continuous compounding

3.27 Compute the (*a*) nominal and (*b*) effective annual rate of return of an investment of $18,000 now which will return $25,500 in 5 years if the continuous-compounding assumption is made.

3.28 How much money would be accumulated in 8 years if an investor deposits $2500 now at a nominal interest rate of 8% per year compounded semiannually?

3.29 If a person buys a car for $5500 and must make monthly payments of $200 for 36 months, what are the nominal and effective interest rates per year for this transaction?

3.30 How much should you be willing to pay for an annuity that will provide $300 every 3 months for 6 years starting 3 months from now if you want to make a nominal 12% per year compounded quarterly?

3.31 If a person deposits $75 into a savings account every month, how much money will be accumulated after 10 years if the interest rate is a nominal 12% per year compounded monthly?

3.32 If a person borrows $3000 and must repay the loan in 2 years with equal monthly installments, how much is the monthly payment if the interest rate is 1% per month?

3.33 If a person made a lump-sum deposit 12 years ago which has accumulated to $9500 now, how much was the original deposit if the interest rate received was a nominal 4% per year compounded semiannually?

3.34 What is the present worth of $10,000 now, $6000 eight years from now, and $9000 twelve years from now if the interest rate is a nominal 6% per year compounded quarterly?

3.35 In Prob. 3.34, what would be the value of the deposits 25 years from the date of the original deposit?

3.36 A man has been presented with the opportunity to buy a second mortgage note valued at $1500 for $1300. The note is due 4 months from now. If he purchases the note, what nominal and effective rates of return will he make? Assume interest is compounded monthly.

3.37 What monthly interest rate is equivalent to an effective semiannual rate of 4%?

3.38 If a pants manufacturing company spends $14,000 in order to improve the efficiency of a sewing operation, how much must it save each month in reduced work force costs in order to recover its investment in $2\frac{1}{2}$ years if the effective interest rate is 12.68% per year compounded monthly?

3.39 A woman who has just won $45,000 in a lottery wants to deposit enough of her winnings into a savings account so that she will have $10,000 for her son's college education. Assume that her son just turned 3 years old and will begin college when he is 18 years of age. How much must the woman deposit to earn 7% per year compounded quarterly on her investment?

3.40 How much money can be withdrawn semiannually for 20 years from a retirement fund which earns 5% per year interest compounded semiannually and has a present amount of $36,000 in it?

3.41 What year-end payment is equivalent to the monthly payments that would be paid on a $2000 loan that must be repaid in 1 year at an interest rate of 1% per month?

3.42 Compute the monthly deposit required to accumulate $5000 in 5 years at a nominal 6% per year compounded daily.

3.43 Draw two cash-flow diagrams illustrating the timing of cash flow if banks paid interest on interperiod deposits but not on withdrawals. Assume that interest is payable quarterly. Represent deposits by *X*'s and withdrawals by *Y*'s, with at least one deposit and one withdrawal per interest period. Draw one diagram for actual deposits and withdrawals and another illustrating deposits and withdrawals from the bank's point of view.

3.44 How much money would be in a savings account in which a person had deposited $100 every month for 5 years at an interest rate of 5% per year compounded quarterly? Assume simple interperiod interest.

3.45 How much money would the person in Prob. 3.44 have if interest were compounded (*a*) monthly? (*b*) Daily?

3.46 Calculate the future worth of the transactions shown in Fig. 3.4*a* if interest is 6% per year compounded semiannually and interperiod interest is not paid.

3.47 Rework Prob. 3.46, except assume that simple interest is paid on interperiod deposits but not withdrawals.

3.48 A tool-and-die company expects to have to replace one of its lathes in 5 years at a cost of $18,000. How much would the company have to deposit every month in order to accumulate $18,000 in 5 years if the interest rate is 6% per year compounded semiannually? Assume simple interperiod interest.

3.49 What monthly deposit would be equivalent to a deposit of $600 every 3 months for 2 years if the interest rate is 6% per year compounded semiannually? Assume simple interperiod interest on all deposits.

3.50 How much money would be in the savings account of a person who had deposited $150 every month and withdrawn $300 every 6 months for 2 years? Use an interest rate of 4% per year compounded semiannually and assume that interperiod interest is not paid.

3.51 How many monthly deposits of $75 would a person have to make in order to accumulate $15,000 if the interest rate is 6% per year compounded semiannually? Assume that simple interperiod interest is paid.

4 Use of Multiple Factors

Because many of the cash-flow situations encountered in real-world engineering problems do not fit the cash flows from which the equations in Chap. 2 were developed, it is often necessary to combine the equations in order to solve the problem at hand. For a given set of cash-flow amounts, there are usually many ways to determine the equivalent cash flow desired. In this chapter, you will learn how to combine several of the engineering-economy factors in order to satisfy the more complex cash-flow situations.

Section Objectives

After completing this chapter, you should be able to do the following:

4.1 Determine the year in which the present worth or future worth is located, given a statement describing a randomly placed uniform series.

4.2 Determine the present worth, equivalent uniform annual worth, or future worth of a uniform series of disbursements or receipts which begin at a time other than year 1, given the amount and time of each payment and the interest rate.

4.3 Calculate the present worth or future worth of randomly distributed single amounts and uniform-series amounts, given the times and amounts of the payments and the interest rate.

4.4 Calculate the equivalent uniform annual worth of randomly distributed single amounts and uniform-series amounts, given the times, amounts, and the interest rate.

4.5 Calculate the present worth and equivalent uniform annual worth of cash flows involving shifted gradients or escalating series, given the interest rate and statement of the problem.

4.6 Calculate the present worth or equivalent uniform annual worth of cash flows which include a decreasing gradient, given the interest rate and statement of the problem.

Study Guide

4.1 Location of Present Worth and Future Worth

When a uniform series of payments begins at a time other than at the end of interest period 1, several methods can be used to find the present worth. For example, the present worth of the uniform series of disbursements shown in Fig. 4.1 could be determined by any of the following methods:

1. Use the single-payment present-worth factor $(P/F, i\%, n)$ to find the present worth of each disbursement at year 0 and add them.
2. Use the single-payment compound-amount factor $(F/P, i\%, n)$ to find the future worth of each disbursement in year 13, add them, and then find the present worth of the total using $P = F(P/F, i\%, 13)$.
3. Use the uniform-series compound-amount factor $(F/A, i\%, n)$ to find the future amount by $F = A(F/A, i\%, 10)$, and then find the present worth using $P = F(P/F, i\%, 13)$.
4. Use the uniform-series present-worth factor $(P/A, i\%, 10)$ to compute the "present worth" (which will *not* be located in year 0!), and then find the present worth in year 0 by using the $(P/F, i\%, 3)$ factor. (Present worth is enclosed in quotation marks to represent the present worth as determined by the uniform-series present-worth factor and to differentiate it from the present worth in year 0.)

This and the next section illustrate the fourth method for calculating the present worth of a uniform series that does not begin at the end of period 1.

For the cash-flow diagram shown in Fig. 4.1, the "present worth" that would be obtained by using the $(P/A, i\%, n)$ factor would be located in *year 3*, not year 4. This is shown in Fig. 4.2, with the present-worth arrow representing an equivalent amount in year 3. Note that P is located *1 year prior* to the beginning of the first annual disbursement. Why? Because the P/A factor was derived with the P in time period 0 and the A beginning at the end of interest period 1; that is, the P must

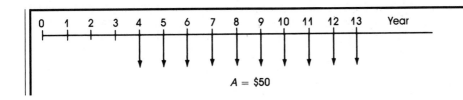

$A = \$50$

Figure 4.1 A randomly placed uniform series.

Figure 4.2 Loca-
tion of *P* for the
randomly placed
uniform series in
Fig. 4.1.

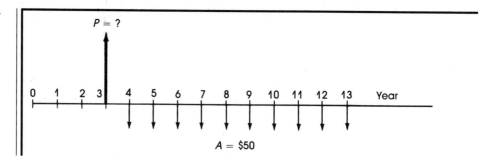

always be *one interest period ahead* of the first *A* value (Sec. 2.2). The most common mistake made in working problems of this type is improper placement of *P*. Therefore, it is extremely important that you remember the following rule. *The present worth is always located one interest period prior to the first uniform payment when using the uniform-series present-worth factor, (P/A, i%, n).*

On the other hand, the uniform-series compound-amount factor, $(F/A, i\%, n)$, was derived (Sec. 2.3) with the future worth *F* located in the *same period* as the last payment. Figure 4.3 shows the location of the future worth when F/A is used for the cash flow shown in Fig. 4.1. Thus, *the future worth is always located in the same period as the last uniform payment when using the uniform-series compound-amount factor $(F/A, i\%, n)$.*

It is also important to remember that the number of periods *n* that should be used with the P/A or F/A factors is equal to the number of payments. It is generally helpful to *renumber* the cash-flow diagram to avoid errors in counting. Figure

Figure 4.3 Placement of *F* for the uniform series of Fig. 4.1.

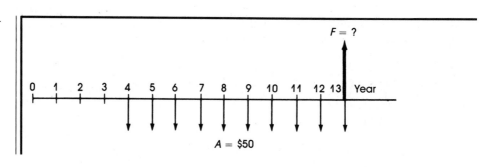

Figure 4.4 Renumbering of payments in Fig. 4.1 to show that $n = 10$ in the P/A or F/A factors.

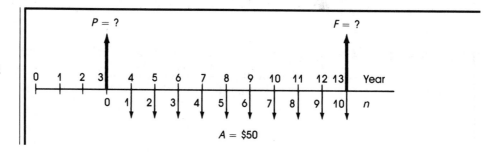

4.4 shows the cash-flow diagram of Fig. 4.1 renumbered for determination of n. Note that in this example $n = 10$.

Additional Example 4.13
Prob. 4.1

4.2 Calculations for a Uniform Series that Begins after Period 1

As stated in Sec. 4.1, there are many methods that can be used to solve problems having a uniform series that begins at a time other than the end of period 1. However, it is generally much more convenient to use the uniform-series formulas than it is to use the single-payment formulas.

There are specific steps which should be followed in solving problems of this type in order to avoid errors:

1. Draw a cash-flow diagram of the receipts and disbursements of the problem.
2. Locate the present worth or future worth on the cash-flow diagram.
3. Determine n by renumbering the cash-flow diagram.
4. Draw the cash-flow diagram representing the desired equivalent cash flow.
5. Set up and solve the equations.

These steps are illustrated in the following two examples.

Example 4.1

A person buys a piece of property for $5000 down and deferred annual payments of $500 a year for 6 years starting 3 years from now. What is the present worth of the investment if the interest rate is 8% per year?

Solution   The cash-flow diagram is shown in Fig. 4.5. The nomenclature P_A is used throughout this chapter to represent the present worth of a uniform annual series A, and P'_A represents the present worth at a time other than period 0. Similarly, P_T represents the total present worth at time 0. The correct placement of P'_A and diagram renumbering to obtain n are also indicated in Fig. 4.5. Note that P'_A is located in year 2, not year 3, and $n = 6$, not 8. To solve this problem, first find the value of P'_A:

$$P'_A = \$500(P/A, 8\%, 6)$$

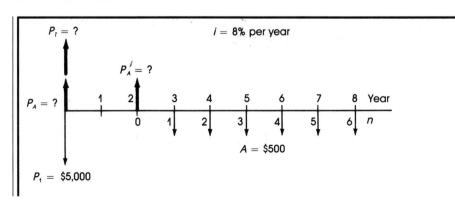

Figure 4.5   Placement of present-worth values, Example 4.1.

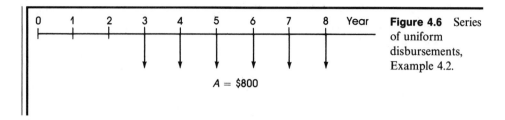

Figure 4.6 Series of uniform disbursements, Example 4.2.

Since P'_A is located in year 2, it is necessary to find P_A in year 0:

$$P_A = P'_A(P/F, 8\%, 2)$$

The total present worth can now be determined by adding P_A and P_1 (the initial investment is labeled P_1):

$$
\begin{aligned}
P_T &= P_1 + P_A \\
&= 5000 + 500(P/A, 8\%, 6)(P/F, 8\%, 2) \\
&= 5000 + 500(4.6229)(0.8573) \\
&= \$6981.60
\end{aligned}
$$

Example 4.2

Calculate the 8-year equivalent uniform annual worth at 16% per year interest for the uniform disbursements shown in Fig. 4.6.

Solution Figure 4.7 shows the original cash-flow diagram and the desired equivalent diagram. In order to convert uniform cash flows that begin sometime after period 1 into an equivalent uniform worth over *all* periods, the first step is to convert the cash flow into a present worth or future worth. Then either the conventional capital-recovery factor $(A/P, i\%, n)$ or the sinking-fund factor $(A/F, i\%, n)$ can be used to determine the equivalent uniform worth. Both of these methods are illustrated below.

1. Present-worth method (refer to Fig. 4.7).

$$P'_A = 800(P/A, 16\%, 6)$$

$$
\begin{aligned}
P_T &= P'_A(P/F, 16\%, 2) = 800(P/A, 16\%, 6)(P/F, 16\%, 2) \\
&= \$2190.78
\end{aligned}
$$

Figure 4.7 Desired equivalent diagram for the series of Fig. 4.6.

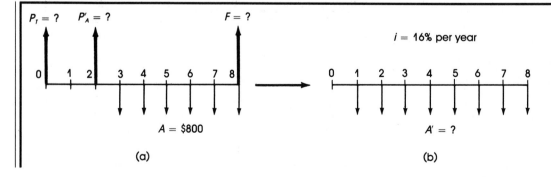

(a)

(b)

where P_T is the total present worth of the cash flow. The equivalent series A' can now be determined with the A/P factor

$$A' = P_T(A/P, 16\%, 8) = \$504.36$$

for 8 years as shown in Fig. 4.7b.

2. Future-worth method (Fig. 4.7): The first step is to calculate the future worth F in year 8:

$$F = 800(F/A, 16\%, 6) = \$7184$$

The sinking-fund factor $(A/F, i\%, n)$ can now be used to obtain A':

$$A' = F(A/F, 16\%, 8) = \$504.46$$

Comment   In the present-worth method, note that P'_A was located in year 2, not year 3. After the present worth was determined, the equivalent series was calculated using $n = 8$. In the future-worth method, $n = 6$ was used to find F, and $n = 8$ was used to find the equivalent series, since the cost must be spread uniformly over *all* of the years.

Additional Example 4.14
Probs. 4.2 to 4.17

4.3   Calculations Involving Uniform-Series and Randomly Distributed Amounts

When a uniform series of payments is included in a cash flow that also contains randomly distributed single amounts, the procedures learned in Sec. 4.2 should be applied to the uniform-series amounts and the single-payment formulas applied to the single-payment amounts. This type of problem, illustrated below, is merely a combination of previous types.

Example 4.3

A couple owning 50 hectares of valuable land decided to sell the mineral rights on their property to a mining company. Their primary objective was to obtain long-term investment income and sufficient money to finance the college education of their two children. Since the children were 12 and 2 years of age at the time the couple was negotiating the contract, they knew that the children would be in college 6 and 16 years, respectively, from the present. They therefore made a proposal to the company that it pay them $20,000 per year for 20 years beginning 1 year hence plus $10,000 six years from now and $15,000 sixteen years from now. If the company wanted to pay off its lease immediately, how much would it have to pay now if the interest rate is 16% per year?

Solution   The cash-flow diagram for this problem is shown in Fig. 4.8. This problem is solved by finding the present worth of the uniform series and adding it to the present worth of the two individual payments. Thus,

$$P = 20,000(P/A, 16\%, 20) + 10,000(P/F, 16\%, 6) + 15,000(P/F, 16\%, 16)$$
$$= \$124,075$$

Comment   In this example, note that the uniform series started at the end of year 1 so that the present worth obtained with the P/A factor represented the present worth at year 0. It was not necessary to use the P/F factor on the uniform series.

Figure 4.8 Diagram including a uniform series and single amounts, Example 4.3.

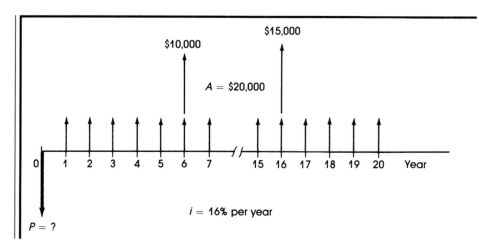

Example 4.4

If the uniform payments described in Example 4.3 did not begin until 3 years from the time the contract was signed, what would be the present worth of the amounts at $t = 0$?

Solution The cash-flow diagram is shown in Fig. 4.9, with the n scale shown above the time axis. The number of years for the uniform series is still 20.

$$P'_A = 20,000(P/A, 16\%, 20)$$

$$P_T = P'_A(P/F, 16\%, 2) + 10,000(P/F, 16\%, 6) + 15,000(P/F, 16\%, 16)$$
$$= 20,000(P/A, 16\%, 20)(P/F, 16\%, 2) + 10,000(P/F, 16\%, 6) + 15,000(P/F, 16\%, 16)$$
$$= \$93,625$$

Comment Displacement of the annual series by 2 years has decreased the present worth of the cash flows by $30,450.

Figure 4.9 Diagram from Fig. 4.8 with the A series shifted 2 years.

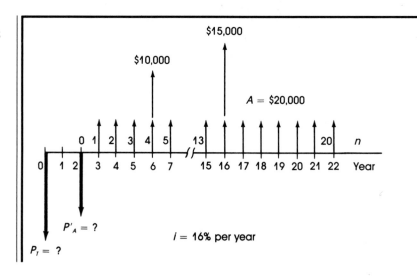

Example 4.5

Calculate the future worth of the receipts shown in Fig. 4.9 using $i = 16\%$ per year.

Solution The future worth of the uniform series and the single random receipts can be calculated as follows:

$$F = 20{,}000(F/A, 16\%, 20) + 10{,}000(F/P, 16\%, 16) + 15{,}000(F/P, 16\%, 6) = \$2{,}451{,}626$$

Comment Although the determination of n is straightforward, make sure that you fully understand how the values 20, 16, and 6 were obtained for calculating the future worth. Also, in order to get a feel for the effect that interest rates have on the time value of money, rework Examples 4.3 through 4.5 using $i = 6\%$ per year instead of 16%. You should get the results shown below.

i	Example 4.3	Example 4.4	Example 4.5
6%	$P = \$242{,}352$	$P_T = \$217{,}118$	$F = \$782{,}381$
16%	$P = \$124{,}025$	$P_T = \$93{,}625$	$F = \$2{,}451{,}626$

Obviously, a 10%-per-year difference in i has a significant effect on P and F.

Additional Example 4.15
Probs. 4.18 to 4.25

4.4 Equivalent Uniform Worth of Both Series and Single Payment Amounts

Whenever it is necessary to calculate the equivalent uniform annual series of randomly distributed single payments and/or uniform amounts, the most important fact to remember is that the payments *must first be converted to a present worth or future worth*. Then the equivalent uniform series can be obtained with the appropriate A/P or A/F factor.

Example 4.6

Calculate the 20-year equivalent uniform annual worth for the receipts described in Example 4.3 (Fig. 4.8).

Solution The desired equivalent cash-flow diagram is shown in Fig. 4.10. From the cash-flow diagram shown in Fig. 4.8, it is evident that the uniform-series receipts (that is, $20,000) are already distributed through all 20 years of the diagram. It is therefore necessary to convert only the single amounts to an equivalent uniform annual series and add the value obtained to the $20,000. This can be done by either (*a*) the present-worth or (*b*) the

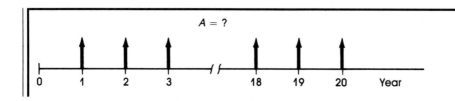

Figure 4.10
Desired equivalent series, Example 4.6.

future-worth method:

(a) $A = 20{,}000 + 10{,}000(P/F, 16\%, 6)(A/P, 16\%, 20) + 15{,}000(P/F, 16\%, 16)(A/P, 16\%, 20)$

$= 20{,}000 + [10{,}000(P/F, 16\%, 6) + 15{,}000(P/F, 16\%, 16)](A/P, 16\%, 20)$

$= \$20{,}928$ per year   (present-worth method)

(b) $A = 20{,}000 + 10{,}000(F/P, 16\%, 14)(A/F, 16\%, 20) + 15{,}000(F/P, 16\%, 4)(A/F, 16\%, 20)$

$= 20{,}000 + [10{,}000(F/P, 16\%, 14) + 15{,}000(F/P, 16\%, 4)](A/F, 16\%, 20)$

$= \$20{,}928$ per year   (future-worth method)

Comment   Note that it was necessary to take the single payments to either end of the time scale before annualizing. Failure to do so would result in unequal receipts in some years.

Example 4.7

Convert the cash flow shown in Fig. 4.9 to an equivalent uniform annual series over 22 years. Use $i = 16\%$ per year.

Solution   Since the uniform-series receipts are not distributed through all 20 years of the time scale, it is first necessary to find the present worth or future worth of the series. This was done in Examples 4.4 and 4.5, respectively. The equivalent uniform annual worth can now be determined by multiplying the values previously obtained by the $(A/P, 16\%, 22)$ factor or the $(A/F, 16\%, 22)$ factor as follows:

$$A = P_T(A/P, 16\%, 22) = 93{,}625(A/P, 16\%, 22) = \$15{,}575$$

or $A = F(A/F, 16\%, 22) = 2{,}451{,}626(A/F, 16\%, 22) = \$15{,}568$

Comment   When a uniform series begins at a time other than at the end of period 1, or when intermediate single amounts are involved, it is important to remember that the equivalent present or future worth of the uniform series must be determined before the equivalent uniform series can be obtained.

Probs. 4.26 to 4.34

4.5   Present Worth and Equivalent Uniform Series of Shifted Gradients

In Sec. 2.5, Eq. (2.12) was derived for calculating the present worth of a uniform gradient. You will recall that the equation was derived for a present worth in year 0 with the gradient starting between periods 1 and 2 (see Fig. 2.5). Therefore, the present worth of a uniform gradient will always be located *2 periods before the gradient starts.* The examples that follow illustrate where the present worth of the gradient is located.

Figure 4.11 Diagram of gradient, Example 4.8.

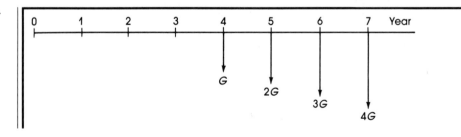

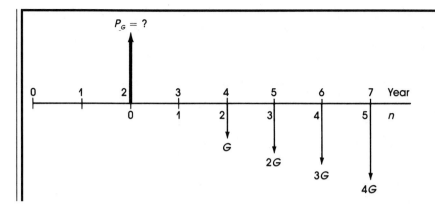

Figure 4.12 Diagram locating the present worth of gradient in Fig. 4.11.

Example 4.8

For the cash-flow diagram shown in Fig. 4.11, locate the gradient present worth.

Solution   The present worth of the gradient, P_G, is shown in Fig. 4.12. In the derivation of the present worth of a gradient series, the present worth was located 2 periods before the start of the gradient. Therefore, for the gradient in Fig. 4.11, the present worth would be located at the end of year 2. It is usually advantageous to renumber the cash-flow diagram so that the gradient year 0 and the number of years n of the gradient can be determined. The best method for accomplishing this is to determine where the gradient begins and label that time as year 2 and then work backward and forward. In this example, since the gradient started between years 3 and 4, gradient year 2 was placed below year 4 on the original diagram (Fig. 4.12). Year 0 for the gradient was then located by moving back 2 years.

Example 4.9

For the cash-flow diagram shown in Fig. 4.13, explain why the present worth of the gradient is located in year 3.

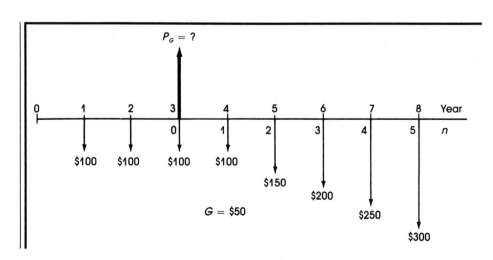

Figure 4.13 Location of the present worth of a gradient.

Solution The gradient is $50 and begins between years 4 and 5 of the original cash-flow diagram. Therefore, year 5 represents year 2 of the gradient; the present worth of the gradient would then be located in year 3. If Fig. 4.13 is divided into two cash-flow diagrams, the location of the gradient becomes quite clear, as in Fig. 4.14.

Figure 4.14 Partitioned cash flow of Fig. 4.13: $(a) = (b) + (c)$.

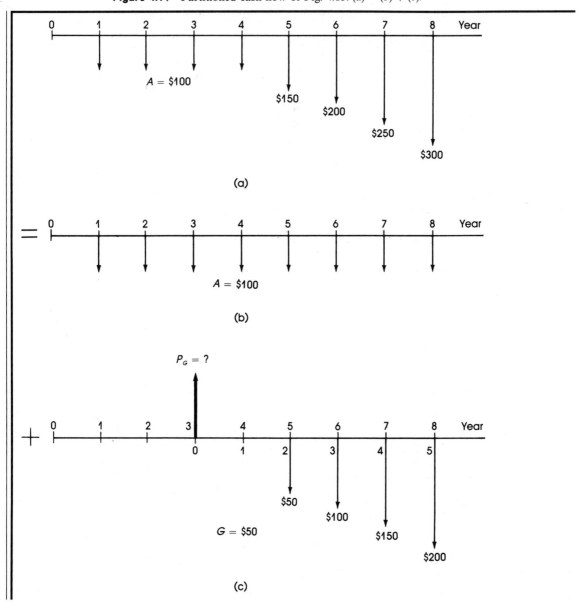

When the gradient of a cash-flow sequence starts between periods 1 and 2, it is called a conventional gradient, as discussed in Sec. 2.5. When a gradient begins at a time before or after period 2, it is referred to as a *shifted gradient*. To determine n, the same renumbering procedure used to determine where the present worth of the gradient is located is necessary. For the cash-flow diagrams shown in Fig. 4.15a to c, the gradients G, number of years n, and gradient factors used to calculate the present worth and annual series of the gradients are shown on each cash-flow diagram, assuming an interest rate of 6% per year.

It is important to note that the A/G factor *cannot* be used to find an equivalent A value in periods 1 through n for cash flows involving a shifted gradient. Consider the cash-flow diagram of Fig. 4.16. To find the equivalent annual disbursement in years 1 through n, you must first find the present worth of the gradient, take this present worth back to year 0, then annualize the present worth from year 0 with the A/P factor. If you used the annual-series gradient factor $(A/G, i\%, n)$, the gradient would be converted into an equivalent annual worth over years 3 through 7 only. For this reason, the first step in a problem wherein a uniform series through all of the periods is desired is always to find the present worth of the gradient at actual year 0. The steps involved in handling problems of this type are illustrated in Example 4.10.

Example 4.10

Compute the equivalent annual worth for the payments of Fig. 4.16.

Solution   The solution steps are:

1. Consider the $50 base amount as an annual amount for all 7 years (Fig. 4.17).
2. Find the present worth of the gradient P_G that occurs in year 2 as shown by the gradient-year time scale.

$$P_G = 20(P/G, i\%, 5)$$

3. Bring the gradient present worth back to actual year 0.

$$P_0 = P_G(P/F, i\%, 2)$$

4. Annualize the gradient present worth from year 0 through year n.

$$A = P_0(A/P, i\%, 7)$$

5. Finally, add the remaining annual worths to the gradient annual worth. Here the base amount is $50 for all 7 years and the annual series is

$$A = 20(P/G, i\%, 5)(P/F, i\%, 2)(A/P, i\%, 7) + 50$$

For cash flows involving an escalating series E which starts at a time other than between interest periods 1 and 2, the same considerations apply as for shifted gradients. In this case, P_E would be located on the diagram in a manner similar to the way P_G is located for shifted gradients. Example 4.11 shows these calculations.

Figure 4.15
Determination of
G and *n* values
used in gradient
factors.

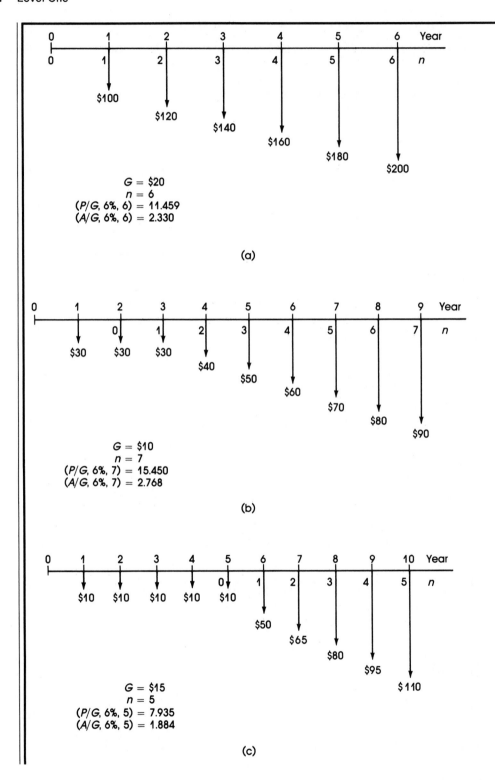

$G = \$20$
$n = 6$
$(P/G, 6\%, 6) = 11.459$
$(A/G, 6\%, 6) = 2.330$

(a)

$G = \$10$
$n = 7$
$(P/G, 6\%, 7) = 15.450$
$(A/G, 6\%, 7) = 2.768$

(b)

$G = \$15$
$n = 5$
$(P/G, 6\%, 5) = 7.935$
$(A/G, 6\%, 5) = 1.884$

(c)

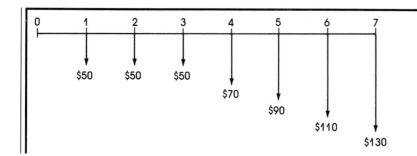

Figure 4.16 Diagram illustrating a shifted gradient.

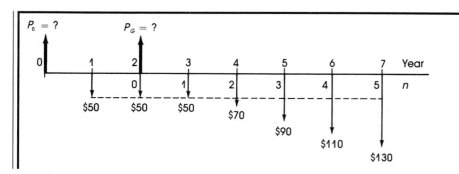

Figure 4.17 Completed diagram, Example 4.10.

Example 4.11

Calculate the equivalent present cost of a $35,000 expenditure now and $7000 per year for 5 years beginning 1 year from now with increases of 12% per year thereafter for the next 8 years. Use an interest rate of 15% per year.

Solution
(a) Figure 4.18 presents the cash flows. The present worth P is found using $i = 15\%$ and Eq. (2.18) for the escalating series which has its present worth P_E in year 4.

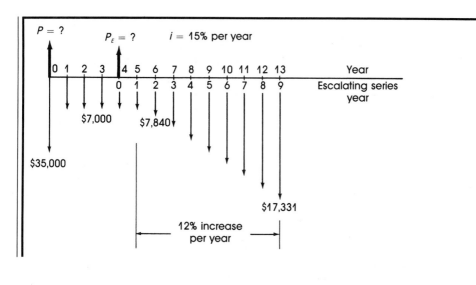

Figure 4.18 Cash-flow diagram, Example 4.11.

$$P = 35{,}000 + 7000(P/A, 15\%, 4) + \left\{ 7000 \frac{[(1.12/1.15)^9 - 1]}{0.12 - 0.15} \right\}(P/F, 15\%, 4)$$

$$= 35{,}000 + 19{,}985 + 28{,}247$$

$$= \$83{,}232$$

Note that $n = 4$ in the P/A factor because the $7000 in year 5 is D in Eq. (2.18). The last term in the expression for P is P_E in year 4 (Fig. 4.18), which is moved to time 0 with the $(P/F, 15\%, 4)$ factor.

Comment   You can check the result of the escalating-cost factor by multiplying each of the amounts in years 5 through 13 by the appropriate P/F factor for $i = 15\%$.

Additional Examples 4.16 and 4.17
Probs. 4.35 to 4.54

4.6   Decreasing Gradients

The use of the gradient factors is the same for increasing and decreasing gradients, except that in the case of decreasing gradients the following are true:

1. The base amount is equal to the *largest* amount attained in the gradient series.
2. The gradient term is *subtracted* from the base amount instead of added; thus, the term $-G(A/G, i\%, n)$ or $-G(P/G, i\%, n)$ must be used in the computations.

The present worth of the gradient will still take place 2 periods before the gradient starts and the A value will start at period 1 and continue through period n.

Example 4.12

Find the (a) present worth and (b) annual series of the receipts shown in Fig. 4.19 for $i = 7\%$ per year.

Solution
(a) The cash flows of Fig. 4.19 may be separated as in Fig. 4.20. The dashed line in Fig. 4.20a indicates that the gradient is subtracted from an annual receipt of $900. The present

Figure 4.19
Diagram including a decreasing gradient.

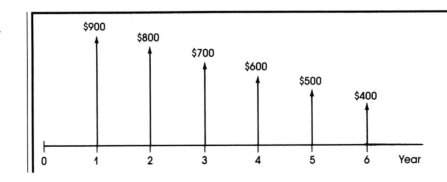

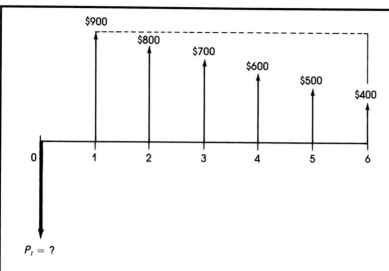

$P_I = ?$

(a)

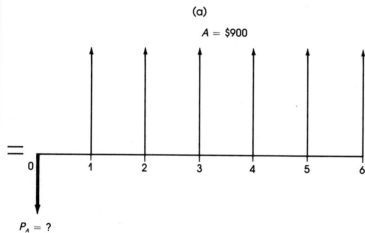

$A = \$900$

$P_A = ?$

(b)

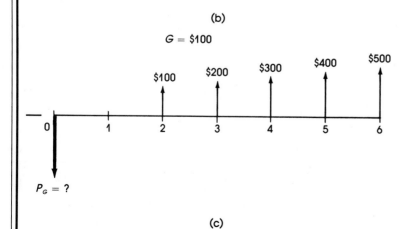

$G = \$100$

$P_G = ?$

(c)

Figure 4.20 Partitioned cash flow of Fig. 4.19: $(a) = (b) - (c)$.

worth is computed as

$$P_T = P_A - P_G = 900(P/A, 7\%, 6) - 100(P/G, 7\%, 6)$$
$$= 900(4.7665) - 100(10.978)$$
$$= \$3192.05$$

(b) The annual series is made up of two components: the base amount and the equivalent uniform-gradient amount. The annual-receipt series ($A_1 = \$900$) is the base amount, and the annual series A_G, which is equivalent to the gradient, is subtracted from A_1.

$$A = A_1 - A_G = 900 - 100(A/G, 7\%, 6)$$
$$= 900 - 100(2.303)$$
$$= \$669.70 \text{ per year for years 1 to 6}$$

Comment Shifted decreasing gradients are handled in a fashion similar to shifted increasing gradients. For an example that combines conventional increasing and shifted decreasing gradients, see the Additional Examples.

Additional Example 4.18
Probs. 4.55 to 4.61

Additional Examples

Example 4.13

A family decides to buy a new refrigerator on credit. The payment scheme calls for a $100 down payment now (the month is March) and $55 a month from June to November with interest at $1\frac{1}{2}\%$ per month compounded monthly. Construct the cash-flow diagram and indicate P in the month in which you can compute an equivalent value using one P/A and one F/P factor. Give the n values for all computations.

Solution Since the payment period, months, equals the compounding period, the 1.5% interest tables of Appendix A can be used. Figure 4.21 solves the problem by placing P in May. The relation using only the two factors is $P = 100(F/P, 1.5\%, 2) + 55(P/A, 1.5\%, 6)$, where $n = 2$ for the F/P factor and $n = 6$ for the P/A factor.

Figure 4.21
Placement of an equivalent amount using only P/A and F/P factors, Example 4.13.

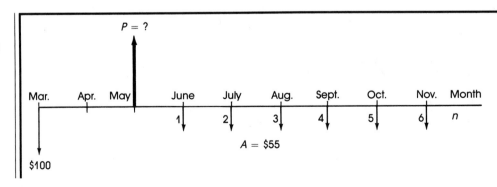

Comment The placement of P is controlled by the uniform series, since the P/A factor is inflexible in this regard.

Example 4.14

Consider the two uniform series shown in Fig. 4.22. Compute the present worth at 15% per year using three different methods.

Solution There are numerous ways to find the present worth. The two simplest are probably the future-worth and present-worth methods. For a third method, the use of the *intermediate-year method* at year 7 is demonstrated.

(*a*) Present-worth method (see Fig. 4.23*a*): The use of P/A factors for the uniform series and the use of P/F factors to obtain the actual present worths allow us to find P_T.

$$P_T = P_{A1} + P_{A2}$$

where

$$P_{A1} = P'_{A1}(P/F, 15\%, 2) = A_1(P/A, 15\%, 3)(P/F, 15\%, 2)$$
$$= 1000(2.2832)(0.7561)$$
$$= \$1726$$

$$P_{A2} = P'_{A2}(P/F, 15\%, 8) = A_2(P/A, 15\%, 5)(P/F, 15\%, 8)$$
$$= 1500(3.3522)(0.3269)$$
$$= \$1644$$

Then

$$P_T = 1726 + 1644 = \$3370$$

(*b*) Future-worth method (Fig. 4.23*b*): Using the F/A, F/P, and P/F factors, we have

$$P_T = (F_{A1} + F_{A2})(P/F, 15\%, 13)$$

where

$$F_{A1} = F'_{A1}(F/P, 15\%, 8) = A_1(F/A, 15\%, 3)(F/P, 15\%, 8)$$
$$= 1000(3.472)(3.0590)$$
$$= \$10,621$$

$$F_{A2} = A_2(F/A, 15\%, 5) = 1500(6.742)$$
$$= \$10,113$$

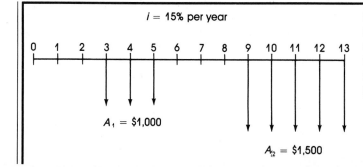

$i = 15\%$ per year

$A_1 = \$1,000$

$A_{l2} = \$1,500$

Figure 4.22 Uniform series to compute a present worth by several factors, Example 4.14.

Figure 4.23
Computation of
the present worth
of Fig. 4.22 by
three methods.

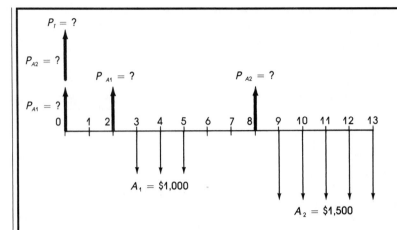

(a) Present-worth method

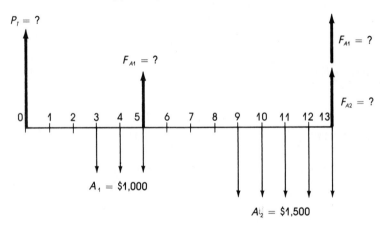

(b) Future-worth method

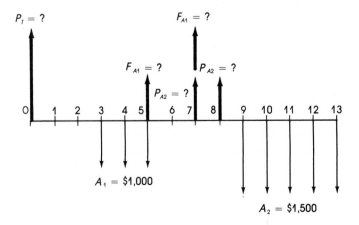

(c) Intermediate-year method

Then

$$P_T = F_{A1} + F_{A2}(P/F, 15\%, 13)20{,}734(0.1625)$$
$$= \$3369$$

(c) Intermediate-year method (Fig. 4.23c): If we find the present worth of both series at year 7 and use the P/F factor, we have

$$P_T = (F_{A1} + P_{A2})(P/F, 15\%, 7)$$

The P_{A2} value is computed as a present worth; but to find the total value P_T at year 0, it must be treated as an F value. Thus,

$$F_{A1} = F'_{A1}(F/P, 15\%, 2) = A_1(F/A, 15\%, 3)(F/P, 15\%, 2)$$
$$= 1000(3.472)(1.3225)$$
$$= \$4592$$

$$P_{A2} = P'_{A2}(P/F, 15\%, 1) = A_2(P/A, 15\%, 5)(P/F, 15\%, 1)$$
$$= 1500(3.3522)(0.8696)$$
$$= \$4373$$

Then

$$P_T = (F_{A1} + P_{A2})(P/F, 15\%, 7)$$
$$= 8965(0.3759)$$
$$= \$3370$$

Example 4.15

Calculate the present worth of the following series of cash flows if $i = 18\%$ per year.

Year	0	1	2	3	4	5	6	7
Cash flow, $	+460	+460	+460	+460	+460	+460	+460	−5,000

Solution The cash-flow diagram is shown in Fig. 4.24. Since the disbursement in year 0 is equal to the disbursements of the A series, the P/A factor can be used for either 6 or 7 years. The problem is worked both ways below.

1. Using P/A and $n = 6$: For this case, the disbursement in year 0 (P_1) is added to the present worth of the remaining payments, since the P/A factor for $n = 6$ will place P_A

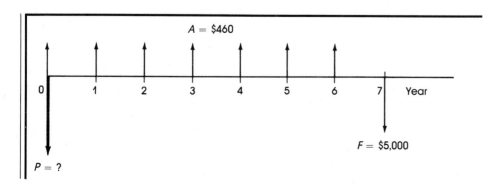

Figure 4.24
Cash-flow diagram, Example 4.15.

in year 0. Thus,

$$P = P_1 + P_A - P_F$$
$$= 460 + 460(P/A, 18\%, 6) - 5000(P/F, 18\%, 7)$$
$$= \$499.40$$

Note that the present worth of the -5000 cash flow (P_F) is negative, since it is a negative cash flow.

2. Using P/A and $n = 7$: By using the P/A factor for $n = 7$, the "present worth" is located in year 1, not year 0, because the P must always be 1 year ahead of the first A when the P/A factor is used. It is therefore necessary to move the P_A value 1 year forward with the F/P factor. Thus,

$$P = 460(P/A, 18\%, 7)(F/P, 18\%, 1) - 5000(P/F, 18\%, 7)$$
$$= \$499.38$$

Comment Rework the problem by first finding the future worth of the series and then solving for P. You should obtain the same answer as above, if you work it correctly.

Example 4.16

Determine the amount of the gradients, the location of the present worth of the gradients, and the n values of the cash flow of Fig. 4.25.

Solution You should construct your own cash-flow diagram upon which you locate the gradient present worths and determine the n values. If we call the series from year 1 to year 4 G_1, the base amount is $25, G_1 is $15, n_1 equals 4 years, and P_{G1} occurs in year 0. For the second series, the base amount is also $25, but G_2 is $5, n_2 equals 7 years, and P_{G2} takes place in year 5.

Comment Even though Fig. 4.25 shows a series of disbursements and receipts, both gradients are increasing. Decreasing gradients are discussed in Sec. 4.6.

Example 4.17

Using $i = 8\%$ for the cash flows of Fig. 4.26, compute the (*a*) equivalent annual worth and (*b*) present worth.

Figure 4.25
Cash flow of two gradients, Example 4.16.

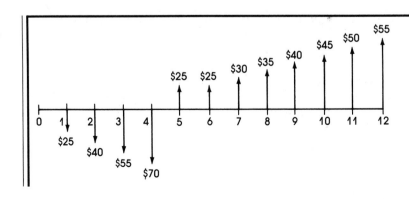

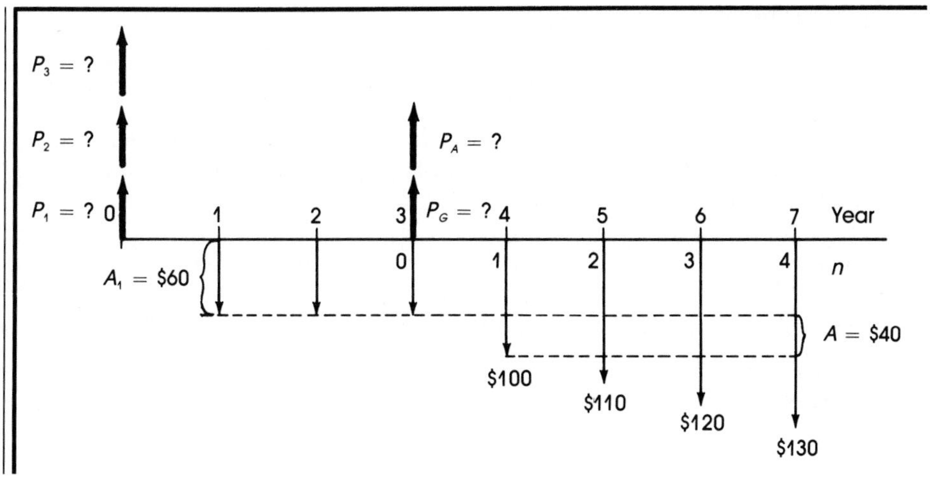

Figure 4.26
Shifted gradient,
Example 4.17.

Solution

(a) The dashed lines of Fig. 4.26 should help you in the solution for present worth and equivalent annual worth. For the annual series, use the steps outlined in Example 4.10.

1. $A_1 = \$60$ for 7 years

    $A = \$40$ base amount of gradient for 4 years

    $A_3 =$ equivalent series of base amount for 7 years

    $\quad = P_2(A/P, 8\%, 7)$

    where

    $P_2 =$ present worth of $A = \$40$ series

    $\quad = 40(P/A, 8\%, 4)(P/F, 8\%, 3)$

    $\quad = \$105.17$

    Then,

    $A_3 = 105.17(A/P, 8\%, 7)$

    $\quad = \$20.20$

2. $P_G =$ present worth of gradient in year 3

    $\quad = G(P/G, 8\%, 4) = 10(4.650)$

    $\quad = \$46.50$

3. $P_1 =$ present worth of P_G in year 0

    $\quad = P_G(P/F, 8\%, 3) = 46.50(0.7938)$

    $\quad = \$36.91$

4. $A_2 =$ equivalent 7-year A value of gradient

    $\quad = P_1(A/P, 8\%, 7) = 36.91(0.19207)$

    $\quad = \$7.09$

5. The equivalent annual worth is

$$A = A_1 + A_2 + A_3 = 60.00 + 7.09 + 20.20$$
$$= \$87.29$$

(b) To find the present worth of the cash flows shown in Fig. 4.26, note that the present worth P_1 of the gradient is the same as that calculated in step 3 above. Then the $40 series has a present worth of

$$P_2 = 40(P/A, 8\%, 4)(P/F, 8\%, 3) = 40(3.3121)(0.7938)$$
$$= \$105.17$$

The $60 annual series has a present worth of

$$P_3 = 60(P/A, 8\%, 7) = 60(5.2064)$$
$$= \$312.38$$

The total present worth is

$$P_T = P_1 + P_2 + P_3 = 36.91 + 105.17 + 312.38$$
$$= \$454.46$$

which is equivalent to $87.29 per year, as in part (a).

Example 4.18

Assume that you are planning to invest money at 7% per year shown by the increasing gradient of Fig. 4.27. In addition, you plan to withdraw according to the decreasing gradient shown. Find the net present worth and equivalent annual worth for the cash flow sequence.

Solution For the investment sequence, G is $500, the base amount is $2000, and n equals 5; for the withdrawal sequence, G is $-1000, the base amount is $5000, and n equals 5; there is also a 2-year annual series with A equal to $1000 in years 11 and 12. For the

Figure 4.27 An investment and withdrawal sequence, Example 4.18.

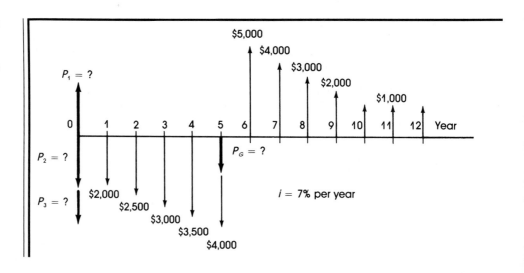

investment sequence,

P_I = present worth of investments

$= 2000(P/A, 7\%, 5) + 500(P/G, 7\%, 5)$

$= 2000(4.1002) + 500(7.646)$

$= \$12,023.40$

For the withdrawal sequence,

P_W = present worth of withdrawal gradient

+ present worth of withdrawals in years 11 and 12

$= P_2 + P_3 = P_G(P/F, 7\%, 5) + P_3$

$= [5000(P/A, 7\%, 5) - 1000(P/G, 7\%, 5)](P/F, 7\%, 5) + 1000(P/A, 7\%, 2)(P/F, 7\%, 10)$

$= [5000(4.1002) - 1000(7.646)](0.7130) + 1000(1.8080)(0.5084)$

$= \$10,084.80$

Since P_I is actually a negative cash flow and P_W is positive, the resultant present worth P is

$P = P_W - P_I = 10,084.80 - 12,023.40$

$= \$ - 1938.60$

The A value may be computed using

$A = P(A/P, 7\%, 12)$

$= \$ - 244.07$

Comment   Thus, in present-worth equivalent, you will invest $1938.60 more than you plan to withdraw. This is equivalent to an annual savings of $244.07 per year for the 12-year period.

4.1   Construct a net cash-flow diagram for each of the following and locate the present **Problems** worth and future worth of each uniform series *separately*.
(a) A $600 deposit for 7 years starting 4 years from now.
(b) A $2000-per-month deposit for 6 months starting next month and a $1500-per-month withdrawal for 4 months starting 3 months from now.
(c) The following transaction in your Christmas club account for the past 12 months:

Month	Deposit	Withdrawal	Month	Deposit	Withdrawal
January	$20	$ 0	July	$75	$ 25
February	20	0	August	75	25
March	20	10	September	25	10
April	20	10	October	25	10
May	75	10	November	25	10
June	75	25	December	0	250

4.2   Determine the amount of money a person must deposit now in order to be able to make ten $3600-per-year withdrawals starting 20 years from now if the interest rate is 14% per year.

4.3   What is the equivalent uniform annual worth in years 1 through 16 of $250 per year for 12 years, with the first payment starting 5 years from now, if the interest rate is 20% per year?

4.4   A couple purchases an insurance policy which they plan to use to finance their child's college education. If the policy provides $10,000 twelve years from now, how much can be withdrawn every year for 5 years if the child starts college 15 years from now? Assume $i = 16\%$ per year.

4.5   A woman deposited $700 per year for 8 years. Starting in the ninth year she increased her deposits to $1200 per year for 5 more years. How much money did she have in her account immediately after she made her last deposit if the interest rate was 15% per year?

4.6   How much money will the woman in Prob. 4.5 have in her account 30 years from the present time if she makes no deposits after the one in year 13?

4.7   What is the present worth 1 year prior to the first deposit for the investment specified in Prob. 4.5?

4.8   A man plans to begin saving for his retirement such that he will be able to withdraw money every year for 30 years starting 25 years from now. He estimates that he will be able to start saving money 1 year from now and plans to deposit $500 per year through year 24. What uniform annual amount will he be able to withdraw when he retires, if the interest rate is 12% per year?

4.9   A businessman purchased an existing building and found that the ceiling was poorly insulated. He estimated that with 6 inches of foam insulation, he could cut the heating bill by $25 per month and the air conditioning cost by $20 per month. Assuming that the winter season is the first 6 months of the year and the summer season is the next 6 months, how much can he afford to spend on insulation if he expects to keep the building for only 2 years? Assume $i = 1\frac{1}{2}\%$ per month.

4.10   A couple plans to make an investment now in order to finance their child's college education. If the child should be able to withdraw $1000 per year in years 15 through 20 such that the last withdrawal will close the account, how much must be invested now if interest is computed at 9% per year?

4.11   If the couple in Prob. 4.10 wanted to make a uniform deposit for 14 years instead of the lump-sum investment, how much would they have to deposit every year starting 1 year from now?

4.12   How much money would the heirs of the man in Prob. 4.8 receive if he died 4 years after his retirement?

4.13   How much money would the woman in Prob. 4.5 have if the interest rate increased from 15 to 16% after 5 years?

4.14   Giovanni's Pizza Palace has a 10-year lease on 200 square meters of space in an enclosed shopping center. The rent is paid yearly at a rate of $100 per square meter. At the end of the fourth year of the lease, the owner of the pizza shop decides to purchase a building and relocate the business. How much must the owner of the shopping center be paid for the remainder of the lease if the interest rate is 12% per year?

4.15   In order to compare leasing versus purchasing a microprocessor, an engineer was told to convert beginning-of-period payments (for leasing) to end-of-period payments. The beginning-of-period amount was $1000 per month, and the computer was to be leased for 3 years. If the company's nominal interest rate is 18% per year compounded monthly, what is the equivalent end-of-period amount?

4.16 Determine the beginning-of-year payments which would be equivalent to the cash flow shown below. Use an interest rate of 15% per year.

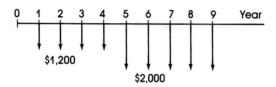

4.17 Calculate the value of x in the cash flow shown below such that the equivalent total value in month 7 is $8000, using an interest rate of $1\frac{1}{2}$% per month.

Month	Cash flow
0	200
1	200
2	200
3	200
4	200
5	x
6	x
7	x
8	x
9	500
10	500
11	500

4.18 If a couple opens a savings account by depositing $1500 now and deposits $1500 every year for 14 years, how much will be in the account after the last deposit if the interest rate is 12% per year?

4.19 How much will be in the account in Prob. 4.18 if the interest rate changes to 14% per year after the first 5 years?

4.20 Find the value of x such that the positive cash flows will be exactly equivalent to the negative cash flows if the interest rate is 15% per year.

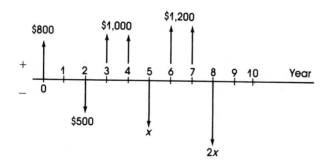

4.21 Find the value of x in the diagram below that would make the equivalent present worth of the cash flow equal to $22,000, if the interest rate is 15% per year.

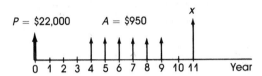

4.22 Calculate the amount of money in year 7 that would be equivalent to the following cash flows if the interest rate is a nominal 13% per year compounded semiannually.

Year	0	1	2	3	4	5	6	7	8	9
Amount, $	900	900	900	900	1,300	1,300	1,300	0	900	900

4.23 The GRQ Company purchased a machine for $12,000 with an expected salvage value of $2000. Operating expenses for the machine will be $1800 per year. In addition, a major overhaul will be required every 5 years at a cost of $2800. What is the equivalent present cost of the machine if it will have an 18-year life and the interest rate is 15% per year?

4.24 A petroleum company is planning to sell a number of existing oil wells. The wells are expected to produce 100,000 barrels of oil per year for 11 more years. If the selling price per barrel of oil is currently $35, how much would you be willing to pay for the wells if the price of oil is expected to increase by $3 per barrel every 3 years, with the first increase to occur 2 years from now? Assume that the interest rate is 12% per year for the first 4 years and 15% per year thereafter, and that oil sales are made at the end of each year.

4.25 Calculate the (a) present worth (year 0) and (b) future worth in year 10 of the following series of disbursements:

Year	Disbursement	Year	Disbursement
0	$3,500	6	$ 5,000
1	3,500	7	5,000
2	3,500	8	5,000
3	3,500	9	5,000
4	5,000	10	15,000
5	5,000		

Assume that $i = 16\%$ per year compounded semiannually.

4.26 A building contractor purchased a dirt scraper for $35,000. He maintained the scraper at a cost of $2500 per year. He overhauled the machine 4 years after the purchase at a cost of $4000. He sold the scraper for $18,000 two years after the overhaul. What was his equivalent uniform annual cost if the interest rate was 10% per year?

4.27 How much money would you have to deposit for 6 consecutive years starting 1 year from now if you want to be able to withdraw $45,000 eleven years from now? Assume that the interest rate is 15% per year.

4.28 What is the equivalent uniform annual worth of the disbursements shown in Prob. 4.25?

4.29 A large manufacturing company purchased a semiautomatic machine for $13,000. Its annual maintenance and operation cost was $1700. After 5 years from the initial purchase, the company decided to purchase an additional unit for the machine which would make it fully automatic. The additional unit had a first cost of $7100. The cost for operating the machine in the fully automatic condition was $900 per year. If the company used the machine for a total of 16 years and then sold the automatic addition for $1800, what was the equivalent uniform annual worth of the machine at an interest rate of 9% per year?

4.30 Calculate the number of $15,000 payments that would be required in the cash-flow diagram shown below in order for the annual payments to be equivalent to the initial $37,000 saving. Use an interest rate of an effective 17% per year compounded monthly.

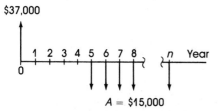

4.31 A company borrows $8000 at an interest rate of a nominal 12% per year compounded monthly. The company desires to repay the loan in 14 equal monthly payments, with the first payment starting 1 month from now.
(a) What should be the size of each payment?
(b) If after making eight payments the company decides to pay off the balance of the loan in the ninth month, how much must the company pay?

4.32 If you want to have $125,000 for your child's college education 21 years from now, how much will you have to deposit each year if your first deposit is now and the last deposit is 18 years from now? Assume that the interest rate is 15% per year.

4.33 A woman plans to make a total of eight deposits, with the first deposit now and succeeding deposits at 1-year intervals, so that she will be able to withdraw $4000 per year for 10 years, the first withdrawal starting 16 years from now. How much must she deposit each year if the interest rate is a nominal 12% per year compounded quarterly?

4.34 An elderly man starts a retirement account by depositing $10,000 now and $300 each month for 10 years. How much money can he withdraw per month for 5 years if he makes his first withdrawal 3 months after his last deposit? Assume that the interest rate is a nominal 18% per year compounded monthly.

4.35 A couple expects to borrow $500 each year for the next 2 years to cover Christmas expenses. Due to increasing costs, the couple expects to have to borrow $550 three years from now, $600 the next year, and $650 the following year. However, due to their children's ages, they hope to have to borrow only $300 per year after that. Calculate (a) the present worth and (b) the equivalent uniform annual worth of the disbursements for a total of 15 years using an interest rate of 13% per year.

4.36 The UR-OK Company is considering two types of machines, both of which will do the same job. The net cash flows for each are tabulated below. Calculate the present worth of each machine using an interest rate of 18% per year.

	Cash flows				Cash flows	
Year	Machine A	Machine B	Year	Machine A	Machine B
0	$+2,000	$+2,000	7	$+3,000	$+2,500
1	2,000	2,000	8	3,500	2,500
2	2,000	2,500	9	4,000	2,500
3	2,500	3,000	10	4,500	3,000
4	2,500	3,500	11	3,000	3,500
5	2,500	4,000	12	3,000	4,000
6	2,500	2,500			

4.37 A person borrows $8000 at a nominal 7% per year compounded quarterly. It is desired to repay the loan in 12 semiannual payments, with the first payment to start 3 months from now. If the payments are to increase by $50 each time, determine the size of the first payment.

4.38 Find the value of G in the diagram below that would make the income stream equivalent to the disbursement stream, using an interest rate of 20% per year.

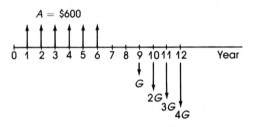

4.39 For the diagram shown below, find the value of the last receipt in the income stream that would make the receipts equivalent to the $500 initial investment at time 0. Use an interest rate of 15% per year.

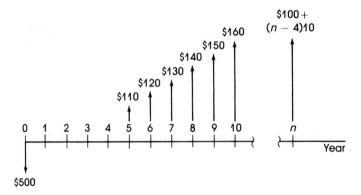

4.40 Calculate the value of x for the cash-flow series shown below such that the equivalent total value in month 5 is $25,000, using an interest rate of a nominal 12% per year compounded monthly.

Month	Cash flow
0	100
1	$100 + x$
2	$100 + 2x$
3	$100 + 3x$
4	$100 + 4x$
5	$100 + 5x$
6	$100 + 6x$
7	$100 + 7x$
8	$100 + 8x$
9	$100 + 9x$
10	$100 + 10x$
11	$100 + 11x$
12	$100 + 12x$
13	$100 + 13x$
14	$100 + 14x$

4.41 Solve for the value of G such that the left cash-flow diagram is equivalent to the one on the right. Use an interest rate of 13% per year.

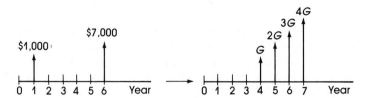

4.42 Solve for the value of x, using an interest rate of 12% per year.

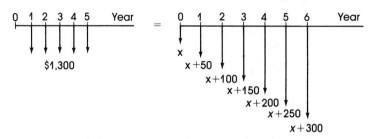

4.43 If a machine costs $15,000 to purchase and the operating costs are $1000 at the end of the first year, $1200 at the end of the second, and amounts increasing by $200 per year through year 12, what is the present worth of the machine if the interest rate is a nominal 15% per year compounded semiannually?

4.44 If the operating costs in Prob. 4.43 start now instead of at the end of the first year, what will the present worth be at time 0 for the 12 years (i.e., 13 payments)?

4.45 For the diagram shown below, find the value of x that will make the negative cash flows equal to the positive cash flow of $800 at time 0. Assume $i = 15\%$ per year.

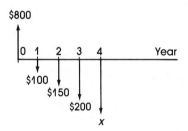

4.46 Rework Prob. 4.45 above, using an interest rate of a nominal 15% per year compounded quarterly.

4.47 Find the future worth (in month 9) of the cash flow shown in the figure below, using an interest rate of 1% per month.

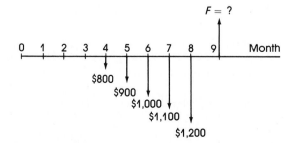

4.48 Mr. Alum Nye is planning to make a contribution to the junior college from which he graduated. He would like to donate an amount of money now so that the college can support students. Specifically, he would like to provide financial support for tuition for five students per year for a total of 20 years (i.e., 21 grants), with the first tuition grant to be made immediately and continuing at 1-year intervals. The cost of tuition at the school is $3800 per year and is expected to stay at that amount for 4 more years. After that time, however, the tuition cost will increase by $90 per year. If the school can deposit the donation and earn interest at a rate of a nominal 8% per year compounded semiannually, how much must Mr. Alum Nye donate?

4.49 Find the present worth of the cash flow shown below using an interest rate of 16% per year.

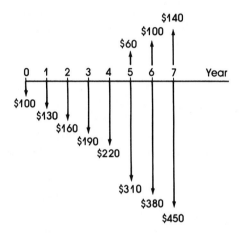

4.50 Calculate the present worth (time 0) of a lease which requires a payment now of $20,000 and amounts increasing by 6% per year. Assume the lease payments are beginning-of-year payments and the lease is for a total of 10 years. Use an interest rate of 14% per year.

4.51 Calculate the present worth of a machine which has an initial cost of $29,000, a salvage value of $5000 after 8 years, and an annual operating cost of $13,000 for the first 3 years, increasing by 10% per year thereafter. Use an interest rate of 15% per year.

4.52 Calculate the present worth to the A-1 Box Company of leasing a computer if the yearly cost is $15,000 for year 1 and $16,500 for year 2, with costs increasing by 10% each year thereafter. Assume that the lease payments must be made at the beginning of the year and that a 7-year study period is to be used. The company's minimum attractive rate of return is 16% per year.

4.53 Calculate the present worth of a machine which costs $55,000 and has an 8-year life with a $10,000 salvage value. The operating cost of the machine is expected to be $10,000 in year 1 and $11,000 in year 2, with amounts increasing by 10% per year thereafter. Use an interest rate of 15% per year.

4.54 Calculate the equivalent annual worth of a machine which costs $73,000 initially and will have a $10,000 salvage value after 9 years. The operating cost is $21,000 in year 1, $22,050 in year 2, and amounts increasing by 5% each year. The minimum attractive rate of return is 19% per year.

4.55 Find the present worth (time 0) of the cash flow shown in the figure below. Assume $i = 12\%$ per year.

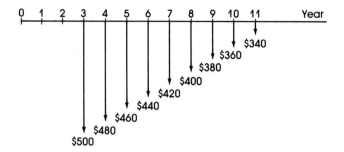

4.56 If you start a bank account now by depositing $2000, how long will it take for you to deplete the account if you start withdrawing money $1\frac{1}{2}$ years from now by withdrawing $500 the first month, $450 the second month, $400 the next month, and amounts decreasing by $50 per month until the account is depleted? Assume that the account earns interest at a rate of a nominal 12% per year compounded monthly.

4.57 Compute the present worth of the following cash flows at $i = 12\%$ per year.

Year	0	1–4	5	6	7	8	9	10
Amount, $	5,000	1,000	900	800	700	600	500	400

4.58 Compute the present worth and equivalent uniform annual worth at $i = 10\%$ per year for the cash flows below.

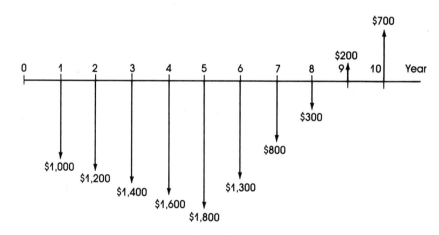

4.59 Find the present worth of the cash flows shown below at $i = 20\%$ per year.

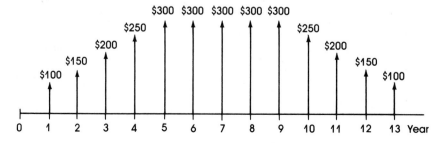

4.60 The Chisel-Em Company plans to purchase a new piece of construction equipment now. Realizable income is $15,000 the first year, $12,000 in year 2, $9000 in year 3, and so on. If the company plans to sell the equipment after 7 years and interest is 15% per year, compute the present worth and equivalent annual worth of the incomes.

4.61 Compute the present worth and future worth of the cash flows shown below if $i = 16\%$ per year.

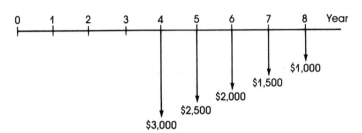

LEVEL TWO

Now that you have learned to correctly account for the time value of money and the effective interest rate in a project, you are ready to evaluate complete alternatives. In this level you will learn how to perform an economic-based analysis of one or more alternatives, and you will learn how to make a choice of the economically best method from two or more alternatives.

There are three basic methods used to perform an economic analysis: present worth (PW), equivalent uniform annual worth (EUAW), and rate of return (ROR). All three methods will give identical decisions for alternative selection when applied to the same set of cost and revenue estimates and when the comparisons are properly conducted.

Chapter	Subject
5	Present-worth and capitalized-cost evaluation
6	Equivalent-uniform-annual-worth evaluation
7	Rate-of-return computations for a single project
8	Rate-of-return evaluation for multiple alternatives

Present-Worth and Capitalized-Cost Evaluation

5

When a future amount of money is converted into its equivalent present value, the magnitude of the present amount is always less than the magnitude of the cash flow from which it was calculated. This is because for any interest rate greater than zero, all P/F factors have a value which is less than 1.0. For this reason, present-worth calculations are often referred to as *discounted cash-flow* (DCF) methods. Similarly, the interest rate used in making the calculations is referred to as the *discount rate*. Other terms frequently used in reference to present-worth calculations are present value (PV), present worth (PW), and net present value (NPV). Regardless of what they are called, present-worth calculations are routinely used for making economic-related decisions. Up to this point, present-worth computations have been made from cash flow associated with only a single project. In this chapter, techniques for comparing alternatives by the present-worth method are discussed. While the examples presented in this chapter (and the next) will be based on only two alternatives, the same procedures should be followed in a present-worth evaluation of more than two.

Section Objectives

After completing this chapter, you should be able to do the following:

5.1 Select the better of two alternatives using present-worth calculations for alternatives that have *equal lives*, given the cash flows and their respective dates, the lives and salvage values of the alternatives, and the interest rate.

5.2 Select as in objective 5 1, except for alternatives that have *different lives*.

5.3 Define *capitalized cost* and calculate the capitalized cost of a series of disbursements, given the disbursements, their respective dates, and the interest rate.

5.4 Select the better of two alternatives on the basis of capitalized cost, given the cash flow for each alternative and the interest rate.

Study Guide

5.1 Present-Worth Comparison of Equal-Lived Alternatives

The present-worth (PW) method of alternative evaluation is very popular because future expenditures or receipts are transformed into *equivalent dollars now*. That is, all of the future cash flows associated with an alternative are converted into present dollars. In this form, it is very easy, even for a person unfamiliar with economic analysis, to see the economic advantage of one alternative over one or more other alternatives.

The comparison of alternatives having equal lives by the present-worth method is straightforward. If both alternatives are used in identical capacities for the same time period, they are termed *equal-service* alternatives. Frequently, the cash flow involves disbursements only, in which case it is generally convenient to omit the minus sign from the disbursements. Then the alternative with the *lowest* present-worth value (i.e., cost) should be selected. On the other hand, when disbursements *and* incomes must be considered, it is generally more convenient to consider income as positive and disbursements as negative, in which case the alternative selected will be the one with the *highest* present worth. While it does not matter which sign convention is adopted for the cash flow, it is important to be consistent in assigning the proper sign to each cash-flow element and then interpreting the results in accordance with that sign convention. Example 5.1 illustrates a present-worth comparison.

Example 5.1

Make a present-worth comparison of the equal-service machines for which the costs are shown below, if $i = 10\%$ per year.

	Type A	Type B
First cost, P	$2,500	$3,500
Annual operating cost, AOC	900	700
Salvage value, SV	200	350
Life, years	5	5

Solution   The cash-flow diagram is left to the reader. The present worth of each machine is calculated as follows:

$$P_A = 2500 + 900(P/A, 10\%, 5) - 200(P/F, 10\%, 5) = \$5788$$

$$P_B = 3500 + 700(P/A, 10\%, 5) - 350(P/F, 10\%, 5) = \$5936$$

Type A should be selected, since $P_A < P_B$.

Comment Note the minus sign on the salvage value, since it is a negative cost. Also, when alternatives are evaluated by the present-worth method, it is common to use PW rather than *P*. In this case, then, $PW_A = \$5788$ and $PW_B = \$5936$.

An assumption inherent in all present-worth analyses is that any funds received through a project are reinvested immediately after they become available at the rate of return (interest rate) used to compute the present worth.

Additional Example 5.5
Probs. 5.1 to 5.8

5.2 Present-Worth Comparison of Different-Lived Alternatives

When the present-worth method is used for comparing alternatives that have different lives, the procedure of the previous section is followed with this exception: *The alternatives must be compared over the same number of years.*

This is because, by definition, a present-worth comparison involves calculating the equivalent present value of all of the future cash flow for each alternative. Obviously, a fair comparison can be made only when the present worths represent costs (or income) associated with equal service, as described in the preceding section. Failure to compare equal service would always favor the shorter-lived alternative (for costs), even if it were not the most economical one, because fewer periods of costs would be involved. The equal-service requirement can be satisfied by either of the following two methods: (1) compare the alternatives using a planning horizon (*n*) which does not take into consideration the lives of the alternatives, or (2) compare the alternatives over a period of time equal to the least common multiple (LCM) for their lives.

For the first method, a time horizon is chosen over which the economic analysis is to be conducted, and only those cash flows occurring through that time period are considered relevant. Any cash flows occurring beyond the stated horizon, whether income or disbursement, are not considered as part of the alternative and are ignored in the present-worth calculation. The time horizon chosen might be relatively short, as when short-term goals are most important, or vice versa. In any case, once the horizon has been selected and the cash flows identified for each alternative, the present worths are determined and the most economical one is chosen. The planning-horizon concept is especially useful in replacement analysis as discussed in Chap. 10.

For the second method, equal service is achieved by making the comparison over the least common multiple of lives between the alternatives, which automatically makes their cash flows extend through the same time period. That is, the cash flow for one "cycle" of an alternative must be duplicated for the least common multiple of years, so that service is compared over the same total life for each alternative. For example, if it is desired to compare alternatives which have lives of 3 years and 2 years, respectively, the alternatives must be compared over a period of 6 years. Such a procedure obviously requires that some assumptions be made about the alternatives in their subsequent life cycles. Specifically, these assumptions are (1) that the alternatives under consideration (processes, machines, services, etc.) will be needed for as long as the least common multiple of years, and (2) that the respective costs of the alternatives will be the same in all subsequent life cycles as they were in the first one. As will be shown in Chap. 12, this second assumption is valid as long as the cash flows are expected to change by exactly the inflation or deflation rate that is applicable through the LCM time period. If the cash flows are expected to change by any other rate, then a planning-horizon-type present-worth analysis must be conducted. This also holds true when the first assumption (about the length of time the alternatives are needed) cannot be made. It is important to remember that when an alternative has a terminal salvage value, this must also be included and shown as an income on the cash-flow diagram at the time reinvestment is made.

Since a planning-horizon type of analysis is relatively straightforward and readily understandable, we will use the least-common-multiple method in the examples and problems in this text unless otherwise specified. Example 5.2 shows evaluations for least-common-multiple and specified-planning horizons.

Example 5.2

A plant superintendent is trying to decide between the machines detailed below.

	Machine A	Machine B
First cost	$11,000	$18,000
Annual operating cost	3,500	3,100
Salvage value	1,000	2,000
Life, years	6	9

(a) Determine which one should be selected on the basis of a present-worth comparison using an interest rate of 15% per year.
(b) If a planning horizon of 5 years is specified and the salvage values are not expected to change, which alternative should be selected?

Solution
(a) Since the machines have different lives, they must be compared over their least common multiple of years, which is 18 years in this case. The cash-flow diagram is shown in Fig. 5.1. Thus, if costs are considered positive

$$PW_A = 11{,}000 + 11{,}000(P/F, 15\%, 6) - 1000(P/F, 15\%, 6) + 11{,}000(P/F, 15\%, 12)$$
$$- 1000(P/F, 15\%, 12) - 1000(P/F, 15\%, 18) + 3500(P/A, 15\%, 18)$$
$$= \$38{,}559$$

$$PW_B = 18{,}000 + 18{,}000(P/F, 15\%, 9) - 2000(P/F, 15\%, 9) - 2000(P/F, 15\%, 18) +$$
$$3100(P/A, 15\%, 18)$$
$$= \$41{,}384$$

Machine A should be selected, since $PW_A < PW_B$.
(b) For a 5-year planning horizon, the present-worth equations are:

$$PW_A = 11{,}000 + 3500(P/A, 15\%, 5) - 1000(P/F, 15\%, 5)$$
$$= \$22{,}236$$

$$PW_B = 18{,}000 + 3100(P/F, 15\%, 5) - 2000(P/F, 15\%, 5)$$
$$= \$27{,}397$$

Machine A is still the better choice.

Figure 5.1 Cash-flow diagram for unequal-life assets, Example 5.2a.

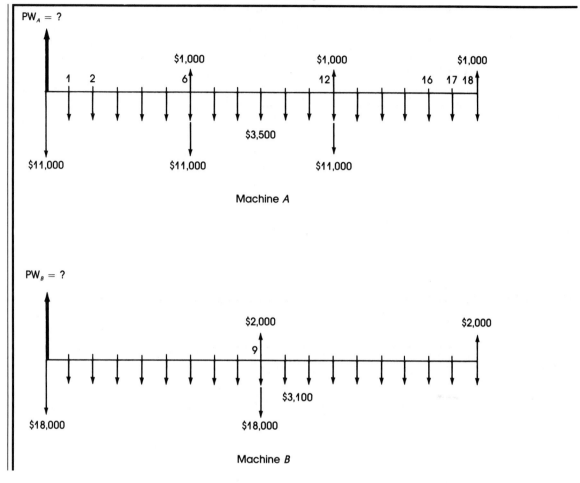

Machine A

Machine B

Comment Note that in part (*a*) the salvage value of each machine must be recovered *after each life cycle* of the asset. In this example, the salvage value of machine *A* was recovered in years 6, 12, and 18, and that for machine *B* was recovered in years 9 and 18.

In part (*b*), the costs after year 5 do not matter as far as the economic analysis is concerned. It is assumed that the machines can be disposed of after year 5 with no additional economic consequences.

When a decision maker is not interested in realizing a return on an investment, an evaluation procedure sometimes used is *payback analysis*. The *payback period n*, the number of years necessary to exactly recover the initial investment *P*, is computed by summing the annual cash-flow values and estimating *n* through the relation

$$0 = -\text{initial investment} + \text{sum of annual cash flows}$$

When the time value of money is considered and each cash flow is discounted to time 0, this technique can be used as supplemental to the present-worth analysis. However, it still has the additional disadvantage that it automatically neglects any cash flows which occur after the payback period and which may increase the actual rate of return of the project. Payback analysis and its usefulness is discussed in detail in Sec. 16.3.

The term *life-cycle cost* is frequently used in present-worth evaluation studies of large projects, especially defense-related projects. In its simplest form, life-cycle cost means that *all* costs associated with an alternative must be included in the evaluation. These costs should include, for example, expenditures for research and development, support costs, production costs, etc. The subject of life-cycle costing is further developed in Sec. 16.4.

Additional Examples 5.6 and 5.7
Probs. 5.9 to 5.24

5.3   Capitalized-Cost Calculations

Capitalized cost refers to the present-worth value of a project that is assumed to last forever. Certain public works projects such as dams, irrigation systems, and railroads fall into this category. In addition, permanent university or charitable-organization endowments must be handled by capitalized-cost methods.

In general, the procedure that should be followed in calculating the capitalized cost of an infinite sequence of cash flows is as follows:

1. Draw a cash-flow diagram showing all nonrecurring (one-time) expenditures or receipts and at least two cycles of all recurring (periodic) expenditures or receipts.
2. Find the present worth of all nonrecurring expenditures (receipts).
3. Find the equivalent uniform annual worth (i.e., *A*) through one life cycle of all recurring expenditures and add this to all other uniform amounts occurring in years 1 through infinity to obtain a total equivalent uniform annual worth (EUAW).
4. Divide the EUAW obtained in step 3 by the interest rate to get the capitalized cost of the EUAW.
5. Add the value obtained in step 2 to the value obtained in step 4.

The purpose for beginning the solution by drawing a cash-flow diagram should be evident from previous chapters. However, the cash-flow diagram is probably more important in this calculation than it is anywhere else, because it facilitates the differentiation between nonrecurring and recurring (periodic) expenditures. In step 2, the present worth of all nonrecurring expenditures (receipts) should be determined. Since the capitalized cost is the *present worth* of a perpetual project, the reason for this step should be obvious. In step 3 the EUAW (which has been called *A* thus far) of all recurring and uniform annual expenditures should be calculated. This is done to compute the present worth of a perpetual annual cost (capitalized cost) using the following equation:

$$\text{Capitalized cost} = \frac{\text{EUAW}}{i} \qquad\qquad (5.1)$$

The validity of Eq. (5.1) can be illustrated by considering the time value of money. If $10,000 is deposited into a savings account at 20%-per-year interest compounded annually, the maximum amount of money that can be withdrawn at the end of every year for *eternity* is $2000, or the amount equal to the interest that accumulated in that year. This leaves the original $10,000 deposit to earn interest so that another $2000 will be accumulated in the next year. Mathematically, the amount of money that can be accumulated and withdrawn in each consecutive interest period for an infinite period of time is

$$A = Pi \tag{5.2}$$

Thus, for the example,

$$A = 10000(0.20) = \$2000 \text{ per year}$$

The capitalized-cost calculation proposed per Eq. (5.1) is the reverse of the one just made; that is, Eq. (5.2) is solved for P to obtain:

$$P = \frac{A}{i} \tag{5.3}$$

For the example just cited, if it were desired to withdraw $2000 every year for eternity at an interest rate of 20% per year, from Eq. (5.3),

$$P = \frac{2000}{0.20} = \$10,000$$

After the present worths of all cash flows have been obtained, the total capitalized cost is simply the sum of these present worths. Capitalized-cost calculations are illustrated in Example 5.3.

Example 5.3

Calculate the capitalized cost of a project that has an initial cost of $150,000 and an additional investment cost of $50,000 after 10 years. The annual operating cost will be $5000 for the first 4 years and $8000 thereafter. In addition, there is expected to be a recurring major rework cost of $15,000 every 13 years. Assume that $i = 15\%$ per year.

Solution The format outlined above will be used.

1. Draw cash flows for two cycles (Fig. 5.2).

2. Find the present worth (P_1) of the nonrecurring costs of $150,000 now and $50,000 in year 10:

   $$P_1 = 150,000 + 50,000(P/F, 15\%, 10) = \$162,360$$

3. Convert the recurring cost of $15,000 every 13 years into an EUAW (A_1) for the first 13 years:

   $$A_1 = 15,000(A/F, 15\%, 13) = \$437$$

   Note that the same A value (i.e., $437) applies to all of the other 13 periods as well.

4. The capitalized cost for the annual-cost series can be computed through either of the following two ways: (a) Consider a series of $5000 from now to infinity and find the present worth of $8000 − $5000 = $3000 from year 5 on, or (b) find the present worth of

Figure 5.2 Diagram used to compute capitalized cost, Example 5.3.

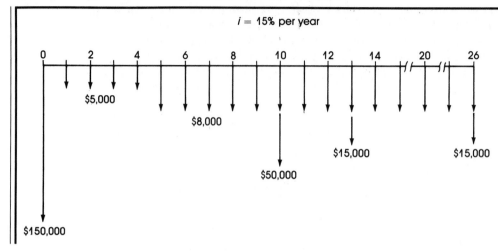

$5000 for 4 years and the present worth of $8000 from year 5 to infinity. Using the first method, we find that the annual cost (A_2) is $5000, and the present worth (P_2) of $3000 from year 5 to infinity, using Eq. (5.3) and the P/F factor, is

$$P_2 = \frac{3000}{0.15}(P/F, 15\%, 4) = \$11,436$$

The two annual costs are converted into a capitalized cost (P_3):

$$P_3 = \frac{A_1 + A_2}{i} = \frac{437 + 5000}{0.15} = \$36,247$$

5. The total capitalized cost (P_T) can now be obtained by addition:

$$P_T = P_1 + P_2 + P_3 = \$210,043$$

Comment In calculating P_2, $n = 4$ was used in the P/F factor because the present worth of the annual $3000 cost is computed in year 4, since P is always one period ahead of the first A. You should rework the problem using the second method suggested for calculating P_2.

A capitalized cost can also be obtained for alternatives which have a finite life. In this case, simply calculate the EUAW for one life cycle of the alternative and then divide the resulting A value by the interest rate. Example 5.8 illustrates these calculations.

Probs 5.25 to 5.32

5.4 Capitalized-Cost Comparison of Two Alternatives

When two or more alternatives are compared on the basis of their capitalized cost, the procedure of Sec. 5.3 is followed for each alternative. Since the capitalized cost represents the present total cost of financing and maintaining a given alternative forever, the alternatives will automatically be compared for the same number of years (i.e., infinity). The alternative with the smaller capitalized cost will represent the most economical one. As in present-worth and all other alternative evaluation methods, it is only the differences in cash flow between the alternatives which must be considered for comparative purposes. Therefore, whenever possible, the calculations should be simplified by eliminating the elements of cash flow which are common to both alternatives. On the other hand, if true capitalized-cost values are needed instead of just comparative ones, the actual cash flows rather than the differences should be used. True capitalized-cost values would be needed, for example, if one wanted to know the actual or true financial obligations associated with a given alternative. Example 5.4 shows the procedure for comparing two alternatives on the basis of their capitalized cost.

Example 5.4

Two sites are currently under consideration for a bridge to cross the Ohio River. The north site would connect a major state highway with an interstate loop around the city and would alleviate much of the local through traffic. The disadvantages of this site are that the bridge would do little to ease local traffic congestion during rush hours, and the bridge would have to stretch from one hill to another to span the widest part of the river, railroad tracks, and local highways below. This bridge would therefore be a suspension bridge. The south site would require a much shorter span allowing for construction of a truss bridge, but would require new road construction.

The suspension bridge would have a first cost of $30 million with annual inspection and maintenance costs of $15,000. In addition, the concrete deck would have to be resurfaced every 10 years at a cost of $50,000. The truss bridge and approach roads are expected to cost $12 million and would have annual maintenance costs of $8000. The bridge would have to be painted every 3 years at a cost of $10,000. In addition, the bridge would have to be sandblasted and painted every 10 years at a cost of $45,000. The cost of purchasing right-of-way is expected to be $800,000 for the suspension bridge and $10.3 million for the truss bridge. Compare the alternatives on the basis of their capitalized cost if the interest rate is 6% per year.

Solution Construct the cash-flow diagrams before you attempt to solve the problem.

Capitalized cost of suspension bridge

P_1 = present worth of initial cost = $30.0 + 0.8 = \$30.8$ million

The recurring operating cost is $A_1 = \$15,000$, while the annual equivalent of the resurface cost is

$A_2 = 50,000(A/F, 6\%, 10) = \$3794$

P_2 = capitalized cost of recurring costs = $\dfrac{A_1 + A_2}{i}$

$= \dfrac{15,000 + 3794}{0.06}$

$= \$313,233$

Finally, the total capitalized cost (P_S) is

$$P_S = P_1 + P_2 = \$31,113,233 \qquad (\$31.1 \text{ million})$$

Capitalized cost of truss bridge

$$P_1 = 12.0 + 10.3 = \$22.3 \text{ million}$$

$$A_1 = \$8000$$

$$A_2 = \text{annual cost of painting} = 10,000(A/F, 6\%, 3)$$
$$= \$3141$$

$$A_3 = \text{annual cost of sandblasting} = 45,000(A/F, 6\%, 10)$$
$$= \$3414$$

$$P_2 = \frac{A_1 + A_2 + A_3}{i} = \$242,583$$

The total capitalized cost (P_T) is

$$P_T = P_1 + P_2 = \$22,542,583 \qquad (\$22.5 \text{ million})$$

Since $P_T < P_S$, the truss bridge should be constructed.

Additional Example 5.8
Probs. 5.33 to 5.38

Additional Examples

Example 5.5

A traveling saleswoman expects to purchase a used car this year. She has collected or estimated the following data: first cost is \$4800; trade-in value will be \$500 after 4 years; annual maintenance and insurance costs are \$350; and additional annual income due to ability to travel is \$1500. Will the woman be able to make a rate of return of 20% per year on her investment?

Solution   Compute the PW value of the investment at $i = 20\%$. (A cash-flow diagram will aid you.)

$$PW = -4800 - 350(P/A, 20\%, 4) + 1500(P/A, 20\%, 4) + 500(P/F, 20\%, 4)$$
$$= \$-1582$$

Indeed, she would not make 20%, since the PW is much less than zero.

Comment   If the PW value had been greater than zero, an excess of 20% would be returned. In Chap. 7, calculations similar to those above will be made to determine the actual rate of return on project investments.

Example 5.6

A cement plant plans to open a new rock pit. Two plans have been devised for movement of raw material from the quarry to the plant. Plan A requires the purchase of two earth

Table 5.1 Details of plans to move rock from quarry to cement plant

	Plan A		Plan B
	Mover	Pad	Conveyor
P	$45,000	$28,000	$175,000
AOC	6,000	300	2,500
SV	5,000	2,000	10,000
n, years	8	12	24

movers and construction of an unloading pad at the plant. Plan B calls for construction of a conveyor system from the quarry to the plant. The costs for each plan are itemized in Table 5.1. Which plan should be selected if money is presently worth 15% per year?

Solution Evaluation will take place over 24 years, since we plan to use a present-worth (PW) analysis. Reinvestment in the two movers will occur in years 8 and 16 and the unloading pad must be repurchased in year 12. No reinvestment is necessary for plan B. You are advised to construct your own cash-flow diagram for each plan to follow the PW analysis.

To simplify computations, we can use the fact that plan A will have an extra AOC in the amount of $12,300 − $2500 = $9800 per year.

PW of plan A

$$PW_A = PW_{movers} + PW_{pad} + PW_{AOC}$$

$$PW_{movers} = 2(45,000)[1 + (P/F, 15\%, 8) + (P/F, 15\%, 16)]$$
$$- 2(5000)[(P/F, 15\%, 8) + (P/F, 15\%, 16) + (P/F, 15\%, 24)]$$
$$= \$124,355$$

$$PW_{pad} = 28,000[1 + (P/F, 15\%, 12)] - 2000[(P/F, 15\%, 12) + (P/F, 15\%, 24)]$$
$$= \$32,790$$

$$PW_{AOC} = 9800(P/A, 15\%, 24) = \$63,051$$

$$PW_A = \$220,196$$

PW of plan B

$$PW_B = PW_{conveyor} = 175,000 - 10,000(P/F, 15\%, 24)$$
$$= \$174,651$$

Since $PW_B < PW_A$, the conveyor should be constructed.

Example 5.7

A restaurant owner is trying to decide between two different garbage disposals. A regular steel (RS) disposal has an initial cost of $65 and a life of 4 years. The alternative is a corrosion-resistant disposal constructed primarily of stainless steel (SS). The initial cost of the SS disposal is $110, but it is expected to last 10 years. Because the SS disposal has a slightly larger motor, it is expected to cost about $5 per year more to operate than the RS disposal. If the interest rate is 16% per year, which disposal should be selected, assuming both have a negligible salvage value?

Figure 5.3
Present-worth
comparison of two
unequal-life assets,
Example 5.7.

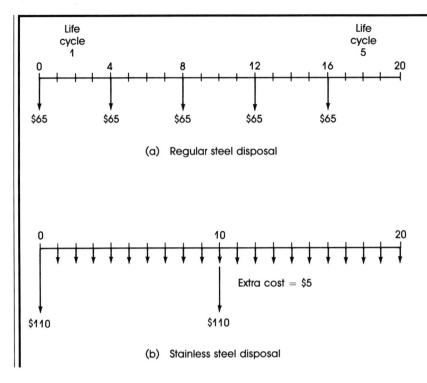

(a)   Regular steel disposal

(b)   Stainless steel disposal

Solution   The cash-flow diagram (Fig. 5.3) uses a comparison period of 20 years with reinvestment in year 10 for the SS disposal and in years 4, 8, 12, and 16 for the RS disposal. The present-worth calculations are as follows:

$$PW_{RS} = 65 + 65(P/F, 16\%, 4) + 65(P/F, 16\%, 8) + 65(P/F, 16\%, 12)$$
$$+ 65(P/F, 16\%, 16) = \$137.72$$

$$PW_{SS} = 110 + 110(P/F, 16\%, 10) + 5(P/A, 16\%, 20) = \$164.58$$

The regular steel disposal should be purchased, since $PW_{RS} < PW_{SS}$.

Comment   In the solution presented, the extra operating cost of $5 per year was regarded as an expense for the SS disposal. However, the same decision would have been reached if the $5 per year had been shown as an income for the RS disposal, but the present worths of both would have been lower by $5(P/A, 16\%, 20)$. This illustrates that unless the absolute money values are sought, it is only important to consider *differences* in cash flow for alternative evaluation.

Example 5.8

A city engineer is considering two alternatives for the local water supply. The first alternative would involve construction of an earthen dam on a nearby river, which has a highly variable flow. The dam would serve as a reservoir so that the city would have a dependable source of water indefinitely. The initial cost of the dam is expected to be $8 million and will require annual upkeep costs of $25,000. The dam is expected to last indefinitely.

Alternatively, the city can drill wells as needed and construct pipelines for transporting the water to the city. The engineer estimates that an average of 10 wells will be required initially at a cost of $45,000 per well, including the pipeline. The average life of a well is expected to be 5 years with an annual operating cost of $12,000 per well. If the city uses an interest rate of 15% per year, determine which alternative should be selected on the basis of their capitalized costs.

Solution The cash-flow diagram is left to the reader. The capitalized cost of the dam is calculated as follows:

$$PW_{dam} = 8,000,000 + \frac{25,000}{0.15} = \$8,166,667$$

The capitalized cost of the wells can be calculated by first converting the recurring costs and annual operating costs to an EUAW and then dividing by the interest rate. Thus,

$$EUAW_{wells} = EUAW \text{ of investment} + \text{annual operating costs}$$
$$= 45,000(10)(A/P, 15\%, 5) + 12,000(10)$$
$$= \$254,244$$

The capitalized cost of the wells, using Eq. (5.3), is

$$PW_{wells} = \frac{254,244}{0.15} = \$1,694,960$$

The wells should be constructed instead of the dam.

Comment The capitalized cost of the wells could also have been obtained by using the A/F factor for calculating the EUAW of future wells. The value obtained should then be divided by i and added to the initial investment cost P_1 of $450,000. Thus,

$$P_A = \frac{EUAW}{i} = \frac{450,000(A/F, 15\%, 5) + 120,000}{0.15}$$
$$= \$1,244,960$$

$$PW_{wells} = P_A + P_1$$
$$= 1,244,960 + 450,000 = \$1,694,960$$

Problems

5.1 Two machines are under consideration by a metal fabricating company. Machine A will have a first cost of $15,000, an annual maintenance and operation cost of $3000, and a $3000 salvage value. Machine B will have a first cost of $22,000, an annual cost of $1500, and a $5000 salvage value. If both machines are expected to last for 10 years, determine which machine should be selected on the basis of present-worth values using an interest rate of 12% per year.

5.2 A public utility is trying to decide between two different sizes of pipe for a new water main. A 250-millimeter line will have an initial cost of $35,000, whereas a 300-millimeter line will cost $55,000. Since there is less head loss through the 300-millimeter pipe, the pumping cost for the larger line is expected to be $3000 per year less than for the 250-millimeter line. If the pipes are expected to last for 20 years, which size should be selected if the interest rate is 15% per year? Use a present-worth analysis.

5.3 A couple is trying to decide between purchasing a house and renting one. They can purchase a new house with a down payment of $15,000 and a monthly payment of $750, beginning 1 month from now. Taxes and insurance are expected to amount to $100 per month. In addition, they expect to paint the house every 4 years at a cost of $600. Alternatively, they can rent a house for $700 per month payable in advance with a $600 deposit, which will be returned when they vacate the house. The utilities are expected to average $135 per month whether they purchase or rent. If they expect to be able to sell the house in 6 years for $10,000 more than they paid down, should they buy a house or rent one, if the interest rate is a nominal 12% per year compounded monthly? Use a present-worth analysis.

5.4 A consulting engineer is trying to determine which of two methods should be specified for screening sewage. A manually cleaned bar screen will have an initial installed cost of $400. The labor cost for cleaning is expected to be $800 the first year, $850 the second year, and $900 the third year and to increase by $50 each year. An automatically cleaned bar screen will have an initial cost of $2500 with an annual power cost of $150. In addition, the motor will have to be replaced every 2 years at a cost of $40 per motor. General maintenance is expected to cost $100 the first year and increase by $10 per year. If the screens are expected to last for 10 years, which method should be selected if the interest rate is 10% per year? Use the present-worth method.

5.5 A consulting engineering firm is trying to decide between purchasing and leasing cars. It estimates that medium-sized cars will cost $8300 and will have a probable trade-in value in 4 years of $2800. The annual cost of such items as fuel and repairs is expected to be $950 the first year and to increase by $50 per year. Alternatively, the company can lease the same cars for $3500 per year payable at the beginning of each year. Since some maintenance is included in the rental price, the annual maintenance and operation expenses are expected to be $100 lower if the cars are leased. If the company's minimum rate of return is 20% per year, which alternative should be selected?

5.6 A building contractor is trying to determine if it would be economically feasible to install rainwater drains in a large shopping center currently under construction. Since the project is being built in the arid southwest, the total annual amount of rainfall is slight, but the rain that does occur is in the form of brief but heavy thundershowers. The thundershowers tend to cause erosion of soil at the project site, which was formed by filling in a large arroyo. In the 3 years required for construction, 12 heavy thundershowers are expected. If no drains are installed, the cost of refilling the washed-out area is expected to be $1000 per thunderstorm. Alternatively, a corrugated steel drain-pipe could be installed which will prevent the soil erosion. The installation cost of the pipe would be $6.50 per meter, with a total length of 2000 meters required. After the 3-year construction period, some of the pipe could be recovered with an estimated value of $3000. Assuming that the thunderstorms occur at 3-month intervals, determine which alternative should be selected, if the interest rate is a nominal 20% per year compounded quarterly.

5.7 A southwestern university is considering installing electric valves with automatic timers on some of their sprinkler systems. It is estimated that 45 valves and timers at a cost of $85 per set will be needed. The initial installation cost is expected to be $2000. At the present time, there are four employees who are in charge of maintaining these lawns. These employees, each of whom earns $12,000 per year, spend 25% of their time in watering. The present cost of water for these lawns is $2200 per year. If the automatic system is installed, the work force cost for watering could be reduced by 80% and the water bill by 35%. However, extra maintenance on the automatic system is expected to cost $450 per year. If the timers and valves are expected to last for 8 years, which system should be used if the interest rate is 16% per year? Use a present-worth analysis.

5.8 A manufacturing company is in need of 1000 square meters of storage space for 3 years. The company is considering the purchase of land for $8000 and erecting a temporary metal structure on it at a cost of $70 per square meter. At the end of the 3-year use period, the company expects to be able to sell the land for $9000 and the building for $12,000. Alternatively, the company can lease storage space for $1.50 per square meter per month payable at the beginning of each year. If the company's minimum attractive rate of return is 20% per year, which type of storage space should be used? Use the present-worth method of analysis.

5.9 Machines that have the following costs are under consideration for a continuous production process.

	Machine G	Machine H
First cost	$62,000	$77,000
Annual operating cost	15,000	21,000
Salvage value	8,000	10,000
Life, years	4	6

Using an interest rate of 15% per year, determine which alternative should be selected on the basis of a present-worth analysis.

5.10 Rework Prob. 5.9 assuming that machine G requires an extensive overhaul at the end of 2 years that costs $10,000.

5.11 Which screen should be selected in Prob. 5.4 if the manually cleaned screen will last 20 years and the automatically cleaned screen will last only 10 years? Assume the automatic screen maintenance cost will increase by $10 per year through year 20.

5.12 Compare the machines below on the basis of their present worths, using an interest rate of 18% per year.

	Machine P	Machine Q
First cost	$29,000	$37,000
Salvage value	4,000	5,000
Life, years	3	5
Annual maintenance cost	3,000	3,500
Overhaul every 2 years	3,700	2,000

5.13 A small strip-mining coal company is trying to decide whether it should purchase or lease a new clamshell. If purchased, the shell will cost $150,000 and is expected to have a $65,000 salvage value in 8 years. Alternatively, the company can lease the clamshell for $30,000 per year, but the lease payment will have to be made at the *beginning* of each year. If the clamshell is purchased, it will be leased to other strip-mining companies whenever possible, an activity that is expected to yield revenues of $10,000 per year. If the company's minimum attractive rate of return is 22%, should the clamshell be purchased or leased? Make calculations on the basis of a present-worth analysis.

5.14 A production plant manager has been presented with two proposals for automating an assembly process. Proposal A involves an initial cost of $15,000 and an annual operating cost of $2000 per year for the next 4 years. Thereafter, the operating cost is expected to be $2700 per year. This equipment is expected to have a 20-year life with no salvage value. Proposal B requires an initial investment of $28,000 and an annual operating cost of $1200 per year for the first 3 years. Thereafter, the operating cost is expected to increase by $120 per year. This equipment is expected to last for 20 years and have a $2000 salvage value. If the company's minimum attractive rate of return is 10%, which proposal should be accepted on the basis of a present-worth analysis?

5.15 An environmental engineer is trying to decide between two operating pressures for a wastewater irrigation system. If a high-pressure system is used, fewer sprinklers and less pipe will be required, but the pumping cost will be higher. The alternative is to use lower pressure with more sprinklers. The pumping cost is estimated to be $3 per 1000 cubic meters of wastewater pumped at the high pressure. Twenty-five sprinklers will be required at a cost of $30 per unit. In addition, 1000 meters of aluminum pipe will be required at a cost of $9 per meter. If the lower pressure system is used, the pumping cost will be $2 per 1000 cubic meters of wastewater. Also required will be 85 sprinklers and 4000 meters of pipe. The aluminum pipe is expected to last 10 years and the sprinklers 5 years. If the volume of wastewater is expected to be 500,000 cubic meters per year, which pressure should be selected if the company's minimum attractive rate of return is 20% per year? The aluminum pipe will have a 10% salvage value.

5.16 The owner of the Good Flick Drive-In Theatre is considering two proposals for upgrading the parking ramps. The first proposal involves asphalt paving of the entire parking area. The initial cost of this proposal would be $35,000, and it would require annual maintenance of $250 beginning 3 years after installation. The owner expects to have to resurface the theater in 15 years. Resurfacing will cost only $8000, since grading and surface preparation are not necessary, but the $250 annual maintenance cost will continue. Alternatively, gravel can be purchased and spread in the drive areas and grass planted in the parking areas. The owner estimates that 29 metric tons of gravel will be needed per year starting 1 year from now at a cost of $90 per metric ton. In addition, a riding lawn mower, which will cost $800 and have a life of 10 years, will be needed. The cost of labor for spreading gravel, cutting grass, etc., is expected to be $900 the first year and $950 the second and will increase by $50 per year thereafter. The owner figures that a gravel surface would not be used for more than 30 years. If the interest rate is 12% per year, which alternative should be selected? Use a present-worth analysis and a 30-year study period.

5.17 An automobile owner is trying to decide between purchasing four new radial tires or having the worn-out tires recapped. Radial tires for the car will cost $85 each and will last 60,000 kilometers. The old tires can be recapped for $25 each, but they will last for only 20,000 kilometers. Since this is a second car, it probably will register only 10,000 kilometers per year. If the radial tires are purchased, the gasoline mileage will increase by 10%. If the cost of gasoline is assumed to be $0.42 per liter and the car gets 10 kilometers per liter, what type tires should be purchased if the interest rate is 12% per year? Use the present-worth method and assume that the salvage value of the tires is zero.

5.18 A state highway department is trying to decide between "hot patching" an existing road and resurfacing it. If the hot-patch method is used, approximately 300 cubic meters of material will be required at a cost of $25 per cubic meter (in place). Additionally, the shoulders will have to be improved at the same time at a cost of $3000. The annual cost of routine maintenance on the patched-up road would be $4000. These improvements will last 2 years, at which time they will have to be redone. Alternatively, the state could resurface the road at a cost of $65,000. This surface will last 10 years if the road is maintained at a cost of $1500 per year beginning 4 years from now. No matter which alternative is selected now, the road will be completely rebuilt in 10 years. If the interest rate is 13% per year, which alternative should the state select on the basis of a present-worth comparison?

5.19 The Bee-Low Mining Company is considering purchasing a machine which costs $30,000 and is expected to last 12 years, with a $3000 salvage value. The annual operating expenses are expected to be $9000 for the first 4 years, but owing to decreased use, the operating costs will decrease by $400 per year for the next 8 years. Alternatively, the company can purchase a highly automated machine at a cost of $58,000. This machine will last only 6 years because of its high technology and delicate design, and its salvage value will be $15,000. Because it is so automated, its operating cost will be only $4000 per year. If the company's minimum attractive rate of return is 20% per year, which machine should be selected on the basis of a present-worth analysis?

5.20 Two metal fabricating machines are presently under consideration by the Heat 'N' Beat Metal fabricating company. The manual model will cost $25,000 to buy with an 8-year life and a $5000 salvage value. Its annual operating cost will be $15,000 for labor and $1000 for maintenance. A computer-controlled model will cost $95,000 to buy and it will have a 12-year life if upgraded at the end of year 6 for $15,000. Its terminal salvage value will be $23,000. The annual costs for the computer-controlled model will be $7500 for labor and $2500 for maintenance. If the company's minimum attractive rate of return is 25%, which machine would be preferred on the basis of the equivalent present cost of each?

5.21 The Board-Stiff Lumber Company is considering whether it should provide disposable or reusable plates and utensils for its employee cafeteria. Disposable utensils will cost $4700 for a 2-year supply. Because of the trash created by the throwaway items, the refuse disposal costs will be $48 per month higher. Alternatively, the company could purchase reusable utensils which will cost initially $10,000. Their life will be 8 years, but due to breakage, another $2000 will have to be spent in 5 years for replacements. After 8 years, the usable items remaining can be sold for $1500. The cost of hiring a part-time dishwasher, buying detergents, obtaining hot water, etc., is expected to be $350 per month. If the interest rate is a nominal 18% per year compounded monthly, should the company purchase the disposable or the reusable items? Use the present-worth method of analysis.

5.22 A company is considering the purchase of one of two processes identified as Q and Z. Process Q will have a first cost of $43,000, a *monthly* operating cost of $10,000 and a $5000 salvage value at the end of its 4-year life. Process Z will have a first cost of $31,000 with a *quarterly* operating cost of $39,000. It will have an 8-year life with a $2000 value at that time. If the interest rate is a nominal 12% per year compounded monthly, which alternative would be preferred on the basis of a present-worth analysis?

5.23 Compare the alternatives shown below on the basis of a present-worth analysis. Use an interest rate of 12% per year compounded monthly.

	Alternative YEA	Alternative TEAM
First cost	$20,000	$31,000
Annual operating cost	4,000	5,000
Monthly income	600	900
Salvage value	3,000	6,000
Life, years	4	5

5.24 Compare the alternatives shown below on the basis of a present-worth comparison. The interest rate is 20% per year.

	Alternative BUY	Alternative STOCK
First cost	$47,000	$56,000
Annual cost	11,000 in year 1; increases by 5% per year	30,000 in year 1; increases by 3% per year
Salvage value	5,000	2,000
Life, years	6	3

5.25 A local planning commission has estimated the first cost of a new city-owned amusement park to be $35,000. They expect to improve the park by adding new rides every year for the next 5 years at a cost of $6000 per year. Annual operating costs are expected to be $12,000 the first year; these will increase by $2000 per year until year 5. After that time, the operating expenses will remain at $20,000 per year. The city expects to receive $11,000 in profits the first year, $14,000 the second, and amounts increasing by $3000 per year until year 8, after which the net profit will remain the same. Calculate the capitalized cost of the park if the interest rate is 6% per year.

5.26 How much additional uniform annual cost can the city incur for the amusement park in Prob. 5.25 to break even?

5.27 What is the capitalized cost of $75,000 now, $60,000 five years from now, and a uniform annual amount of $700 per year for year 10 and every year thereafter, if the interest rate is 8% per year?

5.28 What is the capitalized cost of $200,000 now, $300,000 four years from now, $50,000 every 5 years, and a uniform annual amount of $8000 beginning 15 years from now, if the interest rate is 16% per year?

5.29 A wealthy alumnus of a small university wants to establish a permanent fund for tuition scholarships. He wants to support three students for the first 5 years after the fund is established and five students thereafter. If tuition alone is expected to cost $1000 per year, how much money must the alumnus donate now if the university can earn 10% per year on the fund?

5.30 If the tuition in Prob. 5.29 increases by $20 per year for the first 20 years, how much money must the alumnus donate?

5.31 A donor wishes to endow a scholarship to a certain university in the name of a certain professor. The scholarship is to provide $40,000 per year for the first 5 years and $100,000 per year thereafter. If the university expects to be able to earn 10% per year on the endowment, how much must the donor give now if the first scholarship is to be given 1 year from now?

5.32 As a wealthy university grad, you plan to set up an account which can be drawn upon by a person teaching engineering economy. You plan to deposit $1 million now with the stipulation that only the interest can be withdrawn for the first 10 years. After that time, the remaining balance will be given to the professor. If the money is deposited into an account which earns 14% per year compounded annually, (a) how much will the professor get each year, and (b) how much will he or she get at the end of the 10-year period?

5.33 Machines with costs shown below are presently under consideration by the Go-For-It Company. Using an interest rate of 15% per year, compare the alternatives on the basis of their capitalized costs.

	Machine WHY	Machine NOT
First cost	$31,000	$43,000
Annual operating cost	18,000	19,000
Salvage value	5,000	7,000
Life, years	4	6

5.34 Compare the machines shown below on the basis of their capitalized cost using an interest rate of 20% per year.

	Machine X	Machine Z
First cost	$50,000	$200,000
Annual operating cost	62,000	24,000
Salvage value	10,000	0
Overhaul after 6 years	. . .	4,000
Life, years	7	∞

5.35 Compare the machines in Prob. 5.34 on the basis of their capitalized costs if the life of machine Z is 10 years instead of infinite. Use an interest rate of a nominal 14% per year compounded semiannually.

5.36 A city planning commission is considering two proposals for a new civic center. Proposal F requires an initial investment of $10 million now and an expansion cost of $4 million 10 years from now. The annual operating cost is expected to be $250,000 per year. Income from conventions, shows, etc., is expected to be $190,000 the first year and to increase by $20,000 per year for 4 more years and then remain constant until year 10. In year 11 and thereafter income is expected to be $350,000 per year. Proposal G requires an initial investment of $18 million now and an annual operating cost of $300,000 per year. However, income is expected to be $260,000 the first year and increase by $30,000 per year to year 7. Thereafter, income will remain at $400,000 per year. Determine which proposal should be selected on the basis of capitalized cost if the interest rate is 6% per year.

5.37 Compare the alternatives shown below on the basis of a present-worth comparison, using an interest rate of 14% per year.

	Alternative PAY LATER	Alternative KNOT NOW
First cost	$ 160,000	$25,000
Annual operating cost	15,000	3,000
Overhaul every 4 years	12,000	. . .
Salvage value	1,000,000	4,000
Life, years	∞	7

5.38 Compare the alternatives shown below on the basis of a present worth comparison. Use $i = 14\%$ per year compounded quarterly.

	Alternative U.R.	Alternative O.K.
First cost	$8,500,000	$50,000,000
Annual operating cost	8,000	7,000
Salvage value	5,000	2,000
Life, years	5	∞

6 Equivalent-Uniform-Annual-Worth Evaluation

The objective of this chapter is to teach you the primary methods of calculating the equivalent uniform annual worth (EUAW) of an asset and how to select the better of two alternatives on the basis of an annual-worth comparison. Although the word "annual" is included in the name of the method, the procedures developed in this chapter can be used to find an equivalent uniform series over any interest period desired, as per Chap. 3. Additionally, the word "cost" is often used interchangeably with "worth" in describing a series so that EUAC and EUAW really mean the same thing. However, EUAW more properly describes the cash flow because oftentimes the uniform series developed represents an income rather than a cost. Regardless of which term is used to describe the resulting uniform cash flow, the alternative selected as best will be the same as that chosen by the present-worth or any other evaluation method when the comparisons are properly conducted.

Section Objectives

After completing this chapter, you should be able to do the following:

6.1 State why the EUAW needs to be calculated for only one cycle of each alternative when the alternatives have different lives.

6.2 Calculate the EUAW of an asset having a salvage value, using the *salvage sinking-fund* method, given the asset initial cost, salvage value, life, and interest rate.

6.3 Calculate the EUAW as in objective 6.2, except using the *salvage present-worth* method.

6.4 Calculate the EUAW as in objective 6.2, except using the *capital-recovery-plus-interest* method.

6.5 Select the better of two alternatives on the basis of their EUAW, given their initial costs, salvage values, lives, cash flows, and the interest rate.

6.6 Calculate the EUAW of a perpetual investment, given the initial cost of the asset, the cash flows, and the interest rate.

Study Guide

6.1 Study Period for Alternatives Having Different Lives

The EUAW (equivalent uniform annual worth) is another method that is commonly used for comparing alternatives. As illustrated in Chap. 4, the EUAW means that all incomes and disbursements (irregular and uniform) must be converted into an equivalent uniform annual amount, (that is, an end-of-period amount) which is the *same each period*. The major advantage of this method over all the other methods is that it does not require making the comparison over the least common multiple of years when the alternatives have different lives. That is, the equivalent uniform annual worth of the alternative needs to be calculated for *one life cycle only*. Why? Because, as its name implies, the EUAW is an equivalent annual worth over the life of the project. If the project is continued for more than one cycle, the equivalent annual worth for the next cycle and all succeeding cycles would be exactly the same as for the first, assuming all cash flows were the same for each cycle.

The repeatability of the uniform annual series through various life cycles can be demonstrated by considering the cash-flow diagram shown in Fig. 6.1. The diagram shows cash flow which represents two life cycles of an asset which has a first cost of $20,000, an annual operating cost of $8000, and a 3-year life.

The EUAW for one life cycle (i.e., 3 years) would be calculated as follows:

$$EUAW = 20,000(A/P, 22\%, 3) + 8000$$
$$= \$17,793$$

The EUAW for two life cycles would be calculated as follows:

$$EUAW = 20,000(A/P, 22\%, 6) + (20,000)(P/F, 22\%, 3)(A/P, 22\%, 6) + 8000$$
$$= \$17,793$$

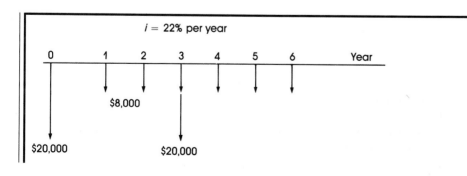

Figure 6.1 Cash flow diagram for two life cycles of an asset.

Note that the EUAW for the first life cycle is exactly the same value as that obtained when two life cycles are considered. This same EUAW will be obtained when three, four, or any other number of life cycles are evaluated. Thus, the EUAW for one life cycle of an alternative represents the equivalent uniform annual worth of that alternative *every time the cycle is repeated.*

When information is available which indicates that the costs will not be the same in succeeding life cycles (or more specifically, that they will change by an amount other than the inflation rate), then a planning-horizon type of approach must be used as discussed in Sect. 6.5. In this text, unless otherwise specified, it will be assumed that all future costs will change exactly in accordance with the applicable inflation or deflation rate for that time.

Prob. 6.1

6.2 Salvage Sinking-Fund Method

When an asset of a given alternative has a terminal salvage value (SV), there are several ways by which the EUAW can be calculated. This section presents the salvage sinking-fund method, probably the simplest of the three discussed in this chapter. This is the method that will be used in this text hereafter. In the salvage sinking-fund method, the initial cost (P) is first converted to an equivalent uniform annual amount using the A/P (capital-recovery) factor. The salvage value, after conversion to an equivalent uniform amount via the A/F (sinking-fund) factor, is *subtracted* from the annual equivalent of the first cost. The calculations can be represented by a general equation:

$$EUAW = P(A/P, i\%, n) - SV(A/F, i\%, n) \tag{6.1}$$

Naturally, if the alternative has any other cash flows, they must be included in the EUAW computation. An EUAW computation is illustrated in Example 6.1.

Example 6.1

Calculate the EUAW of a machine that has an initial cost of $8000 and a salvage value of $500 after 8 years. Annual operating costs (AOC) for the machine are estimated to be $900, and an interest rate of 20% per year is applicable.

Solution The cash-flow diagram (Fig. 6.2) requires us to compute

$$EUAW = A_1 + A_2$$

where A_1 = annual cost of initial investment less salvage value, Eq. (6.1)

A_2 = annual maintenance cost = $900

$A_1 = 8000(A/P, 20\%, 8) - 500(A/F, 20\%, 8) = \$2055$

$EUAW = 2055 + 900 = \$2955$

Comment Since the maintenance cost was already expressed as an annual cost over the life of the asset, no conversions were necessary.

Figure 6.2   (*a*) Diagram for machine costs, and (*b*) conversion to an EUAW.

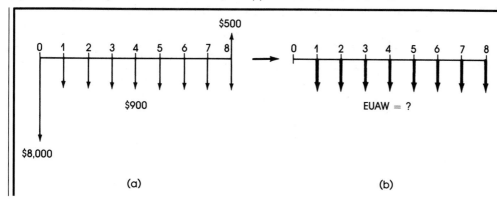

(a)   (b)

The simplicity of the salvage sinking-fund method should be obvious from the straightforward calculations shown in the above example. The steps can be summarized as follows:

1. Annualize the initial investment cost over the life of the asset using the *A/P* factor.
2. Annualize the salvage value using the *A/F* factor.
3. Subtract the annualized salvage value from the annualized investment cost.
4. Add the uniform annual amounts to the value from step 3.
5. Convert any other cash flows into equivalent uniform annual worths and add them to the value obtained in step 4.

Additional Example 6.7
Probs. 6.2 and 6.3

6.3   Salvage Present-Worth Method

The salvage present-worth method is the second method by which investments having salvage values can be converted into an EUAW. The present worth of the salvage value is subtracted from the initial investment cost, and the resulting difference is annualized for the life of the asset. The general equation is

$$\text{EUAW} = [P - \text{SV}(P/F, i\%, n)](A/P, i\%, n) \qquad (6.2)$$

The steps that must be followed in this method are the following:

1. Calculate the present worth of the salvage value via the *P/F* factor.
2. Subtract the value obtained in step 1 from the initial cost *P*.
3. Annualize the resulting difference over the life of the asset using the *A/P* factor.
4. Add the uniform annual worths to the result of step 3.
5. Convert all other cash flows to annual equivalents and add them to the value obtained in step 4.

Example 6.2

Compute the EUAW for the machine detailed in Example 6.1 using the salvage present-worth method.

Solution   Using the steps outlined above and Eq. (6.2),

EUAW = [8000 − 500(*P/F*, 20%, 8)](*A/P*, 20%, 8) + 900 = $2955

Prob. 6.4

6.4   Capital-Recovery-Plus-Interest Method

The final procedure that will be presented here for calculating the EUAW of an asset having a salvage value is the capital-recovery-plus-interest method. The general equation for this method is

$$\text{EUAW} = (P - SV)(A/P, i\%, n) + SV(i) \tag{6.3}$$

In subtracting the salvage value from the investment cost *before* multiplying by the *A/P* factor, it is recognized that the salvage value will be recovered. However, the fact that the salvage value will not be recovered for *n* years must be taken into account by adding the interest (SV*i*) lost during the asset's life. Failure to include this term would assume that the salvage value was obtained in year 0 instead of year *n*. The steps to be followed for this method are as follows:

1. Subtract the salvage value from the initial cost.
2. Annualize the resulting difference with the *A/P* factor.
3. Multiply the salvage value by the interest rate.
4. Add the values obtained in steps 2 and 3.
5. Add the uniform annual amounts to the result of step 4.
6. Add all other uniform amounts for additional cash flows.

Example 6.3

Use the values of Example 6.1 to compute the EUAW using the capital-recovery-plus-interest method.

Solution   From Eq. (6.3) and the steps above,

EUAW = (8000 − 500)(*A/P*, 20%, 8) + 500(0.20) + 900 = $2955

While it makes no difference which method is used to compute the EUAW, it would be good procedure hereafter to use only one method in order to avoid errors caused by mixing various techniques. We will use the salvage sinking-fund method (Sec. 6.2).

Prob. 6.5

6.5 Comparing Alternatives by EUAW

The equivalent-uniform-annual-worth method of comparing alternatives is probably the simplest of the alternative evaluation techniques presented in this book. Selection is made on the basis of EUAW, with the alternative having the lowest cost being the most favorable. Obviously, as discussed in later chapters, nonquantifiable data must also be considered in arriving at the final decision, but in general, the alternative having the lowest EUAW should be selected.

Perhaps the most important rule to remember when making EUAW comparisons is that *only one life cycle* of each alternative must be considered. This is because the EUAW will be the same for any number of life cycles as it is for one, as shown in Sec. 6.1. This procedure is, of course, subject to the assumptions underlying this method. These assumptions are similar to those applicable to a present-worth analysis, namely: (1) the alternatives will be needed for their least common multiple of years or, if not, the equivalent uniform annual worth will be the same for any portion of the asset's life cycle as it is for the entire cycle, (2) the cash flows in succeeding life cycles will change by exactly the inflation or deflation rate, and (3) any funds generated by the project will be reinvested at the interest rate used in making the calculations. When information is available that would indicate that one or more of these assumptions would not be valid, then a planning horizon–type approach should be followed. That is, the costs actually expected through a specified time period (i.e. planning horizon) must be identified and converted into their equivalent uniform annual worths. Examples 6.4 and 6.5 illustrate these procedures.

Example 6.4

The following costs are estimated for two equal-service tomato-peeling machines in a food-canning plant:

	Machine A	Machine B
First cost	$26,000	$36,000
Annual maintenance cost	800	300
Annual labor cost	11,000	7,000
Extra income taxes	. . .	2,600
Salvage value	2,000	3,000
Life, years	6	10

If the minimum required rate of return is 15% per year, which machine should be selected?

Solution The cash-flow diagram for each alternative is shown in Fig. 6.3. The EUAW of each machine using the salvage sinking-fund method, Eq. (6.1), is calculated as follows:

$$\text{EUAW}_A = 26,000(A/P, 15\%, 6) - 2000(A/F, 15\%, 6) + 11,800 = \$18,442$$

$$\text{EUAW}_B = 36,000(A/P, 15\%, 10) - 3000(A/F, 15\%, 10) + 9900 = \$16,925$$

Select machine B, since $\text{EUAW}_B < \text{EUAW}_A$.

Figure 6.3 Cash flows for two alternative tomato-peeling machines, Example 6.4.

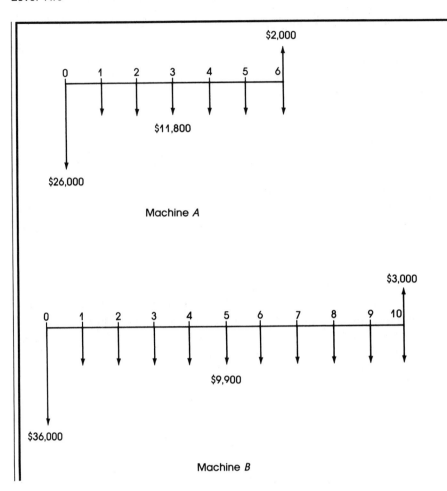

Figure 6.3 Cash flows for two alternative tomato-peeling machines, Example 6.4.

Example 6.5

(a) Assume the company in Example 6.4 is planning to get out of the tomato-canning business in 4 years. At that time, the company expects to be able to sell machine A for $12,000 and machine B for $15,000. All other costs are expected to remain the same. Which machine should the company purchase under these conditions?

(b) If all costs, including the salvage values, will be as originally estimated, which machine should the company purchase using the 4-year horizon?

Solution

(a) The planning horizon is now 4 years, so only 4 years' worth of costs are relevant to the decision. The equivalent uniform annual worths are now calculated as follows:

$$EUAW_A = 26{,}000(A/P,\ 15\%,\ 4) - 12{,}000(A/F,\ 15\%,\ 4) + 11{,}800 = \$18{,}504$$

$$EUAW_B = 36{,}000(A/P,\ 15\%,\ 4) - 15{,}000(A/F,\ 15\%,\ 4) + 9900 = \$19{,}506$$

Select machine A, since $EUAW_A < EUAW_B$.

(b) The only change from part (a) is in the salvage values of the machines. The EUAWs are now:

$$\text{EUAW}_A = 26,000(A/P, 15\%, 4) - 2000(A/F, 15\%, 4) + 11,800 = \$20,506$$

$$\text{EUAW}_B = 36,000(A/P, 15\%, 4) - 3000(A/F, 15\%, 4) + 9900 = \$21,909$$

Again, select machine A.

Comment   Note that in using a 4-year planning horizon, the decision as to which machine should be purchased has changed from that recommended when no planning horizon was specified. This illustrates the importance of recognizing the assumptions inherent in the alternative evaluation methods.

Additional Example 6.8
Probs. 6.6 to 6.27

6.6 EUAW of a Perpetual Investment

Evaluation of flood-control, irrigation, bridge, or other large-scale projects requires the comparison of alternatives which have very long lives, which may be considered infinite in economic analysis terms. For this type of analysis, it is important to recognize that the annual worth of the *initial* investment is simply equal to the annual interest earned on the lump-sum investment, as expressed by Eq. (5.2), that is, $A = Pi$. This is clearly shown by considering the capital recovery relation $A = P(A/P, i\%, n)$. If the numerator and denominator of the A/P factor are divided by $(1 + i)^n$, the following relations are derived.

$$A = P\left(\frac{i(1 + i)^n}{(1 + i)^n - 1}\right) = P\left(\frac{i}{1 - \dfrac{1}{(1 + i)^n}}\right)$$

As the value of n increases toward infinity, this expression for A simplifies to $A = Pi$. The amount A is an EUAW value since it recurs each year in the future. This is, of course, the same result obtained deductively in Chap. 5.

Costs recurring at regular or irregular intervals are handled exactly as in conventional EUAW problems. That is, they must be converted into equivalent uniform annual amounts for *one cycle*. They are thus automatically annual for each succeeding life cycle as well, as discussed in Sec. 6.1. Example 6.6 illustrates EUAW calculations for a perpetual project.

Example 6.6

The U.S. Bureau of Reclamation is considering two proposals for increasing the capacity of the main canal in their Lower Valley irrigation system. Proposal A would involve dredging the canal in order to remove sediment and weeds which have accumulated during previous years' operation. Since the capacity of the canal will have to be maintained near its design peak flow because of increased water demand, the bureau is planning to purchase the dredging equipment and accessories for $65,000. The equipment is expected to have a 10-year life with a $7000 salvage value. The annual labor and operating costs for the dredging operation is estimated to be $22,000. In order to control weeds in the canal itself and along the banks, herbicides will be sprayed during the irrigation season. The yearly cost of the weed-control program, including labor, is expected to be $12,000.

Proposal B would involve lining the canal with concrete at an initial cost of $650,000. The lining is assumed to be permanent, but minor maintenance will be required every year at a cost of $1000. In addition, lining repairs will have to be made every 5 years at a cost of $10,000. Compare the two alternatives on the basis of equivalent uniform annual worth using an interest rate of 5% per year.

Solution The cash-flow diagrams are left to the reader. The EUAW of each proposal is determined as follows:

Proposal A

EUAW of dredging equipment:	
$65,000(A/P, 5\%, 10) - 7000(A/F, 5\%, 10)$	$ 7,861
Annual cost of dredging	22,000
Annual cost of weed control	12,000
	$41,861

Proposal B

EUAW of initial investment: $650,000(0.05)$	$32,500
Annual maintenance cost	1,000
Lining repair cost: $10,000(A/F, 5\%, 5)$	1,800
	$35,310

Proposal B should be selected.

Comment For proposal A, it was necessary to consider only one cycle. No calculations were necessary for the dredging and weed-control costs since they were already expressed as annual costs. For proposal B, the EUAW of the initial investment was obtained by multiplying by the interest rate, which is nothing more than Eq. (5.2), that is,
$EUAW = A = Pi$.

If *nonrecurring* single or series costs are involved, they must be converted to a present worth and then multiplied by the interest rate. Note the use of the A/F (sinking-fund) factor for the lining repair cost. The A/F factor is used instead of the capital-recovery (A/P) factor because the lining repair cost began in year 5 instead of year 0 and continued indefinitely at 5-year intervals.

Additional Examples 6.9 and 6.10
Probs. 6.28 to 6.42

Additional Examples

Example 6.7

A drugstore chain has just purchased a fleet of five pickup trucks to be used for delivery in a particular city. Initial cost was $4600 per truck and the expected life and salvage value is 5 years and $300, respectively. The combined insurance, maintenance, gas, and lubrication

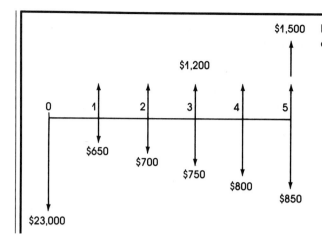

$1,500   **Figure 6.4** Diagram used to compute EUAW, Example 6.7.

costs are expected to be $650 the first year and to increase by $50 per year thereafter, while delivery service will bring an extra $1200 per year for the company. If a return of 10% per year is required, use the EUAW method to determine if the purchase should have been made.

Solution   The cash-flow diagram is shown in Fig. 6.4. If we compute the EUAW by the salvage sinking-fund method, we can first use Eq. (6.1).

A_1 = annual cost of fleet purchase

$$= -5(4600)(A/P, 10\%, 5) + 5(300)(A/F, 10\%, 5)$$
$$= \$-5822$$

The minus signs are used for costs since incomes and disbursements are involved. The annual disbursement and income can be combined into an annual-income equivalent (A_2) so that the net income conveniently follows a decreasing gradient.

$$A_2 = 550 - 50(A/G, 10\%, 5) = \$460$$

Now the EUAW is equal to the *algebraic* sum of the annual disbursement and annual income.

$$\text{EUAW} = -5822 + 460 = \$-5362$$

Since EUAW < 0, a return less than 10% per year will be made and the purchase is therefore not justified.

Comment   Try one of the other EUAW methods of Secs. 6.3 and 6.4 to solve the problem. Obviously, you should get the same answer.

Example 6.8

Compare the two plans proposed in Example 5.6 using the EUAW method.

Solution   Even though the two component parts of plan A, movers and pad, have different lives, the EUAW analysis must be conducted for only one life cycle. For the salvage sinking-fund method, Eq. (6.1),

$$\text{EUAW}_A = \text{EUAW}_{\text{movers}} + \text{EUAW}_{\text{pad}} + \text{EUAW}_{\text{AOC}}$$

where $EUAW_{movers} = 90,000(A/P, 15\%, 8) - 10,000(A/F, 15\%, 8) = \$19,328$

$EUAW_{pad} = 28,000(A/P, 15\%, 12) - 2000(A/F, 15\%, 12) = \$5096$

$EUAW_{AOC} = \$9800$

Then,

$EUAW_A = 19,328 + 5096 + 9800$

$= \$34,224$

$EUAW_B = EUAW_{conveyor}$

$= 175,000(A/P, 15\%, 24) - 10,000(A/F, 15\%, 24)$

$= \$27,146$

As was also shown in the present-worth analysis of Example 5.6, select plan B.

Comment   You should recognize a fundamental relation between the PW and EUAW values for the two examples discussed here. If you have the PW of a given plan, you can get the EUAW by $EUAW = PW(A/P, i\%, n)$ or with an EUAW, $PW = EUAW(P/A, i\%, n)$. The question is: What value does n assume? What would you use? We vote for the least-common-multiple value used in the present-worth method, since this method of evaluation must take place over an equal time period for each alternative. Therefore, the present-worth values are

$PW_A = EUAW_A(P/A, 15\%, 24) = \$220,190$

$PW_B = EUAW_B(P/A, 15\%, 24) = \$174,652$

as found in Example 5.6.

Example 6.9

If an investor deposits $1000 now, $3000 three years from now, and $600 per year for 5 years starting 4 years from now, how much money can be withdrawn every year forever beginning 12 years from now, if the rate of return on the investment is 18% per year?

Solution   The cash-flow diagram is shown in Fig. 6.5. The uniform amount of money that can be withdrawn every year forever is equal to the amount of interest that accumulates each

Figure 6.5   Diagram to determine perpetual annual withdrawal, Example 6.9.

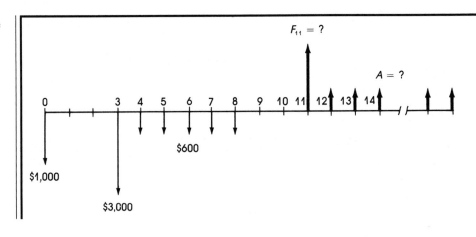

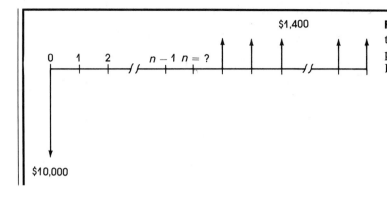

Figure 6.6 Diagram to determine n for a perpetual withdrawal, Example 6.10.

year on the principal amount. To solve the problem, therefore, it is necessary to determine the total amount that would be accumulated in year 11 (*not* year 12) and then multiply by the interest rate i to obtain A. The future amount in year 11 would be

$$F_{11} = 1000(F/P, 18\%, 11) + 3000(F/P, 18\%, 8) + 600(F/A, 18\%, 5)(F/P, 18\%, 3)$$
$$= \$24,505$$

The perpetual withdrawal can now be found by multiplying F_{11} (which is now a P value with respect to the perpetual withdrawal) by the interest rate.

$$A = Pi = 24,505(0.18) = \$4411$$

This A value is the EUAW.

Example 6.10

If an investor deposits \$10,000 now at an interest rate of 7% per year, how many years must the money accumulate before the investor can withdraw \$1400 per year forever?

Solution The cash-flow diagram is shown in Fig. 6.6. The first step is to find the total amount of money that must be accumulated in year n (P_n), which is 1 year prior to the first withdrawal, to permit the perpetual \$1400-per-year withdrawal.

$$P_n = \frac{A}{i} = \frac{1400}{0.07} = \$20,000$$

When \$20,000 is accumulated, the investor can withdraw \$1400 per year forever. The next step is to determine when the initial \$10,000 deposit will accumulate to \$20,000. This can be done with the F/P factor:

$$20,000 = 10,000(F/P, 7\%, n)$$

By interpolation, $(F/P, 7\%, n)$ is 2.0000 when $n = 10.24$ years.

Problems

6.1 Why must the least common multiple of years be used for assets that have different lives when a present-worth comparison is made, whereas an EUAW analysis requires annualization of costs over only one life cycle of each asset?

6.2   If a woman purchased a new car for $6000 and sold it 3 years later for $2000, what was the equivalent uniform annual worth if she spent $750 per year for upkeep and operation? Use an interest rate of 15% per year and the salvage sinking-fund method.

6.3   An entrepreneur purchased a dump truck for the purpose of offering a short-haul earth-moving service. He paid $14,000 for the truck and sold it 5 years later for $3000. His operation and maintenance expense while he owned the truck was $3500 per year. In addition, he had the truck engine overhauled for $220 at the end of the third year. Calculate his equivalent uniform annual worth using the salvage sinking-fund method, if the interest rate was 14% per year.

6.4   Work the following problems by the salvage present-worth method: (a) 6.2 and (b) 6.3.

6.5   Work the following problems by the capital-recovery-plus-interest method: (a) 6.2 and (b) 6.3.

6.6   The manager in a canned-food processing plant is trying to decide between two labeling machines. Their respective costs are as follows:

	Machine A	Machine B
First cost	$15,000	$25,000
Annual operating cost	1,600	400
Salvage value	3,000	6,000
Life, years	7	10

(a) Determine which machine should be selected using a minimum attractive rate of return of 12% per year and an EUAW analysis.

(b) If a present-worth analysis were used, over how many years would you make the comparison?

6.7   Compare the following machines on the basis of their equivalent uniform annual worth. Use $i = 18\%$ per year.

	New machine	Used machine
First cost	$44,000	$23,000
Annual operating cost	7,000	9,000
Annual repair cost	210	350
Overhaul every 2 years	...	1,900
Overhaul every 5 years	2,500	...
Salvage value	4,000	3,000
Life, years	15	8

6.8   Compare the two processes below on the basis of their equivalent uniform annual worth at an interest rate of 18% per year.

	Process M	Process R
First cost	$80,000	$120,000
Salvage value	10,000	18,000
Life, years	10	15
Annual operating cost	15,000	13,000
Annual revenues	39,000	55,000 for year 1, decreasing by $2,000 per year

6.9   A moving and storage company is considering two possibilities for warehouse operations. Proposal 1 requires the purchase of a forklift for $5000 and 500 pallets that cost $5 each. The average life of a pallet is assumed to be 2 years. If the forklift is purchased, the company must hire an operator for $9000 annually and spend $600 per year in maintenance and operation. The life of the forklift is expected to be 12 years, with a $700 salvage value.

Alternatively, proposal 2 requires that the company hire two people to operate power-driven hand trucks at a cost of $7500 per person. One hand truck will be required at a cost of $900. The hand truck will have a life of 6 years with no salvage value. If the company's minimum attractive rate of return is 12% per year, which alternative should be selected?

6.10 The supervisor of a country club swimming pool is trying to decide between two methods used for adding chlorine. If gaseous chlorine is added, a chlorinator, which has an initial cost of $800 and a useful life of 5 years, will be required. The chlorine will cost $200 per year and the labor cost will be $400 per year. Alternatively, dry chlorine can be added manually at a cost of $500 per year for chlorine and $800 per year for labor. If the interest rate is 6% per year, which method should be used?

6.11 A carpenter is trying to determine how much insulation should be put into a ceiling. The higher the R rating of the insulation, the better the insulation. The choices are limited to either R-11 or R-19 insulation. The R-11 insulation costs $2.50 per square meter and the R-19 costs $3.50 per square meter. The annual saving in heating and cooling costs is estimated to be $25 per year greater with R-19 than with R-11. If the house has 250 square meters and the owner expects to keep the house for 25 years, which insulation should be installed at an interest rate of 10% per year?

6.12 In Prob. 6.11, how much would the saving have to be per year in order for the R-19 insulation to be just as economical as the R-11?

6.13 A meat-packing plant manager is trying to decide between two different methods for cooling cooked hams. The spray method involves spraying water over the hams until the ham temperature is reduced to 30 degrees Celsius. With this method, approximately 80 liters of water are required for each ham.

Alternatively, an immersion method can be used in which only 16 liters of water are required per ham. However, this method will require an initial extra investment of $2000 and extra overhaul expenses of $100 per year, with the equipment expected to last 10 years. The company cooks 10 million hams per year and pays $0.12 per 1000 liters for water. The company must also pay $0.04 per 1000 liters for wastewater discharged. If the company's minimum attractive rate of return is 15% per year, which method of cooling should be used?

6.14 Two environmental chambers (A and B) are being considered for a government project which is to last for 6 years. Pertinent data are listed below.

	Chamber *A*	Chamber *B*
First cost	$4,000	$2,500
Annual operating cost	400	300
Salvage value	1,000	−100
Estimated life, years	3	2

(*a*) What chamber should be selected if money is worth 12% per year?
(*b*) What must the difference in annual operating cost be to make the equivalent annual worth of both chambers equal?

6.15 Compare the two plans below at $i = 15\%$ per year.

| | | Plan *B* | |
	Plan *A*	Machine 1	Machine 2
First cost	$10,000	$30,000	$5,000
Annual operating cost	500	100	200
Salvage value	1,000	5,000	−200
Life, years	40	40	20

6.16 The Mighty Mouse Company is considering the purchase of a trap system to rid the plant of stray cats. Compare the two systems below at 10%-per-year interest.

	Scram-um	Catch-um
First cost	$25,000	$50,000
Annual operating cost	500	200
Salvage value	1,000	500
Life, years	20	40

6.17 The Toe-Main Food Processing Company is evaluating various methods for disposing of the sludge from the wastewater treatment plant. Currently under consideration is land disposal of the sludge by spraying or incorporation into the soil. If the spraying alternative is selected, an underground distribution system will be constructed at a cost of $600,000. The salvage value after 20 years is expected to be $20,000. Operation and maintenance of the system is expected to cost $26,000 per year.

Alternatively, the company can use Big Foot trucks to transport and dispose of the sludge by incorporation below the soil surface. Three trucks will be required at a cost of $220,000 per truck. The operating cost of the trucks, including driver, routine maintenance, overhauls, etc., is expected to be $42,000 per year. The used trucks can be sold after 10 years for $30,000 each. If the trucks are used, field corn can be planted and sold for $20,000 per year. For spraying, grass must be planted and harvested, and because of the presence of the "contaminated" sludge on the cuttings, the grass will have to be landfilled at a cost of $14,000 per year. If the company's minimum attractive rate of return is 20% per year, which method should be selected based on an equivalent-uniform-annual-worth analysis?

6.18 Two methods can be used for producing a certain machine part. Method 1 costs $20,000 initially and will have a $5000 salvage value after 3 years. The operating cost with this method is $8500 per year. Method 4 has an initial cost of $15,000, but it will last only 2 years. Its salvage value is $3000. The operating cost for method 4 is $7000 per year. If the minimum attractive rate of return is 16% per year, which method should be used on the basis of an equivalent-uniform-annual-worth analysis?

6.19 A transport company on the U.S. border is trying to decide between purchasing a diesel- or gasoline-powered truck. A diesel-powered truck will cost $1000 more to purchase, but the kilometers per liter will be a respectable 6. Diesel fuel can be purchased in Mexico for 4 cents a liter, but one time per year the fuel system will have to be cleaned of paraffins at a cost of $190 per cleaning. The gasoline-powered truck will average 4 kilometers per liter and gasoline can be purchased for 14 cents a liter (also in Mexico). The diesel-powered truck can be used for 270,000 kilometers if the engine is overhauled for $5500 after 150,000 kilometers. The gasoline-powered truck can be used for 180,000 kilometers if overhauled at a cost of $2200 after 120,000 kilometers. If the company's trucks average 30,000 kilometers per year, which type should be purchased at a minimum attractive rate of return of 20% per year? Assume zero salvage values for both trucks.

6.20 Two types of materials can be used for roofing a commercial building which has 1500 square meters of roof. Asphalt shingles will cost $14 per square meter installed and are guaranteed for 15 years. Fiberglass shingles will cost $17 per square meter installed, but they are guaranteed for 20 years. If the fiberglass shingles are selected, the owner will be able to sell the building for $1500 more than if the asphalt shingles are used. If the owner plans to sell the building in 8 years, which shingles should be used if the minimum attractive rate of return is 17% per year and the equivalent-uniform-annual-worth method of analysis is to be used?

6.21 The R-Sun Specialty Company is considering two types of siding for its proposed new building (the previous building was destroyed by fire). Anodized metal siding will require very little maintenance and minor repairs will cost only $500 every 3 years. The initial cost of the siding will be $250,000. If a concrete facing is used, the building will have to be painted now at a cost of $80,000 and every 5 years at a cost of $8000 more than the previous time. The building is expected to have a useful life of 23 years, and the "salvage value" will be $25,000 greater if the metal siding is used. Compare the equivalent uniform annual worths of the two methods at an interest rate of 15% per year.

6.22 The warehouse for a large furniture manufacturing company currently requires too much energy for heating and cooling because of poor insulation. The company is trying to decide between urethane foam and fiberglass insulation. The initial cost of the foam insulation will be $35,000, with no salvage value. The foam will have to be painted every 3 years at a cost of $2500. The energy saving is expected to be $6000 per year.

Alternatively, fiberglass batts can be installed for $12,000. The fiberglass batts would not be salvageable either, but there would be no maintenance costs. If the fiber-glass batts would save $2500 per year in energy costs, which method of insulation should the company use at an interest rate of 15% per year? Use a 24-year study period and an equivalent-uniform-annual-worth analysis.

6.23 Compare the alternatives below on the basis of an equivalent-uniform-annual-worth analysis, using an interest rate of 20% per year.

	Plan A	Plan B
First cost	$28,000	$36,000
Installation cost	3,000	4,000
Annual maintenance cost	1,000	2,000
Annual operating cost	$2,200 + 75k$*	$800 + 50k$
Life, years	10	10

    * k = years, 1 through 10.

6.24 Compare the alternatives shown below on the basis of an equivalent-uniform-annual-worth analysis. Use an interest rate of 1% per month.

	Alternative WHY	Alternative ME
First Cost	$70,000	$90,000
Monthly operating cost	1,200	1,400
Salvage value	7,000	10,000
Life, years	3	6

6.25 Compare the alternatives shown below on the basis of an equivalent-uniform-annual-worth analysis, using an interest rate of 15% per year compounded continuously.

	Alternative PAY LATER	Alternative KNOT NOW
First cost	$18,000	$25,000
Annual cost	4,000	3,600
Salvage value	3,000	2,500
Life, years	3	4

6.26 A company is considering the purchase of one of two processes identified as E and Z. Process E will have a first cost of $43,000, a monthly operating cost of $10,000, and a $5,000 salvage value at the end of its 4-year life. Process Z will have a first cost of $31,000 and a quarterly operating cost of $39,000. It will have an 8-year life with a $2,000 value at the end of that time. If the interest rate is a nominal 12% per year compounded monthly, which alternative would be preferred on the basis of an equivalent-uniform-annual-worth analysis?

6.27 Data for machines X and Y are shown below. If the interest rate is 12% per year compounded quarterly, which machine should be selected on the basis of an equivalent-uniform-annual-worth analysis?

	Machine X	Machine Y
First cost	$25,000	$55,000
Annual operating cost	8,000	6,000
Annual increase in operating cost	5%	3%
Salvage value	12,000	9,000
Life, years	5	10

6.28 Calculate the perpetual equivalent uniform annual worth of $14,000 now, $55,000 six years from now, and $5000 per year thereafter if the interest rate is 8% per year.

6.29 Rework Prob. 6.28 using an interest rate of a nominal 18% per year compounded semiannually.

6.30 The first cost of a small dam is expected to be $3 million. The annual maintenance cost is expected to be $10,000 per year; a $35,000 outlay will be required every 5 years. If the dam is expected to last forever, what will be its equivalent uniform annual worth at an interest rate of 12% per year?

6.31 A city that is attempting to attract a professional football team is planning to build a new football stadium costing $12 million. Annual upkeep is expected to amount to $25,000 per year. In addition, the artificial turf will have to be replaced every 10 years at a cost of $150,000. Painting every 5 years will cost $65,000. If the city expects to maintain the facility indefinitely, what will be its equivalent uniform annual worth? Assume that $i = 12\%$ per year.

6.32 An alumnus of Watsa Matta University desires to establish a permanent scholarship in his name. He plans to donate $20,000 per year for 10 years starting 1 year from now and leave $100,000 when he dies. If the university expects the alumnus to die 15 years from now, how much money can be given to each of five students beginning 1 year from now and continuing forever if the interest rate is 8% per year?

6.33 As a grateful alumnus of Ima-Wanta University, Ms. B. G. Spender would like to make an endowment which will provide scholarships to needy skydivers who want to study to become engineers. Ms. Spender would like the scholarships to be in the amount of $13,000 per year, with the first scholarship to be given 15 years from now. Ms. Spender wants to deposit enough money into the fund in 14 years so that scholarships can be given in her name forever thereafter (i.e., from year 15 on). If Ms. Spender plans to make her first deposit 1 year from now, how much should she deposit each year if the fund will earn interest at a rate of 12% per year?

6.34 Another grateful alumnus of Watsa Matta University, Mr. I. S. Rich, would also like to establish a perpetual scholarship fund for would-be engineers. Mr. Rich would like to provide one scholarship per year in the amount of $20,000 for an infinite time, with the first scholarship to be given 10 years from now. Mr. Rich plans to make his first deposit 1 year from now and then increase each succeeding one by $1000 through year 9, at which time no more money will be donated. If the fund earns interest at a rate of 14% per year, how much must Mr. Rich deposit in each of the first 2 years?

6.35 For the cash-flow sequence shown below, determine the amount of money that can be withdrawn annually for an infinite time if the first withdrawal is to be made in year 13 and the interest rate is 15% per year.

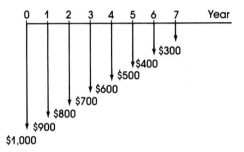

6.36 For the cash-flow sequence shown below, how much time must elapse between the last payment (in year 9) and the first withdrawal of $4000 per year for an infinite time, if the interest rate is 13% per year?

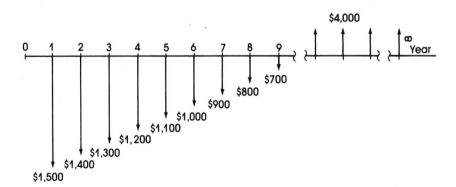

6.37 For the deposits shown in Prob. 6.36, how much money could be withdrawn for an infinite time beginning in year 12, if the interest rate is 13% per year?

6.38 For the cash-flow diagram shown below, determine the equivalent uniform annual series in years 1 through infinity. Use an interest rate of effective 20% per year compounded continuously.

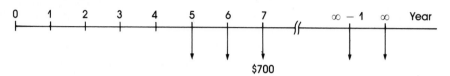

$700

6.39 Compare the alternatives shown below on the basis of an equivalent-uniform-annual-worth comparison. Use $i = 14\%$ per year compounded quarterly.

	Alternative U.R.	Alternative O.K.
First cost	$8,500,000	$50,000,000
Annual operating cost	8,000	7,000
Salvage value	5,000	2,000
Life, years	5	∞

6.40 Compare the alternatives shown below on the basis of their perpetual equivalent uniform annual worth, using an interest rate of 14% per year.

	Alternative PAY NOW	Alternative NOT LATER
First cost	$ 160,000	$25,000
Annual operating cost	15,000	3,000
Overhaul every 4 years	12,000	. . .
Salvage value	1,000,000	4,000
Life, years	∞	7

6.41 Compare the alternatives shown below on the basis of their equivalent uniform annual worths, using an interest rate of an effective 11% per year compounded semiannually.

	Alternative MAX	Alternative MIN
First cost	$150,000	$ 900,000
Annual operating cost	50,000	10,000
Salvage value	8,000	1,000,000
Life, years	5	∞

6.42 Compare the alternatives shown below on the basis of their perpetual equivalent uniform annual worths. Use an interest rate of 1.3% per month.

	Project C	Project D
First cost	$8,000	$99,000
Monthly operating cost	1,500	400
Increase per month in operating cost	10	. . .
Salvage value	1,000	47,000
Life, years	10	∞

Rate-of-Return Computations for a Single Project 7

In this chapter, the procedures to correctly compute the rate of return for one project using the present-worth and equivalent-uniform-annual-worth methods are discussed. Since rate-of-return calculations frequently require trial-and-error solutions, a method for estimating the interest rate that will satisfy the rate-of-return equation is discussed. One of the problems with a rate-of-return analysis is that in some cases, multiple values will satisfy the rate-of-return equation. We discuss how to recognize when this possibility exists. We also discuss the composite (or external) rate of return and how this calculation overcomes the multiple rate-of-return dilemma. While only one project is considered in this chapter, the next chapter is an application of the principles discussed here for comparison of alternatives.

Section Objectives

After completing this chapter, you should be able to do the following:

7.1 Write the definition of *rate of return* for a project and state the general equations used to compute the rate of return, given the receipts and disbursements.

7.2 Calculate the rate of return for a single project using the present-worth method and the steps to initially estimate the rate of return, given the amounts and times of project disbursements and receipts.

7.3 Calculate the rate of return as in objective 7.2, except using the equivalent-uniform-annual-worth method.

7.4 State whether a given cash-flow sequence is *conventional* or *nonconventional* and determine the *multiple rates of return*, given the amounts and times of nonconventional cash flows.

7.5 State the definitions of *internal* and *composite rate of return* and compute both rates, given the amounts and times of the cash flows and a reinvestment rate for all receipts (positive cash flows) resulting from the project.

Study Guide

7.1 Overview of Rate-of-Return Computation

If money is borrowed, the interest rate is applied to the *unpaid balance* so that the total loan amount and interest are paid in full exactly with the last payment. If money is lent or invested in a project, there is an *unrecovered balance* at each time period. The interest rate is the return on this unrecovered balance so that the total loan and interest are recovered exactly with the last receipt. Rate of return defines both of these situations.

> *Rate of return* (ROR) is the rate of interest paid on the unpaid balance of borrowed money or the rate of interest earned on the unrecovered balance of an investment (loan) so that the final payment or receipt brings the balance to zero with interest considered.

The rate of return is expressed as a percent per period, for example, $i = 10\%$ per year and is always positive, $i > 0$; that is, the fact that interest paid on a loan is actually a "negative" rate of return is not considered. Note that the definition above does not state that the rate of return is on the initial amount of the investment but is rather on the *unrecovered* balance, which varies with time. The example below illustrates the difference between these two concepts.

Example 7.1

A $1000 investment is expected to produce a net cash flow of $315.47 for each of 4 years. This represents a 10% per year rate of return on the unrecovered balance. Compute the amount of the unrecovered investment for the 4 years using (*a*) the rate of return on the unrecovered balance and (*b*) the rate of return on the initial $1000 investment. (*c*) Explain why all of the investment is not recovered in part (*b*).

Solution
(*a*) Table 7.1 presents the unrecovered balance figures for each year using the 10% rate on the unrecovered balance at the beginning of the year. After 4 years the total $1000 investment is recovered and the balance in col. 6 is zero.

Table 7.1 Unrecovered balances using a rate of return of 10%

(1) Year	(2) Beginning unrecovered balance	(3) = 0.10(2) Interest on unrecovered balance	(4) Cash flow	(5) = (4) − (3) Removal of unrecovered balance	(6) = (2) − (5) Ending unrecovered balance
0	...	...	$ −1,000.00	...	$ −1,000.00
1	$ −1,000.00	$100.00	+315.47	$ 215.47	−784.53
2	−784.53	78.45	+315.47	237.02	−547.51
3	−547.51	54.75	+315.47	260.72	−286.79
4	−286.79	28.68	+315.47	286.79	0
		$261.88		$1,000.00	

Table 7.2 Unrecovered balances using a 10% return on the initial investment

(1)	(2) Beginning unrecovered	(3) = 0.10($1,000) Interest on initial	(4) Cash	(5) = (4) − (3) Removal of unrecovered	(6) = (2) − (5) Ending unrecovered
Year	balance	investment	flow	balance	balance
0	. . .	. . .	$−1,000.00	. . .	$−1,000.00
1	$−1,000.00	$100	+315.47	$215.47	−784.53
2	−784.53	100	+315.47	215.47	−569.06
3	−569.06	100	+315.47	215.47	−353.25
4	−353.25	100	+315.47	215.47	−138.12
		$400		$861.88	

(b) Table 7.2 shows the unrecovered balance figures if the 10% return is always figured on the initial investment of $1000. Column 6 in year 4 shows a remaining unrecovered amount of $138.12, because only $861.88 is recovered in the 4 years.

(c) A total of $400 in interest must be earned if the 10% return each year is figured on the initial investment. However, only $261.88 in interest must be earned if a 10% return on the unrecovered balance is used. There is more of the annual cash flow available to reduce the remaining investment when the rate is applied to the unrecovered balance.

Comment As defined, rate of return is the interest rate on the unrecovered balance; therefore, the computations in Table 7.1 for part (a) present a correct interpretation of a 10% rate of return. Obviously, an interest rate of 10% per year applied to the principal only, actually represents a higher rate than it appears. The so-called add-on interest rates are frequently based on principal only.

To determine the rate of return value i of a project, the present worth of disbursements P_D is equated to the present worth of receipts P_R. That is,

$$P_D = P_R$$

Equivalently,

$$0 = -P_D + P_R \qquad (7.1)$$

In this analysis, investments are disbursements and incomes are receipts. The equivalent-uniform-annual-worth method can also be used.

$$EUAW_D = EUAW_R$$

or

$$0 = -EUAW_D + EUAW_R \qquad (7.2)$$

The i value which makes the relations correct may be referred to by several titles— rate of return, internal rate of return, breakeven rate of return, profitability index, or return on investment (ROI)—and is customarily represented as i^* (i star).

Probs. 7.1 and 7.2

7.2 Rate-of-Return Calculations by the Present-Worth Method

In Sec. 2.11 the method for calculating the rate of return on an investment was illustrated when only one factor was involved. In this section the present-worth method for calculating the rate of return on an investment when several factors are involved is demonstrated. To understand rate-of-return calculations more clearly, remember that the basis for engineering-economy calculations is *equivalence*, or time value of money. In previous chapters, we have shown that a present sum of money is equivalent to a larger sum of money at some future date when the interest rate is greater than zero. In rate-of-return calculations, the objective is to find the interest rate at which the present sum and future sum are equivalent; in other words, the calculations that will be made here are simply the reverse of calculations made in previous chapters, where the interest rate was known.

The backbone of the rate-of-return method is a rate-of-return relation, such as Eq. (7.1) or (7.2), which is simply an expression equating a present sum of money to the present worth of future sums. For example, if you invest $1000 now and are promised receipts of $500 three years from now and $1500 five years from now, the rate-of-return equation is

$$1000 = 500(P/F, i^*\%, 3) + 1500(P/F, i^*\%, 5) \qquad (7.3)$$

where the value of i^* to make the equality correct is to be computed (see Fig. 7.1). If the $1000 is moved to the right side of Eq. (7.3), we have

$$0 = -1000 + 500(P/F, i^*\%, 3) + 1500(P/F, i^*\%, 5) \qquad (7.4)$$

Equation (7.4) is in the general form of Eq. (7.1), which will be used in setting up all rate-of-return calculations by the present-worth method. The equation must then be solved for i by trial and error to obtain $i^* = 16.95\%$. Since there are always receipts and disbursements involved in any project, a definite value of i^* can be found; however, the rate of return will be greater than zero only if the total amount of receipts is greater than the total amount of disbursements.

It should be evident that rate-of-return calculations are merely the reverse of present-worth calculations. That is, if the above interest rate were given (16.95%) and it was desired to find the present worth of $500 three years from now and $1500

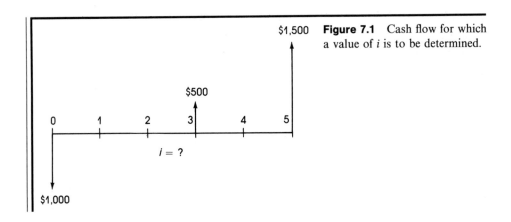

Figure 7.1 Cash flow for which a value of i is to be determined.

five years from now, the equation would be

$$P = 500(P/F, 16.95\%, 3) + 1500(P/F, 16.95\%, 5) = \$1000$$

which is easily rearranged to the form of Eq. (7.4). This illustrates that rate-of-return and present-worth equations are set up in exactly the same fashion. The only difference is in what is given and what is sought.

The general procedure used to make a rate-of-return calculation by the present-worth method is the following:

1. Draw a cash-flow diagram.
2. Set up the rate-of-return equation in the form of Eq. (7.1).
3. Select values of i by trial and error until the equation is balanced. It will probably be necessary to find i* using linear interpolation.

When using the trial-and-error method to determine i*, it is advantageous to get fairly close to the correct answer on the first trial. If the cash flows are combined in such a manner that the income and disbursements can be represented by a *single factor* such as P/F, P/A, and so forth, it is possible to look up the interest rate (in the tables) corresponding to the value of that factor for n years as discussed in Chap. 2. The problem, then, is to combine the cash flows into the format of only one of the standard factors. This may be done through the following procedure:

1. Convert all *disbursements* into either single amounts (P or F) or uniform amounts (A) by neglecting the time value of money. For example, if it is desired to convert an A into an F value, simply multiply the A by the number of years n. The scheme selected for movement of cash flows should be the one which minimizes the error caused by neglecting the time value of money.
2. Convert all *receipts* to either single or uniform values, as in step 1.
3. Having combined the disbursements and receipts so that either a P/F, P/A, or A/F format would apply, use the interest tables to find the approximate interest rate at which the P/F, P/A, or A/F value, respectively, is satisfied for the proper n value. The rate obtained is a good ball-park figure to use in the first trial.

It is important to recognize that the rate of return obtained in this manner is only an *estimate* of the actual rate of return, because the time value of money is neglected. This procedure is illustrated in Example 7.2.

Example 7.2

If $5000 is invested now in common stock that is expected to yield $100 per year for 10 years and $7000 at the end of 10 years, what is the rate of return?

Solution The rate-of-return procedure as described above is used to compute i*.

1. Figure 7.2 shows the cash-flow diagram.
2. Using Eq. (7.1),

$$0 = -5000 + 100(P/A, i*\%, 10) + 7000(P/F, i*\%, 10)$$

3. Use the estimation procedure above to determine the interest rate for the first trial. All income will be regarded as a single F in year 10 so that the P/F factor can be used.

Figure 7.2 Cash flow for a stock investment, Example 7.2.

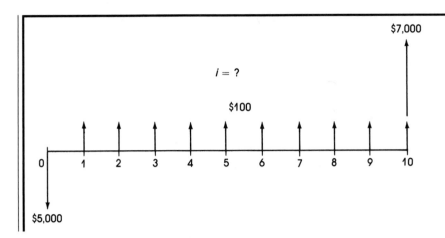

The P/F factor is selected because most of the cash flow (i.e., $7000) already fits this factor and errors created by neglecting the time value of the remaining money will be minimized. Thus

$P = \$5000$

$F = 10(100) + 7000 = 8000$

$n = 10$

Now we can state that

$$5000 = 8000(P/F, i\%, 10)$$

$$(P/F, i\%, 10) = 0.625$$

The approximate interest rate is between 4 and 5%. Therefore, use $i = 5\%$ in the equation of step 2 to get the actual rate of return.

$$0 = -5000 + 100(P/A, 5\%, 10) + 7000(P/F, 5\%, 10)$$

$$0 \neq \$69.46$$

We are too large on the positive side, indicating that the return is more than 5%. Therefore, try $i = 6\%$.

$$0 = -5000 + 100(P/A, 6\%, 10) + 7000(P/F, 6\%, 10)$$

$$0 \neq -\$355.19$$

Since the interest rate of 6% is too high, interpolate (Sec. 2.7) using Eq. (2.21):

$$b \begin{bmatrix} a \begin{matrix} \rightarrow +69.46 \\ \rightarrow 0 \\ \rightarrow -355.19 \end{matrix} \end{bmatrix} \qquad \begin{bmatrix} 5\% \\ i^* \\ 6\% \end{bmatrix} c \Bigg] d$$

$$c = \frac{a}{b}(d) = \frac{(69.46 - 0)}{69.46 - (-355.19)}(1.0) = 0.16$$

$$i^* = 5.00 + 0.16 = 5.16\%$$

Comment   Note that 5% rather than 4% was used for the first trial. The higher value was used because, by assuming that the ten $100 amounts were equivalent to a single $1000 in year 10, the approximate rate estimated from the P/F factor was *lower* than the true value. This is due to the neglect of the time value of money. Therefore, the first-trial i value used was above that indicated by the P/F factor in order to improve the accuracy of the first guess.

Additional Example 7.6
Probs. 7.3 to 7.13

7.3   Rate-of-Return Calculations by the Equivalent-Uniform-Annual-Worth Method

Just as $i*$ can be found by the present-worth method, it may also be determined using the EUAW relation of Eq. (7.2). This method would be preferred, for example, when uniform annual cash flows are involved or when the cash flows increase or decrease by a constant gradient or percentage gradient. The procedure is as follows.

1. Draw a cash-flow diagram.
2. Set up the relations for the EUAW of disbursements $(EUAW_D)$ and receipts $(EUAW_R)$ with $i*$ as an unknown in the relations.
3. Set up the rate-of-return equation in the form of Eq. (7.2), that is,

$$0 = -EUAW_D + EUAW_R$$

4. Select values of i by trial and error until the equation is balanced. If necessary, interpolate to determine $i*$.

The estimation procedure in Sec. 7.2 for the first i value is used here also. Example 7.3 illustrates the EUAW method.

Example 7.3

Use EUAW computations to find the rate of return for the investment situation in Example 7.2.

Solution

1. Figure 7.2 shows the cash-flow diagram.
2. The EUAW relations for disbursements and receipts are

   $EUAW_D = -5000(A/P, i\%, 10)$

   $EUAW_R = 100 + 7000(A/F, i\%, 10)$

3. The EUAW formulation using Eq. (7.2) is

   $0 = -5000(A/P, i*\%, 10) + 100 + 7000(A/F, i*\%, 10)$

4. Trial-and-error solution yields the results: $i = 5\%$, $0 \neq \$+9.02$; and $i = 6\%$, $0 \neq \$-48.26$. Interpolation yields $i* = 5.16\%$, as before.

Thus, for rate-of-return calculations you can choose either the present-worth or equivalent-uniform-annual-worth method. It is generally better to get accustomed to using only one of the two methods in order to avoid errors.

The computer program ROIDS (Return On Investment Determination System) is introduced in the appendix section. It computes the rate of return value (i^*) for any sequence of cash flows.

Probs. 7.3 to 7.13

7.4 Multiple Rate-of-Return Values

In the two previous sections a unique rate of return (ROR) value i^* was determined for the given cash-flow sequences. Investigation shows that the signs on the *net cash flows* changed only once, usually from minus in year 0 to plus for the rest of the investment's life. This is called a *conventional* cash flow. If there is more than one sign change, the series is called *nonconventional*. As shown in the examples in Table 7.3, the runs of net cash-flow signs may be one or more in length.

When there is more than one sign change (that is, when the net cash flow is nonconventional), it is possible that there will be multiple i^* values which will balance the rate-of-return equation [Eqs. (7.1) or (7.2)]. The total number of real ROR values is always less than or equal to the number of sign changes in the sequence. (It is possible to determine that imaginary values or infinity will also satisfy the equation, but these are of little value to the analyst.) Example 7.4 presents the determination and graphical interpretation of multiple rate-of-return values.

Example 7.4

A new synthetic lubricant has been marketed for 3 years, with the following net cash flows in thousands of dollars.

Year	0	1	2	3
Cash flow (In thousands)	$+2,000	−500	−8,100	+6,800

(a) Plot the present worth versus the rate of return for i values of 5, 10, 20, 30, 40, and 45%.
(b) Determine whether the cash-flow series is conventional or nonconventional and estimate the rate of return from the plot in part (a).

Table 7.3 Examples of conventional and nonconventional net cash-flow sequences for six year-lived projects

Type	Sign on net cash flow							Number of sign changes
	0	1	2	3	4	5	6	
Conventional	−	+	+	+	+	+	+	1
Conventional	−	−	−	+	+	+	+	1
Conventional	+	+	+	+	+	−	−	1
Nonconventional	−	+	+	+	−	−	−	2
Nonconventional	+	+	−	−	−	+	+	2
Nonconventional	−	+	−	−	+	+	+	3

Table 7.4 Computation of present worth for several rate-of-return values, Example 7.4

Cash flow (In thousands)	Year				Present worth for different i values (In thousands)
	0	**1**	**2**	**3**	
	$+2,000	**$-500**	**$-8,100**	**$+6,800**	
5%	$+2,000	$-476.20	$-7,346.70	$+5,873.84	$ +51.44
10%	+2,000	-454.55	-6,693.84	+5,108.84	-39.55
20%	+2,000	-416.65	-5,624.64	+3,935.16	-106.13
30%	+2,000	-384.60	-4,792.77	+3,095.36	-82.01
40%	+2,000	-357.15	-4,132.62	+2,477.92	-11.85
45%	+2,000	-344.85	-3,852.36	+2,230.40	+33.19

i (to the left of the rate column)

Solution

(*a*) The present-worth values are found in Table 7.4 using the P/F factor for each i value. Figure 7.3 shows that the present worth has a parabolic shape and crosses the i axis two times, because there are two sign changes in the given cash-flow sequence.

(*b*) The sequence is nonconventional because of two sign changes for the cash flows (plus to minus in years 0 to 1 and minus to plus in years 2 to 3). The two ROR values (i_1^* and

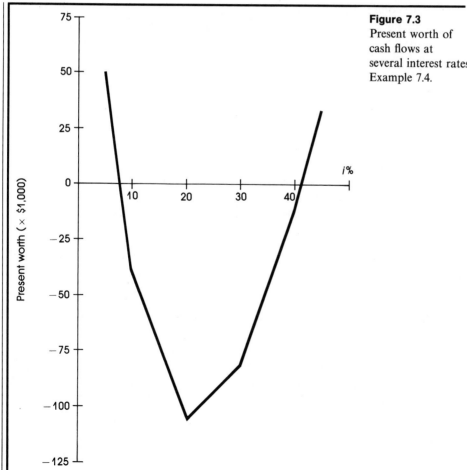

Figure 7.3
Present worth of cash flows at several interest rates, Example 7.4.

i_2^*) may be graphically determined from Fig. 7.3 to be approximately

$$i_1^* = 8\% \text{ and } i_2^* = 41\%$$

Comment If two ROR values are solved mathematically, the more exact values are found to be 7.47 and 41.35%. If there had been three sign reversals in the cash-flow sequence, it would have been possible for there to be three different ROR values.

In many cases some of the multiple ROR values will seem ridiculous because they are too large or too small (negative). For example, values of 10, 150, and 750% for a sequence with three sign changes is difficult to explain. It is common to neglect the large values or to simply never compute them. Alternatively, a composite rate of return can be calculated as described in the next section. Obviously, one advantage of the PW or EUAW (Chaps. 5 and 6) methods for alternative analysis is that unrealistic rates do not enter and confuse the analysis.

For the interested reader, further discussion of multiple rate-of-return computations is presented in the reference texts at the end of this chapter. The appendix devoted to computer applications includes a discussion on ROIDS, which is a program that may be used to determine all real roots for a specified cash-flow sequence. ROIDS allows the user to interactively input the cash-flow sequence and determine all ROR values within a specified range. The procedure for determining a unique rate-of-return value is presented in the next section.

Additional Example 7.7
Probs. 7.14 to 7.16

7.5 Internal and Composite Rates of Return

The rate-of-return values we have computed thus far have assumed that any net positive cash flows (receipts) are reinvested immediately at the rate of return that balances the rate-of-return equation. Therefore, if the balancing rate is, say, 40%, any receipt prior to the end of the project is assumed to earn 40% for the remaining years. Of course, this assumption may be unrealistic when the balancing rate is much greater or less than the minimum attractive rate of return (MARR). The balancing rate, which is computed by Eqs. (7.1) and (7.2), is called the *internal rate of return* (IRR) because it does not consider any of the economic factors external to the project. By definition,

> The *internal rate of return* i^* is a rate of return for a project which assumes that all positive cash flows are reinvested at the same rate of return as that which balances the rate-of-return equation.

It is the reinvestment assumption of the internal rate of return, coupled with the reversals in net cash-flow signs, that allows there to be multiple rates of return for nonconventional cash-flow sequences.[1] However, *if a specific reinvestment rate is*

---

[1] It is actually possible to perform tests of the cash-flow signs, the accumulated cash-flow signs, the algebraic sum of all cash flows, and the project unrecovered investment balance to determine the number and values of rates of return for any nonconventional cash-flow sequence. A good summary of these tests is presented in Newnan [1].

used to compute the future worth of all positive cash flows that can be invested external to the project, a return to a conventional cash-flow sequence is accomplished and the problem of multiple rates of return is eliminated. (The cumulative cash-flow sequence, which is obtained by successively adding cash-flow values, must also be conventional to ensure a single rate of return.)

The *reinvestment rate*, symbolized by c, is often set equal to the MARR. The interest rate determined in this fashion to satisfy the rate-of-return equation will be called the *composite rate of return* and will be symbolized by i'. This rate of return is also known as the *external rate of return*. By definition

> The *composite rate of return* i' is the rate of return of a project which assumes that net positive cash flows which represent funds not immediately needed in the project are reinvested at the rate c.

The term *composite* is used to describe this rate of return because it is derived from more than one interest rate. If c happens to equal any one of the IRR values, then the composite rate will equal that IRR value.

The reinvestment rate is applied to all net positive cash flows to obtain the composite rate of return for the project. Net positive cash flows are considered over-recovery of the project investments, which are the negative cash flows. The correct i' value is that which makes the overall net project investment equal to exactly zero at the end of the project. One procedure for doing so, the *project-net-investment* technique, is summarized here. (This procedure, which has several names, is explained in greater detail in Bussey [2] and in others [3, 4, 5].)

The process involves finding the future worth F of the net investment amount 1 year (period) into the future. That is, find the cash flow next year (period), F_{t+1}, from F_t by using the F/P factor for one period. The interest rate in the F/P factor is c if the net investment F_t is positive and it is i' if F_t is negative. Mathematically, each year sets up the relation

$$F_{t+1} = F_t(1 + i) + C_{t+1} \tag{7.5}$$

where $t = 1, 2, \ldots, n - 1$
  n = total years in the project
  C_t = cash flow in period t
  $i = \begin{cases} c & \text{if } F_t > 0 \quad \text{(net positive investment)} \\ i' & \text{if } F_t < 0 \quad \text{(net negative investment)} \end{cases}$

The equation for F_n obtained using this procedure is set equal to zero and solved for i' by trial and error. The i' value obtained is unique for a given reinvestment rate c.

The development of F_1 through F_3 for the cash-flow sequence below, which is also shown in Fig. 7.4a, is illustrated for $c = 15\%$ as the reinvestment rate.

Year	Cash flow
0	$ 50
1	−200
2	50
3	100

Figure 7.4 Cash-
flow sequence for
which the com-
posite rate of
return i' is com-
puted: (*a*) original
form, (*b*) equiv-
alent form in year
1, and (*c*) equiv-
alent form in year
2.

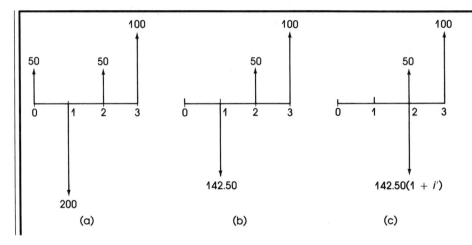

The net investment for year $t = 0$ is \$50, which is positive, so it returns the 15% during the first year. Thus, by Eq. (7.5), F_1 is

$$F_1 = 50(1 + 0.15) - 200 = -\$142.50$$

This result is shown in Fig. 7.4*b*. Since the net investment is now negative, the value F_1 earns interest at the i' rate for year 2. Therefore,

$$F_2 = -142.50(1 + i') + 50$$

The i' value is to be found. Since F_2 will be negative for all $i' > 0$, use i' to set up F_3 as shown in Fig. 7.4*c*.

$$F_3 = F_2(1 + i') + C_3 = [-142.50(1 + i') + 50](1 + i') + 100 \qquad (7.6)$$

Setting Eq. (7.6) equal to zero and solving for i' will result in the unique composite rate of return for the cash-flow sequence. If there had been more F_t expressions, i' would be used in all subsequent equations (see Example 7.8).

This project-net-investment procedure to find i' may be summarized as follows:

1. Draw a cash-flow diagram of the original cash-flow sequence.
2. Develop the series of project net investments using Eq. (7.5) and the stated c value. The result is the F_n expression in terms of i'.
3. Set the F_n expression equal to 0 and find the i' value to balance the equation. If necessary interpolate to determine i'.

Several comments are in order before we show an example. If the reinvestment rate c equals a specific i^*, the value i' will be the same as the internal rate of return i^*, that is, $c = i^* = i'$ because the i^* value assumes reinvestment at the internal rate (see the definition of IRR). The closer the c value is to i', the smaller the difference between the composite and internal rates of return. It is common to use $c =$ MARR, realizing that all receipts can realistically be reinvested at the minimum acceptable rate of return.

Example 7.5

Compute the composite rate of return for the synthetic lubricant investment in Example 7.4 if the reinvestment rate is (a) 7.47% and (b) 20%.

Solution
(a) Use the procedure stated above to determine i' for $c = 7.47\%$.

1. Figure 7.5 shows the original cash flow.
2. The first project-net-investment expression is $F_0 = \$2000$. Since $F_0 > 0$, use $c = 7.47\%$ to get F_1 by Eq. (7.5).

$$F_1 = 2000(1.0747) - 500 = \$1649.40$$

Again $F_1 > 0$, so use $c = 7.47\%$ for determining F_2.

$$F_2 = 1649.40(1.0747) - 8100 = -\$6327.39$$

Figure 7.6 shows the equivalent cash flow at this time. Since $F_2 < 0$, use i' to express F_3.

$$F_3 = -6327.29(1 + i') + 6800$$

3. Let $F_3 = 0$ and solve for i'

$$-6327.39(1 + i') + 6800 = 0$$

$$1 + i' = \frac{6800}{6327.39} = 1.0747$$

$$i = 0.0747 \quad (7.47\%)$$

The composite rate of return is 7.47%, which is the same as c, the reinvestment rate, and the i_1^* value discussed in Example 7.4. Note that $i = 41.35\%$, which is the other IRR value, no longer balances the rate-of-return equation. The equivalent future-worth result for the cash flow in Fig. 7.6 at $i' = 41.35\%$ is

$$6327.39(F/P, 41.35\%, 1) = 8943.77 \neq 6800$$

which indicates a return much lower than 41.35%.

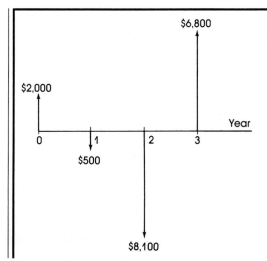

Figure 7.5   Original cash flow (in thousands), Example 7.5.

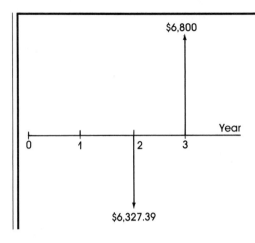

Figure 7.6 Equivalent cash flow (in thousands) of Fig. 7.5 with reinvestment at 7.47%.

(b) For $c = 20\%$, the net-investment series is

$F_0 = 2000$ $\qquad\qquad\qquad\qquad$ ($F_0 > 0$, use c)

$F_1 = 2000(1.20) - 500 = \$1900$ $\qquad$ ($F_1 > 0$, use c)

$F_2 = 1900(1.20) - 8100 = -\$5820$ $\qquad$ ($F_2 < 0$, use i')

$F_3 = -5820(1 + i') + 6800$

Set $F_3 = 0$ and solve for i' directly

$$(1 + i') = \frac{6800}{5820} = 1.1684$$

$\qquad i' = 0.1684 \qquad (16.84\%)$

The composite rate of return is $i' = 16.84\%$ at a reinvestment rate of 20%, which is a marked increase from $i' = 7.47\%$ at $c = 7.47\%$.

Comment If the value $c = 41.35\%$ is used, the equation will be balanced with $i' = 41.35\%$, which is the second internal rate-of-return value found in Example 7.4. This is possible because the assumption that receipts are reinvested at the IRR value is correct if $c = i^* = i' = 41.35\%$.

It is possible to summarize the relations between c, i', and any i^* as follows:

Relation between reinvestment rate c and IRR i^*	Relation between i' and IRR i'^*
$c = i^*$	$i' = i^*$
$c < i^*$	$i' < i^*$
$c > i^*$	$i' > i^*$

These relationships were demonstrated in Example 7.5.

Additional Example 7.8
Probs. 7.17 to 7.23

Additional Examples

Example 7.6

Assume that a couple invests $10,000 now and $500 three years from now and will receive $500 one year from now, $600 two years from now, and amounts increasing by $100 per year for a total of 10 years. They will also receive lump-sum payments of $5000 in 5 years and $2000 in 10 years. Calculate the rate of return on their investment.

Solution The cash-flow diagram in Fig. 7.7 is used to set up a rate-of-return relation using the present-worth method.

$$0 = -10{,}000 - 500(P/F, i*\%, 3) + 500(P/A, i*\%, 10) + 100(P/G, i*\%, 10)$$
$$+ 5000(P/F, i*\%, 5) + 2000(P/F, i*\%, 10)$$

Solving by trial and error and interpolating between $i = 7\%$ and $i = 8\%$, we find $i* = 7.8\%$.

Comment Note that the single values at years 3, 5, and 10 were handled separately so that the P/A and P/G factors could be used on the remaining cash flows. As an exercise, use the procedure of Sec. 7.2 to estimate the interest rate by the P/A factor for $n = 10$ years. Assume that the gradient term is a uniform series with an average value of $500. Your estimate should show that i is in the neighborhood of 7%.

Example 7.7

Assume a series of cash flows as shown for an ongoing project. (Data are adapted from an article by McLean [6] and some results of Barish and Kaplan [7].) The negative net cash flow in year 4 is the result of a major alteration to the project. Compute the rate of return values for the project.

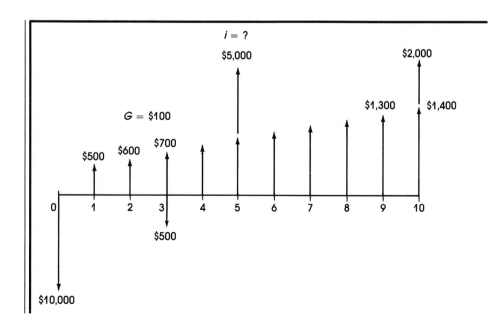

Figure 7.7 Cash-flow diagram for Example 7.6.

Figure 7.8 Plot
of present worth
versus i, Example
7.7.

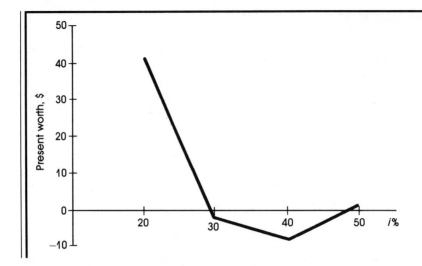

Year	Net Cash flow	Year	Net Cash flow
1	$ 200	6	$500
2	100	7	400
3	50	8	300
4	−1,800	9	200
5	600	10	100

Solution The present-worth method, Eq. (7.1), can be used to set up the rate-of-return equation.

$$0 = 200(P/F, i\%, 1) + 100(P/F, i\%, 2) + \cdots + 100(P/F, i\%, 10) \qquad (7.7)$$

The results obtained for several values of i are shown in the following table and are plotted in Fig. 7.8.

i, %	10	20	30	40	50
Results of Eq. (7.7), $	+198	+42	−2	−8	+1

As you can see, the two values which satisfy Eq. (7.7) are at approximately $i_1^* = 29\%$ and $i_2^* = 49\%$. There are two i^* values because the net cash flow is nonconventional and two sign changes are present (one between years 3 and 4, the second between years 4 and 5).

Example 7.8

Determine the composite rate of return in Example 7.7 if the stated reinvestment rate is 15% per year.

Solution Use the steps summarized in Sec. 7.5 to write the net investment series for $t = 1$ to $t = 10$.

$F_1 = \$200$ $\qquad\qquad\qquad\qquad$ ($F_1 > 0$, use c)

$F_2 = 200(1.15) + 100 = \$330$ $\qquad\qquad$ ($F_2 > 0$, use c)

$F_3 = 330(1.15) + 50 = \$429.50$ $\qquad$ ($F_3 > 0$, use c)

$$F_4 = 429.50(1.15) - 1800 = -\$1306.08 \qquad (F_4 < 0, \text{ use } i')$$

$$F_5 = -1306.08(1 + i') + 600$$

Since we do not know if F_5 is greater than zero or less than zero, all remaining expressions use i'.

$$F_6 = F_5(1 + i') + 500 = [-1306.08(1 + i') + 600](1 + i') + 500$$

$$F_7 = F_6(1 + i') + 400$$

$$F_8 = F_7(1 + i') + 300$$

$$F_9 = F_8(1 + i') + 200$$

$$F_{10} = F_9(1 + i') + 100$$

To find i', the rather cumbersome expression $F_{10} = 0$ must be solved. Computerized solution using ROIDS or its equivalent should be used, but manual solution is possible to find $i' = 21.25\%$.

Comment   At this time you might want to rework this problem with a reinvestment rate of 29 or 49%, as found in Example 7.7, to see that the i' value will be the same as these reinvestment rates; that is, if $c = 29\%$, then $i' = 29\%$. In the example above, $c = 15\%$ is less than $i^* = 29\%$ so $i' = 21.25\% < i^*$, as discussed at the end of Sec. 7.5.

References

1. D. G. Newnan, *Engineering Economic Analysis*, rev. ed., Engineering Press, San Jose, CA, 1980, pp. 373–389.
2. L. E. Bussey, *The Economic Analysis of Industrial Projects*, Prentice-Hall, Englewood Cliffs, NJ, 1978, pp. 220–238.
3. J. Lorie and L. J. Savage, "Three Problems in Capital Rationing," *Journal of Business*, vol. 28, no. 4, October 1955, pp. 229–239.
4. D. Teichroew, A. A. Robichek, and M. Montalbano, "Mathematical Analysis of Rates of Return Under Certainty," *Management Science*, vol. 11, no. 3, January 1965, pp. 395–403.
5. D. Teichroew, A. A. Robichek, and M. Montalbano, "An Analysis of Criteria for Investment and Financial Decisions Under Certainty," *Management Science*, vol. 12, no. 3, November, 1965, pp. 151–179.
6. J. G. McLean, "How to Evaluate New Capital Investments," *Harvard Business Review*, November–December 1958, pp. 59–69.
7. N. N. Barish and S. Kaplan, *Economic Analysis for Engineering and Managerial Decision Making*, 2d ed., McGraw-Hill, New York, 1978, p. 204.

7.1   State a definition for rate of return.

Problems

7.2   Derive a formula to compute the beginning unrecovered balance B_{t+1} for year $t + 1$ computed in Table 7.1 in terms of B_t, the interest rate i, and the cash flow C_t for year t. Demonstrate that your formula works.

7.3   A real estate investor purchased a piece of property for $6000 and sold it 17 years later for $21,000. The property taxes were $80 the first year and $90 the second year, and they increased by $10 per year until the property was sold. What was the rate of return on the investment?

7.4   If the property taxes in Prob. 7.3 increased by $10 per year for the first 6 years and then increased by $20 per year thereafter, what was the rate of return on the investment?

7.5   A family purchased a run-down house for $25,000 with the idea of making major improvements and then selling for a profit. In the first year that they owned the house, they spent $5000 on improvements. They spent $1000 the second year and $800 the third year. In addition, they paid property taxes of $500 per year for 3 years and then sold the house for $35,000. What rate of return did they make on their investment?

7.6   The Karry-Mor Trucking Company purchased a new dump truck for $54,000. The total operating expenses were $36,000 the first year, $39,000 the second, and amounts increasing by $3000 per year thereafter. The income for the first year was $66,000 and decreased by $500 per year thereafter. If the company kept the truck for 10 years and then sold it for $15,000, what was the rate of return on the investment?

7.7   If a company spends $6000 now and $900 per year for 17 years, with the first disbursement 5 years from now, what rate of return would the company receive if the income during the 21 years was $5000 at the end of year 3 and $1200 per year thereafter?

7.8   A real estate investor purchased a building for $130,000 and received $2000 per month in rent. The taxes were $2500 per year and maintenance costs were $700 every 3 years. If the building was sold for $180,000 twelve years after it was bought, what nominal annual and effective monthly rate of return was made on the investment?

7.9   Rework Prob. 7.3 assuming that the property taxes increased by $50 per year.

7.10   An investor purchased three types of stocks (identified here as A, B, and C). The investor purchased 200 shares of A at $13 per share, 400 shares of B at $4 a share, and 100 shares of C at $18 per share. The dividends were $0.50 a share from stock A for 3 years and then the stock was sold for $15 a share. There were no dividends from stock B, but the stock was sold for $5.50 a share 2 years after it was purchased. Stock C resulted in dividends of $2.10 a share for 10 years, but due to a depressed market at the time it was sold, the stock sold for only $12 a share. Calculate the rate of return on each stock as well as the overall rate of return on the stock investments.

7.11   Firewood can be purchased during July for $55 a cord. If the purchaser waits until November, the cost of the same kind of wood is $70 a cord. What rate of return would the purchaser receive if the firewood was purchased in July instead of November?

7.12   A careful shopper is trying to decide between buying an artificial Christmas tree and continuing to buy cut trees. The artificial tree costs $34 and can be used for 8 years, at which time it will be thrown away for no salvage value. Alternately, the shopper can continue to buy cut trees at a cost of $8 now, $9 next year, $10 two years from now, and so on for 8 years. If the artificial tree is purchased, what rate of return is made on the investment?

7.13   A homeowner who plans to build a large brick patio is expecting the price of bricks to increase considerably in the next 2 years. The cost is expected to increase from the present price of $160 per 1000 bricks to $175 per 1000 next year and $195 the following year. If 3000 bricks are purchased now instead of 1000 now and 1000 each year for the following 2 years, what will be the rate of return on the investment?

   Note: For Probs. 7.14 to 7.21 a computer program may be developed to solve the problem more rapidly. (The ROIDS program introduced in the appendix section may be used or the reader may develop his or her own program.)

7.14   A currently funded project is expected to have the following cash flow for the next 5 years.

Year	0	1	2–3	4	5
Cash flow, $	−10,000	10,000	0	30,000	−30,000

(*a*) Compute and plot the present worth at the following interest rates: 0, 10, 20, 30, and 40%.

(*b*) Determine the approximate rate-of-return values from the graph in part (*a*).

(*c*) Use the rate-of-return equation to find the rate of return for the cash-flow sequence. Which value is correct?

7.15 Find all rate-of-return values between 0 and 100% for the following cash flows.

Year	0	1	2	3–6
Cash flow, $	500	−1,000	50	200

If hand solution rather than computerized solution is used, factor values for $i > 50\%$ must be computed from the formulas since tables do not extend above 50%.

7.16 Write a computer program to find multiple rates of return and use it to find the $i*$ values for the cash flows in Example 7.4.

7.17 A commonly referenced multiple rate-of-return problem is the "pump problem" [3]. The cash flow generated by installing a new pump is observed to be

Year	0	1	2
Cash flow, $	1,600	10,000	−10,000

(*a*) Determine the rate-of-return equation solutions, and (*b*) interpret their meaning in terms of internal and composite rates of return.

7.18 Compute (*a*) the internal rates of return, and (*b*) the composite rate of return if $c = 15\%$, for the following cash flow (same data as used in the discussion of Sec. 7.5).

Year	0	1	2	3
Cash flow, $	50	−200	50	100

7.19 (*a*) A company invested $5000 in auxiliary equipment and procedure changes for ore removal 1 year ago. Increased income was observed to be $25,000 this year. What is the calculated rate of return on the investment?

(*b*) If the cash flow is −$23,000 next year, the cash-flow sequence changes signs twice. Compute the two rate-of-return values for the cash-flow and interpret their meaning.

(*c*) If a reinvestment rate of 30% is used, find the composite rate of return and discuss its relation to the answers in part (*b*).

(*d*) Find the composite rate of return if a reinvestment rate of 15% is used.

7.20 Explain why the composite rate of return (i') will be equal to a multiple rate-of-return value ($i*$) if the assumption is made that the reinvestment rate (c) is equal to $i*$.

7.21 (*a*) Use a computer program to find i' for different values of the reinvestment rate c between 15 and 50% for the cash-flow values in Prob. 7.19, parts (*a*) and (*b*).

(*b*) Construct a plot of i' versus c values using the results of part (*a*).

7.22 A certain project, if implemented, would generate income of $40,000 per year for an infinite time. If the initial investment required is $2 million with maintenance costs of $80,000 every 2 years, what is the rate of return for the project?

7.23 The capitalized cost of a certain project is $5 million. If the project has an initial cost of $3.5 million, a salvage value of $2 million, an annual cost of $140,000, and a monthly cost of $5000, find the effective interest rate per year, assuming interest is compounded monthly. The life of the project is infinite.

8 Rate-of-Return Evaluation for Multiple Alternatives

This chapter presents the methods by which two or more alternatives can be evaluated using a rate-of-return comparison. This type of evaluation will result in the same selection as the present-worth and EUAW analysis, but the computational procedure is considerably different. When the alternatives under consideration are mutually exclusive, selection of the best one is necessary. The procedure to select the best is discussed in this chapter.

Section Objectives

After completing this chapter, you should be able to do the following:

8.1 Define what is meant by *independent* and *mutually exclusive* alternatives and state why an incremental analysis is needed when comparing alternatives by the rate-of-return method.

8.2 Prepare a tabulation of net cash flow for two alternatives having equal or different lives, given the details of each alternative.

8.3 State the rationale used to evaluate two alternatives using the incremental-rate-of-return method.

8.4 Select the better of two alternatives on the basis of the *incremental rate of return* computed by the *present-worth method*, given the initial cost, life, and salvage value of each alternative, times and amounts of cash flows, and the minimum attractive rate of return.

8.5 Select the alternative as in objective 8.4, except using the *EUAW method*.

8.6 State the criteria used to select one alternative from several *mutually exclusive alternatives*, compute the incremental rates of return, and select the one best alternative, given the initial cost, salvage value, life, cash flows, and the minimum attractive rate of return.

Study Guide

8.1 The Need for Incremental Analysis

Multiple alternative evaluation generally refers to a situation involving more than two alternatives. These alternatives can be either independent or mutually exclusive. When more than one alternative can be selected from the alternatives which are available, such as when an investor wants to purchase all stocks which are expected to yield a rate of return of at least 25% per year, the alternatives are said to be *independent*. When only one alternative is to be selected from the group of alternatives available (i.e., the best alternative), then the alternatives are said to be *mutually exclusive*. As an example of mutually exclusive alternatives, a contractor wanting to purchase a bulldozer would probably have several models from several companies as possible choices, but only one of the alternatives would ultimately be selected.

Independent alternatives are usually evaluated against a predetermined standard (such as the company's minimum attractive rate of return) and, therefore, do not usually need to be compared against each other. In this case, the procedures discussed in Chap. 7 could profitably be used to identify the appropriate alternatives. Chapter 17 discusses the selection techniques for independent projects. When the alternatives under consideration are mutually exclusive, however, it becomes necessary to be able to identify the one alternative which can be considered the best. The techniques discussed in Chaps. 5 and 6 could obviously be used to do so, producing results which are very easy to interpret. However, the results obtained from rate-of-return evaluations are not as readily understood.

Let us assume that the GRQ Company uses a minimum attractive rate of return of 16% per year, that the company has $90,000 available for investment, and that two mutually exclusive alternatives (*A* and *B*) are being investigated for robotic intervention. Alternative *A* will require an investment of $50,000 and will yield a rate of return of 35% per year. Alternative *B* will require $85,000 and will yield 29% per year. One would intuitively think that the better alternative would be the one which yields the higher rate of return, alternative *A* in this case. However, this is not necessarily so because, while *A* has the higher rate of return, it also requires an initial investment which is much less than the total capital available ($90,000). In a case such as this, a logical question might be, "What happens to the capital that is left over?" It is generally assumed that excess capital will be invested at the company's minimum attractive rate of return. Using this assumption, it is possible to investigate the consequences associated with each of the alternative investments mentioned above. If alternative *A* is selected, $50,000 will be invested at a rate of return of 35% per year. The $40,000 left over will be invested at the company's MARR of 16% per year. The rate of return on the total capital *available* for investment, then, will be the weighted average of these values. Thus, if alternative A is selected,

$$\text{Overall ROR}_A = \frac{50{,}000(0.35) + 40{,}000(0.16)}{90{,}000} = 26.6\%$$

If alternative B is selected, \$85,000 will be invested at a yield of 29% per year and the remaining \$5000 earns 16% per year. The weighted average in this case is

$$\text{Overall ROR}_B = \frac{85{,}000(0.29) + 5000(0.16)}{90{,}000} = 28.3\%$$

This calculation shows that even though the rate of return for alternative A is higher than that of alternative B, the latter represents the better investment for GRQ.

If either a present-worth or equivalent-uniform-annual-worth comparison had been conducted for these two alternatives using $i = 16\%$ per year, alternative B would similarly have been chosen. This rather simple example was presented to illustrate that a major shortcoming of the rate-of-return method for comparing alternatives is that under some circumstances, rate-of-return values alone would not provide the same ranking as would a PW or EUAW analysis. This problem can be overcome by conducting an *incremental-investment rate-of-return* analysis as described in the following sections.

8.2 Tabulation of Net Cash Flow

The concept of cash flow was discussed in Chap. 7 with respect to rate-of-return calculations for a single alternative. In this chapter, it is necessary to prepare a *net cash-flow tabulation* between two alternatives so that an incremental-rate-of-return analysis can be conducted. The column headings for a cash-flow tabulation involving two alternatives are shown in Table 8.1. If the alternatives have *equal* lives, the year column will go from 0 to n, the life of the alternatives. If the alternatives have *unequal* lives, the year column will go from 0 to the least common multiple of the two lives when present-worth analysis is used. The use of the least-common-multiple rule is necessary because rate-of-return analysis on the net cash-flow values must always be done over the same number of period for each alternative (as is the case with present-worth comparisons). If the least common multiple of lives is tabulated, reinvestment in each alternative is shown at appropriate times (as was done in Chap. 5 for cash flow in present-worth analysis).

Table 8.1 Format for cash-flow tabulation

	(1)	(2)	(3) = (2) − (1)
	Cash flow		
Year	Alternative A	Alternative B	Net cash flow
0			
1			
2			
⋮			

Table 8.2 Cash-flow tabulation for Example 8.1

| | Cash flow | | Net cash flow |
Year	Old mill	New mill	(new-old)
0	$ −15,000	$ −21,000	$ −6,000
1–25	−8,200	−7,000	+1,200
25	+750	+1,050	+300
Total	$−219,250	$−194,950	$+24,300

You will see in this chapter that a cash-flow tabulation is an integral part of the procedure for selecting one of two alternatives on the basis of incremental rate of return. Therefore, a standardized format for the tabulation will simplify interpretation of the final results. In this chapter, the alternative with the *higher initial investment* will always be regarded as *alternative B*. That is,

Net cash flow = cash flow$_B$ − cash flow$_A$

The next two examples demonstrate cash-flow tabulation for equal-life and unequal-lived alternatives.

Example 8.1

A tool and die company is considering the purchase of an additional milling machine. The company has the opportunity to buy a slightly used machine for $15,000 or a new one for $21,000. Because the new machine is a more sophisticated model with some automatic features, its operating cost is expected to be $7000 per year, while the old machine is expected to cost $8200 per year. Each machine is expected to have a 25-year life with a 5% salvage value. Tabulate the net cash flow of the two alternatives.

Solution   Net cash flow is tabulated in Table 8.2. The salvage values in year 25 are separated from ordinary cash flow for clarity. Note that a sign must be included to indicate a disbursement (minus) or an income (plus).

Comment   Note that when the cash-flow columns are subtracted, the difference between the totals of the two alternatives should equal the total of the net cash-flow column. This will provide a check of your addition and subtraction in preparing the tabulation.

When disbursements are the same for a number of consecutive years, it saves time to make a single cash-flow listing, as is done for years 1 to 25 of the example. However, remember that several years were combined when adding to get the column totals.

Example 8.2

The Fresh-Pak Tomato Cannery has under consideration two different types of conveyors. Type *A* has an initial cost of $7000 and a life of 8 years. The initial cost of type *B* is $9500 and has a life expectancy of 12 years. The operating cost for type *A* is expected to be $900, while the cost for type *B* is expected to be $700. If the salvage values are $500 and $1000 for type *A* and type *B* conveyors, respectively, (*a*) tabulate the cash flows of each alternative, and (*b*) tabulate the net cash flow using their least common multiple of lives.

Table 8.3 Cash flow for respective asset life for Example 8.2a

	Cash flow	
Year	Type A	Type B
0	$ −7,000	$ −9,500
1–7	−900	−700
8	−900 + 500	−700
9–11	⋯	−700
12	⋯	−700 + 1,000

Table 8.4 Cash flow for 24 years for unequal-lived assets, Example 8.2b

	Cash flow		Net cash flow
Year	Type A	Type B	(B − A)
0	$ −7,000	$ −9,500	$ −2,500
1–7	−900	−700	+200
8	⎰ −7,000 −900 +500 ⎱	−700	+6,700
9–11	−900	−700	+200
12	−900	⎰ −9,500 −700 +1,000 ⎱	−8,300
13–15	−900	−700	+200
16	⎰ −7,000 −900 +500 ⎱	−700	+6,700
17–23	−900	−700	+200
24	⎰ −900 +500 ⎱	⎰ −700 +1,000 ⎱	+700
	$ −41,100	$ −33,800	$ +7,300

Solution

(a) The tabulation of each asset (Table 8.3) shows the cash flows for the respective lives—8 years for A and 12 for B.

(b) The least common multiple of years between 8 and 12 is 24 years. The net cash-flow tabulation for 24 years is given in Table 8.4. Note that the reinvestment and salvage values are shown in years 8 and 16 for type A and in year 12 for type B.

Probs. 8.1 to 8.4

8.3 Interpretation of Rate of Return on Extra Investment

The first step in calculating the rate of return on the extra investment between two alternatives is the preparation of a cash-flow tabulation similar to that of Table 8.2 or 8.4. The net cash-flow column then reflects the *extra investment* that would be required if the alternative with the larger first cost were selected. Thus, in Example 8.1 the new milling machine would require an extra investment of $6000, as shown in the last column of Table 8.2. Additionally, if the new machine were purchased, there would be a "savings" of $1200 per year for 25 years, plus $300 in year 25 as a result of the difference in salvage values. The decision about whether to buy the old or the new milling machine can be made on the basis of the profitability of investing the extra $6000 in the new machine. If the equivalent worth of the savings is greater than the equivalent worth of the extra investment using the company's minimum acceptable rate of return (MARR), then the extra investment should be made (i.e., the higher first-cost proposal should be accepted). On the other hand, if the equivalent worth of the savings is less than the equivalent worth of the extra investment, then the lower-first-cost proposal should be accepted.

Note that if the new mill in Table 8.2 is selected, there will be a net savings of $24,300. Keep in mind that this figure does not take into account the time value of money, since this total was obtained by adding the values for the various years without using the interest factors and cannot therefore be used as a basis for the decision. The totals at the bottom of the table serve only as a check against the additions and subtractions for the individual years. In fact, the $24,300 is the present worth of the net cash flow at $i = 0\%$.

It is important to recognize that the rationale for making the selection decision is the same as if only *one alternative* were under consideration—that alternative being the one represented by the difference (net cash-flow) column in the cash-flow tabulation. When viewed in this manner, it is obvious that unless this investment yields a rate of return equal to or greater than the MARR, the investment should not be made (meaning that the lower-priced alternative should be selected *to avoid this extra investment*). However, if the rate of return on the extra investment equals or exceeds the MARR, the investment should be made (meaning that the higher-priced alternative should be selected).

As further clarification of the extra investment rationale, consider that the rate of return attainable through the net cash flow can be viewed as an alternative to investing at the company's MARR. Recall that in Sec. 8.1, we stated that any excess funds not invested in the project under consideration are assumed to be invested at the company's MARR. Clearly, if the rate of return available through the net cash flow equals or exceeds that associated with the MARR, the net cash flow rate of return should be accepted (and hence, the alternative which requires this extra investment should be selected).

Prob. 8.5

8.4 Incremental-Rate-of-Return Evaluation
Using a Present-Worth Equation

The information in Chap. 7 and in the previous sections of this chapter is used to evaluate two alternatives by the incremental-rate-of-return method. The basic procedure given here assumes that all cash flows are negative (except salvage values) and that one of the two alternatives must be selected. Under this condition, since all cash flows are disbursements, it would not be possible to calculate a rate of return for the individual alternatives. (The rates of return would be negative!) Therefore, the incremental investment *must be* analyzed. (The method involving alternatives with positive cash flows is detailed in Sec. 8.6.)

The procedure for conducting an incremental investment analysis is as follows:

1. Order the alternatives so that the one with the larger initial investment is under the column labeled *B* in Table 8.3.
2. Prepare the cash-flow and net cash-flow tabulation using the least common multiple of years.
3. Draw a net cash-flow diagram.
4. Set up and find the incremental return i_{B-A}^* using the present-worth equation for the net cash flow, Eq. (7.1), or a computer program such as ROIDS or a spreadsheet software package. [Check for sign changes in the net cash-flow sequence which indicate the possible presence of multiple rates of return (Secs. 7.4 and 7.5).]
5. If $i_{B-A}^* <$ MARR, select alternative *A*. If $i_{B-A}^* \geqslant$ MARR, select alternative *B*.

In the absence of a computer program for calculating rates of return, time may be saved if the i_{B-A}^* value is estimated rather than calculated using precise linear interpretation, provided that an exact rate-of-return value is not required. For example, if the MARR is 15% per year and you have established that i_{B-A}^* is in the 15-to-20% range, an exact value is not necessary to accept *B*, since $i_{B-A}^* \geqslant$ MARR. The procedure for the rate-of-return analysis is illustrated in Examples 8.3 and 8.4.

Example 8.3

A manufacturer of boys' pants is considering purchasing a new sewing machine, which can be either semiautomatic or fully automatic. The estimates for each are as follows:

	Semiautomatic	Fully automatic
First cost	$8,000	$13,000
Annual disbursements	3,500	1,600
Salvage value	0	2,000
Life, years	10	5

Determine which machine should be selected if the MARR is 15% per year.

Table 8.5 Cash-flow tabulation for Example 8.3

	(1)	(2)	(3) = (2) − (1)
	Cash flow		
Year	Semiautomatic	Fully automatic	Difference
0	$ −8,000	$ −13,000	$ −5,000
1–5	−3,500	−1,600	+1,900
5	...	$\begin{cases} +2,000 \\ -13,000 \end{cases}$	−11,000
6–10	−3,500	−1,600	+1,900
10	...	+2,000	+2,000
	$ −43,000	$ −38,000	$ +5,000

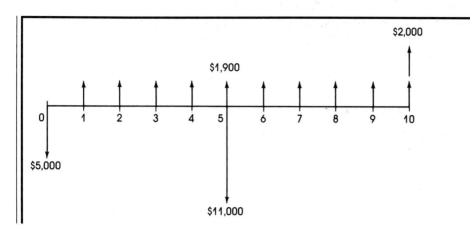

Figure 8.1 Diagram of net cash flow for Example 8.3.

Solution Use the procedure above.

1. Alternative A is the semiautomatic (s) and alternative B is the fully automatic (f) machine.
2. The cash flows for 10 years are tabulated in Table 8:5.
3. The net cash-flow diagram is shown in Fig. 8.1.
4. The incremental-rate-of-return equation for the net cash flow is

$$0 = -5000 + 1900(P/A, i\%, 10) - 11,000(P/F, i\%, 5) + 2000(P/F, i\%, 10)$$

   Solution shows that i^*_{f-s} is between 12 and 15%. By interpolation $i^*_{f-s} = 12.65\%$.
5. Since the rate of return on the extra investment is less than the 15% minimum attractive rate, the lower-cost semiautomatic machine should be purchased. If i^*_{f-s} had been greater than 15%, the fully automatic machine would have been selected.

Comment In step 4, a check of the net cash-flow sign sequence in Table 8.5 indicates that there may be multiple (up to 3) incremental-rate-of-return values. The analysis above, according to the discussion in Sec. 7.5, assumes that the positive net cash flows of $1900 in years 1 to 5 are reinvested at an external rate of $c = 12.65\%$. If this is not a reasonable assumption, the procedure in Sec. 7.5 should be performed using an appropriate reinvestment rate, which produces a different incremental-rate-of-return value to compare with MARR $= 15\%$.

The incremental rate of return obtained above can actually be interpreted as a *breakeven* value, that is, the rate of return at which either alternative might be selected. If the i^* found by the rate-of-return equation is greater than the MARR, the larger-investment alternative is selected. As an illustration, the breakeven rate for the incremental investment in Example 8.3 is 12.65%. Figure 8.2 is a general plot of the present-worth values of net cash flow for different rates of return. At values of $i < 12.65\%$, the present worth for the fully automatic machine is less than that of the semiautomatic. For $i > 12.65\%$, the fully automatic present worth is larger. Thus if MARR $= 10\%$, select the fully automatic machine; whereas if the MARR $= 15\%$, as in Example 8.3, select the semiautomatic, because the breakeven i value is less than 15%.

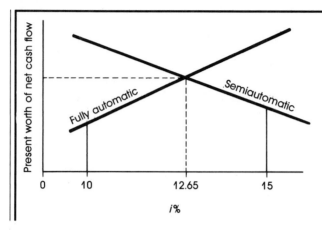

Figure 8.2 Breakeven graph of present worth of net cash flow versus rate of return, Example 8.3.

Example 8.4

Determine which milling machine should be purchased in Example 8.1 using a MARR = 15%.

Solution The alternatives are equal-lived and the net cash flows of Table 8.2 are used with the procedure above to set up the rate-of-return equation.

$$0 = -6000 + 1200(P/A, i\%, 25) + 300(P/F, i\%, 25)$$

The interpolated breakeven rate of return is 19.79%. (Try solving for the value yourself *now*, using the procedure of Sec. 7.2 to estimate the first-trial interest rate.) Since 19.79% > MARR = 15%, the purchase of the new milling machine is justified.

Comment You should realize that you *can* compare two alternatives, A and B, using the actual (rather than the incremental) cash flows with the help of the general relation

$$0 = PW_B - PW_A$$

where alternative B has the larger first cost. In the milling machine trade-off analysis, this approach results in the following:

$$PW_{new} = -21,000 - 7000(P/A, i\%, 25) + 1050(P/F, i\%, 25)$$

$$PW_{old} = -15,000 - 8200(P/A, i\%, 25) + 750(P/F, i\%, 25)$$

Then

$$
\begin{aligned}
0 = PW_{new} - PW_{old} \\
= (-21,000 + 15,000) + (-7000 + 8200)(P/A, i\%, 25) \\
+ (1050 - 750)(P/F, i\%, 25) \\
= -6000 + 1200(P/A, i\%, 25) + 300(P/F, i\%, 25)
\end{aligned}
$$

Note that the reduced form is identical to that used in the solution to this example. This method is not advised for present-worth analysis, since assets with different lives require reinvestment for comparison purposes. Try, for example, the data of Example 8.2. You will find it gets messy and can result in possible mistakes when the rate of return is computed.

Probs. 8.6 to 8.10

8.5 Incremental-Rate-of-Return Evaluation Using an EUAW Equation

Even though the use of present-worth computations to compute the IRR for alternative evaluations is recommended, the conclusions would be identical whether you used a present-worth or an EUAW equation. On some problems you might find EUAW computationally simpler. Remember that for a present-worth rate-of-return equation, the least common multiple of years must always be used in the analysis, no matter whether the difference rate-of-return equation is based on the actual cash flows or net cash flows. For an EUAW analysis, however, this is not necessary when the incremental-rate-of-return equation is obtained from the *actual* cash flows rather than the net cash flows. An EUAW rate-of-return equation on the *net* cash flow must be written over the least common multiple of lives of the alternatives, just as is true for a present-worth equation, as discussed above. The EUAW rate-of-return equation for net cash flow takes the general form

$$0 = \pm \Delta P(A/P, i\%, n) \pm \Delta SV(A/F, i\%, n) \pm \Delta A \tag{8.1}$$

where the Δ (delta) symbol identifies P, SV, and A as differences between the alternatives in the net cash-flow tabulation. Manual interpolation in the tables or a computer program (such as ROIDS) may be used to determine i^*_{B-A}.

   If the lives are *unequal* and the analyst chooses to do the analysis using the alternatives' *actual* cash flows, the EUAW for one cycle of each alternative's cash flows is determined and i^*_{B-A} is found from

$$0 = EUAW_B - EUAW_A \tag{8.2}$$

Note that the net cash flow is not used (and need not be determined) in this analysis, but the rate of return obtained does represent the i^* for the incremental cash flow between the alternatives. It should be emphasized that the net cash flow *can* be used in the EUAW method, but the net cash flow must extend through the least common multiple of lives of the alternatives, just as in the present-worth method.

   The procedure is the same as that in Sec. 8.4 for present worth except that either Eq. (8.1) or (8.2) is used to solve for the rate of return. The next two examples demonstrate the EUAW method.

Example 8.5

Compare the two milling machines in Example 8.1 using the EUAW method to compute the incremental rate of return. Assume that the MARR $= 15\%$ per year.

Solution   Since the lives are equal (25 years), Eq. (8.1) can be used to solve for the incremental rate of return between new (*n*) and old (*o*) mills using the net cash flows of Table 8.2.

$$0 = -6000(A/P, i\%, 25) + 300(A/F, i\%, 25) + 1200$$

Equality occurs at $i^*_{n-o} = 19.79\%$, so the extra investment is justified for the new milling machine (as it was with the present-worth rate of return).

Comment   The value $i = 19.79\%$ is the breakeven rate of return. A graph similar to Fig. 8.2 may be constructed in which EUAW replaces present worth. For values of MARR $\geqslant 19.79\%$, the old milling machines should be purchased; for values of MARR $< 19.79\%$, as is the case here, select the new milling machine.

Example 8.6

Compare the sewing machines of Example 8.3 using an EUAW rate-of-return equation and a MARR of 15% per year

Solution The present-worth rate-of-return equation for the net cash flow in Example 8.3 shows that the semiautomatic machine should be purchased. For the EUAW equation, we may write the equation based on either the *net* cash flow (over the least common multiple of years) or the *actual* cash flow (over one life cycle). Assuming we choose the latter method, the incremental rate of return is found from Eq. (8.2) using the respective lives of 5 years for fully automated (f) and 10 years for semiautomatic (s),

$$\text{EUAW}_f = -13{,}000(A/P, i\%, 5) + 2000(A/F, i\%, 5) - 1600$$

$$\text{EUAW}_s = -8000(A/P, i\%, 10) - 3500$$

From Eq. (8.2),

$$0 = \text{EUAW}_f - \text{EUAW}_s$$

$$0 = -13{,}000(A/P, i\%, 5) + 2000(A/F, i\%, 5) + 8000(A/P, i\%, 10) + 1900$$

At $i = 12\%$, $0 \neq \$+24.33$; for $i = 15\%$, $0 \neq \$-87.52$. Interpolation yields $i^*_{f-s} = 12.65\%$ (as for the present-worth equation); and the semiautomatic should be purchased, since 12.65% is less than the MARR of 15%.

Comment It is important to remember that if an EUAW analysis is to be made on the *net cash flow*, the cash-flow tabulation must be extended for the least common multiple of lives (10 years in this example), as in the present-worth method.

Probs. 8.11 to 8.15

8.6 Selection from Mutually Exclusive Alternatives Using Rate-of-Return Analysis

The analysis in this section deals with *multiple alternatives*, that is more than two, which are mutually exclusive. Acceptance of one alternative automatically precludes acceptance of any others.

As in any selection problem in engineering economics, there are several correct solution techniques. The present-worth and EUAW methods discussed in Chaps. 5 and 6 are the simplest and most straightforward techniques. Using a specified MARR, we could compute the total present worth or EUAW for each mutually exclusive alternative. The alternative that has the most favorable present worth or EUAW would be the one selected. The EUAW method is illustrated in Example 8.9.

When the rate-of-return method is applied, the entire investment must return at least the minimum attractive rate of return (MARR). When the returns on several alternatives equal or exceed the MARR, it is certain that at least one of them (that requiring the lowest investment) is justified. For the others, the incremental capital required must be justified. If the return on the extra investment equals or exceeds the MARR, then the extra investment should be made in order to maximize the total return on the money available for investment as discussed in Sec. 8.1. Thus, for rate-of-return analysis, the following criteria are used to select one mutually exclusive project: select the one alternative that (1) requires the *largest investment* and (2) indicates that the *incremental investment over another acceptable alternative is justified* (i.e., the return is at least the MARR). Therefore, in this regard, an important rule to remember when evaluating alternatives by the incremental-rate-of-return method is that *an alternative should never be compared with one for which the incremental investment has not been justified.* The analysis procedure is:

1. Order the alternatives in terms of increasing *initial* investment cost.
2. For alternatives which have positive cash flows, consider the "do-nothing" (i.e., zero cash flow) alternative as a defender and compute the incremental rate of return $i*$ between the do-nothing alternative and the alternative requiring the lowest initial investment. *For alternatives having only costs,* skip to step 4, using the lowest initial investment cost alternative as the defender and the next-higher one as the challenger.
3. If $i* <$ MARR, remove the lowest-investment alternative from further consideration and compute the overall rate of return for the next-higher-investment alternative. Repeat this step until $i* \geqslant$ MARR for one of the alternatives. When $i* \geqslant$ MARR, that alternative becomes the defender and the next higher-investment alternative is the challenger.
4. Determine the net (incremental) cash flow between the challenger and defender.
5. Calculate the rate of return on the incremental investment required by the challenger using the net cash flow (or the actual cash flow and Eq. (8.2)).
6. If the rate of return calculated in step 5 is greater than the MARR, the challenger becomes the defender and the previous defender is removed from further consideration. Conversely, if the rate of return in step 5 is less than the MARR, the challenger is removed from further consideration and the defender remains as the defender against the next challenger.
7. Repeat steps 4 to 6 until only one alternative remains.

Note that in the incremental analysis (steps 4 to 6), only *two* alternatives are compared at any one time. It is very important, therefore, that the correct alternatives be compared. Unless the procedure is followed as presented above, the wrong alternative may be selected from the rate-of-return analysis. The procedure detailed above is illustrated in Examples 8.7 and 8.8.

Example 8.7

Four different building locations have been suggested, of which only one will be selected. Data for each site are detailed in Table 8.6. Annual cash flow varies because of different

Table 8.6 Four alternative building locations

	Location			
	A	**B**	**C**	**D**
Building cost	$-200,000	$-275,000	$-190,000	$-350,000
Cash flow	+22,000	+35,000	+19,500	+42,000
Life, years	30	30	30	30

Table 8.7 Computation of incremental rate of return for mutually exclusive equal-lived projects

	C	A	B	D
Building cost	$-190,000	$-200,000	$-275,000	$-350,000
Cash flow	19,500	22,000	35,000	42,000
Projects compared	C to none	A to none	B to A	D to B
Incremental cost	-190,000	-200,000	-75,000	-75,000
Incremental cash flow	19,500	22,000	13,000	7,000
$(P/A, i\%, 30)$	9.7436	9.0909	5.7692	10.7143
Incremental i	9.63%	10.49%	17.28%	8.55%
Increment justified?	No	Yes	Yes	No
Project selected	None	A	B	B

tax structures, labor costs, and transportation charges, resulting in different annual receipts and disbursements. If the MARR is 10%, use an incremental-rate-of-return analysis to select a building location.

Solution All alternatives have a 30-year life. The procedure outlined above results in the following analysis:

1. Order the alternatives by increasing initial investment (Table 8.7, first line).
2. Compare location C against do-nothing, since positive cash flows are present in all alternatives. Table 8.7 shows a rate of return of $i_c^* = 9.63\%$, compared with the do-nothing alternative.
3. Since $9.63\% < 10\%$, location C is eliminated. The rate-of-return value $i_A^* = 10.49\%$ eliminates do-nothing so that the defender is now A and the challenger is B.
4. The incremental cash flow (Table 8.7) of B to A for $n = 30$ years is computed.
5. Compute the incremental i_{B-A}^* from

$$0 = -\text{incremental cost} + \text{incremental cash flow } (P/A, i\%, 30) \tag{8.3}$$

Note that $(P/A, 10\%, 30) = 9.4269$; thus any P/A value resulting from Eq. (8.3) greater than 9.4269 indicates that the return is less than 10% and is therefore unacceptable. Comparing B incrementally to A using Eq. (8.3) results in the equation $0 = -75,000 + 13,000 (P/A, i\%, 30)$.

6. A rate of return of 17.28% on the extra investment justifies B, thereby eliminating A.
7. Comparing D to B (steps 4 to 6) results in $0 = -75,000 + 7000 (P/A, i\%, 30)$ with an incremental $i_{D-B}^* = 8.55\%$, which is less than 10%, thereby eliminating location D. Since only alternative B remains, it is selected.

Comment We should mention again as a word of warning that an alternative should *always* be compared with an acceptable alternative, noting that the do-nothing alternative may be the acceptable one. Since C was not justified, location A was *not* compared with C. Thus, *if* the B-to-A comparison had not indicated that B was incrementally justified, then the D-to-A comparison instead of D-to-B would have been made.

It is important to understand the use of incremental-rate-of-return selection, because if it is not properly applied in mutually exclusive alternative evaluation, the wrong alternative may be selected. If the overall rate of return of each alternative in this example is computed (i.e., each alternative compared to the do-nothing alternative), the results are as follows:

Location	C	A	B	D
Overall i^*, %	9.63	10.49	12.40	11.59

If we were now to apply *only* the first criterion stated earlier, that is, make the largest investment that has a MARR of 10% or more, we would choose location *D*. But, as shown above, this is the *wrong selection*, because the extra investment of $75,000 between locations *B* and *D* will not earn the MARR. In fact, it will earn only 8.55% (Table 8.7). Remember, therefore, that incremental analysis is necessary for selection of one alternative from several mutually exclusive ones when the rate-of-return evaluation method is used.

When the alternatives under consideration consist of disbursements only, the "income" is the difference between costs for two alternatives. In this case, there is no need to compare any of the alternatives against the do-nothing alternative as stated in step 2 of the procedure above. (In fact, it can't be done.) The lowest-investment-cost alternative is the defender against the next-lowest-investment-cost alternative (challenger). This procedure is illustrated in Example 8.8.

Example 8.8

Four machines can be used for a certain stamping operation. The costs for each machine are shown in Table 8.8. Determine which machine should be selected if the company's MARR is 13.5% per year.

Solution   The machines are already ordered according to increasing initial investment cost and since no incomes are involved, incremental comparisons must be made beginning with the first two alternatives. Thus, as per step 4 of the incremental rate-of-return procedure, compare machine 2 (challenger) with machine 1 (defender) on an incremental basis to obtain:

$$0 = -1500 + 300 \, (P/A, i\%, 8) + 400 \, (P/F, i\%, 8)$$

Solution of the equation yields $i = 14.6\%$. Therefore, eliminate machine 1 from further consideration. (If you had trouble obtaining the rate-of-return equation above, prepare a tabulation of cash flow for machines 1 and 2, as in Sec. 8.2.) The remaining calculations are summarized in Table 8.9. When machine 3 is compared with machine 2, the rate of return on the increment is less than 0%; therefore, machine 3 is eliminated. The comparison of machine 4 with machine 2 shows that the rate of return on the increment is greater than the MARR, favoring machine 4. Since no additional alternatives are available, machine 4 represents the best selection.

Comment   You should remember that when no incomes are present in the analysis, it is implied that one of the machines *must* be selected. This situation could arise when the alternatives under consideration are part of a larger project which has already been shown

Table 8.8 Four mutually exclusive alternatives

	Machine			
	1	**2**	**3**	**4**
First cost	$-5,000	$-6,500	$-10,000	$-15,000
Annual operating cost	-3,500	-3,200	-3,000	-1,400
Salvage value	+500	+900	+700	+1,000
Life, years	8	8	8	8

Table 8.9 Comparison using incremental rate of return, Example 8.8

	Machine			
	1	2	3	4
Initial investment	$-5,000	$-6,500	$-10,000	$-15,000
Annual operating cost	-3,500	-3,200	-3,000	-1,400
Salvage value	+500	+900	+700	+1,000
Plans compared	...	2 to 1	3 to 2	4 to 2
Incremental investment	...	-1,500	-3,500	-8,500
Incremental annual savings	...	+300	+200	+1,800
Incremental salvage	...	+400	-200	+100
Incremental i^*	...	14.6%	<0%	13.7%
Increment justified?	...	Yes	No	Yes
Alternative selected	...	2	2	4

to be economical regardless of which alternative is selected. You could calculate the present worth and equivalent uniform annual worth of each machine to satisfy yourself that machine 4 would be selected by all of the evaluation methods.

When the alternatives under consideration have different lives, it is necessary to make the comparison over the least common multiple of years between *all* alternatives when using the present-worth method (Chap. 5). This is not necessary in a conventional EUAW analysis as per Chap. 6. In the incremental rate-of-return method, the analysis is done using the least common multiple between only the *two* alternatives compared. Example 8.9 illustrates the calculations for alternatives having different lives. Of course you can use incremental present-worth or incremental-EUAW analysis at the MARR to solve the problem, but it is still necessary to make the comparison over the least common multiple of lives.

Often the lives of the alternatives are so long that they can be considered to be infinite, in which case the capitalized-cost method is used. See Example 8.10 for this special case.

Example 8.9

The three mutually exclusive alternatives in Table 8.10 are available. If the MARR = 15% per year and the alternatives have different lives, as shown, select the one best alternative using (*a*) incremental-rate-of-return analysis and (*b*) conventional EUAW analysis with i = MARR.

Table 8.10 Three mutually exclusive different-lived alternatives

	A	B	C
Initial cost	$-6,000	$-7,000	$-9,000
Salvage value	0	+200	+300
Cash flow	+2,000	+3,000	+3,000
Life, years	3	4	6

Solution

(a) Use the procedure at the beginning of this section to compute the incremental rates of return.

1. Ordering is already done.
2. Comparing A to the do-nothing alternative, $i_A^* = 0\%$ using

$$0 = -6000 + 2000(P/A, i\%, 3)$$

3. Since $i_A^* < \text{MARR} = 15\%$, delete A and compute $i_B^* = 26.4\%$ from

$$0 = -7000 + 3000(P/A, i\%, 4) + 200(P/F, i\%, 4)$$

   Now B is the defender and C is the challenger.
4. The net cash flow between C and B is shown in Table 8.11 for 12 years, the least common multiple between the two.
5. Calculate $i_{C-B}^* = 19.4\%$ using the net cash flow in Table 8.11.
6. Since $19.4\% > 15\%$, select alternative C over B.

(b) For the EUAW analysis, use $i = \text{MARR} = 15\%$ per year and the respective lives. Use of the least common multiple is *not necessary* for unequal lived projects when the EUAW method is utilized.

$$\text{EUAW}_A = -6000(A/P, 15\%, 3) + 2000 = \$-628$$

$$\text{EUAW}_B = -7000(A/P, 15\%, 4) + 3000 + 200(A/F, 15\%, 4) = \$588$$

$$\text{EUAW}_C = -9000(A/P, 15\%, 6) + 3000 + 300(A/F, 15\%, 6) = \$656$$

Alternative C is again selected because it offers the largest positive EUAW, indicating a return in excess of 15%.

Comment   In part (a), when comparing different-lived alternatives by incremental rate-of-return analysis, you must use the least common multiple of years between only the *two* alternatives being compared, and *not* the common multiple of all alternative lives.

Table 8.11 Net cash-flow tabulation for Example 8.9

Year	Cash flow B	Cash flow C	Net cash flow (C − B)
0	$ −7,000	$ −9,000	$−2,000
1	+3,000	+3,000	0
2	+3,000	+3,000	0
3	+3,000	+3,000	0
4	−3,800	+3,000	+6,800
5	+3,000	+3,000	0
6	+3,000	−5,700	−8,700
7	+3,000	+3,000	0
8	−3,800	+3,000	+6,800
9	+3,000	+3,000	0
10	+3,000	+3,000	0
11	+3,000	+3,000	0
12	+3,200	+3,300	+100
	$+15,600	$+18,000	$+3,000

Table 8.12 Capitalized-cost comparison of mutually exclusive dam sites

	C	E	A	B	D	F
P(million)	$ 3	$ 5	$ 6	$ 8	$ 10	$ 11
A($1,000)	125	350	350	420	400	700
Comparison	C to none	E to none	A to E	B to E	D to E	F to E
ΔP(million)	3	5	1	3	5	6
ΔA($1,000)	125	350	0	70	50	350
$\Delta A/i - P$(million)	−0.92	0.83	−1.0	−1.83	−4.17	−0.17
Site selected	None	E	E	E	E	E

Example 8.10

The U.S. Army Corps of Engineers wants to construct a dam on the Sacochsi River. Six different sites have been suggested. The construction costs and average annual benefits (income) to the area are tabulated below. If a MARR of 6% per year is required and dam life is long enough to be considered infinite for analysis purposes, select the best location from the economic point of view.

Site	Construction cost P, millions	Annual income, A
A	$ 6	$350,000
B	8	420,000
C	3	125,000
D	10	400,000
E	5	350,000
F	11	700,000

Solution After ordering the projects by first cost, we can use the capitalized-cost equation, Eq. (5.3), $P = A/i$, in the form of Eq. (8.4) to determine if the incremental investment is justified.

$$0 = \frac{\Delta A}{i} - \Delta P \tag{8.4}$$

Here ΔP is the incremental investment and ΔA is the incremental (net) income or cash flow. Equation (8.4) could be solved for i for each increment of investment, with each value of i then compared with the MARR as per step 6 of the analysis procedure. Alternatively, if the right side of Eq. (8.4) is greater than zero, the incremental investment is justified. Table 8.12 indicates that only site E is justified, so this is the most economical dam site.

Probs. 8.16 to 8.32

Problems

8.1 Prepare a tabulation of net cash flow for the following alternatives:

	Alternative R	Alternative S
First cost	$8,000	$12,000
Annual operating cost	900	1,400
Salvage value	1,000	2,000
Life, years	5	10

8.2 Prepare a tabulation of net cash flow for the following alternatives:

	Alternative *A*	Alternative *B*
First cost	$15,000	$11,000
Annual operating cost	2,300	2,600
Annual income	4,000	3,100
Salvage value	2,000	1,500
Life, years	3	2

8.3 Two different machines are being considered for a certain process. Machine *X* has a first cost of $12,000 and annual operating expenses of $3000 per year. It is expected to have a life of 12 years with no salvage value. Machine *Y* can be purchased for $21,000, and it will have an annual operating cost of $1200 per year. However, a factory overhaul will be required every 4 years at a cost of $2500. It will have a useful life of 12 years with a $1500 salvage value. Prepare a tabulation of net cash flow for the two alternatives.

8.4 The owners of a new home are trying to decide between two types of landscape. They can purchase grass seed for $2.45 per kilogram and plant it themselves at no cost. They estimate that they will need 1 kilogram for 25 square meters. The front yard is 30 × 10 meters and the back yard 30 × 7 meters. They will also have to purchase 3 metric tons of compost when the lawn is seeded and 1 metric ton every 3 years thereafter at a cost of $20 per metric ton. If they elect to plant grass, they will also install a sprinkler system with a $350 installation cost, a $15-per-year maintenance cost, and a life of 18 years. The cost of such items as water, fertilizer, and chemicals is expected to be $120 per year. Alternatively, they can have desert landscaping installed in the front and back yards at a cost of $2800. If they elect this alternative, the annual cost of water, etc., is expected to be only $25 per year. However, they will have to replace the grass barrier under the gravel every 6 years at a cost of $350. Prepare a tabulation of cash flows and net cash flow for the two alternatives if they are to be compared for an 18-year period.

8.5 After preparing a tabulation of cash flow, what would you know immediately if the total of the difference (net cash-flow) column is a negative amount?

8.6 Determine which alternative should be selected in Prob. 8.1 if the company's MARR is 15% per year. Use the incremental-investment method.

8.7 Determine which alternative should be selected in Prob. 8.2 on the basis of incremental rate of return if the company's MARR is 20% per year.

8.8 Determine which alternative should be selected in Prob. 8.3 if the company's MARR is 12% per year. Use the incremental-investment method.

8.9 Determine which alternative should be selected in Prob. 8.4 if the purchaser's MARR is 6% per year. Use the incremental-investment analysis.

8.10 A food-processing company is considering plant expansion. Under the current setup, the company can increase profits $25,000 per year by extending the workday 2 hours through overtime. No investment will be required for this alternative. On the other hand, if additional cookers and freezers are added at a cost of $175,000, the company's profits will increase by $50,000 per year. If the company expects to use the current process for 10 years, would the plant expansion be justified if the company's MARR is 25% per year? Solve this problem two ways.

8.11 Which alternative should be selected in Prob. 8.3 if an EUAW rate-of-return analysis and at a MARR of 12% per year is used? Compare your answer and values with those of Prob. 8.8.

8.12 Select the more economic alternative in Prob. 8.4 by EUAW rate-of-return analysis if the homeowner's MARR is 6% per year and the initial cost of desert landscaping will be $1700 rather than $2800.

8.13 The engineer at the Smoke Ring Cigar Company wants to do a rate-of-return analysis using annual worth for two wrapping machines. The details below are available; however, the engineer does not know what value to use for a MARR figure since some projects are evaluated at 8% and some at 10% per year . Determine whether this difference in MARR would change the decision of which machine to buy. Use rate of return on the incremental-investment method.

	Machine A	Machine M
First cost	$10,000	$9,000
Annual labor cost	5,000	5,000
Annual maintenance cost	500	300
Salvage value	1,000	1,000
Life, years	6	4

8.14 Would the answer to Prob. 8.13 change if the lives of both machines were 6 years?

8.15 A family needs a new roof on their home. They have estimates from roofer A and roofer B. Roofer A wants $2400 for shingle roofing, material, and labor. If the tin on the eaves and valleys and old boards are replaced, an extra $300 cost is incurred. The roof itself will reasonably last 15 years if $20 per year for the first 5 years and $5 per year more each year after year 5 (gradient of $5) is spent on preventive maintenance. The replacement of the tin and board will increase the life to 18 years. Roofer B will put a gravel roof on the house for $2200. It is estimated that the same annual maintenance expenditure as with shingles will give the roof an expected life of 12 years. Use a rate-of-return analysis to compare (a) shingles without tin and board replacement with gravel, and (b) shingles with the replacements and gravel. Assume a MARR of 16% per year.

8.16 What is the basic difference between a problem that requires selection from several mutually exclusive alternatives and one that requires selection from independent projects?

8.17 What criteria are used to select a mutually exclusive alternative by the rate-of-return method?

8.18 Five different methods can be used for recovering by-product heavy metals from a waste stream. The investment costs and incomes associated with each method are shown below. Assuming all methods have a 10-year life with zero salvage value and the company's MARR is 15% per year, determine which one should be selected by (a) the EUAW method and (b) the incremental-rate-of-return method.

	Method				
	1	2	3	4	5
First cost	$15,000	$18,000	$25,000	$35,000	$52,000
Salvage value	+1,000	+2,000	−500	−700	+4,000
Annual income	5,000	6,000	7,000	9,000	12,000

8.19 If method 2 in Prob. 8.18 has a life of 5 years and method 3 has a life of 15 years, which alternative should be selected?

8.20 Select the best alternative using the incremental-rate-of-return method from the proposals shown below if the MARR is 14% per year and the projects will have a useful life of 15 years. Assume that the cost of the land will be recovered when the project is terminated.

			Proposal			
1	2	3	4	5	6	7
Land cost	$ 50,000	$ 40,000	$ 70,000	$ 80,000	$ 90,000	$ 65,000	$ 75,000
Construction cost	200,000	150,000	170,000	185,000	165,000	175,000	190,000
Annual maintenance	15,000	16,000	14,000	17,000	18,000	13,000	12,000
Annual income	52,000	49,000	68,000	50,000	81,000	77,000	45,000

8.21 Any one of five machines can be used in a certain phase of a canning operation. The costs of the machines are shown below and all are expected to have a 10-year life. If the company's minimum attractive rate of return is 18% per year, determine which machine should be selected using (*a*) the incremental-rate-of-return method, and (*b*) the present-worth method.

			Machine		
	1	2	3	4	5
First cost	$28,000	$33,000	$22,000	$51,000	$46,000
Annual operating cost	20,000	18,000	25,000	12,000	14,000

8.22 An oil and gas company is considering five sizes of pipe for a new pipeline. The costs for each size are shown below. Assuming all pipes will last 15 years and the company's MARR is 8% per year, determine which size pipe should be used according to (*a*) the present-worth method and (*b*) the incremental-rate-of-return method.

			Pipe size, millimeters		
	140	160	200	240	300
Initial investment	$9,180	$10,510	$13,180	$15,850	$30,530
Installment cost	600	800	1,400	1,500	2,000
Annual operating cost	6,000	5,800	5,200	4,900	4,800

8.23 An independent dirt contractor is trying to determine which size dump truck to buy. The contractor knows that as the bed size increases, the net income increases, but it is uncertain whether the incremental expenditure required in the larger trucks could be justified. The cash flows associated with each size truck are shown below. If the contractor's MARR is 18% per year and all trucks are expected to have a useful life of 8 years, determine which size truck should be purchased using (*a*) the incremental-rate-of-return method and (*b*) the EUAW method.

			Truck size, square meters		
	8	10	15	20	40
Initial investment	$10,000	$12,000	$18,000	$24,000	$33,000
Annual operating cost	5,000	5,500	7,000	11,000	16,000
Salvage value	2,000	2,500	3,000	3,500	4,500
Annual income	9,000	10,000	10,500	12,500	14,500

8.24 Five processes can be used for producing a certain part. If the company's MARR is 15% per year, determine which process should be selected by (*a*) the present-worth method and (*b*) the incremental-rate-of-return method.

	Process				
	1	**2**	**3**	**4**	**5**
First cost	$15,000	$22,000	$27,000	$31,000	$42,000
Annual operating cost	6,000	5,000	4,500	3,000	2,000
Salvage value	500	1,000	1,100	600	3,000
Life, years	3	4	6	6	6

8.25 One phase of a meat-packing operation requires the use of separate machines for the following functions: pressing, slicing, weighing, and wrapping. All machines under consideration are expected to have a life of 6 years with no salvage value. There are two alternatives for each of the functions as follows:

	Alternative 1		Alternative 2	
	First cost	**Annual cost**	**First cost**	**Annual cost**
Pressing	$ 5,000	$13,000	$10,000	$11,000
Slicing	4,000	10,000	17,000	4,000
Weighing	12,000	15,000	15,000	13,000
Wrapping	3,000	9,000	11,000	7,000

(a) If the company's MARR is 20% per year, use the incremental-rate-of-return method to determine which machine should be selected for each function (identify them as pressing 1, pressing 2, slicing 1, etc.).

(b) For the machines selected in part (a), determine the total investment and operating cost for the entire operation.

8.26 A third alternative can be added in Prob. 8.25: one machine to do the pressing and slicing and another machine to do the weighing and wrapping. The machine that will do the pressing and slicing (identified as pressing-slicing 3) will cost $29,000 and will have an annual operating cost of $9000. The machine that will do the weighing and wrapping (identified as weighing-wrapping 3) will cost $26,000 and will have an annual operating cost of $18,000.

(a) Which machines should be selected for the entire operation?

(b) Determine the total investment and operating cost for the entire operation.

(c) If a fourth alternative can be considered—a single machine to perform all four functions (identified as machine 4) having an initial cost of $45,000 and an annual operating cost of $32,000—which machine(s) should be selected?

8.27 Compare the alternatives below on the basis of a rate-of-return analysis, assuming the MARR is 15% per year.

	Project A	Project B
First cost	$60,000	$90,000
Annual operating cost	15,000	8,000
Annual repair cost	5,000	2,000
Annual increase in repair cost	1,000	1,500
Salvage value	8,000	12,000
Life, years	15	15

8.28 A company is considering the projects shown below, all of which can be considered to last indefinitely. If the company's MARR is 14% per year, determine which should be selected (a) if they are independent, and (b) if they are mutually exclusive.

	A	B	C	D	E
First cost	$10,000	$20,000	$15,000	$70,000	$50,000
Annual income	2,000	4,000	2,900	10,000	6,000
Rate of return	20%	20%	19.3%	14.3%	12%

8.29 Alternative *I* requires an initial investment of $20,000 and will yield a rate of return of 35% per year. Alternative *C*, which requires a $30,000 investment, will yield 25% per year. Which of the following statements is true about the rate of return on the $10,000 increment?
(a) There is no such thing as a rate of return on an increment of investment.
(b) It is greater than 35% per year.
(c) It is exactly 35% per year.
(d) It is between 25 and 35% per year.
(e) It is exactly 25% per year.
(f) It is less than 25% per year.
(g) It is infinity.

8.30 If alternative *C* in Prob. 8.29 yields a rate of return of 35% per year, then which of the statements listed in the problem would be true?

8.31 The four alternatives described below are being evaluated.

Alternative	Initial investment	Overall rate of return for alternative	Incremental rate of return when compared with alternative		
			A	B	C
A	$ 40,000	29%			
B	75,000	15%	1%		
C	100,000	16%	7%	20%	
D	200,000	14%	10%	13%	12%

(a) If the proposals are independent, which should be selected if the MARR is 15% per year?
(b) If the proposals are mutually exclusive, which one should be selected if the MARR is 13% per year?
(c) If the proposals are mutually exclusive, which one should be selected if the MARR is 10% per year?

8.32 A rate-of-return analysis was begun for the alternatives shown below. Fill in the blanks in the incremental-rate-of-return portion of the table.

Alternative	Initial investment	Overall ROR for alternative	Incremental ROR when compared with alternative		
			E	F	G
E	$20,000	20%	—	27%	
F	35,000	23%	27%		
G	50,000	16%			
H	90,000	19%			

LEVEL THREE

In this level you will learn how to perform both a benefit/cost (B/C) analysis for a single project and an incremental B/C analysis to select between two or more alternatives.

The replacement or retention for one or more years in the future of currently-owned assets is examined, and the procedure to determine the number of years to retain an asset so that it has a minimum EUAW is presented.

The understanding of bonds and the calculation of their expected rate of return is discussed. Inflationary effects on present-worth and future-worth computations are covered in association with basic cost indexing and cost-estimation techniques for components and entire systems.

Chapter	Subject
9	Benefit/cost ratio evaluation
10	Replacement analysis
11	Bonds
12	Inflation and cost estimation

Benefit/Cost Ratio Evaluation

9

The objective of this chapter is to teach you how to compare two alternatives on the basis of a benefit/cost (B/C) ratio. This method is sometimes regarded as *supplementary*, since it is used in conjunction with a present-worth, future-worth, or annual-worth analysis. It is, nevertheless, an analytical technique which must be understood because many government projects are analyzed using the benefit/cost ratio method.

Section Objectives

After completing this chapter, you should be able to do the following:

9.1 State the definition used to classify specified expenditures or savings as benefits, costs, or disbenefits.

9.2 Determine whether a single project should be undertaken by comparing its benefits and costs, given values for the benefits, disbenefits, and costs, and the interest rate.

9.3 Select the better of two alternatives on the basis of a benefit/cost analysis, given the initial cost, life, salvage value, and disbursements for each alternative and the required rate of return.

9.4 State the procedure for selecting the best alternative from three or more independent or mutually exclusive projects using a benefit/cost ratio analysis.

9.5 Select one alternative from several *mutually exclusive alternatives* using the incremental B/C ratio method, given initial cost, life, salvage value, and cash flows for each alternative, and the minimum attractive rate of return.

Table 9.1 Examples of benefits, disbenefits, and costs

Item	Classification
Expenditure of $11,000 for new interstate highway	Cost
$50,000 annual income to local residents from tourists because of new reservoir and recreation area	Benefit
$150,000 per year upkeep cost for irrigation canals	Cost
$25,000 per year loss by farmers because of highway right-of-way	Disbenefit

Study Guide

9.1   Classification of Benefits, Costs, and Disbenefits

The method for selecting alternatives that is most commonly used by federal agencies for analyzing the desirability of public works projects is the benefit/cost ratio (B/C ratio). As its name suggests, the B/C method of analysis is based on the ratio of the benefits to costs associated with a particular project. A project is considered to be attractive when the benefits derived from its implementation exceed its associated costs. Therefore, the first step in a B/C analysis is to determine which of the elements are benefits and which are costs. In general, *benefits* are advantages, expressed in terms of dollars, which happen to the *owner*. On the other hand, when the project under consideration involves disadvantages to the owner, these are known as *disbenefits* (*D*). Finally, the *costs* are the anticipated expenditures for construction, operation, maintenance, etc, less any salvage values. Since B/C analysis is used in economy studies by federal, state, or city agencies, it is helpful to think of the *owner* as the *public* and the one who incurs the costs as the *government*. The determination of whether an item is to be considered as a benefit, disbenefit, or cost, therefore, depends on *who is affected* by the consequences. Some examples of each are illustrated in Table 9.1.

While the examples presented in this chapter are straightforward with regard to identification of benefits, disbenefits, or costs, it should be pointed out that in actual situations, judgments must sometimes be made which are subject to interpretation, particularly when it is necessary to determine whether an element of cash flow is a disbenefit or a cost. In other instances, it is not possible to simply place a dollar value on all benefits, disbenefits, or costs that are involved. These nonquantifiable considerations must be included in the final decision, as they are in other methods of analysis. In general, however, dollar values are available, or obtainable, and the results of a proper B/C analysis would agree with the methods studied in preceding chapters (such as present worth, equivalent uniform annual worth, or rate of return on incremental investment).

Probs. 9.1 and 9.2

9.2   Benefits, Disbenefits, and Cost Calculations of a Single Project

Before a B/C ratio can be computed, all of the benefits, disbenefits, and costs that are to be used in the calculation must be converted to common dollar units, as in present-worth or future-worth calculations, or dollars per year, as in annual-worth comparisons. Any method—present worth, future worth, or annual worth—may be used provided the procedures learned in Chaps. 5 and 6 are followed. Regardless of the method used in the B/C analysis, it is important to express both the numerator (benefits, disbenefits) and denominator (costs) in the same terms, such as present dollars or future dollars.

There are several forms of the B/C ratio. The *conventional B/C ratio*, probably the most widely used, will be applied in this text unless specified otherwise. The conventional B/C ratio is calculated as follows:

$$B/C = \frac{benefits - disbenefits}{costs} = \frac{B - D}{C} \tag{9.1}$$

A B/C ratio greater than or equal to 1.0 indicates that the project evaluated is economically advantageous. In B/C analyses, costs are *not* preceded by a minus sign.

Note that in Eq. (9.1) *disbenefits are subtracted from benefits, not added to costs.* It is important to recognize that the B/C ratio could change considerably if disbenefits are regarded as costs. For example, if the numbers 10, 8, and 8 are used to represent benefits, disbenefits, and costs, respectively, the correct procedure results in $B/C = (10 - 8)/8 = 0.25$, while the incorrect regard of disbenefits as costs yields $B/C = 10/(8 + 8) = 0.625$, which is over twice the correct B/C value. Clearly, then, the method by which disbenefits are handled affects the magnitude of the B/C ratio. However, no matter whether disbenefits are subtracted from the numerator or added to costs in the denominator, a B/C ratio of less than 1.0 by the first method, which is consistent with Eq. (9.1), will always yield a B/C ratio less than 1.0 by the latter method, and vice versa.

The *modified B/C ratio*, which is gaining support, includes operation and maintenance (O&M) costs in the numerator and treats them in a manner similar to disbenefits. The denominator, then, contains only the initial investment cost. Once all amounts are expressed in present-worth, annual-worth, or future-worth terms, the modified B/C ratio is calculated as

$$\text{Modified } B/C = \frac{benefits - disbenefits - O\&M \ costs}{initial \ investment} \tag{9.2}$$

Any salvage value is included in the denominator as in the conventional method. The modified B/C ratio will obviously yield a different value than the conventional B/C method. However, as with disbenefits, *the modified procedure can change the magnitude of the ratio, but not the decision to accept or reject.*

A benefit/cost evaluation that does not involve a ratio is based on the *difference* between benefits and costs, that is, B − C. In this case, if B − C is greater than or equal to zero, the project is acceptable. This method has the obvious advantage of eliminating the discrepancies noted above when disbenefits are regarded as costs, since B represents the *net benefits*. Thus, for the numbers 10, 8, and 8 the same result is obtained regardless of how disbenefits are treated.

Subtracting disbenefits: $\quad B - C = [(10 - 8) - 8] = -6$

Adding disbenefits to costs: $\quad B - C = [10 - (8 + 8)] = -6$

Before calculating the B/C ratio, check to be sure that the proposal with the higher EUAW is also the one that yields the higher benefits *after the benefits and costs have been expressed in common units.* Thus, a proposal having a higher initial cost may actually have a lower EUAW, present worth, or future worth when all other costs are considered. Example 9.2 illustrates this point.

Example 9.1

The Wartol Foundation, a nonprofit educational research organization, is contemplating an investment of $1.5 million in grants to develop new ways to teach people the rudiments of a profession. The grants would extend over a 10-year period and would create an estimated savings of $500,000 per year in professor salaries, student tuition, and other expenses. The foundation uses a rate of return of 6% per year on all grant investments. In this case the program would be an addition to ongoing and planned activities. An estimated $200,000 a year would thus have to be released from other programs to support the educational research. To make this program successful, a $50,000 per year operating expense will be incurred by the foundation from its regular O&M budget. Use the following analysis methods to determine if the program is justified over a 10-year period: (*a*) conventional B/C, (*b*) modified B/C, and (*c*) B − C analysis.

Solution The definitions using an equivalent-annual-worth basis are:

Benefit: $500,000 per year

Investment cost: $1,500,000(A/P, 6\%, 10) = \$203,805$ per year

O&M cost: $50,000 per year

Disbenefit: $200,000 per year

(*a*) Using Eq. (9.1) for conventional B/C analysis

$$B/C = \frac{500,000 - 200,000}{203,805 + 50,000} = 1.18$$

The project is justified, since B/C > 1.0.

(*b*) By Eq. (9.2) the modified B/C separates the investment and O&M costs.

$$\text{Modified B/C} = \frac{500,000 - 200,000 - 50,000}{203,805} = 1.23$$

The project is, of course, also justified by the modified method.

(*c*) B is the net benefit, and the annual O&M cost is subtracted as part of C.

$$B - C = (500,000 - 200,000) - (203,805 - 50,000) = \$46,195$$

Since B − C > 0 the investment is again justified.

Comment   In part (*a*), if the disbenefits were added to costs, the *incorrect* B/C value would be

$$B/C = \frac{500,000}{203,805 + 50,000 + 200,000} = 1.10$$

which still justifies the project. However, the disbenefit D = $200,000 is not a direct cost to this program and should be subtracted from B, not added to C.

Example 9.2

Alternative routes are being considered by the state highway department for location of a new highway. Route *A*, costing $4,000,000 to build, will provide annual benefits of $125,000 to local businesses. Route *B* will cost $6,000,000 and will provide $100,000 in benefits. The annual cost of maintenance is $200,000 for *A* and $120,000 for *B*, respectively. If the life of each road is 20 years and an interest rate of 8% per year is used, which alternative should be selected on the basis of a benefit/cost analysis?

Solution   The benefits in this example are $125,000 for route *A* and $100,000 for route *B*. The EUAW of costs for each alternative is as follows:

$$EUAW_A = 4,000,000(A/P, 8\%, 20) + 200,000 = \$607,400$$

$$EUAW_B = 6,000,000(A/P, 8\%, 20) + 120,000 = \$731,100$$

Route *B* has a *higher* EUAW than route *A* by $123,700 per year, and *less benefits* than *A*. Therefore, there would be no need to calculate the benefit/cost ratio for route *B*, since this alternative is obviously inferior to route *A*. Furthermore, if the decision had been made that either route *A* or *B* *must* be accepted (which would be the case if there were no other alternatives), then no other calculations would be necessary and route *A* would be accepted.

Earlier it was stated that in Eq. (9.1), costs are assigned a positive sign. An exception to this occurs when the project or alternative has a lower equivalent cost than is currently being expended; that is, acceptance of the project will result in a reduced cost, or a savings. In this case, the B/C denominator is negative and it is necessary to reverse the interpretation of the B/C ratio. Projects involving reduced (negative) costs should be accepted if the B/C ratio is *less than or equal to 1.0*. If this altered interpretation were not made, projects resulting in savings to a government, with no decrease in benefits to the people, would always be rejected because of the resulting negative B/C ratio. In such a case, it is also possible to expect a negative benefit value (a benefit loss) due to the cost reduction (savings). If the appropriate plus or minus signs are used on benefits and costs, the resulting B/C ratio can be correctly used to accept the project (B/C < 1.0) or reject it (B/C > 1.0). Table 9.2 illustrates acceptance and rejection of alternatives as a function of their conventional B/C ratio and the sign on the costs and benefits.

Probs. 9.3 to 9.8

Table 9.2 Interpretation of conventional B/C ratios for costs, savings, and benefits losses

Change in benefits	Change in costs	B/C ratio	Accept or reject alternative
These ratios use the interpretation B/C > 1.0 to accept			
+$100 (gain)	+$200 (cost)	+0.50	<1.0; reject
+$100 (gain)	+$50 (cost)	+2.0	>1.0; accept
These ratios use the interpretation B/C < 1.0 to accept			
+$100 (gain)	−$200 (saving)	−0.50	<1.0; accept
+$100 (gain)	−$50 (saving)	−2.0	<1.0; accept
−$100 (loss)	−$200 (saving)	+0.50	<1.0; accept
−$100 (loss)	−$50 (saving)	+2.0	>1.0; reject

9.3 Alternative Comparison by Benefit/Cost Analysis

In computing the benefit/cost ratio by Eq. (9.1) for a given alternative, it is important to recognize that the benefits and costs used in the calculation represent the *increments* or *differences* between two alternatives. This will always be the case, since sometimes doing nothing is an acceptable alternative. Thus, when it seems as though only one proposal is involved in the calculation, such as whether or not a flood-control dam should be built to reduce flood damage, it should be recognized that the construction proposal is being compared against another alternative—the do-nothing alternative. Although this is also true for the other alternative evaluation techniques previously presented, it is emphasized here because of the difficulty often present in determining the benefits and costs between two alternatives when only costs are involved. See Example 9.3 for an illustration.

Once the B/C ratio for the differences is computed, a B/C $\geqslant$ 1.0 means that the extra benefits of the higher-cost alternative justify this higher cost. If B/C < 1.0, the extra cost is not justified and the lower-cost alternative is selected. Note that this lower-cost project may be the do-nothing alternative, if the B/C analysis is for only one project.

Example 9.3

Two routes are under consideration for a new interstate highway. The northerly route N would be located about 5 miles from the central business district and would require longer travel distances by local commuter traffic. The southerly route S would pass directly through the downtown area and, although its construction cost would be higher, it would reduce the travel time and distance for local commuters. Assume that the costs for the two routes are as follows:

	Route *N*	Route *S*
Initial cost	$10,000,000	$15,000,000
Maintenance cost per year	35,000	55,000
Road-user cost per year	450,000	200,000

If the roads are assumed to last 30 years with no salvage value, which route should be accepted on the basis of a benefit/cost analysis using an interest rate of 5% per year?

Solution   Since most of the cash flows are already annualized, the B/C ratio will be expressed in terms of equivalent uniform annual worth. The *costs* to be used in the B/C ratio are the initial cost and maintenance cost:

$$EUAW_N = 10,000,000(A/P, 5\%, 30) + 35,000 = \$685,500$$

$$EUAW_S = 15,000,000(A/P, 5\%, 30) + 55,000 = \$1,030,750$$

The *benefits* in this example are represented by the road-user costs, since these are consequences "to the public." The benefits, however, are not the road-user costs themselves but the *difference* in road-user costs if one alternative is selected over the other. In this example, there is a $450,000 - $200,000 = $250,000 per year benefit if route S is chosen instead of route N. Therefore, the benefit B of route S over route N is $250,000 per year. On the other hand, the costs C associated with these benefits are represented by the difference between the annual costs of routes N and S. Thus,

$$C = EUAW_S - EUAW_N = \$345,250 \text{ per year}$$

The route that costs more (route S) is the one that provides the benefits. Hence, the B/C ratio can now be computed by Eq. (9.1).

$$B/C = \frac{250,000}{345,250} = 0.724$$

The B/C ratio of less than 1.0 indicates that the extra benefits associated with route S are less than the extra costs associated with this route. Therefore, route N would be selected for construction. Note that there is no do-nothing alternative in this case, since one of the roads *must* be constructed.

Comment   If there had been disbenefits associated with each route, the difference between the disbenefits would have to be added or subtracted from the net benefits ($250,000) for route S, depending on whether the disbenefits for route S were less than or greater than the disbenefits for route N. That is, if the disbenefits for route S were less than those for route N, the difference between the two would be added to the $250,000 benefit for route S, since the disbenefits involved would also favor route S. However, if the disbenefits for route S were greater than those for route N, their difference should be subtracted from the benefits associated with route S, since the disbenefits involved would favor route N instead of route S. Example 9.5 in the Additional Examples illustrates this calculation.

You should rework this example using the modified B/C method as per Eq. (9.2); the annual maintenance costs are taken out of the EUAW expressions used in the conventional B/C ratios.

Additional Example 9.5
Probs. 9.9 to 9.14

9.4   Benefit/Cost Analysis for Multiple Alternatives

When only one alternative must be selected from three or more mutually exclusive (stand-alone) alternatives, a multiple alternative evaluation is required. In this case, it is necessary to conduct an analysis on the *incremental* benefits and costs similar to the procedure used in Chap. 8 for incremental rates of return. The do-nothing alternative may be one of the considerations.

There are two situations which must be considered in multiple-alternative analysis by the benefit/cost method, as was also true for the other evaluation methods. That is, if *more than one* alternative can be chosen from among several, it is necessary only to compare the alternatives against the do-nothing alternative, because the alternatives are *independent*. For example, if several flood-control dams could be constructed on a particular river and adequate funding is available for all dams, the B/C ratios to be considered should be those associated with a particular dam versus no dam. That is, the result of the calculations could show that three dams along the river would be economically justifiable on the basis of reduced flood damage, recreation, etc., and should be constructed.

On the other hand, when only *one* alternative can be selected from among several, it is necessary to compare the alternatives against each other as well as against the do-nothing alternative. The procedure is the same as that discussed in Chap. 8 for rate of return except that a B/C ratio rather than a rate-of-return equation is written from the net cash-flow column. It is important for you to understand the difference between the procedure to be followed when multiple projects are mutually exclusive and when they are not. In the case of mutually exclusive projects, it is necessary to compare them against each other; in the case of projects that are not mutually exclusive (independent projects), it is necessary to compare them only against the do-nothing alternative.

Prob. 9.15

9.5 Selection from Mutually Exclusive Alternatives Using Incremental Benefit/Cost Ratio Analysis

In order to use the B/C ratio as an evaluation technique for mutually exclusive alternatives, an incremental B/C ratio must be computed in a fashion similar to that used for the incremental rate of return (Chap. 8). The project that has the incremental (conventional) B/C $\geq$ 1.0 and requires the largest *justified* investment is selected. The procedure to be followed is similar to that used for rate-of-return analysis; however, in a B/C analysis it is generally convenient, although not necessary, to compute an overall B/C ratio for each alternative, since the total present-worth or EUAW values must be computed anyway in preparation for the incremental analysis. Those alternatives that have an overall B/C $<$ 1.0 can be eliminated immediately and need not be considered in the incremental analysis. Example 9.4 presents a complete application of the incremental B/C ratio to mutually exclusive alternatives.

Example 9.4

Using the four mutually exclusive alternatives of Example 8.7 (Table 8.6), apply incremental B/C ratio analysis to select the best alternative (MARR = 10% per year).

Solution The alternatives are first ranked from smallest to largest initial investment cost. The next step is to calculate the overall B/C ratio and eliminate those alternatives that have B/C $<$ 1.0. As shown in Table 9.3, location *C* can be eliminated on the basis of its overall

Table 9.3 Incremental B/C ratio analysis for mutually exclusive alternatives, Example 9.4

	C	A	B	D
Building cost	$ − 190,000	$ − 200,000	$ − 275,000	$ − 350,000
Cash flow (CF)	19,500	22,000	35,000	42,000
Present worth of CF	$ 183,826	$ 207,394	$ 329,945	$ 395,934
Overall B/C	0.97	1.03	1.20	1.13
Projects compared	· · ·	· · ·	B to A	D to B
Incremental benefit	· · ·	· · ·	$ 122,551	$ 65,989
Incremental cost	· · ·	· · ·	75,000	75,000
Incremental B/C	· · ·	· · ·	1.64	0.88
Project selected	· · ·	· · ·	B	B

B/C ratio (0.97). All other alternatives are acceptable and must be compared on an incremental basis. The incremental benefits and costs can be determined as follows:

Incremental benefits: increase in present worth between alternatives
Incremental cost: increase in building cost between alternatives

A summary of the incremental B/C analysis is presented in Table 9.3 (bottom half). Using the acceptable alternative that has the lowest investment cost as the defender (A) and the next-lowest acceptable alternative as the challenger (B), the incremental B/C ratio is 1.64, indicating that location B should be selected over location A (therefore eliminating A from further consideration). Using B as the defender and D as the challenger, the incremental analysis yields incremental B/C = 0.88, favoring location B. Since location B has an incremental ratio greater than 1.0 *and* is the largest justified investment, it is selected. This, of course, is the same conclusion reached with the incremental-rate-of-return method in Table 8.7.

Comment   Note that alternative selection should not be made on the basis of the overall B/C ratio, even though location B would still be selected, *coincidentally in this case*. The incremental investment must also be justified in order to select the best alternative.

Although a present-worth ratio was used in this example, an EUAW or future worth ratio could also have been used to compare the investments; in fact, an EUAW ratio is generally simpler if lives are unequal.

Additional Example 9.6
Probs. 9.16 to 9.25

Additional Examples

Example 9.5

Assume the same situation as in Example 9.3 for the routing of a new interstate highway. The B/C analysis showed that the northerly route *N* was to be constructed. However, this route will go through an agricultural region and the local farmers have complained about the great loss in revenue they and the economy will suffer. Likewise, the downtown merchants have complained about the southerly route because of the loss in revenue due to reduced merchandising ability, parking problems, etc. To consider these eventualities, the state highway department has undertaken a study and predicted that the loss to state agriculture for route N will be about $500,000 per year and that route S will cause an estimated reduction in retail sales and rents of $400,000 per year. What effect does this new information have on the B/C analysis?

Solution These new "costs" should be considered as disbenefits. Since the disbenefits of route S are $100,000 less than those of route N, this difference is *added* to the benefits of route S to give a total net benefit of $250,000 + $100,000 = $350,000 to the downtown alternative. Now we have

$$B/C = \frac{350,000}{345,250} = 1.01$$

and route S is to be slightly favored. In this case the inclusion of disbenefits has reversed the earlier decision.

Example 9.6

The U.S. Army Corps of Engineers still wants to construct a dam on the Sacochsi River as in Example 8.10. The construction and average annual dollar benefits (income) are repeated below. If a MARR of 6% per year is required and dam life is infinite for analysis purposes, select the best location using the B/C ratio method.

Site	Construction cost, millions	Annual income
A	$ 6	$350,000
B	8	420,000
C	3	125,000
D	10	400,000
E	5	350,000
F	11	700,000

Solution We make use of the capitalized-cost equation, Eq. (5.2), $A = Pi$, to obtain EUAW values for capital recovery (cost), as shown in the first row of Table 9.4. Since site E is justified and has the largest investment, it is selected. (None of the increments above E was justified.)

Comment Suppose that site G is added with a construction cost of $10 million and an annual benefit of $700,000. What site should G be compared with? What is the ΔB/C

Table 9.4 Use of incremental B/C ratio analysis for Example 9.6

	C	E	A	B	D	F
Capital recovery ($1000)	$180	$300	$360	$480	$600	$660
Annual benefits ($1000)	125	350	350	420	400	700
Comparison	C to none	E to none	A to E	B to E	D to E	F to E
ΔCapital recovery	$180	$300	$ 60	$180	$300	$360
ΔAnnual benefits	125	350	0	70	50	350
ΔB/C ratio	0.70	1.17	0	0.39	0.17	0.97
Site selected	None	E	E	E	E	E

ratio? If you determine that a comparison of G to E is to be made and $\Delta B/C = 1.17$ in favor of G, you are correct! Now site F must be incrementally evaluated with G, but since the annual benefits are the same ($700,000), the $\Delta B/C$ ratio is zero and the added investment is not justified. Therefore, site G is chosen.

9.1 Why should disbenefits be subtracted from benefits rather than added to costs?   **Problems**

9.2 Classify the following cash flows as either benefits, costs, or disbenefits:
  (a) Drive-in theater had to be destroyed because of highway right-of-way
  (b) $10 million expenditure for new highway
  (c) Less disbursement by motorists because of new highway
  (d) Archaeological sites inundated by new reservoir
  (e) $2 million paid for right-of-way for new highway

9.3 The U.S. Army Corps of Engineers is considering the feasibility of constructing a small flood-control dam in an existing arroyo. The initial cost of the project will be $2.2 million, with inspection and upkeep costs of $10,000 per year. In addition, minor reconstruction will be required every 15 years at a cost of $65,000. If flood damage will be reduced from the present cost of $90,000 per year to $10,000 annually, use the benefit/cost method to determine if the dam should be constructed. Assume that the dam will be permanent and the interest rate is 12% per year.

9.4 A state highway department is considering the construction of a new highway through a scenic rural area. The road is expected to cost $6 million, with annual upkeep estimated at $20,000 per year. The improved accessibility is expected to result in additional income from tourists of $350,000 per year. If the road is expected to have a useful life of 25 years, use the (a) $B - C$ method and (b) B/C method at an interest rate of 6% per year to determine if the road should be constructed. (c) What is the modified B/C ratio value?

9.5 If the highway in Prob. 9.4 would result in agricultural-income losses of $15,000 the first year, $16,000 the second, and amounts increasing by $1000 per year, by how much would the tourist income have to increase each year (starting in year 2) in order for the highway to become economically feasible?

9.6 The U.S. Bureau of Reclamation is considering a project to extend irrigation canals into a desert area. The initial cost of the project is expected to be $1.5 million, with annual maintenance costs of $25,000 per year. (a) If agricultural revenue is expected to be $175,000 per year, make a B/C analysis to determine whether the project should be undertaken, using a 20-year study period and an interest rate of 6% per year. (b) Rework the problem using the modified B/C ratio.

9.7 Calculate the B/C ratio for Prob. 9.6 if the canal must be dredged every 3 years at a cost of $60,000 and there is a $15,000-per-year disbenefit associated with the project.

9.8 From the following data, determine the B/C ratio for a project which has an infinite life. Use an interest rate of 12% per year.

Consequences to the people	Consequences to the government
Annual benefits: $90,000 per year	First cost: $750,000
Annual disbenefits: $10,000 per year	Annual cost: $50,000 per year
	Annual savings: $15,000 per year

9.9 Two routes are under consideration for a new interstate highway. The long route would be 22 miles in length and would have an initial cost of $21 million. The transmountain route would be 10 miles long and would have an initial cost of $45 million. Maintenance costs are estimated at $40,000 per year for the long route and $65,000 per year for the transmountain route. Regardless of which route is selected, the volume of traffic is expected to be 400,000 vehicles per year. If the vehicle operating expense is assumed to be $0.12 per mile, determine which route should be selected by (*a*) conventional B/C analysis and (*b*) modified B/C analysis. Assume a 20-year life for each road and an interest rate of 6% per year.

9.10 The U.S. Army Corps of Engineers is considering three sites for flood-control dams (designated as sites *A*, *B*, and *C*). The construction costs are $10 million, $12 million, and $20 million, and maintenance costs are expected to be $15,000, $20,000, and $23,000, respectively, for sites *A*, *B*, and *C*. In addition, a $75,000 expenditure will be required every 10 years at each site. The present cost of flood damage is $2 million per year. If only the dam at site *A* is constructed, the flood damage will be reduced to $1.6 million per year. If only the dam at site *B* is constructed, the flood damage will be reduced to $1.2 million per year. Similarly, if the site *C* dam is built, the damage will be reduced to $0.77 million per year. Since the dams would be built on different branches of a large river, either one or all of the dams could be constructed and the decrease in flood damages would be additive. If the interest rate is 5% per year, determine which ones, if any, should be built on the basis of their B/C ratios. Assume that the dams will be permanent.

9.11 Highway department officials are considering the economics of either resurfacing an existing highway or constructing a new one. The existing highway is 12 miles long and would cost $2 million to resurface. Annual upkeep cost is expected to be $5000 the first year, $10,000 the second, and amounts increasing by $5000 per year until year 10, at which time the road would have to be resurfaced again. If a new road is constructed, the initial cost would be $15 million for a road 10 miles long. The maintenance is expected to cost $5000 the first year, $7000 the second, and amounts increasing by $2000 per year until year 10, after which the cost will be $23,000 per year. If the new road is constructed, the cost of auto accidents is expected to decrease by $500,000 per year. If vehicle operating cost is assumed to be $0.10 per mile and 600,000 vehicles per year travel the road, use the benefit/cost method to determine which road should be constructed at an interest rate of 6% per year.

9.12 The U.S. Forest Service is considering two locations for a new national park. Location *E* would require an investment of $3 million and $50,000 per year in maintenance. Location *W* would cost $7 million to construct, but the forest service would receive an additional $25,000 per year in park-use fees. The operating cost of location *W* will be $65,000 per year. The revenue to park concessionaires will be $500,000 per year at location *E* and $700,000 per year at *W*. The disbenefits associated with each location are $30,000 per year for location *E* and $40,000 per year for location *W*. Use (*a*) the B/C method and (*b*) the modified B/C method to determine which location, if either, should be selected, using an interest rate of 12% per year. Assume that the park will be maintained indefinitely.

9.13 The Bureau of Reclamation is considering the lining of the main canals of its irrigation ditches. The initial cost of lining is expected to be $4 million, with $25,000 per year required for maintenance. If the canals are not lined, a weed-control and dredging operation will have to be instituted, which will have an initial cost of $700,000 and a cost of $50,000 the first year, $52,000 the second, and amounts increasing by $2000 per year for 25 years. If the canals are lined, less water will be lost through infiltration so that additional land can be cultivated for agricultural use. The agricultural revenue associated with the extra land is expected to be $120,000 per year. Use (a) the B/C and (b) the B − C method to determine if the canals should be lined. Assume that the project life is 25 years and the interest rate is 6% per year.

9.14 A city trying to attract professional athletic teams is considering constructing a domed playing arena or a conventional stadium. The domed arena would cost $300 million to construct and would have a useful life of 50 years. The maintenance and operation would be $300,000 the first year, with costs increasing by $10,000 per year. Every 10 years, an expenditure of $800,000 would be required for "remodeling" the interior. The conventional stadium would cost only $50 million to construct and would also have a useful life of 50 years. The cost of maintenance would be $75,000 the first year, increasing by $8000 per year. Periodic costs for repainting, resurfacing, etc., would be $100,000 every 4 years. Revenue from the domed arena is expected to be greater than that from the conventional stadium by $500,000 the first year, with amounts increasing by $200,000 per year through year 15. Thereafter, the extra revenue from the dome would remain the same at $3.3 million per year. Assuming that both structures would have a salvage value of $5 million, use an interest rate of 8% per year and a B/C analysis to determine which structure should be built.

9.15 Why must an incremental B/C analysis be conducted when only one proposal can be selected from three or more mutually exclusive proposals?

9.16 Which dam in Prob. 9.10 should be built if the dams are mutually exclusive (i.e., only one can be constructed)?

9.17 Five methods could be used to recover grease from a rendering plant wastewater stream. The investment costs and incomes associated with each one are shown below. Assuming that all methods have a 10-year life with zero salvage value, determine which one should be selected using a minimum attractive rate of return of 15% per year and the B/C analysis method. Consider operating costs as an O&M cost in the modified B/C method.

	Method				
	1	2	3	4	5
First cost, $	15,000	19,000	25,000	33,000	48,000
Annual operating cost, $	10,000	12,000	9,000	11,000	13,000
Annual income, $	15,000	20,000	19,000	22,000	27,000

9.18 Which alternatives in Prob. 9.17 would be selected if they were not mutually exclusive? Use a conventional B/C analysis.

9.19 Select the best mutually exclusive alternative using the B/C ratio method from the proposals shown below if the MARR is 10% per year and the projects will have a useful life of 15 years. Assume that the cost of the land will be recovered when the project is terminated. Treat maintenance costs as disbenefits.

	Proposal						
	1	2	3	4	5	6	7
Land cost, $	50,000	40,000	70,000	80,000	90,000	65,000	75,000
Construction cost, $	200,000	150,000	170,000	185,000	165,000	175,000	190,000
Annual maintenance, $	15,000	16,000	14,000	17,000	18,000	13,000	12,000
Annual income, $	52,000	49,000	68,000	50,000	81,000	77,000	45,000

9.20 An oil and gas company is considering five sizes of pipe for a new pipeline. The costs for each size are shown below. Assuming that all pipes will last 15 years and the company's MARR is 8% per year, which size of pipe should be used according to the conventional B/C method?

	Pipe size, millimeters				
	140	**160**	**200**	**240**	**300**
Initial investment, $	9,180	10,510	13,180	15,850	30,530
Installation cost, $	600	800	1,400	1,500	2,000
Annual operating cost, $	6,000	5,800	5,200	4,900	4,800

9.21 Which pipe would be selected in Prob. 9.20 if the lives of the 240- and 300-millimeter pipes could be extended to 30 years by installing corrosion protection equipment costing $1200 to purchase and $300 per year to maintain? Use the conventional B/C method of analysis.

9.22 Determine which alternative below should be selected on the basis of a modified B/C analysis. Use an interest rate of 8% per year and assume a project life of 20 years.

	Alternative X	Alternative Y
First cost	$6,000,000	$8,000,000
Annual maintenance cost	40,000	18,000
Annual road-user cost	300,000	170,000

9.23 The bridges shown below are under consideration by a governmental agency. Which would you recommend on the basis of a modified B/C analysis using an interest rate of 10% per year?

	Alternative X	Alternative Y
Initial cost	$400,000,000	$600,000,000
Annual maintenance cost	300,000	200,000
Annual benefits	600,000	500,000
Annual disbenefits	220,000	53,000
Life, years	20	20

9.24 The two highway projects shown below are to be compared using the modified B/C method. Which one, if either, should be built? Use $i = 10\%$ per year.

	2L	3L
First cost	$5,000,000	$8,000,000
Annual maintenance cost	70,000	95,000
Annual benefits	160,000	150,000
Annual disbenefits	30,000	35,000
Life, years	∞	∞

9.25 A state highway department is considering two types of surface coatings for a new road. An armor-coat surface will cost only $800,000 to install, but because of its relatively rough surface, the road users will have to spend more money for gasoline, tire wear, and automobile upkeep. The annual costs for these items are estimated to be $196,000. Additionally, disbenefits of $29,000 per year have been identified for this alternative.

A smooth asphalt coating is an alternative also under consideration. This surface would have an initial cost of $2 million but the annual road user costs will be only $87,000. The asphalt surface will have no disbenefits associated with it. If the life of either surface is expected to be 5 years, determine which should be selected on the basis of a benefit/cost analysis, using an interest rate of 9% per year compounded annually.

10 Replacement Analysis

The result of an alternative evaluation process is the selection and implementation of a project, asset, or service which has a planned economic life. As time passes, it is necessary to determine how the selected and in-place alternative will be replaced. This replacement analysis may be necessary before, at, or after the expected economic life. The basic results of a replacement analysis are answers to the following questions: Has the economic life of this asset or project been reached? Which alternative should be accepted as its replacement?

Whether unplanned or anticipated, replacement is commonly considered because of several reasons. Some are:

Reduced performance—Due to the physical deterioration of parts, the ability to perform at an expected level of reliability (being available and performing correctly when needed) and productivity (performing at a given level of quality and quantity) is not present. This usually results in increased costs of operation, higher scrap and rework costs, lost sales, and larger maintenance expenses.

Altered requirements—New requirements of accuracy, speed, or other specifications have been placed upon the owners. These requirements can not be met by the existing equipment or system. Often the analysis is between complete replacement and enhancement through retrofitting which may result in the new requirements being met.

Obsolescence—The rapidly changing technology of automation, computers, and communications makes currently used systems and assets perform acceptably, but less accurately or productively than equipment coming onto the market. Replacement due to obsolescence is usually possible, but a formal analysis is performed when management determines that competitive forces or newly offered

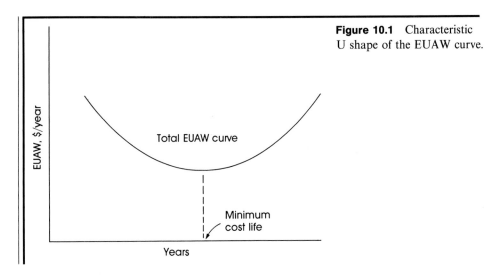

Figure 10.1 Characteristic U shape of the EUAW curve.

equipment may force the company out of markets due to increased require- ments from consumers or contractors. The decrease in time of the development cycle of new products is the cause for much of the replacement analyses per- formed prior to the completion of the expected economic life.

The logic and computations of replacement analysis are discussed in this chap- ter. All evaluations here are before-tax considerations. Since there are often income tax consequences which should be evaluated, after-tax replacement analyses will be presented in Chap. 15.

With each passing year, the following observations usually apply to an asset or project that is currently in place:

- Yearly cost of operation and maintenance increases
- Realizable market or salvage value decreases
- Ownership cost due to the initial investment in terms of equivalent uniform annual worth (EUAW) decreases

These cost factors usually cause the total EUAW to decrease for some years and then increase. The years at which the annualized cost will be a minimum may be found by computing and plotting the EUAW versus time. Figure 10.1 shows the usual U or convex shape of the EUAW plot. This process, called a *minimum-cost-life analy- sis*, is discussed in this chapter.

Section Objectives

After completing this chapter you should be able to do the following:

10.1 Describe the concepts of *consultant's viewpoint* and *sunk cost* for replacement analysis; state the value of the first cost and the other attributes for defender and challenger alternatives; and determine any sunk cost present.

10.2 Select the better of a defender and a challenger plan over a specified planning horizon.

10.3 State the difference between the *conventional* and *cash-flow approaches* to replacement analysis, and use both to perform an analysis.

10.4 Use the *one-additional-year replacement analysis* procedure to determine if a defender should be retained for one more year or replaced, given the data for defender and challenger plans.

10.5 Determine the *minimum-cost life* of an asset using an EUAW analysis, given the first cost or market value, estimated salvage values, operating costs, and required return.

Study Guide

10.1 The Defender and Challenger Concepts in Replacement Analysis

Here, as in previous chapters, we can compare two or more alternatives; however, we now own one of the assets, referred to as the *defender*, and are considering its replacement by one or more *challengers*.

   In the comparison we take the consultant's viewpoint. For evaluation purposes, we assume that we own neither asset. In order to "purchase" the defender, therefore, we must pay the going market value for this used asset. We then use the estimated market value as the first cost P of the defender. There will be an associated salvage value SV, an economic life n, and annual operating cost AOC for the defender. Even though the values may all differ from the original data, it makes no difference for the evaluation because we are using the consultant's viewpoint and are thus making all previous data irrelevant to the present evaluation.

   In replacement analysis this irrelevant data is sometimes used to calculate what is known as a *sunk cost*. The sunk cost of an asset is computed as

Sunk cost = present book value − present realizable value   (10.1)

The present book value is the remaining investment after the total amount of depreciation has been charged; that is, the book value is the current worth of the asset as established by acceptable accounting procedures. (See Chap. 13 for a complete discussion of depreciation.)

   If incorrect estimates have been made about the utility or market value of an asset (as is possible, since no one can be perfect in their estimation of the future), there is a positive sunk cost, *which cannot be recovered*. A sunk cost is a result of a bad decision that was made *at some time in the past*, and past economic decisions must not be allowed to influence decisions of the present. However, some analysts try to "recover" the sunk cost of the defender by erroneously adding it to the first cost of the challenger. This penalizes the challenger, making its costs seem higher than they really are, and thereby jeopardizing the validity of the conclusion. The sunk cost, rather, should be charged to an account entitled "Unrecovered Capital," or the like, which will ultimately be reflected in the company's income statement for the year in which the sunk cost was incurred. Therefore, *for replacement analysis a sunk cost should not be included in the economic comparison*. The following example illustrates the correct data to use in a replacement analysis.

Example 10.1

A dump truck was purchased 3 years ago for $12,000 with an estimated life of 8 years, salvage of $1600, and annual operating cost of $3000. The current book value is $8100.

A challenger is now offered for $11,000 and a trade-in value of $7500 for the old truck. The company estimates challenger life at 10 years, salvage at $2000, and annual operating costs at $1800 per year. New estimates for the old truck are made as follows: remaining life, 3 years; realizable salvage, $2000; same operating costs.

What values should be used for P, n, SV, and AOC for each asset?

Solution When using the consultant's viewpoint, only the most current information is applicable.

Defender	Challenger
$P = \$7{,}500$	$P = \$11{,}000$
AOC = 3,000	AOC = 1,800
SV = 2,000	SV = 2,000
$n = 3$ years	$n = 10$ years

The original cost of $12,000, estimated salvage of $1600, and remaining 5 years of life are not used for the defender; only the current data applies.

A sunk cost is incurred for the defender if it is replaced. By Eq. (10.1),

Sunk cost = 8100 − 7500 = $600

The $600 is not added to the first cost of the challenger, since this action would (1) try to "cover up" past mistakes of estimation, and (2) penalize the challenger since the capital to be recovered each year would be erroneously larger due to the increased first cost.

Probs. 10.1 to 10.5

10.2   Replacement Analysis Using a Specified Planning Horizon

The planning horizon or study period is the number of years used in the economic analysis to compare the defender and challenger. Typically, one of two situations is present: (1) The anticipated remaining life of the defender equals the life of the challenger, or (2) the life of the challenger is greater than that of the defender. We will discuss both possibilities in order.

If the defender and challenger have equal lives, any of the evaluation methods previously discussed can be used with the *most current data*. Example 10.2 compares owning with leasing a van.

Example 10.2

Moore Transfer owns two vans which are deteriorating faster than expected. Owning the vans or leasing on a yearly basis are the replacement options. The two vans were purchased 2 years ago for $60,000 each. The company plans to keep the vans for 10 more years. Fair market value for a 2-year-old van is $42,000 and for a 12-year-old van, $8000. Annual fuel, maintenance, tax, etc., costs are $12,000 per year. Lease cost is $9000 per year (year-end payment) with annual operating charges of $14,000. Should the company lease its vans if a 12% per year rate of return is required?

Solution Consider a 10-year life for the currently owned van (defender) and the leased van (challenger).

Defender	Challenger
$P = \$42,000$	Lease cost = $ 9,000 per year
AOC = 12,000	AOC = 14,000
SV = 8,000	
$n = 10$ years	$n = 10$ years

The EUAW_D calculation for the defender uses the $42,000 as an initial investment.

$$\text{EUAW}_D = P(A/P, i\%, n) - \text{SV}(A/F, i\%, n) + \text{AOC}$$
$$= 42,000(A/P, 12\%, 10) - 8000(A/F, 12\%, 10) + 12,000$$
$$= \$18,977$$

The EUAW_C for the challenger is

$$\text{EUAW}_C = 9000 + 14,000 = \$23,000$$

Clearly, the firm should retain ownership of the two vans.

Comment Recalculate the EUAW of the challenger, assuming the lease payments are beginning-of-year payments. You should get $\text{EUAW}_C = \$23,900$.

In many instances, an asset is to be replaced by another having an estimated life different from that of the defender's remaining life. For analysis, the length of the planning horizon must first be selected, frequently coinciding with the life of the longer-lived asset. The assumption is usually made that the EUAW of the shorter-lived asset is the same throughout the planning horizon. (This assumption will be made in this book unless stated otherwise.) *This implies that the service performed by the shorter-lived asset can be acquired at the same EUAW as presently computed for its expected service life.* For example, evaluating a challenger with a 10-year life and a defender with a 4-year life assumes that the service provided by the defender alternative can be acquired for the same EUAW for the 6 years after the remaining 4 years of defender life. If this assumption is not reasonable, the estimated cost of acquiring the equivalent service beyond year 4 should be included in the defender's cash flow and amortized over its 10-year life.

Example 10.3

A company has owned a particular machine for 3 years. Based on current market value the asset has an EUAW of $5210 per year and an anticipated remaining life of 5 years due to rapid technological growth. The possible replacement for the asset has a first cost of $25,000, salvage value of $3800, life of 12 years, and an annual operating cost of $720 per year. If the company uses a minimum rate of return of 10% per year on asset investments and plans to retain the new machine for its full anticipated life, should the old asset be replaced?

Solution A planning horizon of 12 years to correspond with the challenger's life is selected.

$$EUAW_D = \$5210$$

$$EUAW_C = 25,000(A/P, 10\%, 12) - 3800(A/F, 10\%, 12) + 720$$
$$= \$4211$$

Purchase of the new asset is less costly than is retention of the presently owned machine.

Comment If the planning horizon for different-lived assets is used and a present-worth value is desired, you must realize that a horizon assumes that you plan to purchase a similar asset if the shorter-lived asset is accepted and its EUAW will be the same as during the first life cycle. In other words, in this problem a defender-similar asset would be purchased at the end of 5 and 10 years. Present-worth computations would be:

$$PW_D = 5210(P/A, 10\%, 12) = \$35,500$$

$$PW_C = 4211(P/A, 10\%, 12) = \$28,693$$

Of course, the decision to purchase the new machine is still made.

International competition and the rapid obsolescence of current technology are constant concerns in many industries, especially high-technology. Skepticism about the certainty of the future is often reflected in management's desire to impose abbreviated planning horizons upon economic evaluations. *This often forces the recovery of invested capital and the required return over a shorter period of time than the expected lives of the alternatives.* In such cases, the *n* values in all computations reflect this shortened horizon. The following example explains the consequences of abbreviated planning horizons.

Example 10.4

Consider the data of Example 10.3, except use a 5-year planning horizon. Management specifies 5 years because it is leery of the technological progress being made in this area, progress that has already called into question retention of presently owned, operational equipment. Assume that the challenger's salvage value will remain at $3800.

Solution The approach is as in the preceding example, except a capital-recovery period of only 5 years is used for the challenger.

$$EUAW_D = \$5210$$

$$EUAW_C = 25,000(A/P, 10\%, 5) - 3800(A/F, 10\%, 5) + 720$$
$$= \$6693$$

Now, retention of the defender is less costly, thus reversing the decision made with a 12-year horizon.

Comment By not allowing the full anticipated life of the challenger to be used, management has ruled out its use. However, the decision to not consider the use of this new asset past 5 years is one of management responsibility. The reason why the decision is reversed in this example is quite simple. The challenger is given only 5 years to recover the same investment and a 10%-per-year return, whereas in the previous example 12 years is allowed. Reasonably, $EUAW_C$ must increase. It would be possible to recognize unused value in the challenger by increasing the salvage value from $3800 to the estimated fair market value after 5 years of service, if such a value can be predicted.

Selection of the planning horizon is a difficult decision, one which must be based on sound judgment and data. The use of a short horizon may often bias the economic decision in that the capital-recovery period for the challenger may be abbreviated to much less than the anticipated life. This is the case in Example 10.4, where only 5 years were allowed for recovery of invested capital plus a 10%-per-year return. However, use of a long horizon is also often detrimental due to the uncertainty of the future and its estimate. In this case, the direction of bias is less certain than in the case of a too-short horizon. A common practice is augmentation; that is, the defender is augmented with a newly purchased asset to make it comparable in ability (speed, volume, etc.) with the challenger. Since the analysis is similar to that covered here, a sample solution is included in the Additional Examples.

Additional Example 10.8
Probs. 10.6 to 10.15

10.3 Conventional and Cash-Flow Approach to Replacement Analysis

There are two equally correct and equivalent ways to handle the first cost of alternatives in a replacement analysis. The *conventional approach* (used in Example 10.1) uses the defender's current market value as the first cost of the defender and uses the initial cost of the replacement as the challenger's first cost. Thus, for the defender, P is the highest value attainable through its disposal (sale, trade-in, scrap, etc.) when it is compared with a given challenger. This approach is cumbersome when there is more than one challenger with each offering a different trade-in value for the defender, thus possibly causing a different P value for the defender when compared with each challenger.

The second approach recognizes the fact that when a challenger is selected, the defender's market is a cash inflow to the challenger alternative; and if the defender is selected, there is no actual outlay of cash. This is the *cash-flow approach* to replacement analysis. In this approach, if the defender and challenger have the same life value, set the defender first cost to zero and *subtract* the trade-in value from the challenger first cost. It is important to recognize that this approach can only be used when the lives of the defender and challenger are the same or when the comparison is to be made over a specified or preselected planning horizon. Example 10.5 shows these calculations.

Example 10.5

A 7-year-old asset may be replaced with either of two new assets. Current data for each alternative are given below. Use the cash-flow approach and a MARR = 18% per year to determine the most economical decision.

	Current asset, defender	Possible replacements	
		Challenger 1	Challenger 2
First cost	...	$10,000	$18,000
Defender trade-in	...	3,500	2,500
Annual cost	$3,000	1,500	1,200
Salvage value	500	1,000	500
Life estimate, years	5	5	5

Solution The first-cost value used for the defender (D) in the replacement analysis is different for challenger 1 (C_1) and challenger 2 (C_2). Using the cash-flow approach subtract the trade-in values from the respective challenger's first cost and compute the EUAW over either the respective life of each alternative or the planning horizon, whichever is shorter. The common life of 5 years is the evaluation period for all analyses. With this approach, the defender first cost is zero because the asset is already owned.

Defender: $EUAW_D = 3000 - 500(A/F, 18\%, 5) = \$2930.11$

Challenger 1: $EUAW_{C_1} = (10,000 - 3500)(A/P, 18\%, 5) + 1500$
$$- 1000(A/F, 18\%, 5) = \$3438.79$$

Challenger 2: $EUAW_{C_2} = (18,000 - 2500)(A/P, 18\%, 5) + 1200$
$$- 500(A/F, 18\%, 5) = \$6086.70$$

Since the defender has the smallest EUAW, it should be retained.

Comment Had the conventional approach been used, two analyses would be performed: D versus C_1, with the defender first cost of $3500; and D versus C_2, with the defender first cost of $2500. Results (that you should verify) are as follows:

D versus C_1	D versus C_2
$EUAW_D = \$4049.34$	$EUAW_D = \$3729.56$
$EUAW_{C_1} = \$4558.02$	$EUAW_{C_2} = \$6886.15$

As expected the decision is to retain the defender because it offers the smallest equivalent annual worth.

It is possible to use the conventional approach to determine the replacement value (RV) of the defender that will make its retention or replacement equally attractive. The relation $EUAW_D = EUAW_C$ is used wherein the unknown value RV is substituted for the defender first cost. This is the breakeven market value for the defender that would have to be exceeded to make the challenger more attractive. As an exercise, find RV for the defender in Example 10.2. Your answer should be

$69,250. Since the market value of the vans was estimated at $42,000, the defender should be selected, which is the same conclusion previously reached.

Probs. 10.16 to 10.24

10.4 Replacement Analysis for One-Additional-Year Retention

When a currently owned asset is close to the end of its useful life or has demonstrated deteriorating usefulness to a company, a frequently asked question is whether it should be replaced with a challenger or retained in service for one or more years. Thus, there are essentially three alternatives: Select the challenger, retain the defender for one more year, or retain the defender for its remaining life. In such a case it is not correct to compare only the defender cost and challenger cost over their remaining anticipated lives. Rather, the equivalent-annual worth procedure presented in Fig. 10.2 may be used to first calculate $EUAW_C$ and $C_D(t)$ for $t = 1$, where

$EUAW_C$ = challenger EUAW value

$\quad C_D(1)$ = defender cost for the next year $t = 1$

If $C_D(1) \leqslant EUAW_C$, retain the defender one more year because its cost is less. If $C_D(1) > EUAW_C$, the challenger cost must also be less than the defender $EUAW_D$ for its remaining life. The Fig. 10.2 procedure states that if $EUAW_D \leqslant EUAW_C$, the defender is still retained the next year. In either case it is possible to continue the analysis for future years $t = 2, 3, \ldots$, one year at a time until the challenger is selected or the defender life is reached. Note that only when the challenger cost is less than the next-year defender cost *and* the defender $EUAW_D$, is the challenger selected.

Example 10.6

The U1-Likit Sales Company commonly retains a salesperson's car for 5 years. Due to the purchase of autos exactly 2 years ago that have deteriorated much more rapidly than expected, management has asked you if it is more economical to retain and maintain the fleet for one more year and then replace the fleet, retain it for two more years then replace, or keep it for three more years. Or is it cheaper to replace the fleet this year with new, more reliable autos? Perform the replacement analysis at $i = 20\%$ for a typical defender car and challenger car having the estimated costs detailed here.

	Currently owned (defender)		Possible replacement (challenger)	
	Value at beginning of year	Annual operating cost		
Next year (3)	$3,800	$4,500	First cost	$8,700
Next year (4)	2,800	5,000	Operating cost	3,900 per year
Last year (5)	500	5,500	Life, years	5
Remaining life, years	3		Salvage value	$1,800
Salvage after three more years	$500			

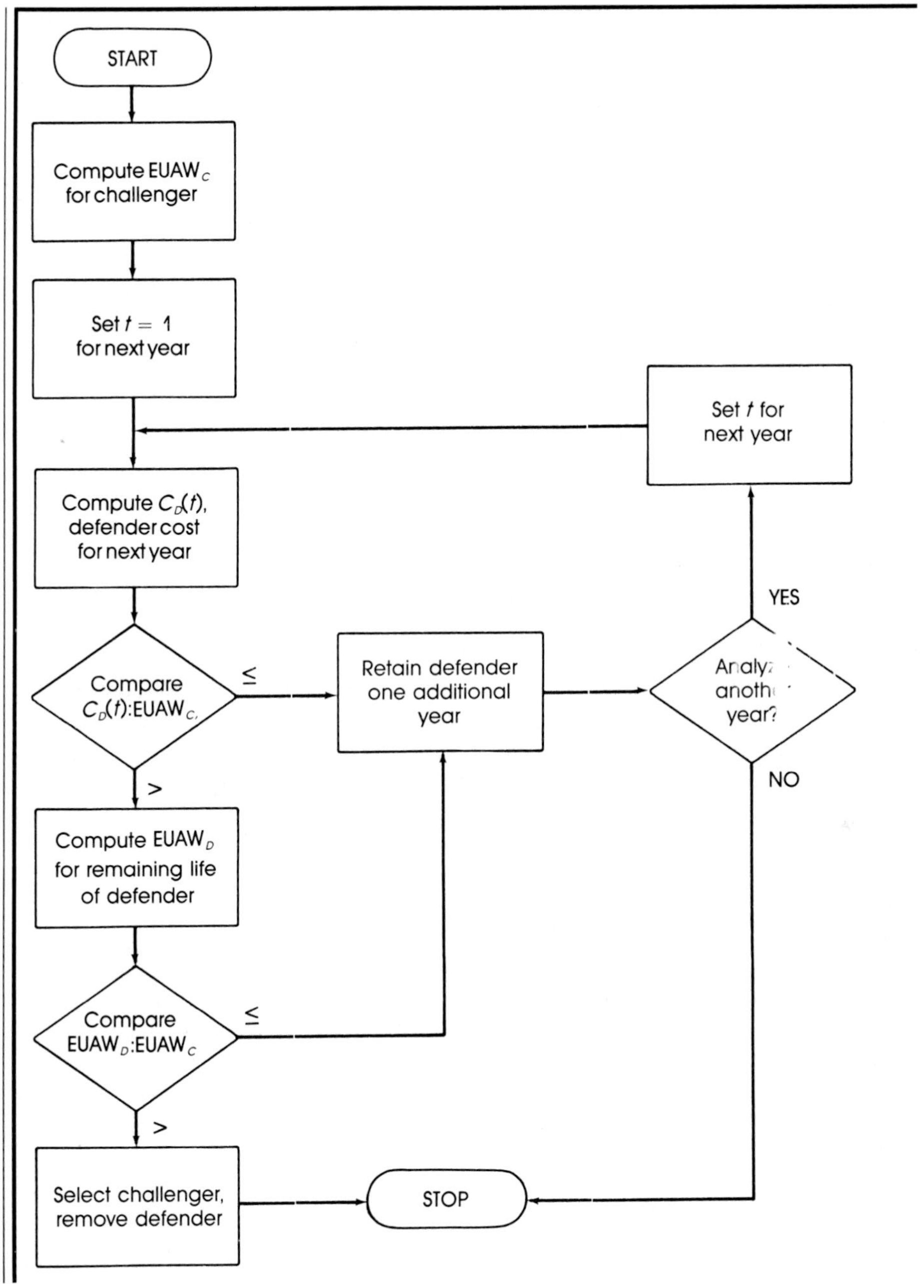

Figure 10.2 Procedure for one-additional-year replacement analysis.

Solution Following the procedure in Fig. 10.2, compute the EUAW for the challenger over 5 years and the defender's cost for next year only ($t = 1$).

$$\text{EUAW}_C = 8700(A/P, 20\%, 5) - 1800(A/F, 20\%, 5) + 3900 = \$6567$$

$$C_D(1) = 3800(A/P, 20\%, 1) - 2800(A/F, 20\%, 1) + 4500 = \$6260.$$

Since $C_D(1) < \text{EUAW}_C$, retain the defending auto for the next year. Note that in $C_D(1)$ the salvage value of $2800 for next year is the expected P value (initial cost) for the year after, that is, for $C_D(2)$. This assumes, of course, that the cost estimates remain the same.

After the next year is over, to determine if the auto should be kept yet another year, follow Fig. 10.2 and set $t = 2$. Then the cost for the defender is

$$C_D(2) = 2800(A/P, 20\%, 1) - 500(A/F, 20\%, 1) + 5000 = \$7860$$

Now $C_D(2) > \text{EUAW}_C = \$6567$, so we compute the EUAW_D value for the remaining 2 years of the defender's life.

$$\begin{aligned} \text{EUAW}_D &= 2800(A/P, 20\%, 2) - 500(A/F, 20\%, 2) + 5000 + 500(A/G, 20\%, 2) \\ &= \$6833 \end{aligned}$$

Since the challenger is also cheaper for the remaining 2 years, select it and replace the defender after one more year of service. Had $\text{EUAW}_C > \text{EUAW}_D$ for 2 years the defender would be retained and a similar analysis done for the last year using $C_D(3)$ and EUAW_D for 1 year.

Comment   If only the EUAW_D value for 3 years were used in the replacement analysis, the wrong decision would be made because the 3-year EUAW_D slightly exceeds the 5-year EUAW_C.

$$\text{EUAW}_D = 3800(A/P, 20\%, 3) - 500(A/F, 20\%, 3) + 4500 + 500(A/G, 20\%, 3) = \$6606$$

$$\text{EUAW}_C = \$6567$$

Here the challenger is purchased immediately, whereas the one-additional-year analysis has shown it is more economical to keep the defender 1 year and then replace it.

Probs. 10.25 to 10.29

10.5 Minimum-Cost Life Analysis

Often an analyst wants to know the amount of time that an asset or project should be kept in service to minimize its total cost with the time value of money and return requirements considered. This time in years is an n value and is referred to by several names including minimum cost life, economic life, retirement life, and replacement life. Up to this point, the life of the assets has been provided with no consideration as to how they were determined. In this section, the determination of asset life is discussed.

Regardless of what it is called, the n value is the number of years which yields a minimum annual cost. The approach to estimate n in the *minimum-cost life analysis* uses the conventional EUAW computations of Chapter 6. To find the minimum-cost life, increase the life value index k from 1 to the largest expected life N, that is, $k = 1, 2, \ldots, N$. For each k determine the value of EUAW_k (or EUAC_k) using

$$\text{EUAW}_k = P(A/P, i\%, k) - \text{SV}_k(A/F, i\%, k) + \left[\sum_{j=1}^{k} \text{AOC}_j(P/F, i\%, j) \right](A/P, i\%, k)$$

$$(10.2)$$

where SV_k = salvage value if the asset is retained k years

   AOC_j = annual operating cost for year j ($j = 1, 2, \ldots, k$)

Table 10.1 Computation of minimum-cost life for a presently owned asset

(1) Life, k years	(2) SV_k	(3) AOC_j ($j = 1, 2, \ldots, k$)	(4) Capital recovery and return	(5) Equivalent operating costs	(6) = (4) + (5) $EUAW_k$
1	$9,000	$2,500	$5,300	$2,500	$7,800
2	8,000	2,700	3,681	2,595	6,276
3	6,000	3,000	3,415	2,717	6,132
4	2,000	3,500	3,670	2,886	6,556
5	0	4,500	3,429	3,150	6,579

The minimum-cost life is the k value for which $EUAW_k$ is the smallest. The corresponding life value $k = n$ and $EUAW_k$ should be used in replacement and alternative evaluation analyses. This approach is illustrated in Example 10.7.

Example 10.7

An asset purchased 3 years ago is now challenged by a new piece of equipment. The market value of the defender is $13,000. Anticipated salvage values and annual operating costs for the next 5 years are given in cols. 2 and 3, respectively, of Table 10.1. What is the minimum-cost life to be used when comparing this defender with a challenger if a 10%-per-year return is required?

Solution Equation (10.2) is used to determine $EUAW_k$ for $k = 1, 2, \ldots, 5$. Column 4 in Table 10.1 gives the capital recovery and return using the first two terms of Eq. (10.2) and col. 5 gives the equivalent operating costs for k years using the last term in the $EUAW_k$ equation. The sum is $EUAW_k$ shown in col. 6. As an example, the computations for $k = 3$ may be determined as follows.

$$EUAW_3 = 13,000(A/P, 10\%, 3) - 6000(A/F, 10\%, 3) + [2500(P/F, 10\%, 1)$$
$$+ 2700(P/F, 10\%, 2) + 3000(P/F, 10\%, 3)](A/P, 10\%, 3)$$
$$= \$6132$$

The minimum cost in Table 10.1 is $6132 per year for $k = 3$, which indicates that 3 years should be the anticipated remaining life of this asset when compared with a challenger. A plot of the results in col. 6 of Table 10.1 will appear much like the U-shaped curve in Fig. 10.1. If several of the $EUAW_k$ values are approximately equal, the curve will be flat on the bottom, which indicates that the cost is relatively insensitive over several candidate life values.

Comment You should realize that the method presented in this example is general. It can be utilized whether the minimum-cost life is to be found for an anticipated purchase or a presently owned asset which may be retained or replaced.

Additional Example 10.9
Probs. 10.30 to 10.35

Additional Examples

Example 10.8

Three years ago the City of Megapolis purchased a new fire truck. Due to expanded growth in a certain portion of the city, new fire-fighting capacity is needed. An additional identical truck can be purchased now or a double-capacity truck can replace the presently owned asset. Data for each asset are presented in Table 10.2. Compare the assets at $i = 12\%$ per year using (a) a 12-year study period and (b) a 9-year period, which the city management believes to be more realistic due to population growth.

Solution Plan A is the retention of the previously owned truck and *augmentation* with the new identical-capacity vehicle; plan B is purchase of the double-capacity truck. Details of each plan are below.

Plan A		Plan B
Presently owned	**Augmentation**	**Double capacity**
$P = \$18,000$	$P = \$58,000$	$P = \$72,000$
AOC = 1,500	AOC = 1,500	AOC = 2,500
SV = 5,100	SV = 6,960	SV = 7,200
$n = 9$ years	$n = 12$ years	$n = 12$ years

(a) For a full-life 12-year horizon,

$$EUAW_A = (\text{EUAW of presently owned}) + (\text{EUAW of augmentation})$$
$$= [18,000(A/P, 12\%, 9) - 5100(A/F, 12\%, 9) + 1500]$$
$$\quad + [58,000(A/P, 12\%, 12) - 6960(A/F, 12\%, 12) + 1500]$$
$$= 4533 + 10,575$$
$$= \$15,108$$

$$EUAW_B = 72,000(A/P, 12\%, 12) - 7200(A/F, 12\%, 12) + 2500$$
$$= \$13,825$$

Purchase the double-capacity truck (plan B) with an advantage of $1283 per year.

(b) The analysis for a truncated 9-year horizon is identical, except that $n = 9$ in each factor; that is, 3 fewer years are given to the augmentation and double-capacity truck to recover the investment plus a 12%-per-year return. Assuming the salvage values remain the same,

$$EUAW_A = \$16,447 \qquad EUAW_B = \$15,526$$

and plan B is again selected but now only by a margin of $921.

If the planning horizon were truncated more severely, at some point the decision would be reversed.

Table 10.2 Data for fire-truck replacement analysis

	Presently owned	New purchase	Double capacity
P	$51,000	$58,000	$72,000
AOC	1,500	1,500	2,500
Trade-in	18,000	. . .	. . .
SV	10% of P	12% of P	10% of P
n	12	12	12

Example 10.9

Assume that an asset can be purchased for $5000 and will have a negligible salvage value. The annual operating costs are expected to follow a gradient of $200 per year with a base amount of $300 in year 1. Find the number of years that the asset should be kept if interest is not considered important.

Solution The main effect of $i = 0\%$ will be to decrease the total annual-cost values and make computations simpler. Cost patterns will be similar to those for $i > 0\%$, except that the minimum-cost life may change. Table 10.3 presents the solution to the problem. Column 3 gives the cumulative AOC according to the gradient, while the average AOC is given in col. 4 $[= (3)/k]$. A sample computation is given below the table, which indicates that $k = 7$ is the minimum-cost life with a total annual cost of $1614. This will be the n value in an economic analysis.

Comment Due to the regularity of this type of problem and the fact that $i = 0\%$, a formula can be derived to find the minimum-cost life directly. Keep in mind that $i = 0\%$ here: We can write

Total annual cost (TAC) = average operating cost + average first cost

Substituting n for k years, TAC for each n value may be expressed as

$$\text{TAC}_n = \frac{\sum\limits_{j=1}^{n} \text{AOC}_j}{n} + \frac{P}{n}$$

where TAC_n = total annual cost for n years of ownership
AOC_j = annual operating cost through year j ($j = 1, 2, \ldots, n$)

For gradients, we can make the substitution

$$\frac{\sum\limits_{j=1}^{n} \text{AOC}_j}{n} = B + \left(\frac{n-1}{2}\right)G$$

where B = base amount of the gradient
G = amount of gradient

Table 10.3 Computation of minimum-cost life for $i = 0\%$

(1)	(2)	(3)	(4)	(5)	(6)
		Operating cost		Average first cost	Total annual cost
Year, k	Annual	Cumulative	Average		
1	$ 300	$ 300	$ 300	$5,000	$5,300
2	500	800	400	2,500	2,900
3	700	1,500	500	1,667	2,167
4	900	2,400	600	1,250	1,850
5	1,100	3,500	700	1,000	1,700
6	1,300	4,800	800	833	1,633
7	1,500	6,300	900	714	1,614*
8	1,700	8,000	1,000	625	1,625
9	1,900	9,900	1,100	555	1,655

* Total annual cost is computed as (6) = (4) + (5) = (3)/k + 5,000/k. For $k = 7$: 6,300/7 + 5,000/7 = $1,614.

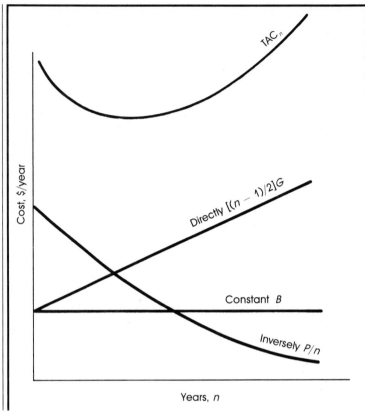

Figure 10.3 Total-annual-cost curve (TAC) for $i = 0\%$ used to solve for minimum-cost life directly.

Then

$$\text{TAC}_n = B + \left(\frac{n-1}{2}\right)G + \frac{P}{n} \qquad (10.3)$$

The general shape of the terms and TAC_n itself is shown in Fig. 10.3. If we take the derivative of Eq. (10.3) and solve for an optimum life value n^*, we have

$$\frac{d\text{TAC}_n}{dn} = \frac{G}{2} - \frac{P}{n^2} = 0$$

$$n^* = \left(\frac{2P}{G}\right)^{1/2}$$

Substitution of $P = \$5000$ and $G = \$200$ for this example yields

$$n^* = \left(\frac{10{,}000}{200}\right)^{1/2} = 7.07 \text{ years}$$

which is, for all practical purposes, the same as $n^* = 7$ obtained in Table 10.3.

10.1 If the difference between present book value and realizable salvage value is negative, how would you treat this sum in a replacement analysis?

10.2 Why is the original asset cost irrelevant in a replacement analysis? What does this fact have to do with the consultant's viewpoint in replacement analysis?

10.3 A new meat display counter was purchased by a supermarket 4 years ago at a cost of $28,000. Book value is presently $10,000 with 5 years remaining before a salvage value of $1000 is reached. Due to lagging sales, the owners wish to trade for a new, smaller counter, which costs $13,000 and has an installation charge of $500. As estimated by the owners, the old counter will last another 8 years and has a trade-in value of $18,000 now. A review of the accounts shows annual repair costs averaging $150 for the old counter. (a) Determine the values of P, n, SV, and AOC for the existing counter to be used in a replacement analysis. (b) Is there a sunk cost involved here? If so, what is its amount?

10.4 The owners of a downtown shoe shop are considering the possibility of moving to a rented shop in a suburban shopping center. They purchased their shop 15 years ago for $8000 cash. They estimate their annual investment in property improvement at $500 and believe the shop should have a current book value that includes the purchase price and these improvements at 8% per year interest. The annual insurance, utility, etc., costs have averaged $1080 per year. If they stay downtown, they hope to retire in 10 years and will give the shop to their son-in-law. They will ask $25,000 for the shop if they sell at this time. If the owners move to the shopping center, they must sign a 10-year lease for $6600 per year with no additional yearly charges. They must pay a $750 deposit when they sign the lease, but this amount is returned at the time of lease expiration. Determine (a) the values of P, n, SV, and AOC for the two alternatives and (b) the amount of the sunk cost if one exists.

10.5 An asset purchased 2 years ago can now be traded for an "improved" version for 40% of the first cost. The asset was purchased for $18,000 and is being depreciated over 5 years for tax purposes with a current book value of $10,125 and 12 years for company income purposes with a current book value of $12,500. Compute the sunk cost (a) to be reported in the tax reports and (b) to be used by company economists.

10.6 City Bus Lines has 20 buses purchased 5 years ago for $22,000 each. The company president plans to have a major overhaul done on these buses next year at a cost of $1800 each. However, the vice president wants to trade these 20 buses in on 25 of a new, smaller model. The trade-in value is $4000 and the new models cost $22,500 each. The president estimates a remaining life of 7 years for the old buses once the overhaul is completed and further states that annual operational costs per bus are $3000, and a $800 salvage is reasonable when sold to an individual as a "vacation van." The vice president interjects the comment that the smaller buses can maneuver in traffic more easily, will cost $1000 per year less to operate, will last for 8 years, and will have a salvage value of $500 when sold to day camps. With this knowledge, you are requested to determine which officer's desire is economically correct at the firm's MARR of 10% per year.

10.7 Rework Prob. 10.6 assuming the new buses have a life of 12 years, which happens to correspond with the planning horizon selected by the vice president.

10.8 Rework Prob. 10.6 using a planning horizon of 5 years.

10.9 Perform a replacement analysis for Prob. 10.3 if the new counter has an expected life of 10 years, annual cost of $30, and a salvage value of $1500. Use an interest rate of 15% per year.

10.10 Determine which alternative is better in Prob. 10.4 at a 10% per year interest rate.

10.11 A new earth mover was purchased by the AAA Cement Company 3 years ago and has been used to transport raw material from the quarry to the crushers. When purchased, the mover possessed the following characteristics: $P = \$55,000$, $n = 10$ years, SV = \$5000, and capacity = 180,000 metric tons per year. With increased construction in industrial parks around the city, an additional mover with a capacity of 240,000 metric tons per year is needed. Such a vehicle can be purchased. If bought, this new asset will have $P = \$70,000$, $n = 10$ years, SV = \$8,000.

However, the company could have constructed a conveyor system to move the material from the quarry. This system will cost \$115,000, have a life of 15 years, no salvage value, and carry 400,000 metric tons per year. The company will need to have some way to move material to the conveyor in the quarry. If the presently owned mover is used, it will more than suffice. However, a new smaller-capacity mover can be purchased. A \$15,000 trade-in on the old mover will be given on any new mover. This one will have $P = \$40,000$, $n = 12$ years, SV = \$3,500, capacity = 400,000 metric tons per year over this short distance. Monthly operating, maintenance, and insurance costs average \$0.01 per ton-kilometer for the movers; similar costs for the conveyor are expected to be \$0.0075 per metric ton. The company wants to make 12% per year on this investment. Records show that the mover must travel an average of 2.4 kilometers from the quarry to the crusher pad. The conveyor will be placed to reduce this distance to 0.32 kilometer. Should the old mover be augmented by a new mover or should the conveyor be considered as a replacement; and if so, which method of moving the material in the quarry should be used?

10.12 Solve Prob. 10.11 under this condition: Only a 4-year planning period is possible because management feels that this spurt in business is very short-lived.

10.13 Machine A, purchased 2 years ago, is wearing out more rapidly than expected. It has a remaining life of 2 years, annual operating costs of \$3000, and no salvage value. To continue the function of this asset, machine B can be purchased and a trade-in value of \$9000 will be allowed for machine A. Machine B has $P = \$25,000$, $n = 12$ years, AOC = \$4000, and SV = \$1000. As an alternative, machine C can be bought to replace A. No trade-in will be allowed for A, but it can be sold for \$7000. This new asset will have $P = \$38,000$, $n = 20$ years, AOC = \$2500, and SV = \$1000. If plan I is the retention of A, plan II is the purchase of B, and plan III is the selling of A and the purchase of C, use a 20-year period and a MARR = 8% to determine which plan is best.

10.14 The new president of Angstrom Technologies feels the company must use the newest and finest equipment in its labs. He has recommended that a 2-year-old piece of precision measurement equipment be replaced immediately. Besides, he feels it can be shown that his proposed equipment is economically advantageous at a 15%-per-year return and a planning horizon of 5 years. (*a*) Perform the replacement analysis for a 5-year period.

	Current	Proposed
Original purchase price	\$30,000	\$40,000
Current market value	15,000	. . .
Estimated useful life, years	5	15
Estimated value, 5 years	\$7,000	\$10,000
Salvage after 15 years	. . .	5,000
Annual operating cost	5,000	3,000

(*b*) Is the decision the same as in (*a*) if a 15-year horizon is used for the replacement analysis? What is the inherent assumption of this analysis for the defender which has only another 5 years of usefulness?

10.15 Resolve Prob. 10.14(*a*) using a 5-year planning horizon so that the challenger is justifiable. To do this assume that the president is able to negotiate the purchase price of the proposed equipment. How much of a "good deal" does he have to negotiate?

10.16 Explain the difference between the conventional and cash-flow approach to replacement analysis computations. Use the cash-flow approach to select the current or proposed equipment in Prob. 10.14(*a*).

10.17 Work Prob. 10.6 using the cash-flow approach to replacement analysis.

10.18 Work Prob. 10.11 using the cash-flow approach to replacement analysis and a planning horizon of 10 years. Assume that all the values in Prob. 10.11 are correct for the 10-year lives.

10.19 Rework Prob. 10.13 using the cash-flow approach to replacement analysis. Assume that machine *A* can be made to last for a total of another 12 years with a $20,000 rework in 2 years. Also let *C* have $n = 12$ years and SV = $1000. Use a 12-year planning horizon.

10.20 Dynamic Computers owns an asset (#101) used in disk-drive construction. This asset has had high annual maintenance costs and may be replaced with one of two new, improved versions. Model A-1 can be installed for a total cost of $155,000 with expected characteristics of $n = 5$ years, AOC = $10,000 and SV = $17,500. Model B-2 has a first cost of $100,000 with $n = 5$ years, AOC = $13,000, and SV = $7000. If the presently owned asset is traded it will bring $31,000 from the A-1 manufacturer and $28,000 from the B-2 producer. Retention of asset #101 has been estimated to be possible for 5 more years at an AOC of $34,000 and a negative salvage value of $2000 after the 5 years. Use the cash-flow approach to determine which is the most economical decision at a required return of 16% per year.

10.21 What is the replacement value of the old display counter described in Prob. 10.3. if the new counter has $n = 10$, AOC = $30, SV = $1500, and $i = 15\%$?

10.22 A construction company bought a 180,000-metric-ton-per-year-capacity earth mover 3 years ago at a cost of $55,000; the expected life at the time of purchase was 10 years with a $5000 salvage value and an annual operating cost of $2700. A 480,000-metric-ton-per-year replacement mover is under consideration. This mover will cost $40,000, have a life of 12 years, a salvage value of $3500, and an annual operating cost of $7200. Compute the required trade-in value of the presently owned mover if the replacement mover is bought and $i = 12\%$.

10.23 (*a*) Solve Prob. 10.22 using a planning horizon of 4 years. (*b*) How does this truncation of the horizon affect the replacement value of the presently owned mover?

10.24 An asset presently owned can last for 6 more years with costs of $24,000 this year and increasing by 10% per year. A desirable challenger would cost $70,000, last for 6 years, and have an annual cost of $12,000 and a salvage value of $4000. What is a trade-in value of the old asset that will make replacement economical if a 5% per year return is desired?

10.25 In Example 10.3 a currently owned machine has an $EUAW_D = \$5210$ for another 5 years of service. The challenger, which has an $EUAW_C = \$4211$ for a 12-year life, is selected because $EUAW_C < EUAW_D$. Management wants to keep the defender another year before replacement. Make a suggestion to management if additional study indicates that the value of the defender is $3000 now with an anticipated value of $1800 one year from now. The projected operating cost for next year is $3000 and the minimum return is still 10% per year.

10.26 The Harvey Paint Company owns an air compressor that should possibly be replaced. A new model which sells for $1500 will last 7 years with annual costs estimated to be

$100 the first year and $50 higher each year and a zero salvage value. Mr. Harvey can sell the old compressor to his brother at the following prices: $400 this year (now), $300 next year, or $50 the third year. Harvey will keep the compressor for a maximum of another 2 years since operating costs are expected to increase to $175 next year and $350 the following year. Should Harvey trade now, next year, or 2 years from now if the new compressor will have the same costs in the future as estimated now? Let $i = 12\%$ per year.

10.27 Last year A. D. Morse MD, made the "gut feeling" decision to keep the aging x-ray equipment used in medical practice for another year (this year) in lieu of purchasing new, equivalent-capability equipment for $15,000.

(a) Use the following information that is now known to determine if the doctor made the correct economic decision at $i = 18\%$ per year.

Defender, last year

Trade-in value last year	$3,000
Market value this year	2,000
Operating cost last year	500

Challenger, last year

$P = \$15,000$	SV after 10 years = $1,000
$n = 10$ years	AOC = $3,000 (constant)

(b) A major price reduction from $15,000 to $8000 has just been announced for the same equipment. Since the AOC of the old equipment is rising substantially this year, Dr. Morse feels this is the year to trade up. Should the doctor trade this year or next year? The following data have been estimated:

Challenger

$P = \$8,000$	SV = $1,000	AOC = $3,000	$n = 10$ years

Defender

This year:	value = $2,000	AOC = $2,000
Next year:	value = 500	AOC = 2,500
Year after next:	value = 0	AOC = 2,500

10.28 A replacement study is to be performed on pressing equipment in an industrial laundry. The challenging asset has a computed $EUAW_C = \$42,000$ for its anticipated 10-year life. Thorough data collection on the defender has resulted in the following projected annual operating costs (AOC) and trade-in values for the next 5 years, after which the currently owned equipment would have to be replaced.

Additional years retained	AOC	Trade-in value
1	$34,000	$28,000
2	30,000	22,000
3	30,000	15,000
4	30,000	5,000
5	30,000	0

If the current equipment is kept for another 5 years, it will cost a net estimated $2000 to remove it from the plant. Perform one-additional-year replacement analysis at a 16%-per-year return to determine how many years to keep this asset before replacing it with the challenger.

10.29 You and your spouse have to make the decision to keep your present car or purchase a new one. A new car will cost $10,000, last you 7 years, have annual maintenance costs of $200 the first year increasing by $100 per year thereafter, and sell for $3000 in 7 years. If you retain the currently owned car, the expected trade-in value and annual maintenance are as follows:

Additional years retained	Annual maintenance cost	Trade-in value
1	$1,800	$2,500
2	1,500	2,000
3	1,500	1,500

You will not consider keeping the car for more than an additional 3 years, at which time you anticipate a $1000 sales price. If all other costs are considered equal for the two cars, use $i = 15\%$ to determine when to purchase a new car. (Neglect financing complications on the new car by assuming that you have just won a contest which gives you a sum of $10,000 after taxes.)

10.30 Ms. Adams just bought a used car for $5800; she paid $400 down and her uncle financed the balance at 5% per year for 3 years. The resale values for the next 6 years are $2200 after the first year, decreasing by $400 per year to year 5, after which the resale value remains at $600. Annual costs of repairs, insurance, gas, etc., are expected to be $1000 the first year, increasing by 10% each year thereafter. If money is worth 7% per year, how many years should the car be retained? Assume that the owner will pay off the entire loan with interest if she sells the car before she has owned it 3 years.

10.31 Rework Prob. 10.30 at $i = 0\%$ rate of return and find the difference between the two answers.

10.32 Machine H was purchased 5 years ago for $40,000 and had an expected life of 10 years. The past and estimated future maintenance and operating costs and salvage values are given below. At a value of $i = 10\%$, determine the number of years the asset should be kept in service before replacement.

Year	Operating cost	Maintenance cost	Salvage value
1	$1,500	$2,000	$25,000
2	1,600	2,000	25,000
3	1,700	2,000	22,000
4	1,800	2,000	22,000
5	1,900	2,000	15,000
6	2,000	2,100	5,000
7	2,100	2,700	5,000
8	2,200	3,300	0
9	2,300	3,900	0
10	2,400	4,500	0

10.33 One year ago the Bullwinkle Pool Company purchased a machine to blow concrete onto the walls of a new swimming pool, thereby greatly reducing the time necessary to construct a pool. The machine cost $800 and is expected to last another 14 years. The owner has already seen newer, improved versions. In order to compare these challengers to the defender, knowing the most economical life of the old version would be of benefit. If operating costs are $500 for the first year and are expected to increase by $100 per year, compute the most economical life for $i = 0\%$. Assume the salvage value is zero for all years.

10.34 Rework Prob. 10.33 at $i = 5\%$ per year and compare the answers.

10.35 The Country Tailors use many sewing machines in their "like grandma used to make" clothes line. The general manager wants to know the minimum-cost life for these machines. (a) Find this value at 20% per year if the first cost is $5000 per machine. (b) Plot the EUAW curve and determine if the equivalent annual worth is insensitive over a range of n values.

Year	Market value	Estimated AOC
0	$5,000	
1	3,000	$1,000
2	1,500	1,500
3	1,000	2,000
4	500	2,500
5	0	3,000
6	0	5,000

Bonds ▌11

A time-tested method of raising capital is through the issuance of an IOU. One form of IOU is a bond. Bonds usually emanate from one of the following three sources: (1) The U.S. government; (2) states and municipalities; (3) corporations. Although these entities could probably raise capital in a number of other ways, bonds are usually used when it would be difficult to borrow a large amount of money from a single source or when repayment is to be made over a long period of time. A major feature that differentiates bonds from other forms of financing is that bonds can be bought and sold in the open market by people other than the original issuer and lender. In this chapter, some of the types of bonds and their characteristics are discussed.

Section Objectives

After completing this chapter, you should be able to do the following:

11.1 Define *mortgage bond, collateral bond, equipment trust bond, debenture bond, convertible debenture, subordinated debenture, municipal bond, general obligation bond, revenue bond, junk bond,* and *bond rating.*

11.2 Calculate the *interest payable* (receivable) per period from the sale (purchase) of a bond, given the face value of the bond, the bond interest rate, and the interest payment period.

11.3 Calculate the *present worth* of a bond, given the face value, bond interest rate, interest payment period, date the bond matures, and the desired rate of return.

11.4 Calculate the *nominal and effective rates of return* that would be received from the purchase of a bond, given the face value, purchase price, bond interest rate, compounding period, and date the bond matures.

Study Guide

11.1 Bond Classifications

A *bond* is a long-term note issued by a corporation or governmental entity for the purpose of financing major projects. In essence, the borrower receives money now in return for a promise to pay later, with interest paid between the time the money was borrowed and the time it was repaid. In general, bonds may be classified as *mortgage bonds*, *debenture bonds*, and *municipal bonds*. These types of bonds can be further subdivided.

A *mortgage bond* is one which is backed by a mortgage on specified assets of the company issuing the bonds. If the company is unable to repay the bondholders at the time the bonds mature, the bondholders have the option of foreclosing on the mortgaged property. Mortgage bonds can be subdivided into first-mortgage and second-mortgage bonds. As their names imply, in the event of foreclosure by the bondholders, the first-mortgage bonds take precedence during liquidation. The first-mortgage bonds, therefore, generally provide the lowest rate of return (less risk). Second-mortgage bonds, when backed by collateral of a subsidiary corporation, are referred to as *collateral bonds*. An *equipment trust bond* is one in which the equipment purchased through the bond serves as collateral. These types of bonds are generally issued by railroads for purchasing new locomotives and cars.

Debenture bonds are not backed by any form of collateral. The reputation of the company is important for attracting investors to this type of bond. As further incentive for investors, debenture bonds sometimes carry a floating interest rate or are often *convertible* to common stock at a fixed rate as long as the bonds are outstanding. For example, a $1000 convertible debenture bond issued by the Get Rich Quick (GRQ) Company may have a conversion option to 50 shares of GRQ common stock. If the value of 50 shares of GRQ common stock exceeds the value of the bond at any time prior to bond maturity, the bondholder has the option of converting the bond to common stock. Debenture bonds generally provide the highest rate of interest because of the increased risk associated with them. *Subordinated debentures* represent debt that ranks behind other debt (senior debt) in the event of liquidation or reorganization of the company. Because they represent even riskier investments, these bonds provide a higher rate of return to investors than regular debentures.

The third general type of bonds are *municipal bonds*. Their attractiveness to investors lies in their income-tax-free status. As such, the interest rate paid by the governmental entity is usually quite low. Municipal bonds can be either *general obligation bonds* or *revenue bonds*. General obligation bonds are issued against the taxes received by the governmental entity (i.e., city, county, or state) that issued the bonds and are backed by the full taxing power of the issuer. School bonds are an example of general obligation bonds. Revenue bonds are issued against the revenue generated by the project financed, as a water treatment plant or a bridge. Taxes cannot be levied for repayment of revenue bonds.

In order to assist prospective investors, all *bonds* are rated by various companies according to the amount of risk associated with their purchase. One such rating is Standard and Poor's, which rates bonds from AAA (highest quality) to DDD (bond in default). In general, first-mortgage bonds carry the highest rating, but it is not un-

Table 11.1 Classification and characteristics of bonds

Classification	Characteristics	Type
Mortgage	Bonds backed by mortgage or specified assets	First mortgage Second mortgage Equipment trust
Debenture	No lien to creditors	Convertible Nonconvertible Subordinated Junk
Municipal	Income tax free	General obligation Revenue

common for debenture bonds of large corporations to carry an AAA rating, or ratings higher than first-mortgage bonds of smaller, less reputable companies. The term *junk bonds* refers to debenture bonds rated lower than BBB. Junk bonds are frequently issued when a corporation wants to raise enough money to purchase another company. The concepts presented in this section are summarized in Table 11.1.

Prob. 11.1

11.2 Bond Terminology and Interest

As stated in the preceding section, a bond is a long-term note issued by a corporation or governmental entity for the purpose of obtaining needed capital for financing major projects. The conditions for repayment of the money obtained by the borrower are specified at the time the bonds are issued. These conditions include the bond face value, the bond interest rate, the bond interest-payment period, and the bond maturity date.

The bond *face value*, which refers to the denomination of the bond, is usually an even denomination starting at $100, with the most common being the $1000 bond. The face value is important for two reasons:

1. It represents the lump-sum amount that will be paid to the bondholder on the bond maturity date.
2. The amount of interest I paid per period prior to the bond maturity date is determined by multiplying the face value of the bond by the bond interest rate (coupon rate) per period as follows:

$$I = \frac{\text{(face value)(bond interest rate)}}{\text{no. of payment periods per year}} = \frac{Vb}{c} \qquad (11.1)$$

Often a bond is purchased at a discount (less than face value) or a premium (greater than face value), but only the face value, not the purchase price, is used to compute bond interest I. Examples 11.1 and 11.2 illustrate the computation of bond interest.

Example 11.1

A shirt manufacturing company planning an expansion issued 4% $1000 bonds for financing the project. The bonds will mature in 20 years with interest paid semiannually. Mr. John Doe purchased one of the bonds through his stockbroker for $800. What payments is Mr. Doe entitled to receive?

Solution　In this example, the face value of the bond is $1000. Therefore, Mr. Doe will receive $1000 on the date the bond matures, 20 years from now. In addition, Mr. Doe will receive the semiannual interest the company promised to pay when the bonds were issued. The interest every 6 months will be computed using $V = \$1000$, $b = 0.04$, and $c = 2$ in Eq. (11.1):

$$I = \frac{1000(0.04)}{2} = \$20 \text{ every 6 months}$$

Example 11.2

Determine the amount of interest you would receive per period if you purchased a 6% $5000 bond which matures in 10 years with interest payable quarterly.

Solution　Since interest is payable quarterly, you would receive the interest payment every 3 months. The amount you would receive is

$$I = \frac{5000(0.06)}{4} = \$75$$

Therefore, you would receive $75 interest every 3 months in addition to the $5000 lump sum after 10 years.

A *zero-coupon bond* is one which does not pay periodic interest. That is, the bond interest rate is zero. Because of this, they often sell at discounts of more than 75% of their face value so that their yield to maturity will be sufficient to attract investors. *Stripped bonds* are simply conventional bonds whose interest payments are sold separately from the face value. The stripped bond then behaves as if it were a zero coupon bond.

Probs. 11.2 to 11.4

11.3　Bond Present-Worth Calculations

When a company or government agency offers bonds for financing major projects, investors must determine how much they are willing to pay for a bond of a given denomination. The amount they pay for the bond will determine the rate of return on the investment. Therefore, investors must determine the present worth of the bond that will yield a specified rate of return. These calculations are shown in Example 11.3.

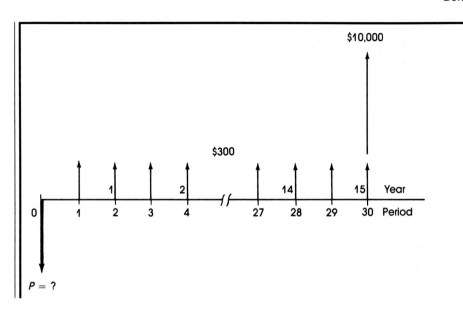

Figure 11.1 Cash flow for a bond investment, Example 11.3.

Example 11.3

Jennifer Jones wants to make a nominal 8%-per-year compounded semiannually on a bond investment. How much should she be willing to pay now for a 6% $10,000 bond that will mature in 15 years and pays interest semiannually?

Solution Since the interest is payable semiannually, Ms. Jones will receive the following payment:

$$I = \frac{10{,}000(0.06)}{2} = \$300 \text{ every 6 months}$$

The cash-flow diagram (Fig. 11.1) for this investment allows us to write a present-worth relation to compute the value of the bond now, using an interest rate of 4% per 6-month period, the same as the interest payment period of the bond. Note in the following equation that I is simply an A value.

$$P = 300(P/A, 4\%, 30) + 10{,}000(P/F, 4\%, 30) = \$8270.60$$

Thus, if Jennifer is able to buy the bond for $8270.60, she will receive a nominal 8%-per-year compounded semiannually on her investment. If she were to pay more than $8270.60 for the bond, the rate of return would be less than 8%, and vice versa.

Comment It is important to note that the interest rate used in the present-worth calculation is the interest rate per period that Ms. Jones *wants to receive*, not the bond interest rate. Since she wants to receive a nominal 8% per year compounded semiannually, the interest rate per 6-month period is 8%/2 = 4%. The bond interest rate is used *only* to determine the amount of the interest payment. If you wish to review nominal and effective interest rates, refer to Secs. 3.1 to 3.3.

When the investor's compounding frequency is either more often or less often than the interest payment frequency of the bond, it becomes necessary to use the techniques learned in Chap. 3. Example 11.4 illustrates the calculations when the investor's compounding period is less than the interest period of the bond.

Example 11.4

Calculate the present worth of a 4.5% $5000 bond with interest paid semiannually. The bond matures in 10 years, and the investor desires to make 8% per year compounded quarterly on the investment.

Solution The interest the investor would receive is

$$I = \frac{5000(0.045)}{2} = \$112.50 \text{ every 6 months}$$

The present worth of the payments shown in Fig. 11.2 can be found in either of two ways:

1. Take each interest payment ($112.50) back to year 0 separately and add to the present worth of $5000. In this case, the interest rate would be 8%/4 = 2% per quarter and the number of periods would be double those shown in Fig. 11.2, since the interest payments are made semiannually while the desired rate of return is compounded quarterly. Thus,

$$P = 112.50(P/F, 2\%, 2) + 112.50(P/F, 2\%, 4) + 112.50(P/F, 2\%, 6)$$
$$+ \cdots + 112.50(P/F, 2\%, 40) + 5000(P/F, 2\%, 40)$$
$$= \$3788$$

2. Determine the effective interest rate compounded *semiannually* (the bond interest payment period) that would be equivalent to the nominal 8% per year compounded quarterly (as stated in the problem), then use the P/A factor to compute the present worth of interest and add to the present worth of $5000. The semiannual rate is 8%/2 = 4%. Since there are two quarters per 6-month period, Table 3.3 indicates that the effective semiannual rate is $i = 4.04\%$. Alternatively, the effective semiannual rate can be computed from

Figure 11.2 Cash flow to calculate the present worth for a bond investment, Example 11.4.

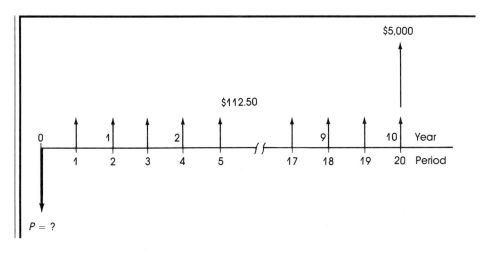

Eq. (3.3):

$$i = \left(1 + \frac{0.04}{2}\right)^2 - 1 = 0.0404$$

The present worth of the bond can now be determined with calculations similar to those in Example 11.3:

$$P = 112.50(P/A, 4.04\%, 20) + 5000(P/F, 4.04\%, 20) = \$3790$$

In summary, the steps that should be followed in calculating the present worth of a bond investment are the following:

1. Calculate the interest payment (I) per period, using the face value (V), the bond interest rate (b), and the number of interest periods (c) per year, by $I = Vb/c$.
2. Draw the cash-flow diagram of the bond receipts to include interest and face value.
3. Determine the investor's desired rate of return per period. When the bond interest period and the investor's compounding period are not the same, it is necessary to use the effective-interest-rate formula to find the proper interest rate per period.
4. Add the present worths of all future cash flows.

Additional Example 11.6
Probs. 11.5 to 11.25

11.4   Rate of Return on Bond Investment

To calculate the rate of return received on a bond investment, the procedures learned in this chapter and Chap. 7 should be followed. That is, the procedures of Secs. 11.1 and 11.2 should be used to establish the timing and the magnitude of the income associated with a bond investment; the rate of return on the investment can then be determined by setting up and solving the rate-of-return equation, Eq. (7.1). The following example illustrates the general procedure for calculating the rate of return on a bond investment.

Example 11.5

In Example 11.1 it was stated that Mr. John Doe paid $800 for a 4% $1000 bond that would mature in 20 years with interest payable semiannually. What nominal and effective interest rates per year would Mr. Doe receive on his investment for semiannual compounding?

Solution   The income Mr. Doe will receive from the bond purchase is the bond interest every 6 months plus the face value in 20 years. The equation for calculating the rate of return using the cash flow of Fig. 11.3 would be

$$0 = -800 + 20(P/A, i\%, 40) + 1000(P/F, i\%, 40)$$

Figure 11.3 Cash flow to compute the rate of return for a bond investment, Example 11.5.

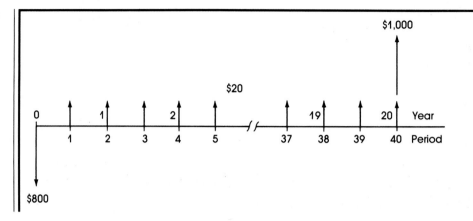

which can be solved to obtain $i = 2.87\%$ compounded semi-annually. The nominal interest rate per year is computed as interest rate per period times the number of periods, which is

Nominal $i = 2.87\%(2) = 5.74\%$ per year

From Table 3.3 or Eq. (3.3), the effective rate is 5.82% per year.

Additional Examples 11.7 and 11.8
Probs. 11.26 to 11.40

Additional Examples

Example 11.6

Ms. VIP wants to invest in some 20-year 4% $10,000 mortgage bonds with interest paid semiannually. If she requires a rate of return of 10% per year compounded semiannually and can purchase the bonds through a broker at a discount price of $8375, (a) should she make the purchase, and (b) if she does purchase the bonds, what will be her total gain in dollars?

Solution
(a) The interest each 6 months is

$$I = \frac{10,000(0.04)}{2} = \$200$$

For the nominal rate of $10\%/2 = 5\%$ per 6 months,

$$P = 200(P/A, 5\%, 40) + 10,000(P/F, 5\%, 40)$$
$$= \$4852$$

If Ms. VIP must pay $8375 per bond, she cannot even come close to 10% per year compounded semiannually, so she should not buy these bonds.

(b) If she buys at \$8375 per bond, we can find the dollars gained by computing the future worth, assuming that Ms. VIP reinvests all interest at 10% per year compounded semiannually:

$$F = 200(F/A, 5\%, 40) + 10,000 = \$34,159$$

Thus, she stands to gain a total of \$34,159 − \$8375 = \$25,784. However, as stated in part (a), the rate of return would be much less than 10% per year.

Example 11.7

In the preceding example, Ms. VIP would naturally be saddened by her inability to make 10% compounded semiannually if she pays \$8375 for a 20-year 4%-per-year \$10,000 bond. Compute the (a) actual nominal and effective returns per year of the bond and (b) actual dollar gain if this rate is used for reinvestment.

Solution
(a) The rate-of-return equation is

$$0 = -8375 + 200(P/A, i\%, 40) + 10,000(P/F, i\%, 40)$$

Solution or interpolation shows that $i = 5.40\%$ per year nominally (2.70% semiannually) and $i = 5.47\%$ effectively.
(b) Using a nominal rate of 5.40% per year compounded semiannually, we find that the future worth of the bond is

$$F = 200(F/A, 2.7\%, 40) + 10,000 = \$24,180$$

which represents a gain of \$15,805.

Example 11.8

An investor paid \$4240 for an 8% \$10,000 bond with interest payable quarterly. The bond was in default and, therefore, it paid no interest for the first 3 years after the investor bought it. If interest was paid for the next 7 years and then the investor sold the bond for \$11,000, what rate of return did he make in the investment? Assume the bond would mature 18 years after the investor bought it.

Solution   The bond interest received by the investor in years 4 through 10 was:

$$I = \frac{(10,000)(0.08)}{4} = \$200 \text{ per quarter}$$

The rate of return *per quarter* can be determined by solving a rate-of-return equation as follows:

$$0 = -4240 + 200(P/A, i\% \text{ per quarter}, 28)(P/F, i\% \text{ per quarter}, 12)$$
$$+ 11,000(P/F, i\% \text{ per quarter}, 40)$$

By trial and error, $i = 4.1\%$ per quarter or a nominal 16.4% per year compounded quarterly.

Comment   The rate-of-return equation obviously could also have been written in terms of dollars per year A or future dollars F.

Problems **11.1** What is the difference between (*a*) mortgage bonds and debenture bonds, and (*b*) general obligation bonds and revenue bonds?

11.2 What would be the interest payment and payment period on a 5% $5000 bond that is payable semiannually?

11.3 What is the frequency and amount of the interest payments on a 6% $10,000 bond for which interest is payable quarterly?

11.4 What are the interest payments and their frequency on a 10% $5000 bond that pays monthly interest?

11.5 How much should you be willing to pay for a 9% $10,000 bond that is due 10 years from now if you want to make a nominal 8% per year compounded semiannually? Assume that the bond interest is payable semiannually.

11.6 What is the present worth of a 5% $50,000 bond that has interest payable semiannually? Assume that the bond is due in 25 years and the desired rate of return is 16% per year compounded semiannually.

11.7 A $5\frac{1}{2}$% $15,000 bond with interest payable quarterly is due 20 years from now. What is the present value of the bond if the purchaser desires to make a nominal 14% per year compounded quarterly?

11.8 You have been offered a 6% $20,000 bond at a 4% discount. If interest is paid quarterly and the bond is due in 15 years, can you make a nominal 8% per year compounded quarterly?

11.9 How much should you be willing to pay for a 9% $15,000 bond which has interest payable semiannually and is due in 20 years, if you desire to make an *effective* 14% per year compounded semiannually?

11.10 How much should an investor be willing to pay for a 15-year $25,000 bond which has a bond interest rate of 8% per year payable semiannually, if the investor desires to make a rate of return of (*a*) a nominal 12% per year compounded quarterly and (*b*) a nominal 18% per year compounded monthly?

11.11 A company is considering issuing bonds for financing a major construction project. The company is presently trying to determine whether it should issue conventional bonds or "put" bonds (i.e., bonds which provide the holder the right to sell the bonds at a certain percentage of face value). If the company issues conventional bonds, it expects the bond purchasers to require an overall rate of return (to maturity) of 14% per year compounded quarterly. If put bonds are issued, however, investors would be satisfied with a 12.5% overall rate of return. If the bond interest rate is 12% payable quarterly and the maturity date is 20 years from now, how much more money will the company receive on each $10,000 bond if the put bonds are issued instead of the conventional bonds?

11.12 A city wanting to repave some local streets needed to acquire $3 million through a bond issue. At the time the bond issue was approved by the voters, the bond interest rate was set at 8%. Between the time the bonds were approved and the time they were sold, however, the interest rate the market required to attract investors changed from 8 to 10%. If the bond interest rate were to remain at 8% per year payable semiannually, how much money would the city receive from the $3-million issue, provided that the bond purchasers required a nominal 10% per year compounded semiannually? Assume that the bonds would mature in 20 years.

11.13 How much would the city receive in Prob. 11.12 if the interest rate required to attract investors were a nominal 10% per year compounded quarterly? Assume that the bond interest rate would remain at 8% per year payable semiannually for 20 years.

11.14 If a manufacturing company needed to raise $2 million in capital to finance a small expansion, what would the face value of the bonds have to be, if the bonds were to have a bond interest rate of 12% per year, payable quarterly, and were to mature in 20 years. Assume that investors would require a rate of return of a nominal 16% per year compounded quarterly.

11.15 Seven years ago, Mr. Hughes purchased a 20-year $10,000 bond having an interest rate of 8% per year payable semiannually for $8000. Mr. Hughes would like to sell the bond now, but he will do so only if he can make a rate of return of a nominal 16% per year compounded semiannually on his investment. How much must he get for the bond in order to achieve his objective?

11.16 A bond was purchased for $900 which has a face value of $1000 and a bond interest rate of 5% per year payable semiannually. The bond is to become due in 9 years. If the purchaser can receive a nominal 18% per year compounded semiannually on any money he invests, how much money will he have accumulated 17 years from now? Assume that all income is invested rather than spent.

11.17 A 4% $1000 bond that has interest payable annually is to become due 6 years from now. However, the company has financial difficulties and has asked the bondholders to defer the due date until 10 years from now. If you could buy the bond now for $800, how much would the company have to pay 10 years from now so that you would make the same rate of return that you would receive if they paid the face value originally scheduled? The bond interest will be paid through the original bond due date, but there will be no interest payments thereafter. Assume that you would have invested the $1000 due in year 6 at the same rate of return you would have made if the company paid the $1000 when originally due.

11.18 A small southern city was authorized by its voters to issue $8 million worth of 25-year bonds. The bonds were to have an interest rate of 10% per year payable semiannually. However, before the bonds were sold, the rate of return required by investors rose to a nominal 12% per year compounded semiannually. If the city needed to receive at least $7.5 million, by how many years would the due date have to be moved up for the bonds to yield the return desired by the investors?

11.19 If a $9\frac{1}{2}$% $10,000 bond that has interest payable semiannually is for sale for $8500, when would the bond have to be due for the purchaser to make a nominal 14% per year compounded semiannually on the investment?

11.20 When would the bond in Prob. 11.19 have to be due if the purchaser wanted to earn a nominal 12% per year compounded continuously on the investment?

11.21 The Wet 'N' Wild Snowmobile Manufacturing Company must raise money for expanding its production facility. If conventional bonds are issued, the bond interest rate will have to be 16% per year compounded semiannually. The face value of the bonds will be $7 million. If convertible bonds are issued, however, the bond interest rate will have to be only 7% per year compounded semiannually. What will the face value of the convertible bonds have to be to make the present worth of both issues the same at an interest rate of 20% per year compounded semiannually? Compounded continuously? The bonds mature in 15 years.

11.22 The Black Belch Smelter is attempting to raise $15 million for air pollution-control equipment by issuing 30-year bonds. With the backing of the local government, industrial development bonds could be issued and, because they are income tax-free, the bond interest rate would have to be only 13% per year payable semiannually to attract investors. If taxable bonds are issued instead of the industrial development bonds, the bond interest rate would have to be $18\frac{1}{2}$% per year payable semiannually. What is the present worth of the extra interest the company will have to pay if the company must issue the taxable instead of the tax-free bonds? The company's MARR is a nominal 20% per year compounded semiannually.

11.23 The underwriters of a $10-million bond issue paid $9.5 million for bonds issued by the Stin-Key Oil Company. The bonds had an interest rate of 15% per year payable semiannually with a maturity date of 20 years. The underwriters thought the bonds could be sold at a yield to investors of a nominal 15% per year compounded semiannually, but they were unpleasantly surprised to discover that the investors required a yield of a nominal 16% per year compounded semiannually. How much money did the underwriters make or lose on their investment?

11.24 A $10,000 zero-coupon bond which matures in 10 years is for sale. An investor who is thinking about buying it thinks that he can sell it for $4000 two years from now. If the investor wants to earn 12% per year compounded monthly on his investment, how much should he pay for the bond? Assume the investor was born 17 years ago.

11.25 An engineer planning for her retirement is considering a zero-coupon bond which has a face value of $50,000 and a maturity date 20 years from now. How much should she pay for the bond if she intends to hold it to maturity and she wants to earn 5% per quarter?

11.26 In Prob. 11.23, what was the "effective" interest paid by the oil company to get the $9.5 million?

11.27 A 9% $10,000 bond is offered for sale for $8000. If the bond interest is payable semiannually and the bond becomes due in 13 years, what nominal rate of return per year will the purchaser make on the investment?

11.28 A 10% $50,000 bond is offered for sale for $43,500. If the bond interest is payable quarterly and the bond becomes due in 15 years, what nominal rate of return per year would the purchaser make on the investment?

11.29 An investor purchased a 5% $1000 bond for $825. The interest was payable semiannually, and the bond was to become due in 20 years. The bond was kept for only 8 years and sold for $800 immediately after the sixteenth interest payment. What nominal rate of return per year was made on the investment?

11.30 What effective rate of return per year would the purchaser of the bond in Prob. 11.29 receive?

11.31 At what bond interest rate will a $10,000 bond yield a nominal 14% per year compounded semiannually if the purchaser pays $8000 and the bond becomes due in 15 years? Assume that the bond interest is payable semiannually.

11.32 At what bond interest rate will a $20,000 bond that has interest payable semiannually yield an effective 8% rate of return per year if the price of the bond is $18,000 and the bond becomes due in 20 years?

11.33 What would the bond interest rate have to be in Prob. 11.31 if the purchaser wanted to make a nominal 15% per year compounded quarterly? Compounded continuously?

11.34 The Sli-Dog Company plans to sell two hundred 4% $1000 bonds. Interest will be paid quarterly and the bonds will be retired after 15 years. If the management wants to set up a fund that will be specifically used to retire the bonds, that is, pay interest and face value, what equivalent annual amount must be placed in the fund? Assume that the fund will earn a nominal 16% per year compounded quarterly. (Note that beginning-of-year deposits are required to have the quarterly interest payments available.)

11.35 Bonds purchased for $9000 have a face value of $10,000 and a bond interest rate of 10% per year payable semiannually. The bonds are due in 3 years. The company that issued the bonds is contemplating a liquidity problem in 3 years and has advised all bondholders that if they will keep their bonds for another 2 years past the original due date, the bond interest for the extended 2-year period will be 16% per year payable semiannually. What nominal rate of return per year would the bondholders receive if they held the bonds for the additional 2-year period?

11.36 What rate of return would the bondholders in Prob. 11.35 have made if the bonds were kept for only 3 years and redeemed on the original due date? Should the bondholders keep the bonds 3 years or 5 years, as the company has advised?

11.37 The GRQ Corporation issued $5 million worth of 25-year bonds, with an interest rate of 10% per year payable semiannually. The company received $4.5 million, but semiannual expenses of $15,000 were expected for servicing the bonds. What nominal interest rate per year did the corporation pay for getting the $4.5 million after all expenses are taken into account?

11.38 An investor purchased a $1000 convertible bond for $850 from the GRQ Company. The bond had an interest rate of 8% per year payable quarterly, and was convertible to 20 shares of GRQ common stock. If the investor kept the bond for $6\frac{1}{2}$ years and then converted it into common stock when the stock was selling for $49 per share, what was the nominal rate of return per year on his investment?

11.39 The Hi-Cee Steel Company issued $5 million worth of 20-year 14%-per-year payable semiannually callable bonds (i.e., bonds which could be called in and paid off at any time). The company agreed to pay a 10% premium on the face value if the bonds were called. Seven years after the bonds were issued, the prevailing interest rate in the marketplace dropped to 11%. (a) What rate of return would the company make by calling the bonds and paying the $5.5 million? (b) Should the company call the bonds?

11.40 If an investor pays $5000 for a zero-coupon bond which has a face value of $20,000 and a maturity date 15 years from now, what rate of return will be made if the investor holds the bond to maturity?

12 | Inflation and Cost Estimation

An increase or decrease in the amount of money or credit without a corresponding increase or decrease in the amount of goods and services causes changes in the price of those goods and services. This occurs because the value of the currency has changed. The terms *inflation* and *deflation* are used to describe price changes brought about by these conditions. In this chapter, the mechanism for conducting an economic analysis under conditions of varying currency values is presented. Additionally, some of the methods for estimating expected plant or equipment costs from past cost information will be presented. While we will be focusing only on inflation, the concepts presented here apply equally well in a deflationary economy.

Section Objectives

After completing this chapter, you should be able to do the following:

12.1 Define *inflated interest rate* and calculate the present worth of specified future sums, given the interest rate, inflation rate, and time period.

12.2 Define what is meant by *real interest rate* and calculate the future worth of a specified present amount, given the interest rate, inflation rate, and time period.

12.3 Calculate the uniform annual amount of money in then-current dollars that would be equivalent to a specified present or future sum, given the interest rate, inflation rate, and time period.

12.4 Define what is meant by a *cost index* and use a specified index to determine an expected present cost from relevant data of previous years.

12.5 Estimate the cost of a component, system, or plant by using a *cost-capacity equation* or the *factor method* of cost estimation.

Study Guide

12.1   Present-Worth Calculations with Inflation Considered

Anyone living today is well aware of the fact that $1 now will not purchase the same amount of goods or services as would $1 in 1930. This is because the value of the dollar has decreased as a result of more dollars chasing fewer goods (inflation). In order to make comparisons between dollar amounts which occur in different time periods, the different-valued dollars must first be converted into dollars which have the same buying power (i.e., constant value dollars).

Currency in one period of time can be brought to the same value as currency in another period of time through the use of the following generalized equation:

$$\text{Dollars in period } t_1 = \frac{\text{dollars in period } t_2}{\text{inflation between } t_1 \text{ and } t_2} \qquad (12.1)$$

If dollars in period t_1 are called today's dollars and dollars in period t_2 are called then-dollars, and f represents the inflation rate per period, Eq. (12.1) becomes

$$\text{Today's dollars} = \frac{\text{then-dollars}}{(1 + f)^n} \qquad (12.2)$$

where n is the number of time periods between t_1 and t_2.

After the dollar amounts in different time periods have been expressed as constant-value dollars per Eq. (12.2), the equivalent present, future, or annual amounts could be determined by using the regular interest rate i in any of the formulas derived in Chap. 2. The calculations involved in this procedure are illustrated in Table 12.1. If an inflation rate of 8% per year is assumed, then col. 2 shows the increase in cost for each of the next 5 years for an item that has a cost of $5000 today. Column 3 shows the cost in then-current dollars with col. 4 showing the cost in constant-value dollars (today's) through the use of Eq. (12.2). Column 5 shows the present-worth calculation at $i = 10\%$ per year. Observe from col. 4 that when the future dollars of col. 3 are converted into today's dollars, the cost is $5000, the same as the cost at the start. This will always be true when the costs are increasing by an amount *exactly equal* to the inflation rate. The actual cost of the item 4 years from now will thus be $6803; but in today's dollars, the cost at that time will be $5000, which at an interest rate of 10% per year has a present worth of $3415.

Table 12.1 Present-worth calculation using today's dollars (rounded)

(1) Year n	(2) Cost increase due to inflation	(3) Future cost in then dollars	(4) = (3)/(1.08)^n Future cost in today's dollars	(5) = (4)(P/F, 10%, n) Present worth at $i = 10\%$
0		$5,000	$5,000	$5,000
1	$5,000(0.08) = $400	5,400	$5,400/(1.08)^1 = $5,000	4,545
2	5,400(0.08) = 432	5,832	5,832/(1.08)^2 = 5,000	4,132
3	5,832(0.08) = 467	6,299	6,299/(1.08)^3 = 5,000	3,757
4	6,299(0.08) = 504	6,803	6,803/(1.08)^4 = 5,000	3,415

An alternative method of accounting for inflation in a present-worth analysis involves adjusting the interest formulas themselves to account for inflation. Let us consider the single-payment present-worth formula:

$$P = F\left[\frac{1}{(1 + i)^n}\right]$$

F (in then dollars) can be converted into today's dollars by using Eq. (12.2). Then:

$$P = \frac{F}{(1 + f)^n}\left[\frac{1}{(1 + i)^n}\right]$$

$$= F\left[\frac{1}{(1 + f)^n(1 + i)^n}\right]$$

$$= F\left[\frac{1}{(1 + i + f + if)^n}\right] \tag{12.3}$$

If the term $i + f + if$ in Eq. (12.3) is defined as i_f, the equation becomes

$$P = F\left[\frac{1}{(1 + i_f)^n}\right] = F(P/F, i_f\%, n) \tag{12.4}$$

The expression i_f is called the inflated interest rate and is defined as

$$i_f = i + f + if \tag{12.5}$$

where i = interest rate
f = inflation rate
i_f = inflated interest rate

For an interest rate of 8% per year and an inflation rate of 10% per year, Eq. (12.5) gives

$$i_f = 0.08 + 0.10 + 0.08(0.10)$$
$$= 0.188 \quad (18.8\%)$$

Table 12.2 shows the use of i_f in the present-worth calculation of the $5000 item considered in Table 12.1. As shown in col. 4 of Table 12.2, the present worth of the item for each year is the same as that calculated in col. 5 of Table 12.1.

The present worth of a uniform-series, gradient, or percentage cash flow can be found similarly. That is, either i or i_f should be used in the P/A, P/G, or P_E equations, depending upon whether the cash flow is expressed in today's dollars or

Table 12.2 Present-worth calculation using an inflated interest rate

(1) Year, n	(2) Future cost in then-current dollars	(3) $(P/F, 18.8\%, n)$	(4) Present worth
0	$5,000	1	$5,000
1	$5,400	0.8418	$4,545
2	$5,832	0.7085	$4,132
3	$6,299	0.5964	$3,757
4	$6,803	0.5020	$3,415

then-current dollars, respectively. If the series is expressed in today's dollars, then its present worth is simply the discounted cash-flow value using the regular interest rate i. If the cash flow is expressed in then-current dollars, however, the amount today that would be equivalent to the inflated then dollars could be obtained by using i_f in the formulas (or by converting the then dollars into today's dollars and then using i). Examples 12.1 and 12.2 illustrate these present-worth calculations

Example 12.1

An alumnus of Taco Tech who "made good" has decided to donate to the college's Excellence Fund and has offered the college any one of the following three plans:

Plan A: $60,000 now
Plan B: $16,000 per year for 12 years beginning 1 year from now
Plan C: $50,000 three years from now and another $80,000 five years from now

The only condition placed on the donation is that the college agree to spend the money on research related to the advancement of robotics. The college would like to select the plan which maximizes the buying power of the dollars received, so it has instructed the engineering professors evaluating the plans to account for inflation in their calculations. If the college can earn 12% per year on its ready-assets account and the inflation rate is expected to be 11% per year, which plan should the college accept?

Solution The simplest method of evaluation is to calculate the present worth of each plan in today's dollars. For plans B and C, the easiest way to obtain the present worth is through the use of the inflated interest rate i_f. By Eq. (12.5),

$$i_f = 0.12 + 0.11 + 0.12(0.11) = 0.243 \qquad (24.3\%)$$

Compute the P value by Eq. (12.4):

$$P_A = \$60,000$$

$$P_B = \$16,000(P/A, 24.3\%, 12) = \$61,003.46$$

$$P_C = \$50,000(P/F, 24.3\%, 3) + 80,000(P/F, 24.3\%, 5) = \$52,995.84$$

Since P_B is the largest in today's dollars, accept plan B.

Comment The present worths of plans B and C could also have been found by first converting the cash flows into today's dollars and then using the regular i. This procedure is obviously tedious and time-consuming, but you may want to satisfy yourself that the answers are the same.

Example 12.2

Calculate the present worth of a uniform series of payments of $1000 per year for 5 years if the interest rate is 10% per year and the inflation rate is 8% per year, assuming that the payments are in terms of (*a*) today's dollars and (*b*) then-current dollars.

Solution
(*a*) Since the dollars are already expressed in today's dollars, the present worth is simply

$$P = 1000(P/A, 10\%, 5) = \$3790.80$$

Figure 12.1 Cash flow, Example 12.2.

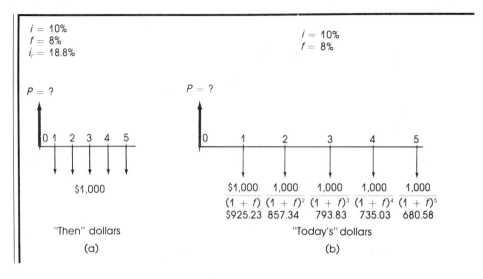

(a) "Then" dollars

(b) "Today's" dollars

(b) Since the dollars are expressed in then-current dollars, as shown in Fig. 12.1a, use the inflated interest rate in the uniform-series present-worth factor.

$$i_f = i + f + if = 18.8\%$$

$$P = 1000(P/A, 18.8\%, 5) = \$3071$$

The present worth can also be obtained by converting the future cash flows into today's dollars and then finding the present worth using the regular i. The calculations associated with this method follow. From Fig. 12.1b,

$$P = 925.93(P/F, 10\%, 1) + 857.34(P/F, 10\%, 2) + 793.83(P/F, 10\%, 3)$$
$$+ 735.03(P/F, 10\%, 4) + 680.58(P/F, 10\%, 5)$$
$$= \$3071$$

This is the same result obtained using i_f.

Comment It is obvious from this example that use of the inflated interest rate i_f in the P/A factor is much simpler than converting the future cash flows into today's dollars and using the P/F factor for each one. Thus, only the former method will be used hereafter.

To summarize, if future dollars are expressed in today's dollars (or have already been converted into today's dollars), the present worth should be calculated using the regular interest rate i in the present-worth formulas. If the future dollars are expressed in then-current dollars, the inflated interest rate i_f should be used in the formulas.

Additional Examples 12.9 and 12.10
Probs 12.1 to 12.14

12.2 Future-Worth Calculations with Inflation Considered

In future-worth calculations, it must be recognized that a future sum of money can represent one of four different amounts:

Case 1: The *actual amount* of money that will be accumulated at time n

Case 2: The *buying power*, in terms of today's dollars, of the actual amount of dollars accumulated at time n

Case 3: The *number of then dollars required* at time n to maintain the same purchasing power as a dollar today (i.e., no interest considered)

Case 4: The number of dollars required at time n *to maintain purchasing power and earn a stated interest rate*.

It should be obvious that for case 1, the actual amount of money accumulated would be obtained by using the regular interest rate i in any of the formulas as has been done before the concept of inflation was introduced. For case 2, the buying power of the future dollars can be determined by using the regular interest rate i to calculate F and then dividing this F by $(1 + f)^n$. Division by $(1 + f)^n$ deflates the inflated dollars which, in effect, recognizes that prices increase during inflation so that 1 dollar then will purchase less goods than 1 dollar now. In equation form, case 2 is

$$F = \frac{P(1 + i)^n}{(1 + f)^n} = \frac{P(F/P, i\%, n)}{(1 + f)^n} \tag{12.6}$$

As an illustration, if $1000 is deposited into a savings account at 10%-per-year interest for 7 years and the inflation rate is 8% per year, the amount of money that will be accumulated with today's buying power will be

$$F = \frac{1000(F/P, 10\%, 7)}{(1 + 0.08)^7} = \$1137$$

If there is no inflation ($f = 0$), in 7 years the 10% rate will accumulate $1948 with today's buying power as follows:

$$F = 1000(F/P, 10\%, 7) = \$1948$$

The future amount of money that would be accumulated with today's buying power could equivalently be determined by using a *real interest rate* i_r in the F/P factor to compensate for the decreased purchasing power of the dollar. This real interest rate can be obtained by equating the single-payment compound-amount formula (F/P factor) with the middle expression in Eq. (12.6), which converts present dollars into future dollars with today's buying power.

$$P(1 + i_r)^n = P\left[\frac{(1 + i)^n}{(1 + f)^n}\right]$$

$$1 + i_r = \frac{1 + i}{1 + f}$$

$$i_r = \frac{i - f}{1 + f} \tag{12.7}$$

The real interest rate i_r represents the rate at which present dollars will expand *with their same buying power* into equivalent future dollars. The use of this interest rate is appropriate when calculating the future worth of a savings account, for example, when the effects of inflation must be taken into consideration. Thus, for the $1000 deposit mentioned previously,

$$i_r = \frac{0.10 - 0.08}{1 + 0.08} = 0.0185 \qquad (1.85\%)$$

$$F = 1000(F/P, 1.85\%, 7) = \$1137$$

Note that the interest rate of 10% per year has been reduced to only 1.85% per year because of the erosive effects of inflation. Also note that an inflation rate larger than the interest rate, $f > i$, leads to a negative real interest rate i_r in Eq. (12.7).

Case 3 also recognizes that prices increase during inflationary periods and, therefore, purchasing something at a future date will require more dollars than would be required now for the same thing. The future dollars are worth less and, therefore, more are needed. This is the type of calculation you would make if someone asks "How much will a car cost in 5 years if its current cost is $15,000 and its price will increase by 6% per year?" (The answer is $20,073.38.) For this calculation, f is used instead of i in the single-payment compound-amount formula as follows:

$$F = P(1 + f)^n = P(F/P, f\%, n)$$

Thus, if the $1000 deposit considered above represented the cost of an item which escalated in price exactly in accordance with the inflation rate of 8% per year, the cost 7 years from now would be:

$$F = 1000(F/P, 8\%, 7) = \$1713.80$$

The calculation associated with case 4, maintaining purchasing power *and* earning interest, takes into account both increasing prices and the time value of money; that is, if real growth of capital is to be obtained, funds must grow at a rate not only equal to the interest rate i, but also at a rate equal to the price increases f. Thus, i_f is used in the equivalence formulas. Using the same $1000 deposit as above,

$$i_f = 0.10 + 0.08 + 0.10(0.08) = 0.188 \qquad (18.8\%)$$

$$F = 1000(F/P, 18.8\%, 7) = \$3340$$

This calculation shows that $3340 of then-current dollars would be equivalent to $1000 now when the interest rate is 10% per year and the inflation rate is 8% per year.

In summary, the calculations made in this section reveal that $1000 now at an interest rate of 10% per year would accumulate to $1948 in 7 years; the $1948 would have the purchasing power of $1137 of today's dollars; an item with a cost of $1000 now would cost $1713.80 in 7 years at an inflation rate of 8% per year; and it would take $3340 of then-current dollars to be equivalent to the $1000 now when inflation is taken into account. Table 12.3 summarizes which i value is to be used in the equivalence formulas as a function of which F value is desired.

Table 12.3 Calculation methods for various future values

Future value desired	Method calculation	Example, using P = $500, n = 5, i = 10%, f = 8%
Case 1: actual dollars accumulated	Use i in equivalence formulas	$F = 500(F/P, 10\%, 5)$
Case 2: buying power of accumulated dollars	Use i in equivalence formulas and divide answer by $(1 + f)^n$ or use i_r in formulas	$F = \dfrac{500(F/P, 10\%, 5)}{(1 + 0.08)^5}$ $F = 500(F/P, i_r\%, 5)$
Case 3: dollars required for same purchasing power	Use f in place of i in equivalence formulas	$F = 500(F/P, 8\%, 5)$
Case 4: maintain purchasing power and earn interest	Use i_f in equivalence formulas	$F = 500(F/P, 18.8\%, 5)$

Example 12.3

The Shur-Kan-Do Sheet Metal Company is trying to decide whether it should pay now or pay later for upgrading its production facilities. If the company selected plan A for "action," the necessary equipment would be purchased now for $20,000. However, if the company selected plan I for "inaction," the equipment purchase would be deferred for 3 years when the cost would be $34,000. The minimum attractive rate of return is 18% per year and the inflation rate is expected to be 12% per year.
(a) Find the real interest rate for these plans.
(b) Determine whether the company should purchase now or purchase later when inflation is not considered, and when inflation is taken into account.

Solution
(a) By Eq. (12.7) the real interest rate is

$$i_r = \frac{0.18 - 0.12}{1.12} = 0.0536 \quad (5.36\%)$$

This means that an effective MARR of 5.36% per year is used when inflation is considered.
(b) *Inflation not considered.* The stated rate is 18% per year, so we can compute P at time 0 or F three years from now and select the plan with the lower cost.
  Computing F as we have previously,

$$F_A = 20,000(F/P, 18\%, 3) = \$32,860.64$$

$$F_I = \$34,000$$

Select action plan A because it costs less.
Inflation considered. Compute the inflated rate by Eq. (12.5).

$$i_f = 0.18 + 0.12 + 0.18(0.12) = 0.322$$

Use i_f to compute the F values 3 years from now to determine the then-current dollars necessary.

$$F_A = 20,000(F/P, 32.2\%, 3) = \$46,208.77$$

$$F_I = \$34,000$$

Now select the inaction plan I because it will require less equivalent then-current dollars.

Comment This example illustrates that when inflation is taken into account, the more economic alternative may be different from the one selected if inflation is ignored.

A present-worth analysis can be used here with the same results:

No inflation considered	Inflation considered
$i = 18\%$	$i_f = 32.2\%$
$P_A = \$20,000$	$P_A = \$20,000$
$P_I = 34,000(P/F, 18\%, 3)$	$P_I = 34,000(P/F, 32.2\%, 3)$
$= \$20,693.45$	$= \$14,715.82$
Select plan A	Select plan I

Probs 12.15 to 12.22

12.3 Capital-Recovery and Sinking-Fund Calculations with Inflation Considered

It is particularly important for capital-recovery calculations to include inflation considerations because present dollars must be recovered with future inflated dollars. There is little significance in considering capital recovery in terms of today's dollars, so only then-current dollars will be considered in this section. Since then-current dollars have less buying power than do today's dollars, it is obvious that more dollars will be required to recover the present investment. This recovering suggests the use of the inflated interest rate in the A/P formula. For example, if \$1000 is invested today when the interest rate is 10% per year and the inflation rate is 8% per year, the annual amount of capital that must be recovered each year for 5 years in then-current dollars will be

$$A = 1000(A/P, 18.8\%, 5) = \$325.59$$

On the other hand, the decreased value of dollars through time means that investors would be willing to spend fewer present (high-value) dollars to accumulate a specified amount of future then-current (inflated) dollars by way of a sinking fund (i.e., A). This suggests the use of a higher interest rate (i_f) to produce a lower A value in the A/F formula. The annual equivalent (when inflation is considered) of $F = \$1000$ five years from now in then-current dollars is thus

$$A = 1000(A/F, 18.8\%, 5) = \$137.59$$

When inflation is not considered, the equivalent annual amount for $F = \$1000$ at $i = 10\%$ is $1000(A/F, 10\%, 5) = \$163.80$. So when F is fixed, uniform future costs should be spread over as long a time period as possible so that inflation will have the effect of reducing the payment involved (\$137.59 versus \$163.80 here). Keep in mind that only the *buying power* of the payment is reduced, not its *amount*.

A situation which sometimes arises in economic calculations involves the determination of the amount of uniform deposit required to accumulate an amount of money with the same buying power as a specified amount *today*. Example 12.4 illustrates the calculations involved.

Example 12.4

What annual deposit will be required for 5 years to accumulate an amount of money that has the same buying power as $680.58 today if the interest rate is 10% per year and the inflation rate is 8% per year?

Solution The actual number of then-current (inflated) dollars required in 5 years is

$F = 680.58(1.08^5) = \$1000$

Therefore, the actual amount of the annual deposit is

$A = 1000(A/F, 10\%, 5) = \$163.80$

Comment This example shows that if $163.80 is deposited each year for 5 years at an interest rate of 10% per year, $1000 will be accumulated after the fifth deposit. However, from an economic analysis point of view, $163.80 for 5 years is *not* equivalent to $1000 in year 5 *when inflation is considered*. As shown above, $137.59 for 5 years is equivalent to $1000 in year 5 at $i_f = 18.8\%$. The discrepancy between the $163.80 and the $137.59 is caused by the use of two different interest rates (i_f and i) in the A/F factor. The use of $i_f = 18.8\%$ accounts for the fact that the dollars spent in years 1 through 4 have greater buying power than dollars spent in year 5. We would therefore not want to spend as many of these. The use of $i = 10\%$, on the other hand, essentially ignores the greater buying power of the dollars spent in earlier years, so the higher "equivalent" amount ($163.80) is obtained. As implied in this example, most of the calculations associated with comparing alternatives in engineering-economic studies require the use of the inflated interest rate when inflation is considered.

Probs. 12.23 to 12.28

12.4 Cost Indexes

Even a cursory study of recent world history reveals that the currency values of virtually every country on earth are in a constant state of change. For engineers involved in project planning and design, this makes the difficult job of cost estimation even more difficult. One method of obtaining preliminary cost estimates is by looking at the costs of similar projects that were completed at some time in the past and updating these past-cost figures. Cost indexes represent a convenient tool for accomplishing this.

A *cost index* is a ratio of the cost of something today to its cost at some time in the past. One such index that most people are familiar with is the consumer price index (CPI), which shows the relationship between present and past costs for many of the things that "typical" consumers must buy. This index, for example, includes such items as rent, food, transportation, and certain services. However, other indexes are more relevant to the engineering profession because they track the costs of goods and services which are more pertinent to engineers. Table 12.4 is a listing of some

Table 12.4 Types and sources of various cost indexes

Type of index	Source
Overall prices	
Consumer (CPI)	Bureau of Labor Statistics
Producer (wholesale)	U.S. Department of Labor
Construction	
Chemical plant overall	*Chemical Engineering*
Equipment, machinery, and supports	
Construction labor	
Buildings	
Engineering and supervision	
Engineering News Record overall	*Engineering News Record (ENR)*
Construction	
Building	
Common labor	
Skilled labor	
Materials	
EPA treatment plant indexes	Environmental Protection Agency, *ENR*
Large-city advanced treatment (LCAT)	
Small-City conventional treatment (SCCT)	
Federal highway	
Contractor cost	
Equipment	
Marshall and Swift (M&S) overall	Marshall & Swift
M&S specific industries	

of the more common indexes and, as shown, some are applicable to a wide variety of goods and services (CPI) while others are more directed toward the needs of a specific user (federal highway index).

Generally, the indexes are made up of a mix of components which are assigned certain weights, with the components sometimes further subdivided into more basic items. For example, the equipment, machinery, and support component of the chemical plant cost index is subdivided further into process machinery, pipes, valves and fittings, pumps and compressors, and so forth. These subcomponents, in turn, are built up from even more basic items like pressure pipe, black pipe, and galvanized pipe. Table 12.5 shows the *Chemical Engineering* plant cost index, the *Engineering News Record (ENR)* construction cost index, and the Marshall and Swift (M&S) equipment cost index for the years 1970 through mid-1987, with the base period of 1957–1959 assigned a value of 100 for the *Chemical Engineering (CE)* plant cost index, 1913 = 100 for the *ENR* index, and 1926 = 100 for the M&S equipment cost index.

The general equation for updating costs through the use of any cost index over a period from time 0 (base) to another time t is

$$C_t = \frac{C_0 I_t}{I_0} \qquad (12.8)$$

Table 12.5 Values for selected indexes

Year	CE plant cost index	ENR construction cost index	M&S equipment cost index
1970	125.7	1,465.07	303.3
1971	132.3	1,665.20	321.3
1972	137.2	1,811.76	332.0
1973	144.1	1,944.50	344.1
1974	165.4	2,103.00	398.4
1975	182.4	2,304.60	444.3
1976	192.1	2,494.30	472.1
1977	204.1	2,672.40	505.4
1978	218.8	2,872.40	545.3
1979	238.7	3,139.10	599.4
1980	261.2	3,378.17	659.6
1981	297.0	3,725.55	721.3
1982	314.0	3,939.25	745.6
1983	316.9	4,108.74	760.8
1984	322.7	4,172.27	780.4
1985	325.3	4,207.84	789.6
1986	318.4	4,347.50	797.6
mid-1987	321.9	4,442.63	803.7

where C_t = estimated cost at present time t
C_0 = cost at previous time t_0
I_t = index value at time t
I_0 = index value at time t_0

Example 12.5 illustrates the use of the *ENR* index to estimate present costs from past values.

Example 12.5

In evaluating the feasibility of a major construction project, an engineer is interested in estimating the cost of skilled labor for the job. The engineer finds that a project of similar complexity and magnitude was completed 5 years ago when the *ENR* skilled labor index was 3496.27. The skilled labor cost for that project was $360,000. If the *ENR* skilled labor index now stands at 4038.44, what would the expected skilled labor cost be for the new project?

Solution The base time t_0 is 5 years ago. Using Eq. (12.8), the present cost estimate is

$$C_t = \frac{C_0 I_t}{I_0}$$

$$= \frac{4038.44}{3496.27} 360,000$$

$$= \$415,825$$

Probs 12.29 to 12.35

Table 12.6 Sample exponent values

Component/system/plant	Size range	Exponent
Activated sludge plant	1–100 MGD	0.84
Aerobic digester	0.2–40 MGD	0.14
Blower	1,000–7,000 ft/min	0.46
Centrifuge	40–60 in	0.71
Chlorine plant	3,000–350,000 tons per year	0.44
Clarifier	0.1–100 MGD	0.98
Compressor	200–2,100 hp	0.32
Cyclone separator	20–8,000 ft^3/min	0.64
Dryer	15–400 ft^2	0.71
Filter, sand	0.5–200 MGD	0.82
Heat exchanger	500–3,000 ft^2	0.55
Hydrogen plant	500–20,000 scfd	0.56
Laboratory	0.05–50 MGD	1.02
Lagoon, aerated	0.05–20 MGD	1.13
Pump, centrifugal	10–200 hp	0.69
Reactor	50–4,000 gal	0.74
Sludge drying beds	0.04–5 MGD	1.35
Stabilization pond	0.01–0.2 MGD	0.14
Tank, stainless	100–2,000 gal	0.67

12.5 Cost Estimating

While the cost indexes discussed above provide a valuable tool for estimating present costs from historical data, they become even more valuable when combined with some of the other cost-estimating techniques. One of the most widely used methods of obtaining preliminary cost information is through the use of *cost-capacity equations*. As the name implies, a cost-capacity equation relates the cost of a component, system, or plant to its capacity. Since many cost-capacity relationships plot as a straight line on log-log paper, one of the most common cost-prediction equations is

$$C_2 = C_1 \left(\frac{Q_2}{Q_1} \right)^x \tag{12.9}$$

where C_1 = cost at capacity Q_1
  C_2 = cost at capacity Q_2
    x = exponent

The value of the exponent for various components, systems, or entire plants can be obtained or derived from a number of sources, including *Chemical Engineers' Handbook*, technical journals (especially *Chemical Engineering*), U.S. Environmental Protection Agency, professional or trade organizations, consulting firms, and equipment companies. Table 12.6 is a partial listing of typical values of the exponent for various units. The next example illustrates the use of the Eq. (12.9).

Example 12.6

The total construction cost for a stabilization pond to handle a flow of 0.05 million gallons per day (MGD) was $73,000 in 1977. Estimate the cost today of a pond 10 times larger. Assume the SCCT index (for updating the cost) was 131 in 1977 and is 225 today. The exponent in Eq. (12.9) has a value of 0.14.

Solution   From Eq. (12.9), the cost of the pond in 1977 dollars would be

$$C_2 = 73,000 \left(\frac{0.50}{0.05} \right)^{0.14}$$

$\quad = \$100,768 \quad$ (\$101,000 in 1977 dollars)

Today's cost can be obtained through the use of Eq. (12.8) as follows:

$$C_t = 101,000 \frac{(225)}{(131)}$$

$\quad = \$173,500 \quad$ (today's dollars)

A different widely used, simplified approach for obtaining preliminary cost estimates of process plants is called the *factor method*. While Eq. (12.9) can be used for estimating both the costs of major items of equipment and the total plant costs, the factor method was developed only for obtaining total plant costs. The method is based on the premise that fairly reliable total plant costs can be obtained by multiplying the cost of the major equipment by certain factors. Since major-equipment costs are readily available, rapid plant estimates are possible if the appropriate factors are known. These factors are commonly referred to as Lang factors after Hans J. Lang, who first proposed the method in 1947.

In its simplest form, the factor method of cost estimation can be expressed as

$$C_T = hC_E \qquad (12.10)$$

where C_t = total plant cost
$\qquad h$ = overall cost factor or summation of individual cost factors
$\qquad C_E$ = summation of cost of major items of equipment

In his original work, Lang showed that construction cost factors and overhead cost factors can be combined into one overall factor for various types of plants as follows: solid process plants, 3.10; solid-fluid process plants, 3.63; and fluid process plants, 4.74. These factors reveal that the total installed-plant cost in many times the purchase cost of the major items of equipment. Example 12.7 illustrates the use of overall cost factors.

Example 12.7

A solid-fluid process plant is expected to have a delivered-equipment cost of \$565,000. If the overall cost factor for this type of plant is 3.63, estimate the plant's total cost.

Solution   From Eq. (12.10), the total plant cost is estimated as

$$C_T = 3.63(565,000)$$

$\quad = \$2,051,000$

Subsequent refinements of the factor method have led to the development of separate factors for various elements of the direct and indirect costs. Direct costs are those which are specifically identifiable with a product, function, or activity. These costs usually include expenditures such as raw materials, direct labor, and specific equipment. Indirect costs are those not directly attributable to a single function but shared by several because they are necessary to perform the overall objective. Examples of indirect costs are general administration, taxes, support functions (such as purchasing), and security. The factors for both direct and indirect costs are sometimes developed from delivered-equipment costs and other times from installed-equipment costs. In this text, we will assume that all factors apply to delivered-equipment costs unless otherwise specified.

Furthermore for indirect costs, some of the factors apply to equipment costs while others apply to the total direct cost. In the former case, the indirect cost factors apply to equipment cost, just as do the direct cost factors. Therefore, the simplest procedure is to add the direct and indirect cost factors before multiplying by the delivered-equipment cost. In the latter case, the direct cost must be calculated first because the indirect cost factor must be applied to the direct cost rather than to the equipment cost. As with direct costs, we will assume that the indirect cost factors apply to the *delivered*-equipment cost. Example 12.11 in the Additional Examples section illustrates these calculations.

It should be pointed out that since the literature-reported values of the factors are decimal fractions of the total equipment costs, the number "1" must be added to their sum in order to obtain the total plant cost estimate from Eq. (12.10). It is *not* necessary to add "1" to the *overall* plant cost factors described earlier (the "1" is already included). Example 12.8 illustrates the use of direct and indirect cost factors for estimating the total plant cost.

Example 12.8

The delivered-equipment cost for a small chemical process plant is expected to be $2 million. If the direct cost factor is 1.61 and the indirect cost factor is 0.25, determine the total plant cost.

Solution
Since all factors apply to the delivered-equipment cost, they can be added to obtain the total cost factor. Remember that 1 must be added to the total because the factors are decimal values. Thus,

$$h = 1 + 1.61 + 0.25 = 2.86$$

From Eq. (12.10), the total plant cost is:

$$C_T = 2.86(2,000,000) = \$5,720,000$$

Comment A more complicated example of cost factors is included in the Additional Example section of this chapter.

Additional Example 12.11
Probs. 12.35 to 12.42

Additional Examples

Example 12.9

A \$50,000 bond which has a bond interest rate of 10% per year payable semiannually is currently for sale. The bond is due in 15 years. If the rate of return required by investors is a nominal 16% per year compounded semiannually and if the inflation rate is expected to be 4.5% per semiannual period, how much should be paid for the bond (a) when inflation is not taken into account and (b) when inflation is considered?

Solution

(a) Without considering inflation, the dividend by Eq. (11.1) is $I = [(50,000)(0.10)]/2 = \$2500$ per semiannual period and the present worth at a nominal 8% per 6 months is

$$P = 2500(P/A, 8\%, 30) + 50,000(P/F, 8\%, 30) = \$33,115$$

(b) To consider inflation, use the inflated rate in the P/A factor.

$$i_f = 0.08 + 0.045 + (0.08)(0.045) = 0.1286 \text{ per semiannual period}$$

$$P = 2500(P/A, 12.86\%, 30) + 50,000(P/F, 12.86\%, 30) = \$20,251$$

Comments The \$12,864 difference in present worth illustrates the tremendous negative effect of inflation on fixed-income investments. On the other hand, the entities which issue the instruments (bonds in this case) are benefactors to the same extent.

Example 12.10

Calculate the equivalent present cost of a \$35,000 expenditure now and \$7000 per year for 5 years beginning 1 year from now with increases of 12% per year thereafter for the next 8 years. Use an interest rate of 15% per year and make the calculations (a) without considering inflation and (b) considering inflation at a rate of 11% per year.

Solution

(a) Figure 12.2 presents the cash flows. The present worth P is found using $i = 15\%$ and Eq. (2.18) for the escalating series which has its present worth P_E in year 4.

$$P = 35,000 + 7000(P/A, 15\%, 4) + \left\{ 7000 \frac{[(1.12/1.15)^9 - 1]}{0.12 - 0.15} \right\}(P/F, 15\%, 4)$$

$$= 35,000 + 19,985 + 28,247$$

$$= \$83,232$$

Note that $n = 4$ in the P/A factor because the \$7000 in year 5 is D in Eq. (2.18). The term in { } is P_E in year 4 (Fig. 12.2), which is moved to time 0 with the $(P/F, 15\%, 4)$ factor.

(b) With inflation taken into account, it is necessary to use the inflated interest rate from Eq. (12.5) in the present-worth equation.

$$i_f = 0.15 + 0.11 + (0.15)(0.11) = 0.2765$$

$$P = 35,000 + 7000(P/A, 27.65\%, 4) + \left\{ \frac{7000[(1.12/1.2765)^9 - 1]}{0.12 - 0.2765} \right\}(P/F, 27.65\%, 4)$$

$$= 35,000 + 7000(2.2545) + 30,945(0.3766)$$

$$= \$62,436$$

Figure 12.2
Cash-flow diagram, Example 12.10.

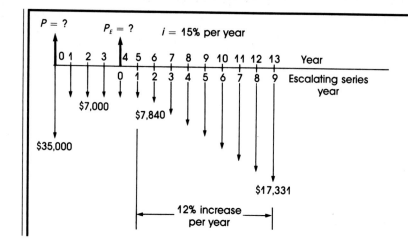

Comment You can check the result of the escalating-series factor in part (a) by multiplying each of the amounts in years 5 through 13 by the appropriate P/F factor for $i = 15\%$. The same thing can be done in part (b) using the inflated interest rate.

Example 12.11

An activated sludge wastewater treatment plant is expected to have the following equipment purchase costs:

Equipment	Cost
Preliminary treatment	$ 20,000
Primary treatment	30,000
Activated sludge	14,000
Clarification	47,000
Chlorination	21,000
Digestion	60,000
Vacuum filtration	17,000
Total cost	$209,000

The multiplication factor for the cost of installation for piping, concrete, steel, insulation, supports, etc. is 0.49. The construction factor is 0.53 and the indirect cost factor is 0.21. Determine the total plant cost if (a) all cost factors are to be applied to the purchase cost of the equipment, and (b) the indirect cost factor is to be applied to the total direct cost.

Solution

(a) The total equipment cost is $209,000. Since both the direct and indirect cost factors are to applied to only the equipment cost, the overall cost factor is

$$h = 1 + 0.49 + 0.53 + 0.21 = 2.23$$

Thus, the total plant cost is

$$C_T = 2.23(209,000) = \$466,000$$

(b) For this part of the problem, the total direct cost must be calculated first. The overall direct cost factor is

$$h = 1 + 0.49 + 0.53 = 2.01$$

The total direct cost is

$$2.01(209,000) = \$420,000$$

Now apply the indirect cost factor to the total direct cost, after adding 1 to the 0.21 factor:

$$C_T = 1.21(420,000)$$
$$= \$508,000$$

Comment   Note the difference in estimated plant cost when the indirect cost is applied to the equipment cost only [part (a)] as opposed to when it is applied to total direct cost [part (b)]. This illustrates the importance of determining what the factors apply to before they are used.

Problems

Note: For the problems below, all costs are in terms of then-current dollars unless otherwise specified.

12.1   Calculate the present worth of $23,000 six years from now if the interest rate is 15% per year and the inflation rate is 10% per year.

12.2   Calculate the present worth of $15,000 eight years from now if the interest rate is 12% per year and the inflation rate is 10% per year.

12.3   Find the present sum of money that would be equivalent to the future sums of $5000 in year 6 and $7000 in year 8 if the interest rate is 15% per year and the inflation rate is (a) 10% and (b) 16% per year.

12.4   How much money could the Kill-Kow Cattle Company afford to spend now for a new tractor trailer in lieu of spending $65,000 three years from now if the interest rate is 13% per year and the inflation rate is 7% per year?

12.5   The manager of the Pick 'N' Pak Food store is trying to determine how much should be spent now to avoid spending $10,000 on freezer equipment 2 years from now. If the interest rate is $1\frac{1}{2}\%$ per month and the inflation rate is 1% per month, what is the maximum amount of money the manager could afford to spend?

12.6   A lucky girl was just told that her great-grandfather died and left her his entire savings account which contained $3 million. If he started the account 50 years ago with a single deposit and never added even a single dollar to the account after the initial deposit, how much did he deposit? Assume that the account earned interest at a rate of 20% per year and that the inflation rate during that time period was 5% per year.

12.7   For the cash flow shown below, calculate the value of *P* by (a) using the inflated interest rate and (b) converting the annual series amounts into today's dollars and then using the appropriate *P/F* factors. Use an interest rate of 10% per year and an inflation rate of 8% per year.

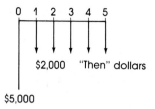

$5,000

12.8 How much money could the GRQ Company afford to spend now in order to avoid spending $5000 per year for 6 years if the interest rate is 15% per year and the inflation rate is 10% per year?

12.9 Two machines under consideration by May-Kit Metal Fabricating Company will have the following costs:

	Machine A	Machine C
First cost	$10,000	$20,000
Annual operating cost	8,000	3,000
Salvage value	3,000	6,000
Life, years	10	10

Which machine should be selected if $i = 15\%$ per year and inflation is 12% per year? Use a present-worth analysis.

12.10 If a company can buy a used computer system for $600,000 with the expectation of spending $48,000 per year for operation and maintenance, how much could it afford to spend on a new one if its annual cost would be only $38,000 per year? Assume that the salvage values will be 10% of the first cost and that both machines will have a 5-year life. Use $i = 15\%$ and $f = 10\%$ per year.

12.11 An engineer is trying to decide which of two machines he should purchase to manufacture a certain part. He obtains estimates from two salesmen, but salesman A gives him the expected costs in constant value dollars (i.e., today's dollars) and salesman B gives him the expected costs in then-current dollars. If the company's minimum attractive rate of return is 15% per year and the company expects inflation to be 10% per year, which salesman's product should he purchase, assuming that their expected costs are correct. Use a present-worth analysis.

	Salesman A (today's dollars)	Salesman B (then dollars)
First cost	$60,000	$95,000
Annual operating cost	25,000	35,000
Life, years	5	10

12.12 An 8% $20,000 bond with interest payable quarterly will mature in 13 years. How much should you pay for the bond if you desire to make a rate of return of a nominal 16% per year compounded quarterly and the inflation rate is 2% per quarter?

12.13 A new pickup truck has a first cost of $9000 with an annual operating cost of $800 the first 2 years, increasing by 9% per year thereafter for years 3 through 8. If the interest and inflation rates are both 0, what is the present worth of the truck?

12.14 A solar hot water heater which is expected to have a 15-year life can be purchased and installed for $3500. The energy savings for the first year are expected to be $300. With an inflation rate of 8% and an interest rate of 10% per year, what percentage increase per year in energy savings must be realized in order for the investment to be justified? Assume that inflation is to be considered.

12.15 Calculate the number of then-current dollars that would be required in 9 years to recover a present investment of $12,000 at an interest rate of 15% per year and an inflation rate of 8% per year.

12.16 If $23,000 is invested now at an interest rate of 20% per year, (*a*) how much money would be *accumulated* in 7 years if the inflation rate were 10% per year and (*b*) how many then-current dollars would be required to preserve the buying power of the original $23,000 and earn interest?

12.17 What future amount of money in then-current dollars 6 years from now would be equivalent to a present sum of $80,000 at an interest rate of 18% per year and an inflation rate of 12% per year?

12.18 Calculate the number of (*a*) today's dollars and (*b*) then-current dollars in year 10 that will be equivalent to a present investment of $33,000 at an interest rate of 15% per year and an inflation rate of 10% per year.

12.19 If the R-Gone Sign Company invests $3000 per year for 8 years beginning 1 year from now in a new production process, how much money must be received in a lump sum in year 8 in then-current dollars in order for the company to recover its investment at an interest rate of 13% per year and an inflation rate of 10% per year?

12.20 A woman deposited $1300 per year for 6 years in a savings account (her first deposit was made at the end of year 1). In years 7 to 12, she deposited $2000. How much money did she have in the account after the last deposit in terms of today's buying power if the interest rate she received was 12% per year and the inflation rate was 9% per year? (*Hint:* Greater buying power of earlier deposits should not be considered in this problem.)

12.21 If a program directed toward reducing product losses saves $6000 per year, what total equivalent saving would the company achieve in 5 years if the interest rate is 15% per year and the inflation rate is 8% per year?

12.22 A bank is offering a 10-year certificate of deposit (CD) at an interest rate of 10% per year. Any person who takes the offer will also receive a free toaster. If the inflation rate during the next 10 years is expected to be 13% per year, how much money must a person have at the end of 10 years just to be able to buy the same things she can buy now for $10,000? Assume the cost of those things will increase by only 8% per year.

12.23 How much money must a meat packing company expect to save each year for 12 years through by-product recovery to justify the expenditure of $35,000 on a flotation system, if the interest rate is 20% per year and inflation rate is 7% per year?

12.24 How much money could the Fix-It Tool Co. afford to spend each year on maintenance if replacement costs 6 years from now will be reduced from $900,000 to $700,000 as a result of the maintenance? Assume that the interest rate is 18% per year and the inflation rate is 8% per year.

12.25 A young couple planning for their daughter's college education 16 years from now would like to have money with an equivalent buying power of $15,000 in today's dollars. How much money will they have to deposit at the end of each year beginning 1 year from now if the interest rate they will earn is 12% per year and the inflation rate is 10% per year?

12.26 A small building will cost $500,000 to construct and will have maintenance costs of $5000 per year for 20 years. If the maintenance costs are expressed in today's dollars, calculate the equivalent uniform annual cost of the building in then-current dollars if the interest rate is 14% per year and the inflation rate is 10% per year.

12.27 Calculate the perpetual equivalent uniform annual cost of $50,000 now, $10,000 five years from now, and $5000 per year thereafter if the interest rate is 12% per year and the inflation rate is 9% per year.

12.28 The GRQ Company is trying to decide whether to spend $10,000 for new office furniture now or wait and purchase it in 3 years. If the company's MARR is 20% per year and the inflation rate is expected to be 9% per year, (a) what equivalent amount could the company spend in 3 years, and (b) how much will the furniture actually cost if its price increases with the inflation rate?

12.29 If a piece of equipment had a cost of $20,000 in 1980 when the M&S equipment index was 659.6, what would it be expected to cost when the index is at 900?

12.30 An item which had a cost of $7000 in 1977 was estimated to cost $13,000 in 1988. If the cost index which was used in the calculation had a value of 326 in 1977, what was its value in 1988?

12.31 A certain labor cost index had a value of 426 in 1960 and 933 in 1980. If the labor cost for constructing a building was $160,000 in 1980, what would the labor cost have been in 1960?

12.32 Use the *ENR* construction index (Table 12.5) to update a cost of $325,000 in 1981 to a 1986 figure.

12.33 If one wanted 1970 to equal 100 for the *CE* plant cost index (Table 12.5), what would be the value of the index in 1983?

12.34 Use the *F/P* or *P/F* factor to calculate the average percent increase per year between 1975 and 1985 for the *CE* plant cost index.

12.35 What will be the value of the M&S equipment cost index in 1996 if it was 797.6 in 1986 and it increases by 6% per year?

12.36 If a 15-horsepower pump has a cost of $2300, what would a 120-horsepower pump be expected to cost if the exponent for the cost-capacity equation is 0.69?

12.37 Mechanical dewatering equipment for a 1-million-gallon-per-day (MGD) plant would cost $400,000. How much would the equipment cost for a 17-MGD plant if the exponent in the cost-capacity equation is 0.63?

12.38 Site work cost for construction of a manufacturing facility which has a capacity of 6000 units per day was $55,000. If the cost for a plant with a capacity of 120,000 units per day was $3 million, what is the value of the exponent in the cost-capacity equation?

12.39 The equipment cost for a wine cooler bottling plant is $160,000. If the direct cost factor is 1.15 and the indirect cost factor is 0.31, estimate the total plant cost.

12.40 The equipment cost for a laboratory dedicated to immunological response research will cost $900,000. If the direct cost factor is 2.10 and the indirect cost factor is 0.26, what is the expected total cost of the laboratory?

12.41 Estimate the cost in 1986 of a 60 square-foot falling-film evaporator if the cost of a 30 square-foot unit was $19,000 in 1970. The exponent in the cost-capacity equation is 0.25. Use the M&S equipment index to update the cost.

12.42 Estimate the cost in 1986 of a 1000-horsepower steam turbine air compressor. A 200-horsepower unit cost $90,000 in 1975. The exponent in the cost-capacity equation is 0.29. Use the M&S equipment index to update the cost.

LEVEL FOUR

The three chapters in this level present the essentials of capital recovery or depreciation as it is utilized in economic analysis studies. The classical and newly introduced methods are covered.

Corporate taxation and the effects of income taxes upon the economic comparison of alternatives is discussed in two chapters. The first of these chapters presents the basic computational requirements, and the second examines the details of after-tax analysis using the alternative evaluation methods of present worth, annual worth, and rate-of-return analysis. The computation of equivalent annual revenue requirements is included in the last chapter of this level.

Chapter	Subject
13	Capital recovery and depletion models
14	Basic of taxation for corporations
15	After-tax analysis

Capital Recovery and Depletion Models 13

The capital investments of a corporation in equipment, vehicles, buildings, and machinery are commonly recovered through a tax-allowed expense deduction called depreciation. The process of depreciating an asset is referred to as capital recovery. An asset loses value for reasons such as age, wear, and obsolescence over its useful life. Even though an asset may be in excellent working condition, the fact that it is worth less through time should be taken into account in economic evaluation studies, especially those which include tax considerations. Income taxes are calculated using the relation

Taxes = (income − deductions)(tax rate)

Depreciation is one of the allowable deductions (along with wages, materials, rent, etc.) and therefore tends to lower the tax burden. Income taxes are described in detail in Chaps. 14 and 15.

    The objective of this chapter is to introduce you to the common methods of recovering the capital invested in assets through different models of depreciation. It teaches you to calculate annual depreciation and book-value amounts for different depreciation methods. The basics of the modified accelerated cost recovery system (MACRS) are presented and illustrated. Additionally, you are introduced to the concept of depletion for reserves of natural resources.

Section Objectives

After completing this chapter, you should be able to do the following:

13.1 Define the terms *depreciation* (*capital recovery*), *book value, market value, basis, recovery period, recovery rate,* and *salvage value.* Identify the common depreciation models and sketch book value versus recovery period for each.

13.2 Given the asset basis and recovery period, calculate annual depreciation, book value, and recovery rate using the *straight-line* method.

13.3 Given the asset basis and recovery period, calculate annual depreciation, book value, recovery rate, and implied salvage value for a specified year using a *declining-balance* method.

13.4 State and define two types of depreciable property and explain the role of the Asset Depreciation Range (ADR) system in recovery period determination.

13.5 State the guidelines and compute the annual depreciation and present worth of depreciation when *switching* from one depreciation model to another. Use the specified procedure to switch from declining-balance to straight-line depreciation.

13.6 Given the basis and recovery period, calculate annual depreciation using the *modified accelerated cost recovery system* (*MACRS*).

13.7 Given the basis, recovery period, and salvage value, calculate annual depreciation and recovery rate using the *sum-of-year-digits* method.

13.8 State the meaning and calculate the amount of the *capital expensing alternative* and an *investment tax credit.*

13.9 Define *depletion* and compute annual depletion using the *factor depletion* and *percentage depletion* methods.

Study Guide

13.1   Terminology

Some depreciation terms commonly used in economic analysis are defined here.

Depreciation or *capital recovery* is the reduction in value of assets owned by the corporation. Depreciation methods are based upon legally approved rules which do not necessarily reflect actual usage patterns of assets during ownership. Annual depreciation charges are tax deductible and indicated in corporate accounts.

Book value represents the remaining undepreciated investment on corporate books after the total amount of annual depreciation charges to date has been subtracted from the initial value. The book value is usually determined at the end of each year, which is consistent with the end-of-year convention used previously.

Market value is the actual amount that could be realized if an asset were sold on the open market. Due to the structure of tax laws, book value and market value may have substantially different numerical values. For example, a commercial building tends to increase in market value, but the book value will decrease as depreciation charges are taken. However, a computer system may have a market value much lower than its book value due to the rapidly changing technology of information systems. In most cases, the market value is used to perform the economic study for asset replacement or upgrading.

Basis or *first cost* is the installed cost of the asset including purchase price, delivery and installation fees, and other depreciable direct costs incurred to ready the asset for use. The term *unadjusted basis* is used when the asset is new, and *adjusted basis* may be used in years after depreciation has been charged.

Recovery period, also called the *depreciable life* or *class life*, is the life of the asset (in years) for depreciation and tax purposes. This may be somewhat different than the expected productive life span due to tax law rulings, management policies, and anticipated rapid obsolescence.

Depreciation rate or *recovery rate* is the fraction of the basis or first cost removed through depreciation from the corporate books each year. This rate may be the same (straight-line rate) or different for each year of the recovery period.

Salvage value is the expected trade-in or market value at the end of the useful life of the asset. The salvage amount may be expressed as a percentage of the first cost and it may be positive, zero, or negative if dismantling costs or carry-away costs are anticipated. Current depreciation methods approved for tax purposes usually assume a salvage value of zero, even though the actual salvage value may be positive. This may force the payment of extra income taxes when an asset is sold with a net realized value greater than the current book value.

Historically there have been several techniques approved for depreciating assets, with the *straight-line (SL)* method the most commonly used. However, *accelerated* methods such as *declining balance (DB)* are used by corporations since the book value decreases to zero (or to the salvage value) more rapidly than it does by the straight-line method, as shown in Fig. 13.1. In 1981 and 1986, respectively, in efforts to standardize accelerated write-off, the *accelerated cost recovery system (ACRS)* and *modified accelerated cost recovery system (MACRS)* were announced in the United States as the prime capital-recovery methods. Prior to these fundamental changes, it was acceptable to utilize the classical straight-line, declining-balance, and sum-of-year digits methods to reduce the book value to the anticipated salvage value. Until further tax law changes, all personal property purchased after 1981 must be depreciated using the ACRS, MACRS, or modified straight-line model explained in this chapter.

Major tax law revisions are occurring more rapidly in recent years, so the exact recovery rates used here may vary slightly from those in effect when you study this material. Nevertheless, the principles and formulas of the recovery methods remain consistent.

Probs. 13.1 and 13.2

13.2 Straight-Line (SL) Depreciation

The straight-line model is a popular method of depreciation and is used as the standard of comparison for most other methods. It derives its name from the fact that the book value decreases linearly with time because the recovery or depreciation

Figure 13.1
Idealized shape of
the book-value
curve for different
depreciation
models.

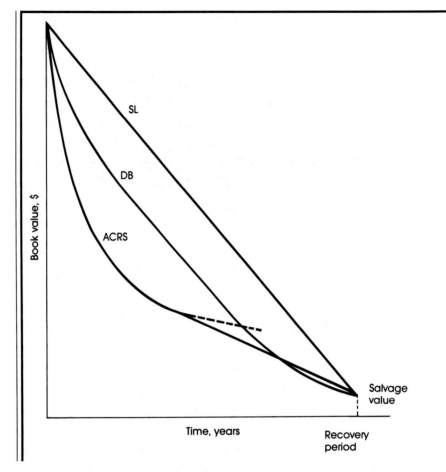

rate is the same each year. The annual depreciation is determined by dividing the
first cost, or unadjusted basis minus any salvage value, by the recovery period of the
asset. In depreciation formulas, we use B for the unadjusted basis because it may be
different for depreciation purposes than the first cost, which has previously been
called P. In equation form,

$$D_t = \frac{B - SV}{n} \tag{13.1}$$

where t = year $(t = 1, 2, \ldots, n)$
    D_t = annual depreciation charge
    B = first cost or unadjusted basis
   SV = salvage value
    n = expected depreciable life or recovery period

Since the asset is depreciated by the same amount each year, the book value
after t years of service, BV_t, will be equal to the first cost of the asset minus the
annual depreciation times t. Thus,

$$BV_t = B - tD_t \tag{13.2}$$

The *recovery rate* d_t is the same for each year t.

$$d_t = \frac{1}{n}$$ (13.3)

The relations above are illustrated in Example 13.1.

Example 13.1

If an asset has a first cost of $50,000 with a $10,000 salvage value after 5 years, (*a*) calculate the annual depreciation and (*b*) compute and plot the book value of the asset after each year using SL depreciation.

Solution

(*a*) The depreciation charged each year can be found by Eq. (13.1).

$$D_t = \frac{B - SV}{n} = \frac{50,000 - 10,000}{5}$$

$$= \$8000 \text{ per year for 5 years}$$

(*b*) The book value of the asset after each year can be found by Eq. (13.2).

$$BV_t = B - tD_t \qquad (t = 1, 2, 3, 4, 5)$$

$$BV_1 = 50,000 - 1(8000) = \$42,000$$

$$BV_2 = 50,000 - 2(8000) = \$34,000$$

$$\dots\dots\dots\dots\dots\dots\dots\dots\dots$$

$$BV_5 = 50,000 - 5(8000) = \$10,000 = SV$$

A plot of BV_t versus t is given in Fig. 13.2.

Probs. 13.3 to 13.6

Figure 13.2 Example 13.1, straight-line-depreciation model of capital recovery.

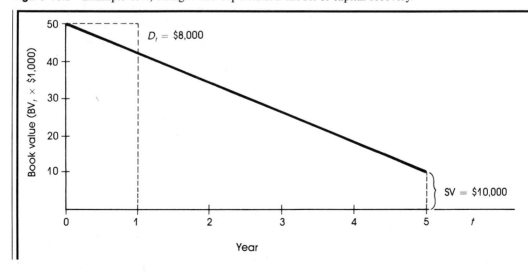

13.3 Declining-Balance (DB) and Double-Declining-Balance (DDB) Depreciation

The declining-balance method of depreciation, also known as the uniform- or fixed-percentage method, is an accelerated write-off technique. Very simply, the depreciation charge for any year is determined by multiplying a *uniform percentage* by the book value (adjusted basis) at the beginning of that year. For example, if the uniform-percentage rate were 10%, then the depreciation write-off for any given year would be 10% of the book value for that year. Obviously, the depreciation charge is largest in the first year and decreases each succeeding year.

    The maximum percentage depreciation that is permitted is 200% (double) the straight-line rate. When this rate is used, the method is known as the double-declining balance (DDB) method. Thus, if an asset had a useful life of 10 years, the straight-line depreciation rate would be $1/n = 1/10$. The uniform rate of 2/10 therefore could be used in the DDB method. The general formula for calculating the maximum DB rate d_M is two times the straight-line recovery rate.

$$d_M = \frac{2}{n} \tag{13.4}$$

This is the rate used for the DDB method. Other commonly used rates for the DB method are 175 and 150% of the straight-line rate, where $d = 1.75/n$ and $d = 1.50/n$, respectively. The actual depreciation rate, relative to the unadjusted basis B for year t is computed as

$$d_t = d(1 - d)^{t-1} \tag{13.5}$$

    When DB or DDB depreciation is used, the estimated salvage value should not be subtracted from the first cost when calculating the depreciation charge. It is important that you remember this, since this procedure further increases the rate of write-off. Even though salvage values are not considered in the depreciation calculation, an asset may not be depreciated below a reasonable salvage value or zero. If salvage value is reached prior to year n, no additional depreciation may be taken thereafter except under the ACRS and MACRS methods (Sec. 13.6).

    The depreciation D_t for year t is the uniform rate d times the book value at the end of the previous year, that is,

$$D_t = (d)BV_{t-1} \tag{13.6}$$

If the BV_{t-1} value is not known, the depreciation charge is

$$D_t = (d)B(1 - d)^{t-1} \tag{13-7}$$

The book value in year t is

$$BV_t = B(1 - d)^t \tag{13-8}$$

Since the salvage value in declining-balance methods does not go to zero, an implied SV after n years may be computed as

$$\text{Implied SV} = BV_n = B(1 - d)^n \tag{13-9}$$

If the implied SV is less than the expected SV, the asset will be totally depreciated before the end of its expected life (n) and vice versa (unless the method of depreciation is switched as discussed in Sec. 13.5). It is possible to compute an implied uniform rate d for comparison with the allowed d by using the expected SV. For SV > 0,

$$\text{Implied } d = 1 - \left(\frac{\text{SV}}{B}\right)^{1/n} \tag{13-10}$$

The allowed range on d is $0 < d < 2/n$. In all DB models, d is stated or calculated by Eq. (13.10); and for the DDB model, $d = 2/n$. Example 13.2 illustrates the DDB model and the Additional Example includes the computations for $d < 2/n$.

Example 13.2

Assume that an asset has a first cost of $25,000 and an expected $4000 salvage after 12 years. Calculate its depreciation and book value for (*a*) year 1, (*b*) year 4, and (*c*) the implied salvage value after 12 years using the DDB method.

Solution   First compute the DDB uniform rate.

$$d = \frac{2}{n} = \frac{2}{12} = 0.1667 \text{ per year}$$

(*a*) For the first year, the depreciation and book value can be calculated using Eqs. (13.7) and (13.8),

$$D_1 = (0.1667)25{,}000(1 - 0.1667)^{1-1} = \$4167.50$$

$$\text{BV}_1 = 25{,}000(1 - 0.1667)^1 = \$20{,}832.50$$

(*b*) For year 4, again using Eqs. (13.7) and (13.8) and $d = 0.1667$,

$$D_4 = 0.1667(25{,}000)(1 - 0.1667)^{4-1} = \$2411.46$$

$$\text{BV}_4 = 25{,}000(1 - 0.1667)^4 = \$12{,}054.40$$

(*c*) Using Eq. (13.9), the salvage value at $n = 12$ is

$$\text{Implied SV} = 25{,}000(1 - 0.1667)^{12} = \$2802.57$$

Since the stated salvage value is anticipated to be $4000, the asset will be completely depreciated before its expected 12-year life is over. Therefore, after the BV reaches $4000, no further depreciation charges would be allowed.

Comment   An important fact to remember about the DB and DDB methods is that salvage value is not subtracted from the first cost when the depreciation is calculated. However, when the book value reaches the expected salvage value ($4000 here), no additional depreciation may be taken. In this case $\text{BV}_{10} = \$4036.02$ and $D_{11} = \$672.80$, making $\text{BV}_{11} = \$3362.22$, which is less than the expected SV of $4000. Therefore, in years 11 and 12 the depreciation is $D_{11} = \$36.02$ and $D_{12} = 0$, respectively. If, on the other hand, a salvage value of zero had been assumed, the entire depreciation could be taken each year.

Additional Example 13.9
Probs. 13.7 to 13.13

13.4 Types of Property and Their Recovery Periods

There are two primary types of property for which depreciation is allowed—*personal* and *real*. *Personal property* is considered to be the income-producing possessions of a corporation used in the conduct of business. Included are property such as vehicles, manufacturing equipment, materials-handling devices, computers, telephone switching equipment, office furniture, refining process equipment, and most other manufacturing and service industry property. *Real property* includes real estate and improvements thereto such as manufacturing and office buildings, warehouses, apartments, and other structures. Land is not considered real property and is not depreciable.

Recovery periods for depreciation models are related to the expected useful life estimates, but for accelerated depreciation a statutory recovery period is established using the Asset Depreciation Range (ADR) system. The ADR is the expected mid-point life of each asset with an accompanying recovery period for capital-recovery purposes. Table 13.1 details several of the recovery-period and ADR relations we will use in later sections and examples. The resulting recovery period is the n value used in a depreciation model. The advantage of the shortened recovery period com-

Table 13.1 Example recovery periods for different classes of property based upon the Asset Depreciation Range (ADR) system

ADR life in years, T	Examples of property	Recovery period, n years
	Personal property classification	
$T \leq 4$	Special manufacturing and handling devices; some motor vehicles; racehorses	3
$4 < T < 10$	Computers; duplicating equipment; trucks; cargo containers; semi-conductor manufacturing equipment	5
$10 \leq T < 16$	Office furniture; many fixtures and equipment; railroad track; agricultural structures	7
$16 \leq T < 20$	Durable-goods manufacturing equipment for castings, forgings; petroleum refining equipment	10
$20 \leq T < 25$	Municipal sewage treatment plants; telephone distribution; barges and tugs	15
$25 \leq T$	Any asset not listed elsewhere having an ADR midpoint life of 25 years or more	20
	Real property classification	
All values	Residential rental property	27.5
All values	Nonresidential rental property	31.5

pared to the anticipated useful life is compounded by the use of accelerated depreciation models which write off more of the basis in the initial years.

Probs. 13.14 and 13.15

13.5   Switching between Depreciation Models

Switching between models for depreciation is done within legal limits to more rapidly reduce the book value and maximize the present value of the total depreciation over the recovery period. Switching also increases the income tax advantage in years where the depreciation is larger by one method than another. The approach you learn in this section is an inherent part of the modified accelerated cost recovery system (MACRS) imposed by the Tax Reform Act of 1986 (Sec. 13.6).

Switching from a DB model to the SL method is most common because it may offer a real advantage, especially if the DB model is the DDB, i.e., the uniform percentage is twice the straight-line rate.

Pertinent rules of switching may be summarized as follows:

1. Switching is recommended when the depreciation for year t using the established method is less than that for a new method. The selected depreciation D_t is the maximum amount.
2. Regardless of the methods used, book value can never go below the salvage value set at purchase time. We will assume a zero salvage value in all switching considerations. BV_t is the adjusted basis.
3. The undepreciated amount or book value, BV_t, is used as the adjusted basis of computation to select D_t when switching is considered.
4. When switching from a declining-balance method, the anticipated salvage value, not the implied salvage, is used to compute the depreciation for the new method.
5. Only one switch can commonly take place during the depreciable life.

In all situations, the criteria of maximizing the present worth of the total depreciation P_D is used in switching determination. The depreciation method, or methods (using switching) which results in the *maximum present worth*, is the best strategy, where

$$P_D = \sum_{t=1}^{n} D_t(P/F, i\%, t) \tag{13.11}$$

This is correct because it minimizes tax liability in the early part of an asset's life (Sec. 14.6 includes a further discussion of the effects of depreciation on taxes).

Virtually all switching occurs from a rapid write-off method to the SL method. The most promising, as mentioned earlier, is the DDB-to-SL switch. This switch is predictably advantageous if the implied salvage computed in Eq. (13.9) exceeds the salvage estimated at purchase time; that is, switch if

$$BV_n = B(1 - d)^n > \text{estimated SV} \tag{13.12}$$

Since we commonly assume that the estimated SV is zero, BV_n will be *greater than zero*; therefore, a switch to SL is advantageous. Depending upon the values of d and *n*, the switch may be best in the last year of the recovery period, which removes the implied SV present in the DDB method.

The procedure to switch from DDB to SL depreciation is:

1. For each year *t*, compute the two depreciation charges.

$$\text{For DDB: } D_D = (d)\, BV_{t-1} \tag{13.13}$$

$$\text{For SL: } \quad D_S = \frac{BV_{t-1}}{n - t + 1} \tag{13.14}$$

2. For each year, select the maximum value. The depreciation for $t = 1, 2, \ldots, n$ is D_t.

$$D_t = \max[D_D, D_S] \tag{13.15}$$

3. If required, compute the present worth of total depreciation charges using Eq. (13.11).

It is acceptable, though not usually financially advantageous, to state that a switch will take place in a particular year, for example, a mandated switch from DB to SL in year 7 of a 10-year recovery period. This approach is usually not taken, but the switching technique will perform correctly for any depreciation models involved with switching in any year $t \le n$.

Example 13.3

A to Z Microelectronics has purchased a $10,000 truck with an estimated 8 years of service. Compute the annual capital recovery and calculate and compare the present worth of total depreciation for (*a*) the straight-line method, (*b*) the double-declining-balance method, and (*c*) the DDB-to-SL switching model. Let $i = 15\%$ per year.

Solution With an expected ADR life of 8 years, Table 13.1 indicates that the recovery period is $n = 5$ years.

(*a*) For $B = \$10,000$, SV $= 0$, and $n = 5$, Eq. (13.1) is used to determine the annual capital recovery.

$$D_t = \frac{10,000 - 0}{5} = \$2000$$

Since D_t is the same for $t = 1, 2, \ldots, 5$, the P/A factor can replace P/F in Eq (13.11) to compute P_D.

$$P_D = 2000(P/A, 15\%, 5) = 2000(3.3522) = \$6704.40$$

(*b*) For DDB depreciation, $d = 2/n = 0.40$. Equation (13.6) or (13.7) can be utilized to compute the results in Table 13.2. The present-worth value $P_D = \$6,991.63$ exceeds the comparable value for SL depreciation above, which indicates that DDB maximizes the present worth of depreciation by an accelerated recovery of the $10,000 investment.

(*c*) Use the DDB-to-SL switching procedure.

Table 13.2 Depreciation and present-worth computations using DDB (For $n = 5$, $B = \$10,000$, and $d = 0.40$)

Year, t	D_t	BV_t	$(P/F, 15\%, t)$	Present worth of D_t
0	—	$10,000.00	—	—
1	$4,000.00	6,000.00	0.8696	$3,478.40
2	2,400.00	3,600.00	0.7561	1,814.64
3	1,440.00	2,160.00	0.6575	946.80
4	864.00	1,296.00	0.5718	494.04
5	518.40	777.60	0.4972	257.75
	$9,222.40			$P_D = \$6,991.63$

Table 13.3 Depreciation and present worth for DDB-to-SL switching, Example 13.3c

Year, t	DDB model		D_s, Eq. (13.14)	D_t	$(P/F, 15\%, t)$	Present worth of D_t
	D_D	B_t				
0	—	$10,000.00	—	—	—	—
1	$4,000.00	6,000.00	$2,000	$4,000	0.8696	$3,478.40
2	2,400.00	3,600.00	1,500	2,400	0.7561	1,814.64
3	1,440.00	2,160.00	1,200	1,440	0.6575	946.80
4	864.00	1,296.00	1,080	1,080	0.5718	617.54
5	518.40	777.60	1,296	1,080	0.4972	536.98
	$9,222.40			$10,000		$7,394.36

1. The DDB values in Table 13.2 are repeated in Table 13.3 as D_D for comparison with the D_S values from Eq. (13.14). Note that the D_S values change each year since the adjusted basis BV_{t-1} is different. Only in year $t = 1$ is $D_S = \$2000$, the same as computed in part (a). For illustration, the following D_S values are computed.

Year $t = 2$: $BV_1 = \$6000$ by DDB method

$$D_S = \frac{6000 - 0}{5 - 2 + 1} = \$1500$$

Year $t = 4$: $BV_3 = \$2160$ by DDB method

$$D_S = \frac{2160 - 0}{5 - 4 + 1} = \$1080$$

2. The D_t values shown in Table 13.3 include a switch to SL in year 4 with $D_4 = D_5 = \$1080$. The amount $D_S = \$1296$ in year 5 would apply only if the switch occurred in year 5. Total depreciation with switching is $10,000 compared to the $9222.40 with DDB alone. Book-value plots with and without switching are shown in Fig. 13.3.
3. With switching, $P_D = \$7394.36$ (Table 13.3), which indicates another increase in depreciation present worth over both SL and DDB with no switch.

Figure 13.3 Comparison of book value for DDB and DDB-to-SL switching in year 4, Example 13.3.

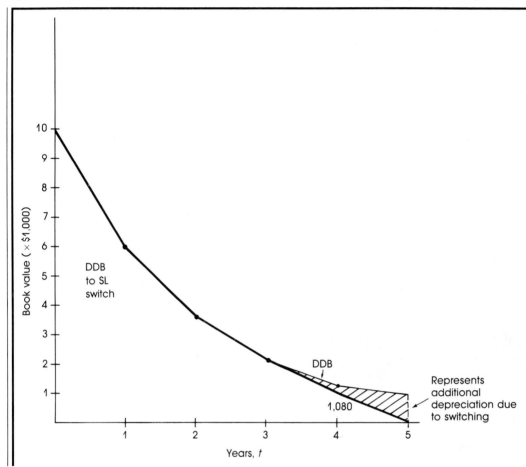

13.6 Modified Accelerated Cost Recovery System (MACRS)

The Economic Recovery Act of 1981 introduced the ACRS system of capital recovery, and the 1986 Tax Reform Act modified it to MACRS. Both systems dictate statutory depreciation rates for all personal and real property while taking advantage of accelerated methods of depreciation. The mechanisms of MACRS as it materially affects economic analysis are presented here. Many aspects of MACRS are more appropriate to depreciation accounting than they are to the evaluation of investment alternatives. If you are interested in additional details and derivation of MACRS depreciation rates, refer to Appendix B.

Both ACRS and MACRS compute the annual depreciation using the relation

$$D_t = d_t B \qquad (13.16)$$

The unadjusted basis B is completely depreciated so the assumption that $SV = 0$ is automatically made, even though there may well be a true $SV > 0$. The MACRS recovery periods are standardized to the Table 13.1 values. Periods of 3, 5, 7, and 10 years for personal property are most commonly used.

Accelerated methods and switching from DB to SL depreciation are inherent components of MACRS for *personal property*. The rates, for example, may start with DDB $(d_t = 2/n)$ and switch to the SL rate when the SL method offers a faster write-off. The SL method *must* be used for *real property* throughout the recovery period.

A *half-year convention* is imposed which assumes that all property is placed in service at the midpoint of the acquisition year. Therefore, only 50% of the first year DB depreciation applies for tax purposes. This removes some of the accelerated depreciation advantage and requires that one-half year of depreciation be taken in year $n + 1$. (We clearly state when the half-year convention is utilized in examples and problems.)

For recovery periods $n = 3$, 5, 7, and 10, DDB with half-year-convention switching to SL applies. When the switch to SL takes place, which is usually in the last 1 to 3 years of the recovery period, any remaining basis is charged off in year $n + 1$ to reach $BV_n = 0$. This is usually 50% of the applicable SL amount after the switch has occurred. For periods of 15 and 20 years, 150% DB with the half-year convention and the switch to SL applies.

For annual depreciation computations on personal property the unadjusted basis is multiplied by the MACRS rate using Eq. (13.16). The d_t values for $n = 3$, 5, 7, 10, and 15 are presented in Table 13.4. (These rates are derived in Appendix B.) The Internal Revenue Service rates may be slightly different due to round-off. Table 13.4 values are used throughout the remainder of this text.

Table 13.4 Recovery rates d_t applied to unadjusted basis B for the MACRS method

Year,	Recovery rate, d_t (%)				
	Recovery period, years				
t	$n = 3$	$n = 5$	$n = 7$	$n = 10$	$n = 15$
1	33.3	20.0	14.3	10.0	5.0
2	44.5	32.0	24.5	18.0	9.5
3	14.8	19.2	17.5	14.4	8.6
4	7.4	11.5	12.5	11.5	7.7
5		11.5	8.9	9.2	6.9
6		5.8	8.9	7.4	6.2
7			8.9	6.6	5.9
8			4.5	6.6	5.9
9				6.5	5.9
10				6.5	5.9
11				3.3	5.9
12–15					5.9
16					3.0

Table 13.5 Depreciation and book value using MACRS, Example 13.4

Year, t	Rate, d_t	D_t	BV_t
0	—	—	$10,000
1	0.20	$ 2,000	8,000
2	0.32	3,200	4,800
3	0.192	1,920	2,880
4	0.115	1,150	1,730
5	0.115	1,150	580
6	0.058	580	0
	1.000	$10,000	

$$P_D = \sum_{t=1}^{6} D_t(P/F, 15\%, t) = \$6901.20$$

Example 13.4

In Example 13.3(c) the DDB-to-SL switching model was applied to a $10,000 eight-year asset resulting in $P_D = \$7394.36$ at $i = 15\%$. Use MACRS to depreciate the same asset using a 5-year recovery period (as indicated by Table 13.1) and compare the P_D values.

Solution Table 13.5 summarizes the computations for depreciation (using Table 13.4), book value, and present worth of depreciation. The P_D values for the different methods are:

Straight-line: $6704.40
Doubling-declining-balance: $6991.63
DDB-to-SL switch: $7394.36
MACRS: $6901.20

Based on the slightly smaller P_D value for MACRS compared to DDB and switching, MACRS provides a less accelerated write-off. This is true because the half-year convention disallows 50% of the first-year depreciation (which amounts to 20% of the basis) compared to other DB-based methods.

The only alternative to the MACRS method is an election to use the SL method with the half-year convention, no salvage value, and a statutory recovery period which is usually longer than the MACRS period. Thus, the advantages of accelerated depreciation and maximization of depreciation present worth are lost. However, an economic analysis may well include the SL method for simplicity of computations. With the half-year convention and a recovery of n years, the SL recovery rates for this alternative to MACRS are:

$$d_t = \begin{cases} \dfrac{1}{2n} & t = 1, n+1 \\ \dfrac{1}{n} & t = 2, 3, \ldots, n \end{cases} \qquad (13.17)$$

Probs. 13.21 to 13.26

13.7   Sum-of-Year-Digits (SYD) Depreciation

The SYD method is a classical accelerated-depreciation technique which removes much of the basis in the first one-third of the recovery period; however, write-off is not as rapid as DDB or MACRS. This technique, though not incorporated or legal in the current MACRS method for real property, is often utilized in economic analyses for rapid depreciation of invested capital and in the actual depreciation of multiple-asset accounts (group and composite depreciation). For the sake of completeness, it is included here.

The mechanics of the method involve initially finding the sum of the year digits from 1 through n. The number obtained represents the sum of year digits. The depreciation charge for any given year is obtained by multiplying the basis of the asset less any salvage value $(B - SV)$ by the ratio of the number of years remaining in the recovery period to the sum of year digits.

$$D_t = \frac{\text{depreciable years remaining}}{\text{sum of year digits}}(\text{basis} - \text{salvage value})$$

$$= \frac{n - t + 1}{S}(B - SV) \tag{13.18}$$

where

$S = $ sum of year digits 1 to n

$$= \sum_{j=1}^{n} j = \frac{n(n + 1)}{2} \tag{13.19}$$

Note that the depreciable years remaining must include the year for which the depreciation charge is desired. That is why the "1" has been included in the numerator of Eq. (13.18). For example, to determine the depreciation for the fourth year of an asset which has an 8-year life, the numerator of Eq. (13.18) is $8 - 4 + 1 = 5$ and $S = 36$.

The book value for any given year can be calculated without making the year-by-year depreciation determinations as follows:

$$\text{BV}_t = B - \left[\frac{t(n - t/2 + 0.5)}{S}\right](B - SV) \tag{13.20}$$

The rate of depreciation d_t, which decreases each year for the SYD method, follows the multiplier in Eq. (13.18); that is,

$$d_t = \frac{n - t + 1}{S} \tag{13.21}$$

Example 13.5

Calculate the depreciation charges for years 1, 2, and 3 for electrooptics equipment with $B = \$25,000$, $SV = \$4000$, and an 8-year recovery period.

Solution The sum of year digits is $S = 36$ and the depreciation amounts for the first 3 years by Eq. (13.18) are:

$$D_1 = \frac{(8 - 1 + 1)}{36}(25{,}000 - 4000) = \$4667$$

$$D_2 = \frac{7}{36}(21{,}000) = \$4083$$

$$D_3 = \frac{6}{36}(21{,}000) = \$3500$$

Figure 13.4 is a plot of the book values for an $80{,}000 asset with $SV = \$10{,}000$ and $n = 10$ years. The MACRS, DDB, and SYD curves track closely except for year 1 and years 9 through 11. You should compute the D_t and BV_t values to confirm the results of Fig. 13.4.

Probs. 13.27 to 13.29

Figure 13.4 Comparison of book value for an asset ($B = \$80{,}000$, $n = 10$ years, and an anticipated $SV = \$10{,}000$) under the SL, SYD, DDB, and MACRS models of depreciation.

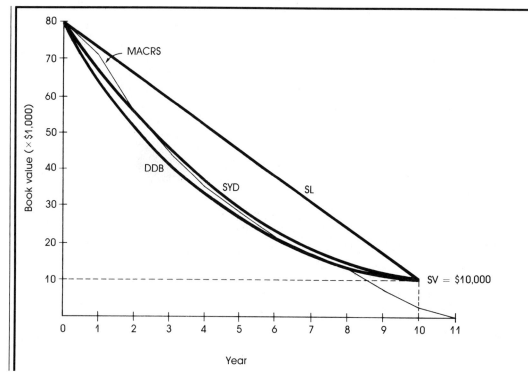

13.8   Capital-Expensing Alternative and Investment Tax Credit

Rather than depreciate the entire basis of a capital investment of no more than $200,000, it is an acceptable alternative to expense up to $10,000 during the first year of service, provided the basis is reduced prior to depreciation by the amount expensed. This is an incentive for smaller corporations to invest in capital goods. This allowance, historically called the *Section 179 expense*, is, however, additionally limited in that any amount invested in excess of $200,000 reduces the $10,000 expensing limit dollar-for-dollar. Thus, any basis over $210,000 prohibits expensing allowance.

As an example, a 3-year recovery property costing $50,000 can have $10,000 expensed in year $t = 1$, and the first-year MACRS depreciation will be $0.333(50,000 - 10,000) = \$13,320$. Therefore, $13,320 is written off in year 1 and the remaining $26,680 is recovered in years 2, 3, and 4. Of course a corporation can claim only a single $10,000 amount in each year, so the allowance is very limited in its overall impact upon most large corporations.

The *investment tax credit* (ITC) has been used in the United States to encourage investment in modern plants and equipment in capital-intensive industries by giving corporations (and in rare cases, individuals) direct tax credits of 6 to 10% of the first cost without reducing the basis for depreciation purposes. That is, the ITC reduces income taxes directly while depreciation is only a deduction from taxable income before taxes are computed. This advantage for ITC was removed by the Tax Reform Act of 1986, which repealed the ITC because it was considered inflationary and unnecessary due to the introduction of MACRS. Since this tax credit has been historically present, you should know how it functions. The ITC guidelines are outlined for reference using the 1981–1986 tax laws (the ACRS-era).

A 10% ITC allows $0.1B$ to be considered a tax credit. Under 1981–1986 law, 100% of this tax was allowable if the ACRS recovery period exceeded 3 years, and 60%, or $0.06B$, for n $\leqslant$ 3 years. The investment tax credit was creditable at the time of asset purchase, that is, year 0. The tax credit claimed in a year is limited by the income tax liability or some specified amount for each year. If the credit exceeds this amount, the unclaimed portion can be carried back for 3 and forward for 15 succeeding tax years. In addition, this tax credit did not reduce the total depreciable value of an asset, but it must be taken on the reduced basis if a Section 179 capital expense was claimed.

Example 13.6

Asset numbers 309 and 318 were purchased in 1 year under ACRS depreciation. Compute the total of first-year depreciation, ITC, and capital expense allowance if both qualify for ITC and a full Section 179 expense was claimed for asset 309.

Asset number	Years, n	Cost, B
309	10	$75,000
318	3	8,000

Solution   Depreciation rates for ACRS are given in the appendix (Table B.1). Asset 309 total is $15,200 in deductions and a $7500 tax credit.

Capital expense: $10,000

ITC   :   0.1(75,000) = $ 7,500

ACRS   :   0.08(75,000 − 10,000) = $ 5,200

Asset 318 total is $2000 in deductions and a $480 tax credit.

$$ITC: 0.06(8,000) = \$ 480$$

$$ACRS: 0.25(8,000) = \$2,000$$

Problems 13.30, 13.31

13.9   Depletion Methods

We have thus far computed depreciation for an asset which has a value that can be recovered by purchasing a replacement. Depletion is similar to depreciation; however, depletion is applicable to natural resources, which, when removed, cannot be "repurchased," as can a machine or building. Therefore, a depletion method is applicable to natural deposits removed from mines, wells, quarries, forests, and the like.

There are two methods of depletion; factor, or cost, depletion and percentage depletion. *Factor depletion* is based on the level of activity or usage, not time, as in the case of depreciation. The depletion factor d'_t for year t is

$$d'_t = \frac{\text{initial investment}}{\text{resource capacity}} \tag{13.22}$$

and the annual depletion charge is d'_t times the year's usage or activity volume. As is the case for depreciation, accumulated depletion by the factor method cannot exceed the total cost of the resource.

Example 13.7

The Wilderness Company has purchased some forest acreage for $350,000 from which an estimated 175 million board feet of lumber are recoverable.
(a) Determine the depletion charges if 15 million and 22 million board feet are removed in the first 2 years, respectively.
(b) If after 2 years the total recoverable board feet is reestimated to be 225 million, compute the new d'_t for years 3 and later.

Solution
(a) Use Eq. (13.22) to compute d'_t for $t = 1, 2, \ldots, n$.

$$d'_t = \frac{350,000}{175} = \$2000 \text{ per million board feet}$$

Apply this rate to the annual usage to obtain depletion of $30,000 in year 1 and $44,000 in year 2. Continue using d'_t until a total of $350,000 is depleted.

(b) After 2 years, a total of $74,000 has been depleted. A new d'_t value must be based on the remaining $350,000 − 74,000 = $276,000 of undepleted investment. Additionally, with the new estimate of 225 million, a total of $225 − 15 − 22 = 188$ million board feet remain. Now, for $t = 3, 4, \ldots$,

$$d'_t = \frac{\$276,000}{188} = \$1468 \text{ per million board feet}$$

The second depletion method, that of *percentage depletion*, is a special consideration given when natural resources are exploited. A flat percentage of the resource's gross income may be depleted each year provided it does not exceed 50% of taxable income (see Chap. 15). Using percentage depletion, total depletion charges may exceed actual costs with no limitation. Since it is possible to use the depletion figure computed either by the factor or by the percentage method, the percentage depletion method is usually chosen, because of the possibility of writing off more than the original cost of the venture. Below are listed some of the percentages for activities that can use the percentage depletion method. The percentages are altered by the federal government when new depletion legislation is enacted.

Activity	Percentage of gross income
Oil and gas wells	15
Coal, sodium chloride	10
Gravel, sand, peat, some stones	5
Sulfur, cobalt, lead, nickel, zinc, etc.	22
Gold, silver, copper, iron ore	15

Example 13.8

A gold mine purchased for $750,000 has an anticipated gross income of $1.1 million per year for years 1 to 5 and $0.85 million per year after year 5. Compute annual depletion charges for the mine.

Solution   A 15% depletion applies to the gold mine. Thus, assuming that depletion charges do not exceed 50% of taxable income, depletion will be $0.15(1.1 \text{ million}) = \$165,000$ years 1 to 5 and $0.15 (0.85 \text{ million}) = \$127,500$ each year thereafter. At this rate, the cost of $750,000 will be recovered in approximately 4.5 years of operation.

A depreciation method similar to factor depletion, but applicable to depreciable assets, is the *unit-of-production* method. The asset can be depreciated using this method only if the rate of use or production is a measure of its rate of deterioration. The depreciation factor is computed in a fashion similar to Eq. (13.22), with the salvage value removed from the initial investment. For example, oil-producing equipment at a lease site costing $300,000 with a salvage of $50,000 may be depreciated on a basis of the estimated 1.0 million barrels of oil to be extracted from the lease. The depreciation factor is then $(300,000 - 50,000)/1,000,000 = \$0.25$ per barrel. This method is not commonly used because of the difficulty of finding a production unit which accurately measures the rate of asset deterioration.

Probs. 13.32 to 13.34

Additional Example

Example 13.9

The Rock Company has just purchased a computer controlled ore-crushing unit for $80,000. The unit has an anticipated life of 10 years and a salvage of $10,000. Use the declining-balance method to develop a schedule of depreciation and book values for each year.

Solution The depreciation rate from Eq. (13.10) using the expected SV of $10,000 is

$$d = 1 - \left(\frac{10,000}{80,000}\right)^{1/10} = 0.1877$$

Note that $0.1877 < 2/n = 0.2$, so this DB model does not exceed twice the straight-line rate. Table 13.6 presents the D_t values using Eq. (13.6) and the BV_t values from $BV_t = BV_{t-1} - D_t$, rounded to the nearest dollar. For example, at $t = 2$,

$$D_2 = (d)BV_1 = 0.1877(64,984) = \$12,197$$

$$BV_2 = 64,984 - 12,197 = \$52,787$$

Due to the round-off to even dollars, $2312 is calculated for depreciation in year 10; but $2318 is deducted to make $BV_{10} = SV = \$10,000$ exactly.

Table 13.6 D_t and BV_t values using declining-balance depreciation, Example 13.9

Year, t	D_t	BV_t
0	—	$80,000
1	$15,016	64,984
2	12,197	52,787
3	9,908	42,879
4	8,048	34,831
5	6,538	28,293
6	5,311	22,982
7	4,314	18,668
8	3,504	15,164
9	2,846	12,318
10	2,318	10,000

13.1   The ABC Company paid $52,000 for an asset in 1985 and installed it at a cost of $3000. **Problems**
The asset was expected to remain in service for 10 years and be sold for 10% of the
original purchase price. If the asset was sold in 1990 for $8700, write the values used in
depreciation analysis for the following: first cost, anticipated life, salvage value, actual
life, and market value in 1990; book value in 1990 if 75% of the first cost had been
written off in depreciation.

13.2   A $100,000 piece of testing equipment was installed and depreciated for 5 years. Each
year the end-of-year book value decreased at a rate of 10% of the book value at the
beginning of the year. The system was sold for $24,000 at the end of the 5 years. (a) Plot
the book-value curve for the 5 years and determine the difference between the book
value and the actual market value at the time of sale. (b) Compute the amount of the
annual depreciation. (c) What is the recovery rate in each year?

13.3   Smoothline Construction, Inc., has just purchased a machine for $275,000. A $75,000
installation charge is required to use the machine. The expected life is 30 years with a
salvage value of 10% of the purchase price. Use the straight-line (SL) method of depreciation to determine (a) first cost, (b) salvage value, (c) annual depreciation, and (d) book
value after 20 years.

13.4   A machine costing $12,000 has a life of 8 years with a $2000 salvage value. Calculate
the (a) depreciation charge and (b) book value of the machine for each year using the
straight-line method. (c) What is the rate of depreciation for this method? Explain the
meaning of this rate.

13.5   Compare accelerated- and decelerated-depreciation methods to the straight-line
method using the definition of depreciation rate.

13.6   A special purpose computer was purchased for $B = \$50,000$ and $n = 4$ years. Tabulate
the values for depreciation, accumulated depreciation, and book value for each year
under the conditions that (a) there is no salvage value, and (b) SV = $10,000 is expected.

13.7   For the declining balance method of depreciation, state the difference between the
uniform percentage rate d used to compute annual depreciation and the recovery rate
d_t, which varies with each year t.

13.8   What is the basic difference between the declining-balance and the double-declining-
balance methods of depreciation?

13.9   Work Prob. 13.4 using the double-declining-balance (DDB) method and plot the book
value for the SL and DDB depreciation methods.

13.10 A building costing $320,000 is expected to have a 30-year life with a 25% salvage value.
Calculate the depreciation charge for years 4, 9, 18, and 26 using the (a) straight-line
method and (b) double-declining-balance method.

13.11 Calculate the book value of the building in Prob. 13.10 for year 13 using the (a) straight-
line method and (b) double-declining-balance method.

13.12 Earth-moving equipment has been purchased. It has $B = \$82,000$, $n = 18$ years, and
SV = $15,000. For the years 2, 7, 12, and 18, compute the annual depreciation charge
using (a) double-declining-balance (DDB) depreciation and (b) DB depreciation with a
recovery rate calculated from the asset data. Compare the book values for these two
sets of computations.

13.13 Declining-balance depreciation at a rate of 1.5 times the straight-line rate is to be used
for automated process-control equipment with $B = \$175,000$, $n = 12$, and expected
SV = $32,000. Compute (a) the depreciation and (b) the book value for years 1 and
12. (c) Compare the expected salvage value and the salvage remaining by the DB
method.

13.14 In terms of asset value write-off, explain why the combination of the ADR system and accelerated-depreciation methods is a significant advantage over the use of straight-line depreciation using the actual life of an asset.

13.15 A newly installed, electronic desk-top printing system has $B = \$25,000$, actual $n = 10$ years, and no salvage value. On a single graph, plot the book-value curves for the following two situations: (a) straight-line depreciation for the actual life and (b) DDB depreciation if the ADR life is 10 years and the Table 13.1 recovery periods are acceptable for depreciation.

13.16 An asset has a first cost of $45,000, a recovery period of 5 years, and a $3000 salvage value. Use the switching procedure from DDB to SL depreciation to maximize the present worth of depreciation. Let $i = 18\%$ per year.

13.17 Management of the Above Board Company has a new piece of machinery with $B = \$110,000$, $n = 10$ years, and SV $= \$10,000$. Determine the depreciation schedule and present worth of depreciation at $i = 12\%$ per year using the 175% DB method for the first 5 years and the SL method for the last 5 years.

13.18 The Electric Company owns a portable building with a first cost of $155,000 and an anticipated salvage of $50,000 after 25 years. (a) Should the switch from DDB to SL depreciation be made? (b) For what rate of depreciation values of the DB method would it be advantageous to switch from DB to SL depreciation some time in the life of the building?

13.19 A company car is purchased for $9000 and is to be depreciated over 8 years and then sold for an estimated $750. (a) Start with DDB depreciation and perform an analysis to determine the maximum depreciation allowed each year if switching to SL is allowed at any time. Let $i = 12\%$ per year. (b) Repeat the analysis above, using the declining-balance method at a rate equal to 150% of the SL rate.

13.20 If $B = \$12,000$, $n = 8$ years, and SV $= \$800$, (a) develop the depreciation schedule and present-worth value for the DDB method if $i = 20\%$ per year. (b) Allow switching to SL depreciation and develop the new schedule and present-worth value. (c) Plot the book values of the asset with and without switching on the same graph.

13.21 Joe's Machine Shop purchased a $30,000 asset and must depreciate it using MACRS over 10 years. If the salvage is zero, compare (a) the plot of book value and (b) the present worth of depreciation at 20% per year for SL and MACRS depreciation over the 10 years.

13.22 If $B = \$45,000$, SV $= \$3000$, and $n = 5$-year recovery period, use $i = 18\%$ per year to maximize the present worth of depreciation using the following methods: DDB-to-SL switching (as done in Prob. 13.15) and MACRS. Comment on the answers and explain why the depreciation amounts are different using these example models.

13.23 An asset is purchased for $20,000. (a) If $n = 5$ years, use MACRS to compute the annual depreciation and present worth of depreciation. (b) Compare these results with DDB depreciation for $n = 10$. Use $i = 15\%$ per year.

13.24 In Prob. 13.23 use a recovery period of 5 years and the SL alternative to MACRS (with the half-year convention) to (a) compute annual depreciation and present worth of depreciation at 15% and (b) compare the two present-worth-of-depreciation values.

13.25 An automated assembly robot costs $45,000 and has an ADR life of 9 years. An analyst in the financial management department used classical SL depreciation to determine end-of-year book values for the robot when the original economic evaluation was performed. You are now performing a replacement analysis after 3 years of service and realize that the robot was actually to be depreciated using the MACRS straight-line alternative with the half-year convention. What is the amount of the error in the book value for the classical SL method?

13.26 The president of a small business wants to understand the difference in the yearly recovery rates for classical SL, MACRS, and the SL alternative to MACRS. Prepare a single graph showing the annual recovery rates (in percent) for the three methods versus the year of the recovery period. Use $n = 3$ years as an illustration.

13.27 A company pick-up truck has a first cost of $12,000, a salvage value of $2000, and a recovery period of 8 years. Use the SYD method to tabulate the recovery rate, depreciation amount, and book value for each year.

13.28 Earth-moving equipment having a first cost of $82,000 is expected to have a life of 18 years. The salvage value at that time is expected to be $15,000. Calculate the depreciation charge and book value for years 2, 7, 12, and 18 using the sum-of-years-digits method.

13.29 If $B = \$12,000$, $n = 6$ years, and SV is 15% of B, use the SYD method to determine the (a) book value after 3 years, (b) rate of depreciation in year 4, and (c) depreciation amount in year 4 using the rate from part (b) rather than Eq. (13.18).

13.30 A janitorial supply company purchased a new truck for $26,500. The owner wants to compute the 3-year MACRS depreciation schedule using the maximum amount of capital expensing allowed. Perform the analysis.

13.31 The Beauty Supply Company acquires a $115,800 mixing machine which is to be capital expensed and depreciated using the 5-year recovery, MACRS method. Determine the depreciation schedule to include the capital expense allowance.

13.32 A coal mining company has owned a mine for the past 5 years. During this time the following tonnage of ore has been removed each year: 40,000; 52,000; 58,000; 60,000; and 56,000 tons. The mine is estimated to contain a total of 2.0 million tons of coal and the mine had an initial cost of $3.5 million. If the company had a gross income for this coal of $15 per ton for the first 2 years and $18 per ton for the last 3 years, (a) compute the depletion charges each year using the larger of the values for the two accepted depletion methods and (b) compute the percent of the initial cost that has been written off in these 5 years.

13.33 If the mine operation explained in Prob. 13.32 is reevaluated after the first 3 years of operation and estimated to contain a remaining 1.5 million tons, answer the two questions posed in Prob. 13.32.

13.34 Assume that the earth-moving equipment described in Prob. 13.28 is to be depreciated by the unit-of-production method. Total tons moved in a lifetime is based on an annual average of 150,000 tons per year. If the tonnage moved in the first 3 years is 200,000, 250,000, and 175,000 tons, respectively, compute the depreciation charge and book value for each year.

14 | Basics of Taxation for Corporations

The prime objectives of this chapter are to help you develop a basic understanding of tax definitions and tax relations for corporations, and to compare depreciation (capital recovery) methods from the tax viewpoint. This chapter presents only an introduction to tax considerations. The details of after-tax economic analysis are presented in the next chapter.

Just as an individual must pay taxes to the federal, state, and local levels of government on income, property, sales, services, etc., a corporation pays taxes on income generated in the process of doing business in sales, services, and the like. When an economic analysis is performed, it is always necessary to determine if it will be a before-tax or after-tax analysis. For tax-exempt organizations (state, university, religious, nonprofit foundations and nonprofit corporations) after-tax analysis is not necessary. For taxed organizations (corporations and partnerships) the after-tax analysis may result in a different economic-based decision than before-tax analysis, but usually the decision is not dramatically altered. However, after-tax analysis does give a much better estimate of the return and the annual cost and revenues expected from the project under the current tax laws. Thus, many analysts prefer an after-tax analysis.

Since we investigate only a simplified version of corporate taxation in this text, more detail may be gained from other texts and IRS (Internal Revenue Service) publications such as Refs. [1], [2], and [3]. Individual tax problems should be resolved by consulting a tax advisor or appropriate IRS publications and personnel.

Section Objectives

After completing this chapter, you should be able to do the following:

14.1. Define *gross income, expenses, taxable income, capital gain, captial loss, operating loss,* and *recaptured depreciation.*

14.2. Compute income taxes using the published tax rates or an effective tax rate.

14.3. Compute the resulting net capital gains or losses and the income tax, given gain and loss amounts, federal tax laws, and the applicable tax rates.

14.4. State the tax law applicable to operating losses.

14.5. State the relations used to compute cash flow before and after taxes.

14.6. Show the tax advantage of one depreciation model over another by computing the *present worth of taxes,* given the model, asset data, tax rate, and after-tax rate of return.

14.7. Show the tax advantage of a shorter recovery period by computing the *present worth of taxes,* given the depreciation model, asset data, tax rate, and after-tax rate of return.

14.8. Compute the before-tax rate of return for a given effective tax rate and after-tax rate of return.

Study Guide

14.1   Definitions Useful in Tax Computations

To help you better understand the tax rates and formulas discussed in this chapter, some basic definitions are presented here.

Gross income. The total of all income from revenue-producing sources, including all items listed in the revenue section of an income statement. Refer to Appendix C for a review of accounting statements.

Expenses. All costs incurred while transacting business.

Taxable income (TI). The dollar value remaining upon which taxes are to be paid; computed as follows:

$$TI = \text{gross income} - \text{expenses} - \text{depreciation} \tag{14.1}$$

Capital gain. A gain incurred when the selling price of depreciable property (assets) or real property (land) exceeds the purchase price (unadjusted basis). Thus, at sale time the computation is

$$\text{Capital gain} = \text{selling price} - \text{unadjusted basis}$$

where the capital gain > 0. If the sales date occurs within a given time of purchase date, the capital gain is referred to as *short-term gain* (STG); if the ownership period is longer, the gain is a *long-term gain* (LTG). Tax law sets the ownership period, commonly 6 months or 1 year.

Capital loss. If the selling price is less than the book value, the loss is

$$\text{Capital loss} = \text{book value} - \text{selling price}$$

The terms *short-term loss* (STL) and *long-term loss* (LTL) are defined in a fashion similar to capital gains, that is, using a time break point. The concept of sunk cost, briefly discussed in Sec. 10.1, results in a capital loss.

Operating loss. When a corporation experiences a year of net loss rather than net profit, it has an operating loss. Special tax considerations are made in an attempt to balance the lean and fat years. Anticipation of operating losses, and thus the ability to take them into account in an economy study, is, of course, virtually impossible, but the tax treatment of past losses may be relevant in an analysis.

Recaptured depreciation (RD). If a depreciable property is sold for an amount greater than the current book value, the excess is depreciation recaptured by the sale and must be considered and taxed as ordinary taxable income. At sale time compute RD as follows:

$$\text{RD} = \text{Selling price} - \text{book value} \tag{14.2}$$

where $\text{RD} \geqslant 0$. If selling price exceeds the unadjusted basis B, a capital gain is also incurred, and all previous depreciation claimed is considered recaptured and taxable.

Tax laws on capital gains and losses and law-change effects on recaptured depreciation computations vary with each new tax bill, but the basic philosophy as applied to after-tax economic analyses remains the same. The tax laws of the late 1980s require that capital gains and losses be classified as long or short term using a 6-month time period, even though there is no tax advantage for corporations or individuals for long-term capital gains as there had been until the 1986 Tax Reform Act. Recaptured depreciation is computed using a salvage value of zero for asset disposal after the recovery period, or Eq. (14.2) for earlier sale. We use this approach in examples and problems. Application of different rules will, of course, alter the final numerical answers.

Prob. 14.1

14.2 Basic Tax Formulas and Computations

Taxes are computed using the general relation

$$\text{Taxes} = (\text{gross income} - \text{expenses} - \text{depreciation})T \tag{14.3}$$

where T is the tax rate. Since Eq. (14.3) uses the definition of taxable income (TI) from Eq. (14.1), we have

$$\text{Taxes} = (\text{TI})T \tag{14.4}$$

However, to give the small businesses a slight assist, corporate taxes are actually computed using the graduated tax-rate schedule in Table 14.1. You can see from this schedule that the nominal tax rate is 34% for TI values above $100,000. The 5% tax surcharge for TI > $100,000 has an upper limit of $11,750 which translates to an effective tax rate of 39% for the $100,001–$335,000 range. If this surcharge is removed or ignored, a federal tax rate of 34% may be used in studies with only a slight overestimate of tax liability. This value of 34% may be compared with 46%,

Table 14.1 Corporation federal income-tax-rate schedule (Effective January 1988 until altered)

(1) TI range	(2) Tax rate, T	(3) Maximum tax for TI range	(4) Total TI	(5) = sum of (3) Maximum tax incurred
$1–$50,000	0.15	$7,500	$50,000	$7,500
$50,001–$75,000	0.25	6,250	75,000	13,750
$75,001–$100,000	0.34	8,500	100,000	22,250
$100,001–$335,000	0.39	91,650[†]	335,000	113,900
All over $335,000	0.34	Unlimited	Above 335,000	Unlimited

[†] Tax surcharge of 5% on TI > $100,000 has an upper limit of $11,750 = 0.05(335,000−100,000).

which was the effective rate for TI > $100,000 in earlier years when more business expenses were deductible from gross income to compute TI by Eq. (14.1). To avoid dependence on current, changing tax law, problems will include a stated effective tax rate, or Table 14.1 rates will be explicitly utilized.

Example 14.1

For a particular year, the software division of the Intelligent Robotics Corp. has a gross income of $2,750,000 with expenses and depreciation totaling $1,950,000. Compute the federal taxes to be paid for the year.

Solution   Compute the TI and taxes using the schedule in Table 14.1.

   TI = 2,750,000 − 1,950,000 = $800,000

Taxes = 113,900 + (800,000 − 335,000)(0.34) = $272,000

Since this represents an overall tax of 34% on the entire $800,000 TI, an effective federal tax of 34% could be estimated.

Comment   If the 34% flat rate is used for all TI over $100,000 and the 5% surcharge is accounted for separately, the tax computation is

Taxes = 22,250 + (800,000 − 100,000)(0.34) + (335,000 − 100,000)(0.05)

   = $272,000

For the sake of simplicity, the tax rate used in an economy study is often a "one-figure" effective tax rate, which serves to account for federal, state, and city taxes. Commonly used effective tax rates are in the range of 35% to 50% but the applicable rate in a particular case is easily approximated. One advantage in applying an effective rate is that state taxes are deductible from federal taxes. Thus, you can use the following relation to compute the effective tax rate as a decimal fraction.

Incremental effective rate = state rate + (1 − state rate)(federal rate)   (14.5)

where the federal rate of 34% may be applicable if TI is already above $100,000.

Example 14.2

Compute the income tax for the data presented in Example 14.1 using an effective tax rate, if the state rate is 8% and the company uses a 34% incremental federal rate.

Solution   First the effective rate is computed using Eq. (14.5):

Incremental effective rate $= 0.08 + (1 - 0.08)(0.34) = 0.3928$

Now, by Eq. (14.4),

Taxes $= $ TI(effective rate)

$\qquad = 800,000(0.3928)$

$\qquad = \$314,240$

Comment   Do not compare this tax value with the results of Example 14.1, since the latter do not include a state tax.

Additional Example 14.7
Probs. 14.2 to 14.7

14.3   Tax Treatment of Capital Gains and Losses

The capital gain and loss definitions and computations presented in Sec. 14.1 are most commonly used in economic analysis for depreciable assets when an after-tax replacement study is conducted (Chap. 15), or the disposal plans for an asset are predictable. Prior to the 1986 tax law alteration, long-term gains (assets retained more than 6 months prior to disposal) were taxed at a lower rate. This text utilizes a tax structure in which long-term and short-term gains are taxed alike, with the distinction between short and long term maintained in case the tax laws are altered and preferential rates are reinstated.

Computationally, long-term gains are offset by long-term losses. Similar computations are necessary for short-term gains and losses. Resulting net gains (long or short) are taxed as ordinary income with a maximum rate of 34%. (We will either use this 34% figure or state an effective tax rate in all examples, but the law may differ when you study this material.) Resulting net losses are used to reduce the tax burden, with the assumption that they are used by the corporation to reduce gains for the same tax year. These gains are not a specific part of the project being analyzed, but can be utilized to offset the anticipated losses. Any net losses above net gains may not be used to reduce taxable income, but they can be carried back for 3 and forward for 5 years. Since most economic studies include only long-term capital investments, the tax treatment used in this text for gains and losses is summarized in Fig. 14.1.

Short-term gains and losses may be important for tax computation purposes as illustrated in Example 14.4, but they do not usually enter into the study of major

Offsetting	Net gain or loss results considered	After-tax treatment
Long-term LTG-LTL	Net long-term gain (NLTG)	Tax as ordinary income
	Net long-term loss (NLTL)	Reduces taxable income (may carry over excess losses)
Short-term STG-STL (neglected here)		

Figure 14.1 Summary of tax treatment of capital gains and losses.

economic investments. Recaptured depreciation may be included in after-tax studies and taxed at the effective rate. If the asset is anticipated to be disposed of with a long-term loss, the net loss is used to reduce taxable income in the year that disposal occurs.

Example 14.3

A new company, Intelligent Control Systems, Inc., expects to record a gross income of $500,000 and business expenses and depreciation of $300,000 in the next tax year. Using an incremental effective tax rate of 35%, compute the expected annual taxes for the following situations: (*a*) Anticipated equipment disposals with long-term gains of $25,000 and long-term losses of $10,000, and (*b*) anticipated long-term gains of $10,000 and long-term losses of $25,000. There are no other gains or losses predicted.

Solution
(*a*) The net long-term gain (NLTG) of $15,000 is taxed as ordinary income.

$$\text{Taxes} = (\text{TI} + \text{NLTG})T$$
$$= (500{,}000 - 300{,}000 + 15{,}000)0.35$$
$$= \$75{,}250$$

(*b*) A net long-term loss (NLTL) of $15,000 is available to reduce TI, but only $10,000 can be applied in this tax year because losses can not reduce TI beyond the extent of gains.

$$\text{Taxes} = (\text{TI} - \text{applicable NLTL})T$$
$$= (500{,}000 - 300{,}000 - 10{,}000)0.35$$
$$= \$66{,}500$$

The remaining NLTL of $5000 is carried over to next year to offset gains.

Comment In economy studies, anticipated losses are usually used to reduce TI with the implied assumption that the full benefit of the incurred loss can be taken in the year it occurs.

Example 14.4

How will the taxable income be affected if the following gains and losses have been incurred in one tax year?

LTG = $40,000   STG = $75,000

LTL = $5000   STL = $90,000

Solution   Offsetting gains and losses results in NLTG = $35,000 and NSTL = $15,000, with a net total of $20,000 for the long-term *gain*. The TI value will be increased by $20,000 when taxes are computed.

Comment   If long-term gains were taxed at a preferentially lower rate, such as the 20% rate used in earlier tax laws, the $20,000 would be taxed at this rate and the TI would incur the higher tax burden.

Probs. 14.8 to 14.11

14.4   Tax Laws on Operating Losses

We have discussed situations in which capital losses can be carried forward for several tax years. Another important tax advantage is the provision that allows an operating loss (Sec. 14.1) to be carried backward and forward until the loss is completely exhausted. The number of years allowed for carry-back and carry-forward may vary (e.g., 3 and 7 years, respectively), but the amount of operating loss claimed in any 1 year cannot exceed taxable income. Since only the amount of the loss is recoverable, this and all carry-back–carry-forward laws present a question of strategy, that is, *when* to utilize the tax advantage.

Prob. 14.12

14.5   Definition of Tax Cash-Flow Terms

In subsequent sections the terms CFBT and CFAT will be used to represent cash flow before taxes and after taxes, respectively. The relations between gross income, TI, taxes, and these terms are:

$$\text{CFBT} = \text{gross income} - \text{expenses} \tag{14.6}$$

$$\text{TI} = \text{CFBT} - \text{depreciation} \tag{14.7}$$

$$\text{Taxes} = (\text{TI})T \tag{14.8}$$

$$\text{CFAT} = \text{CFBT} - \text{taxes} \tag{14.9}$$

Equation (14.7) is simply a rewrite of Eq. (14.1) using the CFBT terminology. When an after-tax analysis is performed and the after-tax return is computed, the CFAT values are used in the computations.

Prob. 14.13

14.6 Effect of Depreciation Models on Taxes

The amount of taxes incurred is affected by the depreciation model chosen. Accelerated methods require less taxes in earlier years due to the reduction in taxable income by the annual depreciation as indicated by Eq. (14.7). We assume that some stated after-tax rate of return is used for the economy study. If we assume (1) a constant tax rate; (2) a gross income that exceeds annual depreciation; (3) capital recovery down to the same salvage value (commonly zero), and (4) the same number of years, the following are correct for all depreciation models:

1. The total taxes paid are equal for any model.
2. The present worth of taxes, P_{tax}, are *less* for accelerated depreciation models.

Example 14.5 examines these facts for three depreciation models.

Example 14.5

In performing an after-tax analysis for a new $50,000 machine for a fiber optics manufacturing line, a CFBT = $20,000 is expected. If a recovery period of 5 years applies, use an effective tax rate of 35% and a return of 8% per year to compare the *classical straight-line model, the DDB model, and the MACRS model.* For each model (*a*) compute and plot the taxes incurred, and (*b*) compute the present worth of taxes. Since MACRS requires 6 years to completely recover the investment, use a 6-year period for comparison purposes.

Solution
(*a*) Table 14.2 details the annual depreciation, taxable income, and taxes for each model. For straight line with $n = 5$, $D_t = \$10,000$ ($t = 1, 2, \ldots, 5$). The CFBT of $20,000 is fully taxed at 35% in year 6 for comparison with other methods.

Table 14.2 Comparison of depreciation models in taxes incurred and present worth of taxes

(1) Year, t	(2) CFBT	Straight line			DDB			MACRS		
		(3) D_t	(4) TI	(5) = 0.35(4) Taxes	(6) D_t	(7) TI	(8) = 0.35(7) Taxes	(9) D_t	(10) TI	(11) = 0.35(10) Taxes
0	$-50,000									
1	+20,000	$10,000	$10,000	$ 3,500	$20,000	$ 0	$ 0	$10,000	$10,000	$ 3,500
2	+20,000	10,000	10,000	3,500	12,000	8,000	2,800	16,000	4,000	1,400
3	+20,000	10,000	10,000	3,500	7,200	12,800	4,480	9,600	10,400	3,640
4	+20,000	10,000	10,000	3,500	4,320	15,680	5,488	5,760	14,240	4,984
5	+20,000	10,000	10,000	3,500	2,592	17,408	6,093	5,760	14,240	4,984
6	+20,000	0	20,000	7,000	0	20,000	7,000	2,880	17,120	5,992
Totals		$50,000		$24,500	$46,112		$25,861[†]	$50,000		$24,500
P_{tax}				$18,386			$18,549			$18,162

[†] Larger than other values since there is an implied salvage value of $3888 not recovered by the DDB model.

Figure 14.2 Taxes incurred by different depreciation models for a 6-year comparison period.

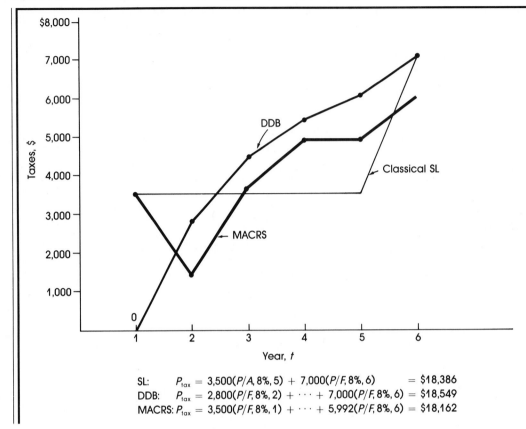

SL: $P_{tax} = 3,500(P/A, 8\%, 5) + 7,000(P/F, 8\%, 6)$ $= \$18,386$
DDB: $P_{tax} = 2,800(P/F, 8\%, 2) + \cdots + 7,000(P/F, 8\%, 6) = \$18,549$
MACRS: $P_{tax} = 3,500(P/F, 8\%, 1) + \cdots + 5,992(P/F, 8\%, 6) = \$18,162$

The DDB percentage of $d = 2/n = 0.40$ is applied for 5 years. An implied salvage value of $3888 is present, as can be computed by Eq. (13.9), which means that not all $50,000 is tax deductible. Thus, the total taxes incurred are $1361 (i.e., 35% of $3888) larger for DDB than SL.

MACRS writes off the $50,000 in 6 years using the rates of Table 13.4. Total taxes are $24,500 for MACRS, the same as classical SL over the 6 years.

The tax cash flows (cols. 5, 8, and 11 in Table 14.2) are compared in Fig. 14.2. Note the low tax values for DDB in years 1 and 2 and in year 2 for MACRS relative to the SL model.

(b) The present worth of taxes, computed in Fig. 14.2, is smallest for MACRS at $18,162. The DDB P_{tax} amount would be smaller than classical SL if all $50,000 were depreciated. These computations indicate that accelerated methods, under the conditions stated in the introduction to this example, have a smaller tax implication using a present-worth-of-tax criterion.

Comment For the DDB method, if the asset is disposed of in year 6 for zero income, there is a long-term loss of $3888, the undepreciated amount. According to the assumptions of Sec. 14.3, this LTL can reduce TI to $16,112 (i.e., $20,000 - 3,888$) in year 6. In this case the total taxes are $24,500, the same taxes as for SL and MACRS. The DDB entries in Table 14.2 would then appear as follows:

Year,		DDB		
t	CFBT	D_t	TI	Taxes
6	$+20,000	0	$16,112	$ 5,639
Totals		$46,112		$24,500
P_{tax}				$17,691

With the LTL accounted for correctly, the DDB method does have the smallest present worth of taxes.

Probs. 14.14 to 14.22

14.7 Tax Effects of Different Recovery Periods

For a specified after-tax rate of return, a constant tax rate and the same depreciation method, a lower n value will offer a tax advantage. Comparison of taxes for different n values will show the following:

1. The total taxes paid are equal for all n values.
2. The present worth of taxes are *less* for smaller n values.

Example 14.6 demonstrates these points for the SL method, but this can be done for any depreciation model.

Example 14.6

Superconductivity, Inc. uses the common practice of keeping two sets of books on depreciable assets, one for its own internal use and one for tax purposes. The company owns an asset with $B = \$9000$ and actual $n = 9$ years; however, a recovery period of 5 years is specified for this equipment. Show the tax advantage afforded the company by the lower n if CFBT = $3000 per year, a tax rate of 35% applies, invested money is returning 10% per year after taxes, and straight-line depreciation is used. Neglect the effect of any salvage value.

Solution Compute the present worth of taxes for both n values and compare to see if there is a tax advantage for $n = 5$ over $n = 9$ years.

$$n = 9 \text{ years:} \quad D_t = \frac{9000}{9} = \$1000 \quad t = 1, 2, \ldots, 9$$

$$TI = 3000 - 1000 = \$2000 \text{ per year}$$

$$\text{Taxes} = 2000(0.35) = \$700 \text{ per year}$$

$$P_{\text{tax}} = 700(P/A, 10\%, 9) = \$4031$$

$$n = 5 \text{ years: } D_t = \begin{cases} \dfrac{900}{5} = \$1800 & t = 1, 2, 3, 4, 5 \\ 0 & t = 6, 7, 8, 9 \end{cases}$$

Using Eqs. (14.7) and (14.8), annual taxes are

$$\text{Taxes} = \begin{cases} (3000 - 1800)0.35 = \$420 & \text{(years 1 to 5)} \\ (3000)0.35 = \$1050 & \text{(years 6 to 9)} \end{cases}$$

$$P_{\text{tax}} = 420(P/A, 10\%, 5) + 1050(P/A, 10\%, 4)(P/F, 10\%, 5)$$
$$= \$3659$$

Note that a total of $6300 in taxes is paid for both the 9- and 5-year period. However, the more rapid write-off for $n = 5$ results in a present-worth tax saving of $372.

Comment   As mentioned in Sec. 13.7, there is a half-year convention necessary when claiming SL depreciation. We have neglected this computational detail for the sake of simplicity, but it allows only a half year of SL depreciation in the first year. However, this amount is recovered by taking it in the year following the end of the recovery period. The net effect is a longer recovery period for the SL method with a tax disadvantage in the first year.

Probs. 14.23 to 14.25

14.8   Before-Tax and After-Tax Rates of Return

If the effect of taxes are important, but the details of an after-tax analysis are not necessary, you may choose to increase or inflate the before-tax rate in an effort to account for the effective tax rates. If incremental tax rates are expressed in decimal form, an approximation of the tax effect is

$$\text{Before-tax rate} = \frac{\text{after-tax rate}}{1 - \text{effective tax rate}} \qquad (14.10)$$

If a company's incremental effective tax rate is 40% (federal and state) and a 12% after-tax return is required, the before-tax rate for the economy study is

$$\frac{0.12}{1 - 0.40} = 0.20$$

Additionally, solution of Eq. (14.10) for the after-tax rate will estimate the expected return on a project, given the effective tax rate, once the before-tax study is complete.

Probs. 14.26 to 14.28

Additional Example

Example 14.7

Biotech One, Inc. is considering the purchase of a cell-analysis machine. Details of two possibilities follow:

	Machine 1	Machine 2
Basis	$150,000	$225,000
Expenses	30,000	10,000
Recovery	5 years	5 years

Use an effective tax rate of 35% and classical straight-line depreciation with no salvage value to determine which machine has a tax advantage. The gross income for both machines is expected to be the same.

Solution   Since the gross income (GI) values are equal, the smaller tax burden will occur for the machine with the larger deductions. Compute straight-line depreciation for machine 1 (D_1) and machine 2 (D_2) using Eq. (13.1) with SV = 0.

$$D_1 = \frac{150,000}{5} = \$30,000$$

$$D_2 = \frac{225,000}{5} = \$45,000$$

Equation (14.3) is used for tax computations.

$$\text{Tax}_1 = (\text{GI} - 30,000 - 30,000)0.35$$

$$\text{Tax}_2 = (\text{GI} - 10,000 - 45,000)0.35$$

Since the GI estimates are equal, machine 1 has the benefit with its $60,000 deductions compared to the $55,000 for machine 2. The size of the tax advantage is 5000(0.35) = $1750.

References

1. *U.S. Master Tax Guide*, Commerce Clearing House, Chicago, published annually.
2. *Tax Information on Corporations*, U.S. Internal Revenue Service Publication 542, published annually.
3. *Sales and Other Dispositions of Assets*, U.S. Internal Revenue Service Publication 544, published annually.

Problems

Note to instructor and students: When current tax rates and law are different from those in this edition and are used in the following problems, minor adjustments may be required in problem statement wording and solution. The answers will also change.

14.1   The situations listed below were recorded for the Pure Milk Company in the past year. For each situation, state which of the following is involved: gross income, taxable income, recaptured depreciation, capital gain, capital loss, or operating loss.
   (a) An asset with a book value of $8000 and a first cost of $15,000 was retired and sold for $8450.
   (b) A milking machine was purchased and had a first-year depreciation of $9600.
   (c) The company estimates that it will report a $-75,000 net profit to the IRS on the tax return.
   (d) The asset in part (b) will have a $4200-per-year interest cost.
   (e) An asset that had a life of 8 years has been owned for 14 years and has a book value of $800. It was sold this year for $275.
   (f) The cost of goods sold in the past year was $468,290.

14.2 Two small businesses have the following data on their tax returns:

	Dough Company	Broke Company
Sales	$1,500,000	$820,000
Interest revenue	31,000	25,000
Expenses	754,000	591,000
Depreciation	48,000	54,000

If both concerns do business in the state of No-Taxes, compute the federal income tax for the year, using the federal tax rates.

14.3 Compute the taxes for the situation in Prob. 14.2 using an effective rate of 34% for the entire TI. What percentage reduction in taxes is allowed by the graduated tax rate?

14.4 A-to-Z Car Dealers will have a $250,000 taxable income (TI) this year. If an advertising campaign is initiated, the TI is estimated to increase to $290,000 for the year. Neglecting any state and local taxes, use the federal tax rates to compute the (a) *effective* federal tax rate on TI = $250,000, (b) *effective* federal tax rate on only the additional taxable income, (c) *effective* federal tax rate on the entire $290,000 taxable income, and (d) after-tax profit on the additional taxable income.

14.5 A company has a gross income of $3.9 million for the year. Depreciation and all expenses amount to $2.45 million. If the combined state and local tax rate amounts to 6.5%, and an effective federal rate of 40% is applicable, compute the income taxes using the effective tax-rate formula.

14.6 Wa lbanger Contractors reported a taxable income of $80,000 last year. If the state tax rate is 8%, compute the (a) federal effective tax rate, (b) overall effective tax rate, and (c) total taxes to be paid by the company based on the tax rate in (b).

14.7 The taxable income for a small partnership business is $150,000 this year. An effective tax rate of 39% is used by the owners. Investment in some new machinery was considered in the first quarter of the year. The equipment would have cost $40,000, have a life of 5 years, be salvaged for an estimated $5000, and be written off using classical straight-line depreciation. The purchase would have increased taxable income by $10,000 and expenses $1000 for the year. Compute the change in income taxes for the year if the purchase had been made.

14.8 The following capital gains and losses are present this year for a small clothing manufacturer.

LTG = $2800

LTL =   500

STL = 2000

If TI = $80,000, compute the income taxes the company must pay if state taxes are 5% and the effective federal rate is 34%.

14.9 Compute the recaptured depreciation, and capital gains and losses for all the asset transactions below and then use them to compute the annual income taxes. Total sales for the year were $80,000, while expenses and accumulated depreciation amounted to $39,400.

(a) A 3-year-old straight-line-depreciated asset was sold for 0.68B. The asset had a first cost of $50,000, no salvage value, and a recovery period of 10 years.

(b) A machine that was only 5 months old was replaced because of extreme technological obsolescence. The asset had $B = \$10,000$, $SV = \$1000$, $n = 3$ years, and was depreciated by the MACRS method. The trade-in deal allowed the company $8000 on a new machine. (Use 50% of annual depreciation for the 5-month period.)

(c) Land purchased 4 months ago for $8000 was sold for a 10% profit.

(d) A 23-year-old asset was sold for $500. When purchased, the asset was entered on the books with $B = \$18,000$, $SV = \$200$, $n = 20$ years. Straight-line depreciation was used for the life of the machine.

(e) A short-term capital loss of $22,000 incurred 4 years ago is not completely exhausted. A total of $3500 remains on the books. The company's tax specialists want to remove one-half this loss from the accounts this year.

14.10 In Prob. 14.7 assume that the asset purchase was made, but in December of the same year, due to lagging sales, the equipment was sacrificed for $28,000. If current tax law requires that assets be retained at least 1 year to qualify for long-term gains and losses, explain how the sale in December will affect income taxes for the year.

14.11 Three years ago Chic Fashions, Inc. purchased land and assets which have recently been transferred to a separate subsidiary of the corporation. Use the information below to determine where recaptured depreciation and gains or losses have been incurred. Compute the amount of each effect.

Category	Purchase price	Life	Current book value	Sales price
Land	$100,000	—	$100,000	$105,000
Asset 1	50,500	5	15,500	17,500
Asset 2	10,000	3	1,000	11,000

14.12 The REALbody Health Club suffered a $25,000 operating loss in 1990. This loss may be carried back through any or all of the previous three tax years (1987, 1988, and 1989). Taxable income has been $110,000, $90,000, and $50,000 for the three previous years with an effective tax rate of 50%. Compute the present worth of taxes in 1987 using the seven potential plans below. If $i = 10\%$ per year, which plan offers the lowest taxes in present worth?

(a–c) Recover entire loss in year 1987, 1988, or 1989.

(d–f) Recover one-half in each of two of the years.

(g) Recover one-third of losses in each year 1987 through 1989.

14.13 The vice president of finance wants to estimate the required CFBT to realize a $1,500,000 CFAT for the year. The effective federal tax rate is 40% and the state tax rate is 8%. Use the additional fact that $1 million of asset value will be written off through depreciation this year. Estimate the CFBT.

14.14 A transportation leasing company bought new trucks for $150,000 and expects to realize a CFBT of $100,000 for each of 3 years. The trucks have a recovery period of 3 years. Assume an effective tax rate of 40% and an interest rate of 15% per year. Show the advantage of accelerated depreciation methods in terms of tax present worth for the MACRS method versus the classical SL method. Since MACRS takes an added year to fully depreciate the basis, assume no CFBT beyond year 3, but include any negative tax as a tax advantage for the trucks.

14.15 Resolve Prob. 14.14 for the MACRS versus DDB methods. Stop any depreciation for the DDB method after 3 years and neglect the remaining implied salvage value.

14.16 What is the difference in the present worth of total taxes paid for the following situations? Asset A can be purchased to produce a CFBT of $65,000, or five of asset B can be purchased to produce the same CFBT for at least 6 years in the future. Money is worth 12% per year and MACRS or the SL alternative to MACRS is used with a 5-year recovery period, as indicated below. Neglect any gains, losses, or recaptured depreciation at sale time, but do use the half-year convention in the SL alternative for years 1 and 6.

	Asset A	Five of asset B
Total first cost	$250,000	$260,000
Total salvage value	25,000	25,000
Total annual CFBT	65,000	65,000
Depreciation method	MACRS	SL alternative
Tax rate	50%	50%
Life, years	5	5

14.17 Rework Prob. 14.16 if asset A is depreciated by the DDB method and asset B uses the SYD method. Use the stated salvage values and 5-year recovery periods for comparison.

14.18 An asset costing $45,000 has a life of 5 years, a salvage value of $3000, and an anticipated CFBT = $15,000 per year. Determine the depreciation schedule for SL and DDB and for switching from DDB to SL to maximize depreciation. (This is the same as Prob. 13.16.) Use $i = 18\%$ and an effective tax rate of 50% to determine how the present worth of taxes decreases when switching is used. (Assume that the asset is sold for $3000 in year 6 and that any negative TI or capital loss at sale time is a tax advantage.)

14.19 Construct the cash-flow diagram for taxes and calculate the present worth of taxes for a $9000, 5-year recovery asset if the effective tax rate is 40%, CFBT is estimated at $10,000 per year, and money is worth 12% per year. Use the 150% DB method of depreciation.

14.20 If the CFBT is not considered, it is possible to compute the *tax savings* TS, in year t due to depreciation alone using

$$TS_t = (\text{tax rate})(\text{depreciation}) \times T(D_t)$$

The present worth of tax savings P_{TS} increases for accelerated depreciation methods and when switching between methods is allowed, as in MACRS. Compute the present worth of tax savings for the depreciation schedules in Prob. 14.18.

14.21 Calculate the present worth of tax savings for the depreciation schedule in Prob. 14.19. Use the relation discussed in Prob. 14.20 to compute tax savings.

14.22 The MACRS rates and the SL depreciation election are being considered for an asset with $B = \$10,000$ and $n = 5$-year recovery. Use CFBT = $4000 to select the better method based on the present worth of taxes. Let $T = 0.35$ and $i = 20\%$ per year. Use the half-year convention.

14.23 The new asset which cost $6000 is expected to last 4 years and produce a CFBT = $3000 for 4 years only. The asset can be straight-line depreciated over 3 or 5 years. If the tax rate is 50% and $i = 5\%$ per year, use the maximum tax present worth to select the 3- or 5-year recovery period. Assume that any depreciation in excess of cash flow is a tax advantage.

14.24 You plan to purchase a $900,000 material handling system for a new manufacturing line. If you have the choice of the classical straight-line method of depreciation ($n = 3$ years) and the DDB method ($n = 5$ years), which method and time period would you choose? Use the criteria of maximum tax savings (Prob. 14.20) to help with the decision. Let the effective tax rate be 45% and assume $i = 12\%$ per year.

14.25 The Leading Edge Microprocessor Co. has just bought an asset for $B = \$88,000$ with an expected life of 10 years. For tax purposes, the company is allowed (a) to use a recovery period of 5 or 10 years and straight-line depreciation with the half-year convention, or (b) to use a period of 5 years and MACRS depreciation. If $i = 10\%$ per year, $T = 52\%$, and the anticipated CFBT is $25,000 per year for only 10 years, determine the recovery period and depreciation method to minimize the time value of taxes. Consider all operating losses to be a tax advantage in the year incurred. (*Note:* Maximum tax savings can also be used to select n and the depreciation method. See Prob. 14.20.)

14.26 Compute the before-tax rate of return for Prob. 14.5 if an after-tax rate of return of 8% per year is required.

14.27 If a company uses a before-tax rate of return of 19% and an after-tax return of 8% per year, what percent of income is assumed to be absorbed in taxes?

14.28 A consultant has worked on new technology justification for the textile industry recently. In both a small business and a large corporation she performed economic evaluations which have a 21%-per-year before-tax return. If the "hurdle rate" (that is, the MARR) for new projects in both companies is 12% per year after taxes, determine if management at both companies would accept the projects provided the before-tax return is used to approximate the after-tax return. Effective incremental tax rates are 48% for the large corporation and 34% for the small company.

15 After-Tax Economic Analysis

This chapter will help you incorporate some of the effects of income taxes into economic-based decision analysis. Rather than utilizing an inflated before-tax rate of return, details of income tax impacts are included in the computation of the rate of return, present worth, or equivalent uniform annual worth through the estimation of cash flow after taxes (CFAT). Additionally, the minimum revenue required to economically justify a project will be computed using after-tax analysis.

As you already realize, the tax laws and guidelines change continuously. Only the impact from the large tax advantages and burdens should be considered in any economic analysis when a capital investment decision is to be made. These major considerations are discussed here.

It is important in economic analysis to bear in mind the different perspectives of tax effects. We are estimating the effect of current tax laws on future streams of cash flows for the purpose of making a sound economic decision *now*. The tax and financial personnel in a corporation are usually more interested in the effect of tax law upon the current year's revenues and costs. The estimation viewpoint toward after-tax analysis—the one we will take—is proactive and includes only the major effects of income tax law. The second, which is reactive, computationally more detailed, and driven by decisions made in past economic decision analyses, includes more detailed tax law effects.

Section Objectives

After completing this chapter, you should be able to do the following:

15.1 Determine the annual income taxes due and the estimated *cash flow after taxes* (*CFAT*) considering depreciation and debt financing, given the cash flow before taxes (CFBT), and the details for taxes, depreciation, and borrowed funds.

15.2 Select the better of two alternatives using *after-tax PW or EUAW* analysis, given the alternatives, tax rate, and required after-tax return.

15.3 Use *after-tax rate-of-return analysis* to compute the return of a single project and select the better of two alternatives, given the alternatives, tax rate, and required after-tax return.

15.4 Choose between challenger and defender alternatives using *after-tax replacement analysis*, given the plans, defender market value, and effective tax rate.

15.5 Compute the *equivalent annual revenue requirement* for a project, using the project data, specified depreciation method, and effective tax rate.

15.1 Effect of Income Taxes on Cash Flow

It is important that you understand and be able to tabulate the effect of income taxes in terms of CFAT. The formulas were first introduced in Chap. 14, but are reviewed here for easy reference. Concentration is upon the estimation of CFAT to be utilized in present worth, EUAW, and rate-of-return computations.

$$\text{CFBT} = \text{gross income} - \text{expenses} \tag{15.1}$$

$$\text{TI} = \text{CFBT} - \text{depreciation} \tag{15.2}$$

$$\text{Taxes} = (\text{TI})T \tag{15.3}$$

$$\text{CFAT} = \text{CFBT} - \text{taxes} \tag{15.4}$$

There are two ways to finance a venture: through the investment of the corporation-owned funds (equity financing) or through the use of borrowed funds which are acquired from some source outside the corporation ownership (debt financing). A combination of these two is the most common mechanism. When any debt financing is involved, the associated interest is tax deductible and Eq. (15.2) for TI will reflect this tax advantage. Additionally, CFAT in Eq. (15.4) must be reduced by interest and principal payments. For debt financing these relations will take the forms:

$$\text{TI} = \text{CFBT} - \text{depreciation} - \text{interest} \tag{15.5}$$

$$\text{CFAT} = \text{CFBT} - \text{taxes} - \text{interest} - \text{principal} \tag{15.6}$$

The aspect of debt and equity financing is introduced here and is examined more thoroughly in Chap. 18.

If Eq. (15.2) or Eq. (15.5) results in a negative TI value, we will assume the resultant negative tax offsets taxes for the same year attributable to other income-producing assets in the corporation. Any negative tax, therefore, increases the CFAT value by a corresponding amount for the year in which the tax advantage occurs. This simplifying procedure is commonly used in lieu of complex carry-forward–carry-back tax laws similar to those overviewed in Sec. 14.4.

Table 15.1 Tabulation of CFAT for Example 15.1

Year	(1) Income	(2) Expenses	(3) CFBT	(4) SL depreciation	(5) = (3) − (4) TI	(6) Taxes	(7) = (3) − (6) CFAT
0	—	$50,000	$ − 50,000	—	—	—	$ − 50,000
1	$27,000	10,000	+ 17,000	$10,000	$7,000	$2,800	+ 14,200
2	26,000	10,500	+ 15,500	10,000	5,500	2,200	+ 13,300
3	25,000	11,000	+ 14,000	10,000	4,000	1,600	+ 12,400
4	24,000	11,500	+ 12,500	10,000	2,500	1,000	+ 11,500
5	23,000	12,000	+ 11,000	10,000	1,000	400	+ 10,600

Example 15.1

A proposal has been made that a new piece of equipment be purchased this year. Characteristics of the purchase plan are:

$$B = \$50,000$$

$$SV = 0$$

$$n = 5 \text{ years}$$

Expected income $= 28,000 - 1000t \quad (t = 1, 2, 3, 4, 5)$

Expected expenses $= 9500 + 500t$

Use an effective tax rate of 40% and straight-line depreciation to tabulate the CFAT.

Solution Table 15.1 details all tax information and CFAT for the asset using Eqs. (15.1) to (15.4). Straight-line depreciation is $50,000 \div 5 = \$10,000$ per year.

Example 15.2

Fifth Ave. Cleaners plans to invest in a new dry cleaner.

$$B = \$15,000$$

$$SV = 0$$

$$\text{Income} = \$7000 \text{ per year}$$

$$\text{Expenses} = \$1000 \text{ per year}$$

Incremental effective tax rate $= 50\%$

$$n = 5 \text{ years}$$

If straight-line depreciation is used, tabulate CFAT for the following conditions: (*a*) All $15,000 is from company funds (100% equity financing) and (*b*) one-half the investment is borrowed from a bank (50% equity − 50% debt financing) at 10% per year interest. Assume that the 10% is simple interest on the total amount borrowed and repayment will be in five equal payments of accrued interest and principal.

Solution
(*a*) For 100% equity financing, CFBT = $6000 per year. Annual depreciation is $15,000/5 = $3000. Table 15.2 details CFAT.

Table 15.2 CFAT for 100% equity financing (Example 15.2a)

Year	CFBT	Depreciation	TI	Taxes	CFAT
0	$-15,000$	—	—	—	$-15,000$
1–5	6,000	$3,000	$3,000	$1,500	4,500
	$ 15,000			$7,500	$ 7,500

Table 15.3 CFAT for 50% equity–50% debt financing (Example 15.2b)

Year	(1) CFBT	(2) Depreciation	(3) Interest	(4) Principal	(5) TI[†]	(6) Taxes	(7) CFAT[‡]
0	$-7,500$	—	—	—	—	—	$-7,500$
1–5	6,000	$3,000	$ 750	$1,500	$2,250	$1,125	+2,625
	$ 22,500		$3,750	$7,500		$5,625	$ 5,625

[†] In column notation, $(5) = (1) - (2) - (3)$, Eq. (15.5).
[‡] $(7) = (1) - (6) - (3) - (4)$, Eq. (15.6).

(b) The 50% debt financing requires that $7500 be borrowed from outside the company and its stockholders. The loan repayment plan will be as follows:

$$\text{Principal:} \qquad \frac{7500}{5} = \$1500 \text{ per year}$$

Interest: $7500(0.10) = \$750$ per year

The $750 interest is tax deductible; however, the principal is *not deductible*. Therefore, the debt-financing formulas for TI and CFAT, Eqs. (15.5) and (15.6), are used. Table 15.3 presents CFAT computation for 50% debt – 50% equity financing. You can see that the annual CFAT has decreased from $4500 to $2625 because of 50% debt financing. The $7500 equity cash outflow in year 0 is used because only 50% of the first cost comes from company funds while the remaining is from the lending bank.

Comment If only equity financing is involved, as in part (a), we can use the relation

$$\text{CFAT} = \text{depreciation} + \text{TI}(1 - T) \tag{15.7}$$

where $\text{TI}(1 - T)$ is the *profit*, that is, the portion of TI not absorbed by taxes. Thus, in Table 15.2

$$\text{CFAT} = 3000 + 3000(1 - 0.50) = \$4500$$

could be used in lieu of the method presented. But you must be careful because, if there is any debt financing whatsoever, as in part (b), this simple approach will not work. Let's try it in Table 15.3.

$$\text{CFAT} = 3000 + 2250(1 - 0.50) = \$4125 \neq \$2625$$

Why doesn't it work? Simple. This method neglects the fact that interest is tax deductible *and* that CFAT is computed in a completely different way in the two cases. You might compare Eqs. (15.7) and (15.6) to verify this.

15.2 PW and EUAW Computations on Cash Flow after Taxes

If the required after-tax rate of return is known, the CFAT values are used to compute the PW or EUAW for a project as in previous chapters. If mutually exclusive alternatives are compared, select the project using the following guidelines:

If income and cost values are estimated, the project with the larger PW or EUAW value offers the better return in excess of the required rate.

If only costs are estimated and a positive sign is used to identify the cost figures, the project with the smaller positive PW or EUAW value offers the better return in excess of the required rate.

In either case, a negative PW or EUAW value indicates that the alternative will not deliver the required return. Additionally, if only costs are involved, the associated tax advantages are assumed to be applied against other expenses.

Example 15.3

For the estimated CFAT values presented below, use a 7%-per-year after-tax return and EUAW analysis to select the economically favorable plan.

Plan X		Plan Y	
Year	CFAT	Year	CFAT
0	$-28,800	0	$-50,000
1–6	5,400	1	14,200
7–10	2,040	2	13,300
10	2,792	3	12,400
		4	11,500
		5	10,600

Solution Use the revenue and cost estimates in EUAW relations and select the larger positive value at $i = 7\%$ per year.

$$\text{EUAW}_X = [-28,800 + 5400(P/A, 7\%, 6) + 2040(P/A, 7\%, 4)(P/F, 7\%, 6)$$
$$+ 2792(P/F, 7\%, 10)](A/P, 7\%, 10)$$
$$= \$421.78 \tag{15.8}$$

$$\text{EUAW}_Y = [-50,000 + 14,200(P/F, 7\%, 1) + 13,300(P/F, 7\%, 2)$$
$$+ 12,400(P/F, 7\%, 3) + 11,500(P/F, 7\%, 4)$$
$$+ 10,600(P/F, 7\%, 5)](A/P, 7\%, 5)$$
$$= \$327.01 \tag{15.9}$$

Plan X is selected since both EUAW values are positive and EUAW_X is larger.

    If present-worth analysis were used, a 10-year horizon would be used to compute $\text{PW} = \text{EUAW}(P/A, 7\%, 10)$. Of course, Plan X would again be selected for its larger PW value.

15.3   Rate-of-Return Computations for Cash-Flow after Taxes

The PW or EUAW relations are used to compute the rate of return of CFAT values using the same procedures as in Chaps. 7 and 8. For a *single project*, set the PW or EUAW of the CFAT sequence equal to zero and solve for the $i*$ value by the most rapid method.

Present worth: $0 = \sum_{t=1}^{n} CFAT_t(P/F, i*\%, t)$ (15.10)

EUAW: $0 = \left[\sum_{t=1}^{n} CFAT_t(P/F, i*\%, t) \right](A/P, i*\%, n)$ (15.11)

Example 15.4

The president of a fiber optics manufacturing company plans to spend $50,000 for a 5-year-life, straight-line-depreciated machine which will have a projected $20,000 annual CFBT and carry an effective incremental tax rate of 50%. Compute the $i*$ value for the president.

Solution   The CFAT in year zero is $-\$50,000$ and for years $t = 1$ through 5 is, according to Eqs. (15.1) through (15.4),

CFAT = CFBT − taxes = 20,000 − (20,000 − 10,000)0.5
    = $15,000

Since the annual revenue CFAT values are equal, Eq. (15.10) is abbreviated to compute $i*$.

   $0 = -50,000 + 15,000(P/A, i*\%, 5)$

$(P/A, i*\%, 5) = 3.3333$

Solution gives $i* = 15.25\%$ as the after-tax rate of return.

Comment   If you use an exaggerated before-tax rate to approximate the tax effect on this type of asset, you can use Eq. (14.10) to obtain $i*/(1 - T) = 0.1525/(1 - 0.50) = 0.305$, or 30.5%. Actual before-tax return using the CFBT figures can be found from the equation

$0 = \$-50,000 + 20,000(P/A, i*\%, 5)$

which gives a value of $i* = 28.7\%$. Comparison shows that the tax effect is slightly overestimated by using a 30.5% before-tax return.

For two (or more) alternatives, use a PW or EUAW relation to compute the *incremental return on the incremental CFAT series* of the larger alternative over the smaller one. These are the same procedures used in Secs. 8.4 and 8.5; they are summarized in Table 15.4. The equations in Table 15.4 are used whether revenues are estimated or not, because the entire analysis is performed on the incremental or net CFAT values,

Table 15.4 Guidelines for computation of the after-tax return using incremental analysis (Alternative B has the larger initial investment)

Method used to compute i	Equal lives $n_B = n_A$	Unequal lives $n_B \neq n_A$
Present worth (PW)	Set PW = 0 for incremental CFAT for n years	Set PW = 0 for incremental CFAT for least common multiple of years, n
Equation used	$$\sum_{t=1}^{n} \Delta CFAT_t (P/F, i^*\%, t) = 0 \qquad (15.12)$$	
Equivalent uniform annual worth (EUAW)	Set EUAW = 0 for incremental CFAT for n years	Set difference of EUAW = 0 over respective lines n_B and n_A
Equation used	(PW of ΔCFAT)$(A/P, i^*\%, n) = 0$ (15.13)	$EUAW_B - EUAW_A = 0$ (15.14)

which we refer to by the Δ (delta) symbol, that is, for each year t and alternatives B and A,

$$\Delta CFAT = CFAT_B - CFAT_A$$

The resulting after-tax return i^* is the breakeven return between the two alternatives, which is compared to the after-tax MARR to select an alternative. (You might want to refer to Secs. 8.3 through 8.5 for a review of incremental rate-of-return analysis.) The following steps outline an after-tax return analysis.

1. Order the alternatives by initial investment cost. Identify as B the one with the larger initial investment.
2. Decide whether a PW or EUAW relation will be used to compute the incremental after-tax return. Select the appropriate equation from Table 15.4.
3. Calculate the net or incremental $\Delta CFAT_t$ values for PW or EUAW (equal lives) analysis. Or, determine the EUAW relations for unequal-lives analysis.
4. Compute the after-tax incremental return i^*_{B-A} using the correct equation by computer or manual trial-and-error. The $B - A$ subscript is usually omitted.
5. Compare the return to the after-tax MARR. Accept alternative B if the return is larger than MARR. Otherwise, select alternative A.

Remember that the acceptance of the larger investment alternative B means that the incremental investment over A is justified because $i^* >$ MARR.

In choosing the method to compute i^* from Table 15.4, the PW method is usually easier for equal lives and the EUAW method is better for unequal lives. This is true simply because less CFAT values and factor manipulations are involved in the computations.

Example 15.5

Use the estimated CFAT values for plans X and Y in Example 15.3 to select the better alternative if a minimum after-tax return of 6% per year is required.

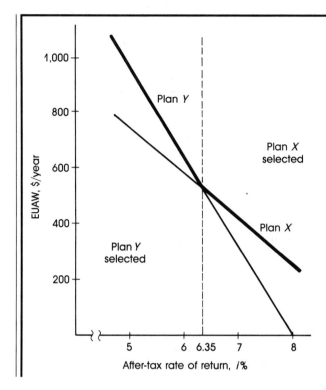

Figure 15.1 EUAW values for different after-tax returns, with a breakeven return of 6.35%. Example 15.4.

Solution Since the lives are unequal, Table 15.4 indicates that either PW analysis over a 10-year period or incremental EUAW analysis may be performed. If an EUAW analysis is selected, plan Y has the larger initial investment and Eqs. (15.8) and (15.9) may be used to estimate i^* from $EUAW_Y - EUAW_X = 0$. The result is $i^* = 6.35\%$ which is greater than 6%. This indicates that plan Y should be selected because the incremental investment is justified.

The EUAW values for each plan are shown in Fig. 15.1 between after-tax return values of 5 and 8%. Any minimum required return less than 6.35% (the breakeven after-tax return between the two plans) favors plan Y because the $EUAW_Y$ value is larger. Similarly, any required return exceeding 6.35% will indicate that plan X is better.

Comment If PW analysis were used to estimate $i^* = 6.35\%$, the 10-year evaluation would involve the assumed repeat of plan Y for years 6 through 10. The $\Delta CFAT$ values would be used to compute the incremental return by Eq. (15.12) in Table 15.4.

Example 15.6

Johnson Enterprises wants to decide between one advanced robotics system for automated printed-wiring-board assembly (system 1), which will require a $100,000 investment now and a 2-year investment program involving two robots purchased for $80,000 now and $60,000 one year from now (system 2). Management plans to implement one of the plans. Use a 20%-per-year after-tax return requirement to select a system.

Year	0	1	2	3	4
System 1 CFAT	$100,000	25,000	25,000	25,000	10,000
System 2 CFAT	$ 80,000	60,000	20,000	20,000	5,000
	Capital investment				

Solution According to the steps for after-tax return analysis, system 1 with the larger investment in year zero is the alternative with the incremental investment which must be justified. Since lives are equal, select PW analysis and compute the incremental CFAT. All values are divided by $1000.

Year, t	0	1	2	3	4
ΔCFAT$_t$	+$20	−35	+5	+5	+5

Equation (15.12) is set up to approximate the after-tax return.

$$20 - 35(P/F, i^*\%, 1) + 5(P/A, i^*\%, 3)(P/F, i^*\%, 1) = 0$$

Interpolation between 25 and 30% indicates a return of 28.77%, which exceeds the 20% required return. The extra investment in system 1 is therefore economically justified.

Comment Notice that the system 2 investment ($140,000) is actually larger than system 1. If we define system 2 as the larger investment alternative and compute the incremental return using PW analysis, the ΔCFAT values are different in sign and the return is zero!

Year, t	0	1	2	3	4
ΔCFAT$_t$	−$20	+35	−5	−5	−5
		$i^* = 0\%$			

A zero return indicates that the incremental investment of $40,000 is not justified. In this case, system 1 is again selected.

The additional example illustrates the differences in after-tax returns for various debt-equity funding percentages on the same project.

Additional Example 15.9
Probs. 15.15 to 15.25

15.4   After-Tax Replacement Analysis

When a defending asset is challenged by a new asset, the effects of income taxes should be considered. To account for all the tax details in *replacement analysis* is sometimes neither time- nor cost-effective; however, it is worthwhile to account for any applicable capital gain or loss which would occur if the defender were replaced. Also important is the future tax advantage stemming from deductible operating and depreciation expenses. The following example will give you an idea of the impact of taxes on replacement analysis.

Example 15.7

Throckmorton Mining purchased ore extraction equipment 3 years ago for $6000 which management now feels is obsolete due to technological change. A challenger has been located. If a $4000 trade-in is offered for the current machine, perform (a) an EUAW analysis before taxes using a 15% return requirement, and (b) a 7%-per-year after-tax analysis. Assume an effective tax rate of 34% on income and recaptured depreciation (RD), and assume straight-line depreciation with no salvage recognized for either asset.

	Defender	Challenger
Basis B	$4,000	$10,000
Salvage value	0	2,000
Annual costs	1,000	150
Recovery period	8 (original)	5

Solution

(a) For the before-tax analysis, $B = \$4000$ and $n = 5$ years for the defender. EUAW analysis uses the conventional approach to replacement analysis (Sec. 10.3).

$$\text{EUAW}_D = 4000(A/P, 15\%, 5) + 1000 = \$2193$$

$$\text{EUAW}_C = 10,000(A/P, 15\%, 5) - 2000(A/F, 15\%, 5) + 150 = \$2837$$

The defender is favored by a $644 margin.

(b) For *retention of the defender* in after-tax analysis, compute the actual current investment.

$$\text{Depreciation} = \frac{6000}{8} = \$750 \text{ for next 5 years}$$

$$\text{Present book value} = 6000 - 3(750) = \$3750$$

$$\text{RD on trade-in} = 4000 - 3750 = \$250$$

$$\text{RD tax savings} = 0.34(250) = \$85$$

$$\text{Actual after-tax investment} = 4000 - 85 = \$3915$$

The after-tax adjusted annual costs for the defender are:

$$\text{Annual costs} = \$1000$$

$$\text{Depreciation} = \$750 \text{ for 5 years}$$

$$\text{Annual tax savings} = (1000 + 750)(0.34) = \$595$$

$$\text{Actual after-tax expenses} = 1000 - 595 = \$405 \text{ per year}$$

After-tax EUAW for the defender is

$$\text{EUAW}_D = 3915(A/P, 7\%, 5) + 405 = \$1360$$

If the trade-in amount is less than the current book value, the resulting capital loss represents a forgone tax advantage since the defender is retained. The loss amount should, therefore, be added to the trade-in value. For example, if the trade-in were $3000, a forgone tax advantage of $(3750 - 3000)(0.34) = \$225$ would be added to the defender basis of $4000.

If *acceptance of the challenger* occurs,

$$\text{Annual depreciation} = \frac{10,000}{5} = \$2000$$

$$\text{Annual tax savings} = (2000 + 150)(0.34) = \$731$$

$$\text{Actual after-tax expenses} = 150 - 731 = -\$581 \text{ per year}$$

The annual expenses are actually a net income. When the asset is sold for $2000 the book value is actually zero and a tax burden is present in the recaptured depreciation for $2000(0.34) = \$680$. The after-tax EUAW for the challenger is

$$\text{EUAW}_C = 10,000(A/P, 7\%, 5) - (2000 - 680)(A/F, 7\%, 5) - 581 = \$1628$$

Select the defender with an advantage of $268 per year. Even though both analyses favor the defender, the after-tax advantage is considerably less. Since the adequacy of the defender is in question, management may well replace it.

Probs. 15.26 to 15.31

15.5   After-tax Evaluation Using Revenue Requirements

The gross revenue expected from an asset or project should recover the investment and all costs associated with its use and ownership or lease, and produce the required return on the investment. Taxes are included in the cost category as are operating and maintenance expenses. The sum of these indicate the equivalent annual amount that an alternative should generate, that is, its *annual revenue requirement (RR)*.

$$\text{RR} = \text{Repayment of initial investment} + \text{return on equity capital} + \text{interest on debt capital} + \text{annual operating costs} + \text{income taxes} \tag{15.15}$$

All the values are expressed in equivalent annual terms. If, for example, annual operating costs (AOC) are expected to rise annually at 10%, economy factors are used to compute the equivalent annual operating cost. If separate equity and debt capital sources are not distinguished, the first three terms of Eq. (15.15) are the after-tax capital recovery and required return on the initial investment. This amount is the equivalent uniform annual worth.

$$\text{EUAW} = P(A/P, i\%, n) - \text{SV}(A/F, i\%, n) \tag{15.16}$$

Since income taxes are affected differently for equity and debt capital, detailing for each finance source requires special, and considerably more complex, relations than those presented here. See the texts by G. T. Stevens (listed in the bibliography) for additional coverage.

The equivalent uniform annual income taxes in Eq. (15.15) may be estimated using the relations in Sec. 15.1 as follows.

$$\begin{aligned}
\text{Taxes} &= (\text{Taxable income})(\text{tax rate}) = (\text{TI})T \\
&= (\text{CFBT} - \text{depreciation})T & \text{by Eq. (15.2)} \\
&= (\text{CFAT} + \text{taxes} - \text{depreciation})T & \text{by Eq. (15.4)}
\end{aligned}$$

Solution for the amount of taxes results in

$$\text{Taxes} = \frac{T}{1-T}(\text{CFAT} - \text{depreciation})$$

The annual CFAT is the amount of the gross revenue that must remain to repay the investment, the equity capital return, and the debt capital interest. This is the same as the after-tax capital recovery and return in Eq. (15.16). Therefore, without considering the differences between equity and debt financing (or by assuming 100% equity financing), taxes are estimated as

$$\text{Taxes} = \frac{T}{1-T}(\text{EUAW} - D)$$

$$= \frac{T}{1-T}[P(A/P, i\%, n) - \text{SV}(A/F, i\%, n) - D] \qquad (15.17)$$

where P is the initial investment or first cost B, and D is the *equivalent annual depreciation*. Any depreciation method can be used, but the differing annual amounts are first converted to an equivalent annual amount, similar to the operating costs in Eq. (15.15).

Substitution of the relations for EUAW and taxes into Eq. (15.15) for RR, and some algebraic simplification, will give the equivalent uniform annual revenue requirement for the asset or project.

$$\text{RR} = \text{EUAW} + \text{AOC} + \text{taxes}$$

$$= P(A/P, i\%, n) - \text{SV}(A/F, i\%, n) + \text{AOC}$$

$$+ \frac{T}{1-T}[P(A/P, i\%, n) - \text{SV}(A/F, i\%, n) - D]$$

$$\text{RR} = \text{AOC} + \frac{1}{1-T}[\text{EUAW} - T(D)] \qquad (15.18)$$

A project is justified if the estimated equivalent annual revenue exceeds or equals this RR value.

It is possible to compute the changing revenue requirement for each year if the individual debt-equity mix is considered (Chap. 18), and the yearly values for depreciation, unrecovered balance on the investment, unpaid debt capital, etc. are estimated. This, of course, is much more time consuming, but more accurate. Historically, the utilities industry has performed annual revenue requirement estimates, as opposed to equivalent annual as we have done here, for its investments. These estimates are submitted to the state government control agencies—the PUCs (Public Utility Commissions)—in partial justification for utility rates set on our electricity.

Example 15.8

A new printed-wiring-board (pwb) assembly system for flat-pack designs in aerospace defense electronics has been acquired by a company. Essential information for the system has been provided to you by the vice president of operations. Use the simplifying assumptions to determine if the system was justified purely on an economic basis.

$$\text{First cost} = \$2,000,000$$

$$\text{Salvage after 5 or 10 years} = 10\% \text{ or } \$200,000$$

$$\text{Planning horizon} = 10 \text{ years}$$

$$\text{Equivalent annual operating cost} = \$150,000$$

$$\text{Depreciation method} = \text{straight line over 5 years}$$

$$\text{Incremental tax rate} = 45\%$$

$$\text{After-tax MARR} = 12\% \text{ per year}$$

$$\text{Approved contract revenue} = \$400,000 \text{ (years 1–4)}$$

$$\text{Anticipated revenue} = \$750,000 \text{ (years 5–10)}$$

Simplifying assumptions: Recognize the salvage value in SL depreciation computations; no debt financing has been used, only equity capital.

Solution Equation (15.18) is used to compute the revenue requirement. Initially it is necessary to determine the EUAW and equivalent depreciation for this system over the 10-year planning horizon.
   The investment recovery and 12% return is

$$\text{EUAW} = 2,000,000(A/P, 12\%, 10) - 200,000(A/F, 12\%, 10) = \$342,564$$

The equivalent depreciation over 10 years recognizing salvage is

$$D = (1,800,000/5)(P/A, 12\%, 5)(A/P, 12\%, 10) = \$229,672$$

and the equivalent annual revenue requirement by Eq. (15.18) is

$$\text{RR} = 150,000 + \frac{1}{1 - 0.45}[342,564 - 0.45(229,672)] = \$584,930$$

A minimum of $584,930 in revenue must be received to economically justify the purchase of the system.
   Using the $400,000 (years 1–4) and $750,000 (years 5–10) values, the estimated equivalent annual revenue is $562,058. Therefore, on an economic basis only, the system is not quite justifiable, but it is close to making the required 12%-per-year return.

Probs. 15.32 to 15.36

Additional Example

Example 15.9

Compute the after-tax rate of return for the two situations in Example 15.2: (*a*) 100% equity financing and (*b*) 50% equity–50% debt financing. Compare the results.

Solution

(a) For 100% equity financing, the CFAT values of Table 15.2 can be used to find i^* in the general form of Eq. (15.10).

$$0 = -15,000 + 4500(P/A, i^*\%, 5)$$

$$(P/A, i^*\%, 5) = 3.3333$$

$$i^* = 15.25\%$$

(b) For the 50%–50% split on financing, Table 15.3 results are used to set up the equation

$$0 = -7500 + 2625(P/A, i^*\%, 5)$$

$$(P/A, i^*\%, 5) = 2.8571$$

$$i^*\% = 22.22\%$$

Comparison of the two return values shows that debt financing increases the rate of return on company investments, simply because less of the firm's capital is tied up in investment.

Comment Why not use close to 100% debt financing and maximize the return? We will let you speculate about this dilemma until Sec. 18.7, where we answer this question. Please venture a guess now. What if every time you wanted to make a purchase you had to borrow? How stable would your business be? What kind of financial reputation would your company have?

Problems

Note: Unless specifically mentioned, problems in this chapter do not use the MACRS or straight-line alternative to MACRS. Classical depreciation methods are utilized to make problems independent of changing tax laws.

15.1 An investment company plans to purchase an apartment complex for $350,000. Annual income before taxes of $28,000 is expected for the next 8 years, after which the property will be sold for an estimated $453,600. The applicable tax rate is 52%, the estimated annual operating cost is $3000, and the gain on property sale is taxed at 40%. Tabulate the cash flow after taxes for the years of ownership, if the property will be straight-line-depreciated over a 20-year life with a 40% salvage value. (Note that both recaptured depreciation and a capital gain are involved at sale time.)

15.2 A new machine was purchased for $40,000 and recovered over a 5-year period using SL depreciation. It had no estimated salvage and produced annual CFBT of $20,000 which was taxed at 40%.

(a) If all funds for the purchase were obtained from corporate retained earnings, and the machine was donated after 5 years, determine the CFAT values for years 0 through 5.

(b) Determine the CFAT values if the machine was sold prematurely after 3 years for $20,000 and no replacement was acquired, but the CFBT for year 4 only was $10,000.

15.3 Determine the CFAT values for the situation in Prob. 15.2(b) if 60% of the investment in the machine was borrowed. You have been given data which shows that $2500 in annual interest was paid. The entire principal of the loan was paid in year 5 and interest continued even though the machine was sold in year 3.

15.4 (a) Derive the following formula for annual CFBT when debt financing is used for capital investment.

$$CFBT = \frac{CFAT - (D + I)T + I + LP}{1 - T} \tag{15.19}$$

where the annual amounts are:

D = depreciation

I = interest payment for the loan principal

LP = loan principal repayment

(b) Use Eq. (15.19) to compute the required cash flow before taxes to result in the CFAT of $2625 in Example 15.2(b) where 50% debt financing is used for the purchase of a new dry cleaner.

15.5 An asset was acquired by AAA Floor Covering. It has

$B = \$10,000 \quad n = 5$ years $\quad$ CFBT = $5000

$SV = 0 \quad$ Straight-line depreciation $\quad T = 48\%$

Tabulate the CFAT if the asset is actually salvaged after 6 years for $3075. Assume the CFBT continued for the year after full depreciation.

15.6 (a) Resolve Prob. 15.5 if MACRS depreciation and a 5-year recovery period is used.
(b) Is there any difference in the total CFAT values for SL and MACRS depreciation? Why?

15.7 Use the same asset data and actual salvage for the depreciation computation in Prob. 15.5 to calculate CFAT if a $5000 loan assisted in acquiring the $10,000 asset. Assume the loan interest was computed at an annual rate of 3% on the initial principal and repayment of the principal and interest took place in five equal installments of $1150 each.

15.8 The following machines are to be economically evaluated.

Machine A: $B = \$15,000 \quad SV = \$3000 \quad AOC = \$3000 \quad n = 10$

Machine B: $B = \$22,000 \quad SV = \$5000 \quad AOC = \$1500 \quad n = 10$

Use classical straight-line depreciation, $T = 50\%$, and an after-tax return of 7% to select the more economical machine. AOC is an expense and a tax advantage.

15.9 In Chap. 6, Prob. 6.7, two systems have been evaluated by the EUAW method using a before-tax return of 18%. The used machine was selected for its lower EUAW value.

Repeat the analysis using an after-tax rate of 10% per year, but use the depreciation recovery periods as the "life" of the machines rather than the expected economic lives. Recovery periods are 6 years for the new machine and 3 years for the used machine. Straight-line depreciation with salvage value considered applies to both machines. The tax rate is 50% and all annual costs are tax savings to the alternative. Base the decision on the EUAW of the annual cost CFAT values, where

Cost CFAT = operating + repair costs $-$ tax savings

$\qquad$ = (operating + repair costs)$(1 - T) -$ depreciation(T)

15.10 Update Prob. 6.6 as follows: Regardless of which machine is selected, a $10,000 loan will be necessary for the purchase. Repayment of this loan will be in five equal annual installments of $2700 each ($2000 principal, $700 interest). If the food-processing company is in the 52% tax bracket, uses straight-line depreciation on all assets, and an after-tax MARR of 6% per year is required, determine which labeling machine is more economical.

15.11 The Siko-so-matic Drug Company has to decide between the two pill-forming machines detailed below.

	Round	Oval
First cost	$24,000	$15,000
Actual salvage	6,000	3,000
Annual CFBT	4,000	2,000
Life, years	12	12

The machines have an anticipated useful life of 12 years, but tax laws require straight-line depreciation over 10 years with a zero salvage value. If an effective tax of 50% applies and an after-tax return of 10% is desired, compare the two machines using (a) present-worth analysis and (b) EUAW analysis over 12 years. Neglect the half-year convention.

15.12 Select the more economical of the two alternatives detailed below if the after-tax MARR is 8% per year, classical straight-line depreciation is used, and $T = 50\%$.

	Alternative A	Alternative B
First cost	$10,000	$15,000
Salvage value	1,000	2,000
Annual cost	1,500	600
Life, years	10	10

15.13 Rework 15.12 if one-half the first cost of each asset will be debt-financed with 10 equal loan repayments of $616.45 for A and $924.68 for B. For each loan, one-tenth of the principal amount is removed each year with the remainder of the payment necessary for interest.

15.14 Compare the two plans below using an after-tax MARR of 10% per year and present-worth analysis.

	Plan A		Plan B
	Machine 1	Machine 2	
First cost	$5,000	$25,000	$40,000
Salvage value (after RD tax)	0	1,000	5,000
Annual savings			
Years 1–4	500	10,000	$15,000
Years 5–8	500	10,000	20,000
Years 9–10	500	5,000	25,000
Tax rate	48%	48%	48%
Life, years	5	10	10

Assume straight-line depreciation (no half-year convention) for all assets and neglect the salvage value in depreciation charges. Further, assume that any operating loss will simply be a tax savings for the year incurred and neglect any recaptured-depreciation tax for the salvage amount.

15.15 Compute the after-tax rate of return for Prob. 15.1.

15.16 (*a*) Compute the after-tax rate of return for Probs. 15.2*b* and (*b*) 15.3. Explain the difference in your answers.

15.17 (*a*) What is the after-tax return for the situation in Prob. 15.7 where partial debt capital is used to finance the purchase? (*b*) What difference does the debt financing make in the after-tax return compared to that for 100% equity financing as described in Prob. 15.5?

15.18 A tractor and farm equipment purchased for $78,000 by Mr. Stimson Farms, Inc. generated $26,080 annually in before-tax cash flow during its 5-year life, which represents a return of 20%. However, the corporate tax expert said the effective CFAT was only $15,000 per year. If the corporation hopes for a return of 10% per year after taxes on all investments, for how many years should the equipment have remained in service to realize the 10% return? What percent decrease or increase from the 5 years does this represent?

15.19 (*a*) Summarize the data in Prob. 15.8 and compute the breakeven return using after-tax analysis. (*b*) Plot the present-worth values and select the better plan for each of the following after-tax MARR values: 6%, 9%, 10%, and 13% per year.

15.20 Compute the after-tax return for each machine in Prob. 15.8, if gross income is $5000 per year.

15.21 Determine the after-tax return at which the machines of Prob. 15.10 are economically indifferent.

15.22 (*a*) Determine the after-tax rate of return for each alternative in Prob. 15.14. (*b*) Find the breakeven after-tax return for the plans in Prob. 15.14 using PW analysis.

15.23 In the fiber optics problem in Example 15.4, $B = \$50,000$, $n = 5$, CFBT $= \$20,000$, and $T = 50\%$. Straight-line depreciation is used to compute $i^* = 15.25\%$. Assume the owner wants an after-tax return of 20%. If the tax rate remains at 50%, compute the value of the (*a*) first cost, (*b*) salvage value, and (*c*) annual depreciation at which this will occur. When determining any one of the values above, assume that the remaining parameters retain the value detailed in the example.

15.24 Compute the after-tax return for the owners of an offshore diving company. They can purchase equipment to handle a special job for $2500. The equipment will have no salvage value and will last 5 years. They will receive revenue of $1500 in year 1 of ownership and an estimated $300 revenue each additional year. Use SL depreciation and $T = 45\%$.

15.25 If the offshore diving equipment in Prob. 15.24 is not purchased, another firm will provide the service for the owners and they will make 5% per year after taxes on the $2500 already invested. If this is done, what percent of the $1500 revenue must they realize in year 1 to make the same return offered by the purchase alternative and additional $300 revenue for years 2 through 5?

15.26 Rework Example 15.7 under the assumptions that the trade-in value of the defender is only $2000 and that new remaining life and salvage value estimates are 10 years and $750, respectively. Salvage is still not considered in depreciation computations.

15.27 Perform an after-tax analysis for Prob. 10.13 using a tax rate of 50% and an after-tax return of 4%. Use classical straight-line depreciation for all assets and assume that asset *A* cost $20,000 when purchased and had an expected life of 8 years.

15.28 If the tax rate is 50%, what is (*a*) the breakeven rate of return between plans I and II and (*b*) the replacement value of asset *A* compared to asset *B* ($i = 4\%$) for Prob. 10.13?

15.29 Gene-splicing equipment placed in service 6 years ago has been depreciated from $B = \$175,000$ to no book value using the ACRS rates initiated in the mid 1980s. The system can be continued in use or replaced by a newer technology system with the following data:

$B = \$40,000 \qquad n = 5 \text{ years} \qquad$ no salvage value

The new system will require $7000 in annual costs, but only $2000 of this is tax deductible. Classical straight-line depreciation is again allowed as a depreciation method.

The currently owned system could be sold to a new upstart firm for $15,000. But, if retained it would be used for five more years with an AOC of $6000 and a $9000 upgrade now. The upgrade investment would be depreciated with $n = 3$ years and no salvage value

(a) Use a 5-year planning horizon, a tax rate of 40%, and after-tax MARR of 12% per year to perform the replacement analysis.

(b) Assume the company owner decides to purchase the new machine now regardless of the replacement analysis recommendation. The owner believes the company can sell the new (challenger) equipment in 5 years for an amount which will make the current decision between the defender and challenger indifferent. Compute the required sales amount (actual salvage value) in 5 years for the defender and challenger to be equally desirable now. Comment on this amount compared to the challenger first cost.

15.30 (a) Compare the two plans detailed below using a tax rate of 48% and an after-tax return of 8%. (b) Is the decision different from the before-tax result? Use $i = 15\%$.

	Defender	Challenger
First cost	$28,000	$15,000
AOC when purchased	—	1,500
Actual AOC	1,200	
Expected salvage when purchased	2,000	3,000
Trade-in value	18,000	
Depreciation	SL	SL
Life, years	10	8
Years owned	2	

15.31 Rework Prob. 10.26 using an after-tax analysis. Assume that the currently owned compressor was purchased 4 years ago for $1000 with $n = 5$ years and no salvage value. The asset has been straight-line-depreciated and the effective tax rate for the company is 50%. When performing the analysis for the challenger assume that the *equivalent* annual costs for maintenance are used as expenses to reduce taxes.

15.32 Mrs. Carmichael owns an engineering consulting firm which is currently in leased space. An offer to purchase a building with the following estimates has been made:

$B = \$650,000 \qquad SV = \$200,000 \qquad n = 10 \text{ years}$

Annual operating costs are $10,000 the first year, increasing $1000 per year thereafter and the equivalent annual depreciation is $89,663. If the owner expects the sales price in 5 years to actually equal the SV, determine the annual revenue requirement for a 15%-per-year return and an effective tax rate of 40%. Neglect the fact that part of the purchase amount would be debt-financed.

15.33 Compute the annual revenue requirement for the offer below. Assume an effective tax rate of 35% and assume you desire a 10% per year return on your investment.

A single-family dwelling and land can be purchased for $30,000. The house, which is valued at $20,000, would be SL-depreciated to $5000 using a 5-year recovery period, and has an equivalent AOC of $5000. This AOC estimate includes the capital gains tax and recaptured depreciation tax on an anticipated sales price of $40,000 five years hence.

15.34 Compute the annual revenue requirement for each of the machines in Prob. 15.8.

15.35 Determine whether the defender or challenger dump truck should be selected in Example 10.1 based on the lower of the annual revenue requirements. Use $T = 50\%$ and a 10%-per-year return requirement. Assume the defender will be depreciated using DDB and the actual salvage is that predicted by DDB. Assume the challenger is SL-depreciated and the estimated $2000 salvage is correct. Also, assume 100% equity financing.

15.36 Describe the difference in meaning between a cash flow before-tax (CFBT) amount and the annual revenue requirement (RR) computed by Eq. (15.18).

LEVEL FIVE

The five chapters in this level illustrate the full range of applications for economic-analysis tools. Breakeven analysis for a single project or between two alternatives is explained. This includes the correct use of payback analysis.

The selection of a portfolio of projects using the capital-budgeting approach is discussed, and the establishment of the MARR is examined for a mixture of debt (borrowed) and equity (owner) capital. A summary of the capital-asset pricing model (CAPM) for equity ownership is included.

Sensitivity and risk are taken into account by varying the value of specific parameters in the economic analysis. Simple probabilistic analysis makes it possible to determine expected values and to use the technique of decision trees to select a decision path from several that are dependent.

The last chapter discusses the role of economic analysis in the evaluation of large capital investment opportunities for integrated systems. Consideration of quantitative and intangible benefits and costs is made via multiple-criteria evaluation techniques.

Chapter	Subject
16	Determination of breakeven values
17	Capital rationing under budget constraints
18	Establishing the MARR
19	Sensitivity analysis and decision trees
20	Decision making for large capital investments

Determination of Breakeven Values | 16

It is not uncommon to hear questions such as the following in relation to investments, marketing strategy, and manufacturing planning.

How many of these do I have to make in order to break even?
How long do we have to keep this machine to recover our money?
Is there a rate of return at which I can be indifferent and select either of the two alternatives?

The answers are determined by estimating the number of units or years, or the rate of return at which revenues and costs are equal. For two alternatives, the circumstances under which PW or EUAW values are equal may answer the last question.

The first question is best addressed using *breakeven analysis*. The level of operation (number of units) or utilized percentage of plant capacity is computed when relatively simple models of revenues and costs are equated. Both linear and nonlinear breakeven analyses are commonly performed as a part of economic studies.

The amount of time that an investment takes to pay for itself is determined by the *payback period*. Usually a specified rate of return is included when the payback is calculated. Also, inflation may be considered in these computations. Payback, which is often misused and misunderstood, is not an alternative evaluation technique, but it can be used to assist decision makers in evaluating the risk present in a project.

We have already seen the use of breakeven points between alternatives in rate-of-return computations. Figures 8.2 and 15.1 show the breakeven returns between alternatives for before-tax and after-tax situations.

Breakeven and payback computations are detailed in this chapter and a brief description of life-cycle costing (LCC) is included.

Section Objectives

After completing this chapter, you should be able to:

16.1 Compute the *breakeven point* and amount of profit associated with a single investment, given revenue and variable-cost and fixed-cost information.

16.2 Calculate the *breakeven point between two alternatives* using present-worth or equivalent-uniform-annual-worth analysis and select one of the alternatives.

16.3 Compute the *payback period* for a single project, given the revenue and cost cash flows and demonstrate the fallacies of utilizing payback analysis to evaluate alternatives.

16.4 Define *life-cycle costing (LCC)* and describe the major cost categories used in performing an LCC analysis.

Study Guide

16.1 Breakeven Value of a Variable

To analyze or estimate profit or loss, it is often necessary to determine the quantity at which revenue and cost will be equal. This value, called the *breakeven point* Q_{BE}, is computed using the best estimates of relations for revenue and cost for different quantities Q. Quantity Q may be expressed in units per year, percentage of capacity, hours per month, and many other dimensions. We will usually use units per year for illustration purposes.

The revenue relation R may be linear or nonlinear as shown in Fig. 16.1a. Linear revenue is commonly assumed, but nonlinear is more realistic since it recognizes that while additional revenues are possible, decreased unit prices are usually necessary. As the unit price is increased, the revenue curve becomes steeper.

Costs, which may also be linear or nonlinear, are usually composed of two components:

Fixed costs (FC) such as building costs, insurance, overhead, some level of labor, capital recovery
Variable costs (VC) such as labor, material, indirect labor, and advertisement

The fixed-cost component can be considered constant. It does not vary with differing production levels or work-force size. Thus, even if no units are produced, the fixed cost is still incurred because the plant must be open and maintained and certain employees paid. Of course, this situation could not last for long before the plant would have to close to eliminate its fixed costs. Variable costs will increase as production and work-force size increase. It is usually possible to decrease the variable cost through better product design and manufacturing efficiency. Fixed costs are reduced through improved plant and work-force utilization, less costly fringe benefits packages, and so on.

There is often a predictable relation between the two cost components. For example, product manufactured at an outdated steel mill may have a high fixed cost due to large maintenance costs, ineffective processes, and lost production time. Also,

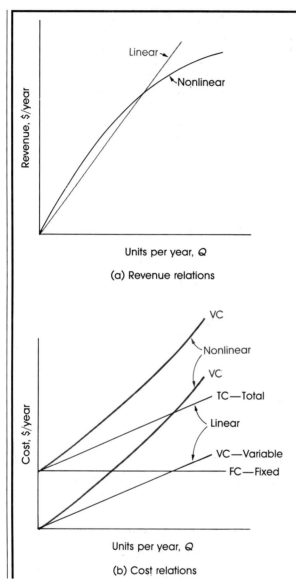

Figure 16.1 Linear and nonlinear revenue and cost relations used in breakeven analysis.

Linear

Nonlinear

Revenue, $/year

Units per year, Q

(a) Revenue relations

VC

Nonlinear

VC

TC—Total

Linear

VC—Variable

FC—Fixed

Cost, $/year

Units per year, Q

(b) Cost relations

the variable cost may be high due to high scrap rates, and to the inability of the process to utilize improved raw materials and improved worker techniques.

Figure 16.1*b* is a plot of linear fixed and variable costs, which together form the total cost (TC).

The quantity at which revenue R and TC intersect is the breakeven point Q_{BE}. As shown in Fig. 16.2 if the variable cost VC is reduced, the TC line will fall and the breakeven point will decrease. This is an advantage because the smaller the value of Q_{BE}, the greater the profit for a constant revenue amount. For linear models of R

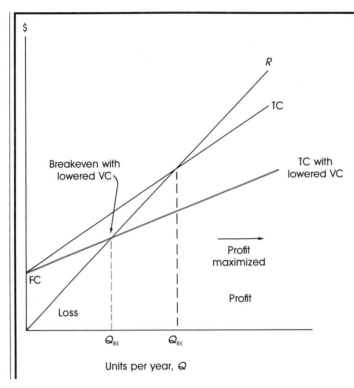

Figure 16.2 Effect on breakeven point when the variable cost per unit is reduced.

and TC, the greater the actual quantity sold (moving to the right of Q_{BE} in Fig. 16.2), the larger the profit will be.

If nonlinear R or TC relations are assumed, there may be more than one breakeven point. Figure 16.3 presents this situation for two breakeven points. The maximum profit is obtained now by operating at a quantity Q_p which has the maximum distance between the R and TC curves.

Of course, no R and TC relations—linear or nonlinear—are able to match exactly the true revenue and cost situations for a company over an extended period of time for any particular product or service. But, it is possible to make estimates accurately enough to calculate breakeven points that can be used as an aim point for planning and design purposes.

Example 16.1

A brake drum plant has historically produced at about 80% of capacity in its production of 14,000 units per month. Decreased demand and a worker slowdown has put production at 8000 units per month for the foreseeable future. Use the information below to determine (*a*) where the 8000-unit level will place production relative to the linear breakeven point and the profit at this reduced level, and (*b*) the variable cost per drum necessary to break even at the 8000-unit level, if revenue per unit and fixed costs remain constant.

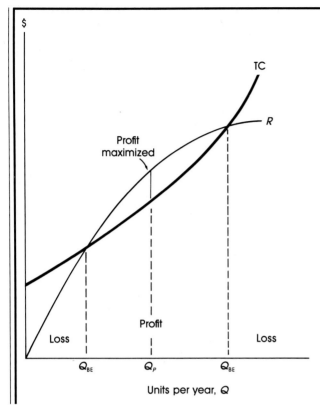

Fixed cost: FC = $75,000 per month

Variable cost: v = $2.50 per unit

Revenue: r = $8.00 per unit

Solution   Before answering the questions, the linear relations for total revenue R and total cost TC must be determined. Let Q be the quantity in units per month.

$R = rQ = 8Q$ ($ per month)

$TC = FC + vQ = 75,000 + 2.5Q$ ($ per month)

At breakeven set $R = TC$ and solve for Q_{BE}.

$rQ = FC + vQ$

$$Q_{BE} = \frac{FC}{r - v} \tag{16.1}$$

(*a*) Use Eq. (16.1) to find Q_{BE}.

$$Q_{BE} = \frac{75,000}{8.00 - 2.50} = 13,636 \text{ units per month}$$

Figure 16.4
Breakeven graph
for Example 16.1.

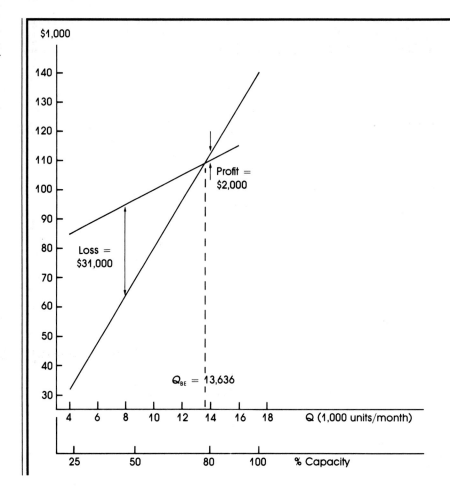

Figure 16.4 indicates that at 14,000 units (80% of capacity) this plant has been producing barely above breakeven, which is actually at 78% of capacity—a very high production requirement to attain breakeven!

To compute profit, p, subtract cost from revenue at a production rate Q of 8000 units.

$$p = R - TC = rQ - FC - vQ \qquad (16.2)$$
$$= (r - v)Q - FC$$
$$= (8.00 - 2.50)8000 - 75,000$$
$$= -\$31,000$$

There is actually a loss of $31,000 per month. At the 14,000-unit-per-month level, only $2000 profit is realized (Fig. 16.4).

(b) A simple way to determine v with $Q_{BE} = 8000$ is to substitute into Eq. (16.2) the values profit $p = 0$ and $Q = 8000$ and solve for v.

$$0 = (8 - v)8000 - 75,000$$

$$v = \frac{-11,000}{8} < 0$$

Since $v < 0$, it is impossible to break even at 8000 units without reducing the fixed cost and/or increasing the revenue. This plant has some serious problems!

Probs. 16.1 to 16.6

16.2 Computation of Breakeven Points Between Two or More Alternatives

In some economic analyses, one or more of the cost components may vary as a function of the number of units produced or consumed in manufacture. In these cases, it is convenient to express the cost relation in terms of the variable and find the value at which the alternative proposals break even. In breakeven analysis, the variable is usually *common to both alternatives*, such as a variable operating cost or production cost. Figure 16.5 graphically illustrates the breakeven concept for two proposals. The fixed cost, which may be simply the equivalent initial investment cost, of proposal 2 is greater than that of proposal 1, but proposal 2 has a lower variable cost as indicated by its smaller slope. The intersection of the two lines represents the breakeven point between the two proposals. Thus, if the units (hours of operation or level of output) are expected to be greater than the breakeven amount, proposal 2 will be selected, since the total cost of the operation will be lower with this alternative. Conversely, an anticipated level of operation below the breakeven point favors proposal 1.

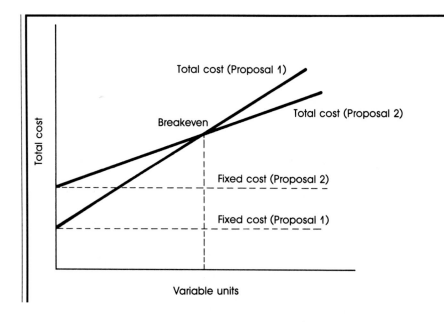

Figure 16.5
Linear breakeven between two alternatives involving costs only.

Instead of plotting the total costs of each alternative and finding the breakeven point graphically, it is generally easier to calculate the breakeven point algebraically. Although the total cost may be expressed as either a present worth or equivalent uniform annual worth, the latter is generally preferable when the variable units are expressed on a yearly basis. Additionally, EUAW calculations are simpler when the alternatives under consideration have different lives. In either case, the following steps may be used to select one of the alternatives:

1. Clearly define the variable and state its dimensional units.
2. Use EUAW or present-worth analysis to express the total cost of each alternative as a function of the defined variable.
3. Equate the two cost relations and solve for the breakeven value of the variable.
4. If the anticipated operating level is below the breakeven value, select the alternative with the higher variable cost (larger slope). If the level is above the breakeven point, select the alternative having the lower variable cost. Refer to Fig. 16.5.

Example 16.2

A sheet metal company is considering the purchase of an automatic feed machine for a certain phase of the finishing process. The machine has an initial cost of $23,000, a salvage value of $4000, and a life of 10 years. If the machine is purchased, one operator will be required at a cost of $12 an hour. The output with this machine will be 8 tons per hour. Annual maintenance and operating cost of the machine is expected to be $3500.

Alternatively, the company can purchase a less sophisticated manual-feed machine for $8000, which has no salvage value and a life of 5 years. However, with this alternative, three workers will be required at a cost of $8 an hour each, and the machine will have an annual maintenance and operation cost of $1500. Output is expected to be 6 tons per hour for this machine. All invested capital must return 10% per year.

(a) How much sheet metal must be finished per year in order to justify the purchase of the automatic-feed machine?
(b) If management anticipates a requirement to finish 2000 tons per year, which machine should be purchased?

Solution
(a) Use the steps above to find the breakeven point.

1. Let x be the number of tons of sheet metal per year.
2. For the automatic-feed machine the annual variable cost may be written as

$$\text{Annual variable cost} = \left(\frac{\$12}{\text{hour}}\right)\left(\frac{1 \text{ hour}}{8 \text{ tons}}\right)\left(\frac{x \text{ tons}}{\text{year}}\right) = \frac{12}{8}x$$

Note that the cost is in dollars per year, which is required for EUAW analysis. The EUAW for the automatic-feed machine is

$$EUAW_{\text{auto}} = 23,000(A/P, 10\%, 10) - 4000(A/F, 10\%, 10) + 3500 + \frac{12}{8}x$$

$$= \$6992 + 1.5x$$

Similarly, the annual variable cost and EUAW for the manual-feed machine is

$$\text{Annual variable cost} = \left(\frac{\$8}{\text{hour}}\right)(3 \text{ persons})\left(\frac{1 \text{ hour}}{6 \text{ tons}}\right)\left(\frac{x \text{ tons}}{\text{year}}\right) = 4x$$

$$EUAW_{manual} = 8000(A/P, 10\%, 5) + 1500 + \frac{3(8)}{6} x$$

$$= \$3610 + 4x$$

3. Equating the two costs and solving for x yields the breakeven value.

$$EUAW_{auto} = EUAW_{manual}$$

$$6992 + 1.5x = 3610 + 4x$$

$$x = 1353 \text{ tons per year}$$

4. If the output is expected to exceed 1353 tons per year, purchase the automatic-feed machine, since its variable cost slope of 1.5 is smaller than the manual-feed slope of 4.

(b) Select the smaller-sloped alternative, which is the automatic-feed machine in this case. Equivalently, it is possible to select an alternative by substituting the expected production level of 2000 tons per year into the EUAW relations. Then $EUAW_{auto} = \$9992$ and $EUAW_{manual} = \$11,610$, and purchase of the automatic-feed machine is justified.

Comment Work the problem on a present-worth basis to satisfy yourself that either method results in the same breakeven value. Remember how to select the correct alternative: As shown in Fig. 16.5, the alternative with the smaller slope (i.e., lower variable cost) should be selected when the variable units are above the breakeven point (and vice versa).

Example 16.3

The Jack n' Jill Toy Company currently purchases the metal parts which are required in the manufacture of certain toys, but there has been a proposal that the company make these parts. Two machines will be required for the operation: Machine A will cost \$18,000 and have a life of 6 years and a \$2000 salvage value; machine B will cost \$12,000 and have a life of 4 years and a \$-500 salvage value. Machine A will require an overhaul after 3 years costing \$3000. The annual operating cost of machine A is expected to be \$6000 per year and for machine B \$5000 per year. A total of four laborers will be required for the two machines at a cost of \$2.50 per hour per worker. In a normal 8-day period, the machines can produce parts sufficient to manufacture 1000 toys. Use a MARR of 15% per year and a purchase price of \$0.50 per toy if the parts are not manufactured.
(a) How many toys must be manufactured each year to justify the purchase of the machines?
(b) If the company expects to produce 75,000 toys per year, what maximum expenditure could be justified for the more expensive machine, assuming its salvage value and all other costs will be the same as stated?

Solution
(a) Use steps 1 to 3 of the procedure above to determine the breakeven point.

1. Let x be the number of toys produced per year.
2. There are variable costs for the workers and fixed costs for the two machines. The annual variable cost is

Variable cost per year = (cost per unit)(units per year)

$$= \left(\frac{4 \text{ persons}}{1000 \text{ units}}\right)\left(\frac{\$2.50}{hour}\right)(8 \text{ hours})x = 0.08x$$

The fixed annual costs for the machines are

$$\text{Fixed EUAW}_A = 18{,}000(A/P, 15\%, 6) - 2000(A/F, 15\%, 6)$$
$$+ 6000 + 3000(P/F, 15\%, 3)(A/P, 15\%, 6)$$

$$\text{Fixed EUAW}_B = 12{,}000(A/P, 15\%, 4) + 500(A/F, 15\%, 4) + 5000$$

3. Equating the annual costs of the purchase option $(0.50x)$ and the manufacture option yields

$$0.50x = \text{fixed EUAW}_A + \text{fixed EUAW}_B + \text{variable cost per year}$$
$$= 18{,}000(A/P, 15\%, 6) - 2000(A/F, 15\%, 6) + 6000$$
$$+ 3000(P/F, 15\%, 3)(A/P, 15\%, 6) + 12{,}000(A/P, 15\%, 4)$$
$$+ 500(A/F, 15\%, 4) + 5000 + 0.08x \tag{16.3}$$

$$0.42x = 20{,}352.43$$

$$x = 48{,}458 \text{ units per year}$$

Therefore, a minimum of 48,458 toys must be produced each year to justify the manufacture proposal.

(b) Substitute 75,000 for the variable x and the unknown first cost P_A for the $18,000 in Eq. (16.3) and solve to obtain $P_A = \$60{,}187$. Thus, as much as $60,187 is justified as the first cost for machine A if 75,000 toys are produced and other costs are as estimated.

Even though the preceding examples dealt with only two alternatives, the same type of analysis can be done for three or more alternatives. In this case, it becomes necessary to compare the alternatives by pairs to find their respective breakeven points. The results reveal the ranges through which each alternative is most economical. For example, in Fig. 16.6, if the output is expected to be less than 40 units per hour, proposal 1 should be selected. Between 40 and 60 units per hour, proposal 2 would be most economical; and above 60 units per hour, proposal 3 is favored.

If the variable-cost relations are nonlinear, analysis is more complicated. If the costs increase or decrease uniformly, mathematical expressions that allow direct determination of the breakeven point can be derived.

Probs. 16.7 to 16.18

16.3 Determination and Use of Payback Period

The payback period for an asset or project is the number of years it must be retained or be economically useful to *recover the initial investment with a stated return*. It should never be used as a method equivalent to PW, EUAW, or rate of return to select between alternatives, but merely as *supplemental information about an alternative*. Payback analysis may be performed using before-tax or after-tax cash flows. To find the payback period n at a stated return, determine the n value in years using

$$0 = -P + \sum_{t=1}^{n} \text{CF}_t(P/F, i\%, t) \tag{16.4}$$

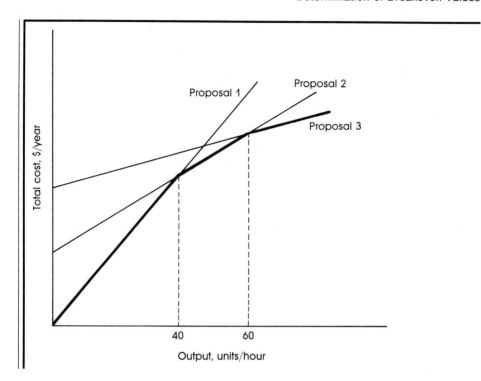

Figure 16.6
Breakeven points
for three
proposals.

where CF_t is the net cash flow at the end of year t. If the cash flows are the same each year, the P/A factor may be used in the relation $0 = -P + CF (P/A, i\%, n)$. After n years (not necessarily an integer), the cash flows will recover the investment and a return of $i\%$. If the asset or project is active for more than n years, a larger return will result; but if the expected retention period is known to be less than n years, there is not enough time to recover the investment and the required return. It should be pointed out that in payback-period analysis, any cash flows occurring after n years are neglected in the Eq. (16.4) computation.

An historically incorrect usage of payback is to neglect the effects of required return ($i = 0\%$) and inherent inflation in the economy and to compute n from

$$0 = -P + \sum_{t=1}^{n} CF_t \tag{16.5}$$

If all CF_t values are assumed equal

$$n = \frac{P}{CF} \tag{16.6}$$

Computation of n from these relations is of little value in an economic study. When $i\% > 0$ is used to estimate the payback-period requirement, the result gives a sense of the risk taken if the project is undertaken. For example, a company which plans to produce a certain product under contract for only the next 3 years should not undertake a project with a payback requirement of 10 years. But, if the same project can be used to launch a new product, a thorough economic and organizational study should be performed. In this example, the payback is only supplemental information, not a substitute for a correct economic study.

Use of the no-return ($i = 0\%$) payback period to make economic alternative decisions is incorrect because:

1. The required return is neglected since the time value of money is omitted.
2. All cash flows which occur after the computed payback period and will contribute to the return on the investment are neglected.

Example 16.4

A semiautomatic machine purchased for $18,000 is expected to generate annual revenues of $3000 and have SV = $3000 at any time during the 10 years of anticipated ownership. If a 15%-per-year required return is imposed on the purchase, compute the payback period.

Solution   The cash flow each year is $3000 with an additional $3000 in year n. Equation (16.4) is slightly altered to

$$0 = -P + CF(P/A, 15\%, n) + SV(P/F, 15\%, n)$$
$$= -18,000 + 3000(P/A, 15\%, n) + 3000(P/F, 15\%, n)$$

The resulting payback period is $n = 15.3$ years. Using the planning horizon of 10 years, the machine will not return the required 15% per year.

Comment   If the unrealistic assumption were made that the company owner required no return on investments, Eq. (16.5) would result in $n = 5$ years as follows:

$$0 = -18,000 + 5(3000) + 3000$$

Note the difference that a required return of 15% makes. At a 15% required return, this asset would have to be retained 15.3 years, while with no return only 5 years of ownership are required. This characteristic of longer required ownership is always present with $i > 0\%$ because of the time value of money. The pattern of cash flows is the primary determinant of the difference in required retention periods.

This result indicates that the payback-period method of analysis is appropriate only when investment capital is in very short supply and management requests recovery of capital in a short period. The payback calculation will give the amount of time required to recover the invested dollars, but from the point of view of the economic and time value of money, the payback-period method of analysis using $i = 0\%$ is not appropriate.

If two or more alternatives are present and payback analysis is used to select one, the second fallacy of payback periods mentioned above may lead to an incorrect decision. When cash flows which would occur after n are neglected, it is possible to favor short-lived assets even when longer-lived assets produce a higher return. In these cases, present-worth or EUAW analysis should always be considered the primary decision technique. Comparison of short- and long-lived assets in Example 16.5 illustrate this incorrect use of payback analysis.

Example 16.5

Two equivalent pieces of farm equipment are being considered for purchase. Machine 2 is expected to be versatile enough for use in future plantings to have longer-term cash flows than machine 1.

Machine 1	Machine 2
$P = \$12,000$	$P = \$8,000$
Net income per year = 3,000	Net income per year = 1,000 (year 1–5)
	= 3,000 (year 6–15)
Maximum $n = 7$ years	Maximum $n = 15$ years

Mr. James used 15%-per-year payback analysis and a microprocessor-based economic analysis package in the local agricultural extension office, which utilizes Eq. (16.4), to choose machine 1 because it has a shorter payback period. The computations are summarized here.

Machine 1: $n = 6.57$ years which is less than the 7-year life.

$$0 = -12,000 + 3000(P/A, 15\%, n)$$

Machine 2: $n = 9.52$ years which is less than the 15-year life.

$$0 = -8000 + 1000(P/A, 15\%, 5) + 3000(P/A, 15\%, n - 5)(P/F, 15\%, 5)$$

Machine 1 was selected since the 15% return will be realized in a shorter number of years.

Use a 15% EUAW analysis to compare the machines and comment on the difference in the results of the analysis.

Solution For the respective number of years

$$\text{EUAW}_1 = -12,000(A/P, 15\%, 7) + 3000 = \$116$$

$$\text{EUAW}_2 = -8000(A/P, 15\%, 15) + 1000$$
$$+ 2000(F/A, 15\%, 10)(A/F, 15\%, 15) = \$485$$

Machine 2 is selected since its return exceeds that of machine 1. The results are contradictory. The EUAW analysis accounts for the increased cash flows for machine 2 in the later years, whereas payback neglects these cash benefits when it uses values of 9 and 10 years to estimate $n = 9.52$ years.

Remember: Use payback only as supplemental, risk assessment information. Rely on PW, EUAW, and rate-of-return analysis to compare alternatives.

Probs. 16.19 to 16.26

16.4 Life-Cycle Costing (LCC)

The technique of life-cycle costing analysis is applied more commonly by large contractors to the government, especially to carry out defense-related projects. LCC may be defined as [1]:

> The total cost of an item including research and development, production, modification, transportation, introduction into inventory, construction, operation, support, maintenance, disposal, salvage revenue, and any other cost of ownership.

The total anticipated costs of an alternative are usually estimated using major categories like:

Research and development costs. All expenditures for design, prototype fabrication, testing, manufacturing planning, engineering services, software engineering, software development, and the like on a product or service.

Production costs. The investment necessary to produce or acquire the product, including expenses to employ and train personnel, transport subassemblies and the final product, build new facilities, and acquire equipment.

Operating and support costs. All costs incurred to operate, inventory, maintain, and manage the product for its anticipated life.

The use of a present-worth analysis to account for the time value of money is necessary when performing LCC. The major difference in LCC and the analyses we have performed thus far is the attempt to include *all* development costs and future maintenance costs, but all computations are the same.

The purpose of such an evaluation is to critically cost-out each alternative for its entire life and select the one with the minimum LCC. Application is usually made to projects which will require research and development time to design and test a product or system intended to perform a specific task. Examples are radar for aircraft range detection, large software and artificial intelligence projects, and computer integrated manufacturing (CIM) efforts.

Actually, an economic study comparison of alternatives with all definable costs estimated for the life of each alternative is the same as the LCC analysis. However, because many applications of LCC are defense-related, the methods of cost estimation and comparison as used for government acquisition are somewhat different in presentation format. For a description of cost-estimation procedure in LCC refer to Seldon [1].

Probs. 16.27 and 16.28

Reference

1. M. R. Seldon, *Life Cycle Costing: A Better Method of Government Procurement*, Westview, Boulder, CO, 1979.

Problems **16.1** The fixed costs for Cameo Imports is $600,000 annually. The average product is sold at a revenue of $2 per unit and has $1.50 variable costs. (*a*) Compute the annual breakeven quantity for Cameo. (*b*) Plot the revenue and total cost relations and estimate from your graph the annual profit if 1.8 million items are sold.

16.2 Average cost per unit (AC) is the total cost divided by the volume. (*a*) Derive a relation for AC and use it to plot AC versus Q for a fixed cost of $60,000 and variable costs of $1 per unit. At what sales volume is a $3-per-unit average cost possible? (*b*) If the fixed cost in 1 year increases to $100,000, plot the new AC curve on the same graph and find the necessary sales volume to maintain a $3 per unit average cost.

16.3 A defense contractor has been able to summarize its fixed and variable cost components for a particular product.

Fixed components ($)		Variable components ($ per unit)	
Administrative	$10,000	Materials	$ 5
Lease cost	20,000	Labor	3
Insurance	7,000	Indirect labor	5
Utilities	3,000	Other overhead	20
Taxes	10,000		
CFAT requirement	$50,000		

(a) Find the revenue per unit to break even if a 5000-unit contract with the government is all that can be expected this year.

(b) If foreign sales of 3000 can be added to the 5000-unit government contract, determine the profit, provided the same price determined in (a) is charged.

16.4 (a) The manufacturer of a toy called Willy Wax has a capacity of 20,000 units annually. If the fixed cost of the production line is $30,000 with a variable cost of $3 and a price of $5 per unit, find the percent of present capacity that must be used to break even.

(b) The toy manufacturer has found a way to utilize 10,000 units of the 20,000-unit capacity currently available on another product. This is expected to reduce the fixed cost for Willy Wax to $15,000. If 10,000 units are produced, what amount of variable cost is allowed for this to be the breakeven volume? How does this compare with the current variable cost?

16.5 It is common to perform nonlinear breakeven analysis using quadratic relations for total cost (TC) and revenue (R) of the form

$$aQ^2 + bQ + c$$

Since profit $p = R - TC$, a general form of R, TC, and p may be written.

$$R = dQ^2 + eQ$$

$$TC = fQ^2 + gQ + h$$

$$p = R - TC = aQ^2 + bQ + c \qquad (16.7)$$

where a through h are constants which must be determined.

As indicated in Fig. 16.3, maximum profit occurs at the quantity Q_p which has the greatest distance between R and TC. This maximum profit point can be found using calculus; it occurs at the quantity

$$Q_p = \frac{-b}{2a} \qquad (16.8)$$

Substitution into Eq. (16.7) yields

$$\text{Maximum profit} = \frac{-b^2}{4a} + c \qquad (16.9)$$

(a) Use calculus to verify the results of Eqs. (16.8) and (16.9).

(b) Plot the following revenue and total-cost relations, determine the maximum profit and quantity at which it occurs, and indicate these values on your graph.

$$R = -0.005Q^2 + 25Q$$

$$TC = 0.002Q^2 + 3Q + 2$$

16.6 (a) A consultant to the Weiner Corporation has determined that the total-cost relation in Prob. 16.5 is correct for the company and that revenue is linear with $r = \$25$. Plot the profit function and determine the value of the maximum profit and the quantity Q_p at which it occurs. (b) What is the difference in the quantity at which maximum profit occurs between this solution (linear revenue) and that in Prob. 16.5?

16.7 Two pumps can be used for pumping a corrosive liquid. A pump with a brass impeller costs $800 and is expected to last 3 years. A pump with a stainless steel impeller will cost $1900 and last 5 years. An overhaul costing $300 will be required after 2000 operating hours for the brass impeller pump while an overhaul costing $700 will be required for the stainless pump after 9000 hours. If the operating cost of each pump is $0.50 per hour, how many hours per year must the pump be required to justify the purchase of the more expensive pump? Use an interest rate of 10% per year.

16.8 The Sli-Dog Company is considering two proposals for improving the employees' parking area. Proposal A would involve filling, grading, and blacktopping at an initial cost of $5000. The life of the parking lot constructed in this manner is expected to be 4 years with annual maintenance costs of $1000.

Alternatively, the parking area would be paved under proposal B, in which case the life would be extended to 16 years. The annual maintenance cost will be negligible for the paved parking area, but the markings will have to be repainted every 2 years at a cost of $500. If the company's minimum attractive rate of return is 12% per year, how much can it afford to spend for paving the parking area so that the proposals would break even?

16.9 The Redi-Bilt Construction Company is considering the purchase of a dirt scraper. A new scraper will have an initial cost of $75,000, a life of 15 years, a $5000 salvage value, and an operating cost of $30 a day. Annual maintenance cost is expected to be $6000 per year.

Alternatively, the company can lease a scraper and driver as needed for $210 per day. If the company's minimum attractive rate of return is 12% per year, how many days per year must the scraper be required to operate in order to justify the purchase?

16.10 The Pant Company is considering the purchase of an automatic cutting machine. The machine will have a first cost of $22,000, a life of 10 years, and $500 salvage value. The annual maintenance cost of the machine is expected to be $2000 per year for the range of usage anticipated. The machine will require one operator at a cost of $24 a day. Approximately 1500 yards of material can be cut each hour with the machine.

Alternatively, if manual labor is used, five workers, each earning $18 a day, can cut 1000 yards per hour. If the company's minimum attractive rate of return is 8% per year, how many yards of material must be cut each year in order to justify the purchase of the automatic machine? Assume an 8-hour work day.

16.11 A couple have an opportunity to buy an old, fire-damaged house for what they believe to be a bargain price of $28,000. They estimate that remodeling the house now will cost $12,000 and annual taxes will be approximately $800 per year. They estimate that utilities will cost $500 per year and that the house must be repainted every 3 years at a cost of $400. At the present time, resale houses are selling for $16 per square foot, but they expect this price to increase by $1.50 per square foot per year. They will lease the house for $2500 per year until it is sold. If the house has 2500 square feet and they want to make 8% per year on their investment, (a) how long must they keep the house before they can break even and (b) what would the selling price be at the time of the sale?

16.12 The Rawhide Tanning Company is considering the economics of furnishing an in-house water-testing laboratory instead of sending samples to independent laboratories for analysis. If the lab were completely furnished so that all tests could be conducted in-house, the initial cost would be $25,000. A technician would be required at a cost of $13,000 per year. The cost of power, chemicals, etc., would be $5 per sample. If the lab is only partially furnished, the initial cost will be $10,000. A part-time technician will be required at an annual salary of $5000. The cost of in-house sample analysis will be $3 per sample, but since all tests cannot be conducted by the company, outside testing will be required at a cost of $20 per sample.

      If the company elects to continue the present condition of complete outside testing, the cost will be $55 per sample. If the laboratory equipment will have a useful life of 12 years and the company's MARR is 10% per year, how many samples must be tested each year in order to justify (*a*) the complete laboratory and (*b*) the partial laboratory? (*c*) If the company expects to test 175 samples per year, which of the three alternatives should be selected?

16.13 A slaughterhouse and packing plant currently pays the city $4000 a month in waste-water discharge fees. The company is considering constructing a treatment plant of its own. The treatment plant will require a part-time operator at $300 per month. In addition, the operating cost is expected to be $500 per month. The treatment plant will require minor repairs every 4 years costing $1500 and is expected to last for 20 years. If the company's MARR is a nominal 12% per year compounded monthly, how much could the company afford to spend on a treatment plant and not exceed its present costs?

16.14 A city engineer is considering two methods for lining water-holding tanks. A bituminous coating can be applied at a cost of $2000. If the coating is touched up after 4 years at a cost of $600, its life can be extended another 2 years. Alternatively, a plastic lining can be installed, which will have a life of 15 years. If the city uses an interest rate of 5% per year, how much money can be spent for the plastic lining so that the two methods just break even?

16.15 A family planning to build a new house is trying to decide between purchasing a lot in the city or in the suburbs. A 1000-square-meter lot in the city will cost $10,000 in the area in which they want to buy. If they purchase a lot outside the city limits, a similar parcel will cost only $2000. For the size of house they plan to build, they expect annual taxes to amount to $1200 per year if they build in the city and only $150 per year in the surburbs. If they purchase the lot outside the city limits, they will have to drill a well for $4000. With their own well, they will save $150 per year in water charges, but they expect the city to provide water to their area in 5 years, after which time they will purchase the city water. They estimate that the increased travel distance will cost $325 the first year, $335 the second year, and amounts increasing by $10 per year. Using a 25-year analysis period and an interest rate of 6% per year, how much extra could the family afford to spend on the house outside the city limits and still have the same total investment? Assume that the land can be sold for the same price as its initial cost.

16.16 The I. M. Rite family is considering insulating their attic to prevent heat loss. They are considering R-11 and R-19 insulation. They can install R-11 for $160 and R-19 for $240. They expect to save $35 per year in heating and cooling with R-11. If the interest rate is 6%, how much money must they be able to save per year in order to justify the R-19 insulation if they want to recover their investment in 7 years?

16.17 A waste-holding lagoon situated near the main plant receives sludge on a daily basis. When the lagoon is full, it is necessary to remove the sludge to a site located 4.95 kilometers from the main plant. At the present time, whenever the lagoon is full, the sludge is removed by pumping it into a tank truck and hauling it away. This requires a portable pump that costs $800 and has an 8-year life. The company supplies the labor to operate the pump at a cost of $25 per day, but the truck and driver must be rented at a cost of $110 per day.

Alternatively, the company can install a pump and pipeline to the remote site. The pump would have an initial cost of $600, and a life of 10 years and would cost $3 per day to operate. The company's MARR is 15%. (a) If the pipeline would cost $3.52 per meter to construct, how many days per year must the lagoon require pumping in order to justify construction of the pipeline? (b) If the company expects to pump the lagoon one time per week, how much money could it afford to spend on the pipeline in order to just break even? Assume a pipeline life of 10 years.

16.18 A building contractor is considering two alternatives for improving the exterior appearance of a commercial building that is being renovated. The building can be completely painted at a cost of $2800. The paint is expected to remain attractive for 4 years, at which time the job would have to be redone. Every time the building is painted, the cost will increase by 20% over the previous time.

Alternatively, the building can be sandblasted now and every 10 years at a cost 40% greater than the previous time. The remaining life of the building is expected to be 38 years. If the company's MARR is 10% per year, what is the maximum amount that could be spent now on the sandblasting alternative so that the two alternatives will just break even? Use present-worth analysis to solve this problem.

16.19 Determine the number of years that an investor must keep a piece of property to make a 15%-per-year return. The purchase price was $6000 with taxes of $80 the first year, increasing by $10 each year until sold. Assume a sales price of $14,000 for the first 10 years and $20,000 thereafter.

16.20 When is it incorrect to use payback-period analysis to choose between two alternatives?

16.21 (a) Determine the payback period for an asset that initially costs $8000, has a salvage of $500 when sold, and has a net profit of $900 per year. The required return is 8% per year.

(b) If the asset will be used for 5 years by the owner, should it be purchased?

16.22 Determine the number of years that the two machines of plan B of Prob. 6.15 must be kept to make a 10% per year return if the annual income of the plan is $4000. Assume that the estimated salvage values apply for all years.

16.23 How many years must the owner of the artificial Christmas tree in Prob. 7.12 use the tree to make 20% per year on the investment?

16.24 The Sundown Detective Agency would like to purchase some new camera equipment. The following data are estimates

$$\text{First cost} = \$1050$$

$$\text{Annual operating cost} = 70 + 5(k) \qquad k = 1, 2, 3, \ldots$$

$$\text{Annual income} = 200 + 50(k) \qquad k = 1, 2, 3, \ldots$$

$$\text{Salvage value} = \$600$$

(a) Compute the payback period to make a return of 10% per year on the investment.
(b) Should the camera be purchased if the expected useful period is 7 years?

16.25 The lease cost of trucks is $3300 per year payable at the beginning of each year. No refund is given for partial-year leases. As an alternative the purchase price per truck is $1700 now with a monthly payment of $300 for 4 years. A truck can be sold for an average of $1200 regardless of the length of time of ownership. The prospective owner, Speede Drug Company, expects to increase net revenue by $400 per month and requires a nominal return of 12% per year. The drug company must pay operation, maintenance, and insurance for the leased or purchased trucks, so this cost is equal and neglected.

(a) How many months must the lease and purchase plans be adhered to in order to make the required return?

(b) If a purchased truck has an expected life of 6 years, should the trucks be leased or purchased?

16.26 (a) Use EUAW analysis to evaluate the alternatives below if the retention years are the maximum life. (b) If a 10%-per-year return, payback-period analysis is used to evaluate the alternatives, with no EUAW computations, is the decision different? Why?

Machine 1	Machine 2
B = $12,000	B = $8,000
Cash flow = $3,000 per year	Cash flow = $1,000 (years 1–5)
	= $3,000 (years 6–15)
Maximum n = 7 years	Maximum n = 15 years

16.27 A medium-sized municipality wants to develop a computerized system which will assist in street-maintenance project selection during the next 10 years. A life-cycle cost approach has been used to categorize costs into development, programming and installation, and annual operating and support costs for the alternatives, each of which may be developed at either of two levels—one level for information collection only using manual project selection, and a second level for data collection and automatic selection of candidate street-maintenance projects.

The first alternative A involves the tailoring of software for the hardware currently owned by the city. This alternative would involve some procedure changes and 1 (level 1) or 2 (level 2) years to get ready. Alternative B is the development of a new system using the current municipal procedures. The same development times are anticipated for this alternative, but the resulting annual costs will be less.

As the final alternative C, the currently used manual system can be upgraded to perform the same function as the level 2 design for $100,000 per year. Assume that a $15,000-per-year cost will be incurred during the development time for each alternative to maintain the current manual system.

The costs are summarized and shown as cash-flow diagrams below. Note that the level 2 programming costs are equally spread over the 2-year development time. Use present-worth analysis and an interest rate of 10% per year to determine which alternative and level has the lowest LCC.

Alternative	Cost component	Level 1 Information only	Level 2 Projects selected
A (tailor software)	Development	$100,000	$200,000
	Programming	175,000	350,000
	Annual	60,000	55,000
	Time, years	1	2
B (new system)	Development	50,000	100,000
	Programming	200,000	500,000
	Annual	45,000	30,000
	Time, years	1	2
C (current)	Annual	⋯	100,000

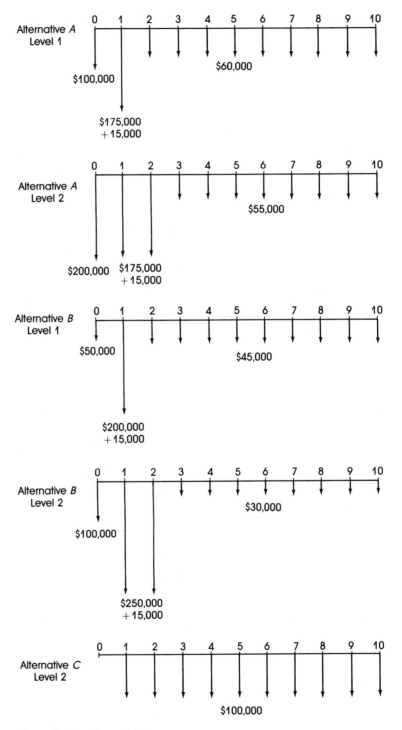

Figure for Problem 16.27

16.28 A window manufacturing company is considering the development of a computerized-decision support system to help reduce the scrap cost of glass. The current manual system allows $150,000 in scrap each year and requires 0.5 work-year to maintain at a cost of $35,000 per work-year. Improvements in the system next year will cost $25,000 and the system will be used for another 10 years. The suggested system has the following projected costs for a life of 10 years:

$125,000 per year for 2 years to develop (years 1 and 2)
$100,000 for a share of a new minicomputer (year 2)
$17,500 per year to maintain the manual system (years 1 and 2)
$8000 per year to maintain hardware and software (years 3 to 10)
$37,000 per year personnel cost (years 3 to 10)
$30,000 per year scrap cost once installed

Sales of the system to other companies is expected to net a $15,000 revenue for years 4 through 10. If a 20%-per-year return is required, which alternative has the lower LCC for the 10-year life?

17

Capital Rationing under Budget Constraints

The limited funds that a company or financial institution have available for capital investments are divided between several proposals (projects) with the anticipation that the rate of return or present value of each proposal will be maximized. Proposals are usually considered independent of each other, not mutually exclusive. That is, if a particular proposal is selected for investment, it does not prohibit the decision maker from selecting any other proposal. This is a major difference from the types of problems we have solved in earlier chapters where projects have generally been mutually exclusive.

Proposal selection under budgetary constraints is referred to as the *capital-budgeting* problem. Besides proposal selection, the solving of the capital-budgeting problem is also called portfolio selection. In this chapter, we investigate the present-worth approach to proposal selection under specified capital rationing and introduce a mathematical-programming (linear-programming) formulation for the solution.

Section Objectives

After completing this chapter, you should be able to:

17.1 Define the *capital-budgeting problem* and state its prominent characteristics.
17.2 Solve a capital-budgeting problem by finding the present worth of project bundles, given the MARR, budget constraint, cash-flow sequence, and life for each project.

17.3 Develop the *integer linear program* for a capital-budgeting problem, given the investments, cash flows, and lives of each project, the MARR, and the budget constraint.

Study Guide

17.1 The Capital-Budgeting Problem

Most corporations have the opportunity to select from several capital-investment proposals which are not mutually exclusive, that is, more than one of the proposals may be selected. However, the money available for investment is limited to some amount which is generally stated by management. This is commonly known as the *capital-budgeting problem* and it has the following characteristics:

1. Several proposals are available that are *independent* of each other; that is, selection of a particular proposal does *not* preclude selection of any other proposal.
2. Each proposal is either selected entirely or not selected; that is, partial investment in a proposal is not possible.
3. A budgetary constraint restricts the total investment possible in the proposals. Budget constraints may be present for several years after the initial investment is made.
4. The objective of financial investment for the corporation is to maximize the value of the investments.

It is also likely that the candidate proposals have nondeterministic cash-flow sequences and different risks associated with them. We will discuss problems which are of equal risk, have only an initial-year budget constraint, and have deterministically predictable cash flows for the life of the proposal.

In the typical capital-budgeting problem, there are several independent proposals, each with a first cost, life, series of yearly independent cash flows, and possibly a salvage value. If any of the proposals are mutually exclusive, this must be specifically accounted for. Likewise, if any are dependent, that is, must be done in conjunction with some other proposal, the dependent components should be combined into *one independent* proposal. (For a review of mutually exclusive alternatives, read Sec. 8.5.)

17.2 Capital Budgeting Using Present-Worth Analysis

To determine the selected proposals for a given budget limitation B, it is necessary first to formulate all mutually exclusive bundles of proposals, each of which has a total investment that does not exceed B. Then the present worth of each bundle is found at the MARR and the bundle with the *largest* PW value is selected.

To illustrate the concept of bundling, consider these four proposals.

Proposal	Investment
A	$10,000
B	5,000
C	8,000
D	15,000

If the budget constraint is $B = \$25,000$ feasible bundles are as follows:

Proposals	Total investment	Proposals	Total investment
A	$10,000	AD	$25,000
B	5,000	BC	13,000
C	8,000	BD	20,000
D	15,000	CD	23,000
AB	15,000	ABC	23,000
AC	18,000		

The number of bundles to be considered increases rapidly according to the relation $2^m - 1$, where m is the number of candidate proposals.

The procedure to solve a capital-budgeting problem using PW analysis is:

1. Develop all feasible mutually exclusive bundles which have a total initial investment that does not violate the budgetary constraint B.
2. Determine the cash-flow sequence CF_{jt} for each bundle j and year t.
3. Compute the PW value P_j for each bundle at the MARR.

$$P_j = \sum_{t=1}^{n_j} CF_{jt}(P/F, i\%, t) \tag{17.1}$$

where n_j is the life of bundle j.
4. Select the bundle with the largest PW value.

Selecting the maximum PW value means that this bundle produces a return larger than any other bundle above the MARR. Any bundle with $P_j < 0$ is discarded because it does not produce a return greater than the MARR. This procedure is illustrated in the next example.

Example 17.1

The Megabuck Company has $20,000 to invest next year on any or all of the proposals detailed in Table 17.1. Select a subset of proposals if the MARR is 15%.

Solution Use the procedure above with $B = \$20,000$ to select one bundle which maximizes present worth. In the solution, a $CF_{jt} > 0$ is a net cash inflow and $CF_{jt} < 0$ is a net outflow. The initial investment for any bundle is CF_{j0}, which is negative.

Table 17.1 Independent proposals considered for investment

Proposal	Initial investment	Expected annual cash flow	Proposal life, years
A	$10,000	$2,870	9
B	15,000	2,930	9
C	8,000	2,680	9
D	6,000	2,540	9
E	21,000	9,500	9

Table 17.2 Present-worth analysis for the capital-budgeting problem in Example 17.1

(1) Bundle, j	(2) Proposals	(3) Investment, CF_{j0}	(4) Cash flow, CF_j	(5) Present worth, P_j at $i = 15\%$
1	A	$-10,000	$2,870	$+ 3,694.49
2	B	$-15,000	2,930	$- 1,019.21
3	C	$- 8,000	2,680	$+ 4,787.89
4	D	$- 6,000	2,540	$+ 6,119.86
5	AC	$-18,000	5,550	$+ 8,482.38
6	AD	$-16,000	5,410	$+ 9,814.36
7	CD	$-14,000	5,220	$+10,907.75

1. All feasible bundles which require less than $20,000 are detailed in cols. 2 and 3 of Table 17.2. Proposal E is eliminated from further consideration because the initial investment exceeds B.
2. The cash-flow sequence, col. 4, is the sum of individual proposal cash flows for the 9-year life of each bundle.
3. Use Eq. (17.1) to compute the present worth of each bundle. Simplify. Use $CF_j = CF_{jt}$ and $n = n_j = 9$, since the cash flows are the same for each bundle and the lives are all 9 years. Use the relation

$$P_j = -CF_{j0} + CF_j(P/A, 15\%, 9)$$

for the present-worth values in Table 17.2. The largest present worth is $P_7 = \$10,907.75$; therefore, invest $14,000 in proposals C and D.

Comment The return on bundle 7 exceeds 15% per year; in fact, solution to $0 = -14,000 + 5220(P/A, i^*\%, 9)$ gives $i^* = 34.8\%$ as the actual return. Note that bundle 2 does not return the MARR since $P_2 < 0$. This analysis also assumes that any funds above the initial investment, for example, the $6000 that is uncommitted by selection of bundle 7, will return the MARR.

In a capital-budgeting problem, it is assumed that the investment is made for the life of the longest-lived proposal, with reinvestment of all positive cash flows at the MARR. There is no renewal of the same proposal at the end of its life, so the use of a least common denominator for the n value for unequal-lived proposals is not appropriate as it is in PW analysis. Rather, Eq. (17.1) and the procedure above is used to select a bundle by present-worth analysis for proposals of any life.

In fact, it is possible to demonstrate that PW evaluation using the factor $(P/A, MARR\%, n_j)$ is the same as reinvesting from year n_j to n_L at the MARR, where n_L is the longest-lived proposal. At the end of a bundle's life, the cash flows, assumed to be the same each year, are worth $CF_j(F/A, MARR\%, n_j)$. Reinvestment of this sum to year n_L and computation of the present worth in year 0 of all future receipts to year n_L results in the following relation:

$$P_j = CF_j(F/A, MARR\%, n_j)(F/P, MARR\%, n_L - n_j)(P/F, MARR\%, n_L) \qquad (17.2)$$

Let MARR $= i$ and use the factor formulas (Chap. 3 or Appendix A) to simplify:

$$P_j = CF_j \left[\frac{(1 + i)^{n_j} - 1}{i} \right] [(1 + i)^{n_L - n_j}] \left[\frac{1}{(1 + i)^{n_L}} \right]$$

$$= CF_j \left[\frac{(1 + i)^{n_j} - 1}{i(1 + i)^{n_j}} \right] = CF_j(P/A, i\%, n_j) \qquad (17.3)$$

Since the bracketed expression in Eq. (17.3) is the $(P/A, i\%, n_j)$ factor, evaluation of each bundle for n_j years by present-worth analysis assumes reinvestment at the MARR of all positive cash flows until the longest-lived proposal is completed in year n_L.

The next example demonstrates capital-budgeting problem selection for unequal-lived proposals.

Example 17.2

Use a MARR $= 15\%$ per year and $B = \$20,000$ to select proposals from the following capital-investment possibilities.

Proposal	Initial investment	Expected annual cash flow	Project life, years
A	$10,000	$2,870	6
B	15,000	2,930	9
C	8,000	2,680	5
D	6,000	2,540	4

Solution   The unequal lives cause the cash flows to vary from year to year, but the present-worth solution procedure is the same as that in Example 17.1. Of the $2^4 - 1 = 15$ bundles, the seven feasible ones and their PW values by Eq. (17.1) are given in Table 17.3. For example, bundle 7 has

$$P_7 = -14,000 + 5220(P/A, 15\%, 4) + 2680(P/F, 15\%, 5) = \$2235.60$$

Select proposals C and D since P_7 is the largest present worth.

This selection coincidentally happens to be the same as that in Example 17.1 where all lives are 9 years. It is the cash-flow values *and* the number of years over which they are received that determine the selected bundle.

Comment   Consider the bundle 7 evaluation in Table 17.3 and its cash-flow diagram in Fig. 17.1. At 15% the future worth at year 9, life of the longest-lived bundle, is

$$F = 5220(F/A, 15\%, 4)(F/P, 15\%, 5) + 2680(F/P, 15\%, 4) = \$57,111.36$$

The present worth is

$$P = -14,000 + 57,111.36(P/F, 15\%, 9) = \$2236.76$$

Allowing for a small round-off error from the multiple use of factors, we find that the P value is the same as P_7 in Table 17.3, thus verifying the reinvestment assumption summarized by Eq. (17.3). We see that it is not necessary to equalize the evaluation periods of bundles in a capital-budgeting problem.

Table 17.3 Present-worth analysis for the capital-budgeting problem in Example 17.2

(1)	(2)	(3)	(4)	(5)	(6)
			Cash flows		
Bundle, j	Proposals	Investment, CF_{j0}	Year, t	CF_{jt}	Present worth, P_j
1	A	$-10,000	1-6	$2,870	$+ 861.52
2	B	-15,000	1-9	2,930	-1,019.21
3	C	- 8,000	1-5	2,680	+ 983.90
4	D	- 6,000	1-4	2,540	1,251.70
5	AC	-18,000	1-5	5,550	+1,845.41
			6	2,870	
6	AD	-16,000	1-4	5,410	+2,113.43
			5-6	2,870	
7	CD	-14,000	1-4	5,220	+2,235.60
			5	2,680	

Proposal selection under budget constraint can be done equally well using the incremental-rate-of-return method you learned earlier. Once all feasible bundles are developed, they are ordered by increasing initial investment. Next, determine or approximate the return on the first entire bundle and the return for each incremental investment and cash-flow sequence on all other bundles. If any bundle has an incremental return less than the MARR, it is removed from consideration. The last justified increment indicates the best bundle.

This approach will result in the same answer as the present-worth procedure. There are a number of incorrect ways that the capital-budgeting problem can be

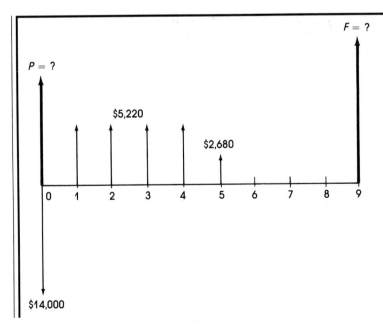

Figure 17.1 Plot of investment and cash flows for bundle 7, proposals C and D, in Example 17.2.

"solved" by the rate-of-return method, but an incremental analysis of mutually exclusive bundles ensures a correct result—one that is the same as the present-worth selection.

Probs. 17.1 to 17.9

17.3 Capital Budgeting Using Mathematical Programming

When there are several alternative proposals (m) in a capital-budgeting problem, the number of mutually exclusive bundles, $2^m - 1$, gets too large to evaluate by hand or calculator. It is then best to use a computerized solution to the capital-budgeting problem by present-worth analysis written in terms of the *integer linear programming* formulation that follows:

$$\text{Maximize: } Z = \sum_{k=1}^{m} P_k x_k$$

$$\text{Constraints: } \sum_{k=1}^{m} |CF_{k0}| x_k \leqslant B \qquad (17.4)$$

$$x_k = 0 \text{ or } 1 \qquad k = 1, 2, \ldots, m$$

where $P_k = \sum_{t=1}^{n_k} CF_{kt} (P/F, i\%, t)$ is the present worth at the MARR of proposal k which has a life of n_k years (investment CF_{k0} is not included here)

B = maximum allowed investment

$$x_k = \begin{cases} 0 & \text{if proposal } k \text{ is not included} \\ 1 & \text{if proposal } k \text{ is included} \end{cases}$$

The objective function Z requires that the combination of proposals included in the solution has the maximum present worth possible. Note that these are proposals, not mutually exclusive bundles, as in previous sections. The first constraint ensures that the sum of the initial investment for all selected proposals does not exceed the budget limitation B. The last constraint, $x_k = 0$ or 1, ensures that each proposal is completely included ($x_k = 1$) or completely excluded ($x_k = 0$) from the solution. This is commonly called the proposal indivisibility constraint and it makes this a 0–1 integer-programming (IP) problem.

Solution is best accomplished by a canned computerized IP solution package. Research in this area is presented in optimization texts and current journal articles.

Since solution to an IP problem is often time-consuming (even for a computer) and therefore quite expensive, it is possible to relax the proposal indivisibility constraint to the form

$$0 \leqslant x_k \leqslant 1 \qquad k = 1, 2, \ldots, m$$

where x_k now represents the fraction of the initial investment of proposal k that is committed in a particular solution. This relaxation assumes that any proposal can be divided for investment purposes, an assumption that may not be correct but does allow the capital-budgeting problem to now be solved using the faster technique of linear programming (LP). The answers by LP and IP will not necessarily be the same because of the removal of the indivisibility constraint to reformulate as an LP problem. Discussion of these aspects is included in advanced texts.

Example 17.3

Develop an integer-programming formulation for the capital-budgeting problem in Example 17.2.

Solution First calculate the P_k values for the four proposals detailed in Example 17.2 using $i = 15\%$.

Proposal, k	Cash flow, CF_{kt}	Life, n_k	$(P/A, 15\%, n_k)$	Present worth, P_k
A	$2,870	6	3.7845	$10,861.52
B	2,930	9	4.7716	13,980.79
C	2,680	5	3.3522	8,983.90
D	2,540	4	2.8550	7,251.70

Substitute the P_k values and $B = \$20,000$ into Eq. (17.4) with subscripts 1, 2, 3, and 4 for proposals A, B, C, and D, respectively.

Maximize: $Z = 10,861.52x_1 + 13,980.79x_2 + 8,983.90x_3 + 7,251.70x_4$

Constraints: $10,000x_1 + 15,000x_2 + 8000x_3 + 6000x_4 \leqslant 20,000$
x_1, x_2, x_3, and $x_4 = 0$ or 1

By inspection of the budget constraint in conjunction with the objective function, it is quite easy to see the feasible combinations of proposals which parallel the mutually exclusive bundles of Example 17.2. You may refer to Table 17.3 for these combinations. The solution is to select proposals C and D for a maximum value of $Z = \$16,235.60$ and an initial investment of $14,000. This solution is written

$x_1 = 0 \quad x_2 = 0 \quad x_3 = 1 \quad x_4 = 1$

Of course, it becomes increasingly difficult to obtain the answer by inspection when there are more proposals, so a computerized IP package should be used.

There are several other constraints that can be used in the problem to accommodate special cases. Situations like dependent proposals, complementary proposals, and investment over several years are easily modeled by additional constraints. For example, if there are budget limitations for several years in the future and proposals require investments for more than the first year, a series of budget constraints is used instead of the single-year restriction in Eq. (17.4). This series may be written

$$\sum_{k=1}^{m} I_{kt}x_k \leqslant B_t \qquad t = 1, 2, \ldots, n_L$$

where I_{kt} = investment in proposal k for year t
B_t = budget limitation in year t
n_L = life of the longest-lived proposal

There are as many constraints of this type as there are B_t values. Additional information on this and other constraint types is available in more advanced texts.

Probs. 17.10 to 17.13

Problems **17.1** Determine which of the following should be selected for investment if $30,000 is available and the required return is 10% per year. Use the present-worth method to make your selection.

Proposal	Investment	Cash flow	Life
A	$10,000	$3,950	8
B	12,000	2,400	8
C	18,000	5,750	8
D	22,000	3,530	8

17.2 (a) Rework Prob. 17.1 using the following proposal-life values:

Proposal	A	B	C	D
Life, years	3	8	5	12

(b) How many mutually exclusive bundles may be formed if the budget restriction is ignored?

17.3 Management of the Hoof Cattle Company has decided it can invest in any three of four available proposals. Each proposal has an initial investment of $10,000 and a specified present-worth value at $i = 18\%$. Select the proposals which offer the best investment opportunity.

Proposal	Life, years	Proposal present worth at 18%
1	13	$1,840
2	5	375
3	10	−1,800
4	8	25

17.4 Use the present-worth procedure to solve the following capital-budgeting problem: Select up to three of four proposals to maximize the return if the MARR after taxes is 10% and the available budget is (a) $16,000 and (b) $24,000. The last year in which a CFAT value is shown indicates the end of the proposal.

Proposal	Investment	CFAT for year				
		1	2	3	4	5
1	$ 6,000	$1,000	$1,700	$2,400	$ 3,100	$ 3,800
2	10,000	500	500	500	500	10,500
3	8,000	5,000	5,000	2,000		
4	10,000	0	0	0	15,000	

17.5 New investment funds for The System Company are restricted to $100,000 for next year. Select any or all of the following proposals using $i = 15\%$ for calculating the present-worth values.

Proposal	Initial investment	Annual cash flow	Life, years	Salvage value
1	$25,000	$ 6,000	4	$ 4,000
2	30,000	9,000	4	−1,000
3	50,000	15,000	4	20,000

17.6 Use the proposal bundle 34 for the data in Prob. 17.4 to demonstrate that equalization of evaluation periods is not necessary for present-worth comparison if reinvestment at the required return (10%) is assumed until the life of the longest-lived bundle is reached.

17.7 Rework Example 17.1 using a rate-of-return approach to solving the capital-budgeting problem.

17.8 Use (a) the rate-of-return method and (b) the present-worth method to solve the following capital-budgeting problem for a budget of $5000. Let MARR = 14%.

Proposal	Investment	Cash flow	Life
I	$3,000	$1,000	5
II	4,500	1,800	5
III	2,000	900	5

17.9 Reconsider the proposals in Prob. 17.5 and select from them using a minimum return criterion of 15%.

17.10 Set up the capital-budgeting problem for Example 17.1 using integer linear programming. Solve the problem manually or using a prepared integer-programming package.

17.11 Set up the integer-programming formulation for Prob. 17.1.

17.12 (a) Use the data in Prob. 17.5 to develop the integer program to solve the capital-budgeting problem. (b) What change is necessary if management has decided to allow partial investment in the proposals assuming that cash flows can also be partitioned?

17.13 Locate an integer-programming package for your computer and solve Prob. 17.4 using it.

18 Establishing the Minimum Attractive Rate of Return

All previous chapters have used a stated minimum attractive rate of return (MARR) for alternative evaluation. Now we will study the basis for setting the MARR using an estimate of the cost of funds for a corporation's daily operations, capital investments and acquisitions, and programs of growth.

After-tax analysis is important for funds obtained from outside the corporation (debt financing) since the interest paid on these loans is tax deductible. This fact tends to make financing from outside the corporate ownership more attractive than from within. We will study the relative advantages of each, why a balance between internal and external financing is important, and how they contribute to the computation of the *cost of capital*.

This cost of capital is translated into the MARR based upon the risk category of a project alternative. We briefly examine the aspect of risk in this chapter.

Section Objectives

After completing this chapter, you should be able to:

18.1 Describe the sources of *debt* and *equity capital* available to a corporation, and define the term *cost of capital*.
18.2 State reasons why the minimum attractive rate of return (MARR) varies.
18.3 Compute the *weighted average cost of capital* and the *debt-to-equity mix*, given the cost of capital and fraction of financing for each source of funds.

18.4 Estimate the *cost of debt capital* for a before-tax and after-tax analysis, given the details of the financing arrangement.

18.5 Estimate the *cost of equity capital* for common stock, preferred stock, and retained earnings. Also, compute the cost of capital for common stock using the *capital asset pricing model (CAPM)*.

18.6 State the relationship between the MARR and the cost of capital when different project risk categories are present.

18.7 Compute the rate of return on equity funds for different debt-equity mixes, and state why a large fraction of debt financing increases the risk level of all company investments.

Study Guide

18.1 Types of Financing and the Cost of Capital

A corporation accumulates funds (capital) by different methods which may be classified as two sources—*debt financing* and *equity financing*. These correspond to the balance sheet sections of liabilities and owner's equity, respectively (Appendix C).

Debt financing represents capital borrowed from others that will be paid back at a stated interest rate by a specified date. The original owner (lender) takes no direct risk on the repayment of the funds and interest, nor does the lender share in the profits the borrowing firm makes on the funds. Debt financing includes borrowing via bonds, mortgages, and loans and may be classed as long-term or short-term liability.

Equity financing represents capital owned by the corporation and used to generate revenue. Equity capital is developed from *owner's funds* and *retained earnings*. Owner's funds are classified as funds obtained from (1) the sale of common and preferred stock for a public corporation, or (2) company owners for a private (non-stock-issuing) company. Retained earnings are funds which have been previously retained by the corporation for future investment and expansion purposes. These funds are owned by the stockholders, not the corporation per se.

To set the MARR for capital budgeting and alternative evaluation, the cost to the corporation of each type of financing is computed *independently*. The proportion from debt and equity sources is weighted to estimate the actual interest rate paid for the investment capital. The procedure for weighting is explained later in this chapter. The resulting rate is called the *cost of capital*. The MARR is then set relative to this cost, sometimes equal to it, but usually higher, depending upon the *risk* that must be taken when any of the available projects are accepted for investment.

If we assume a computer project will be financed by an $800,000 bond issue (debt financing) and the interest on the bond is 8%, the cost of capital is 8%. If the project actually returns 6%, money has been lost. If management requires that any project return 5% above cost, and believes that the risk associated with this computer investment is substantial enough to warrant an additional 10%-return requirement, the MARR is $8 + 5 + 10 = 23\%$. If management considers all computer projects risk-free and requires no return above cost, a MARR of 8% may be used for justification of the project.

Probs. 18.1 to 18.3

18.2 Variations in MARR

The MARR is set relative to the cost of capital, which is a weighted rate dependent upon the mix of debt and equity financing. This rate is not usually determined exactly because the mix will change over time from project to project. Figure 18.1 presents a relationship between several rates used in economic analysis and financial management. A risk-free return is often considered to be the rate offered by U.S. Treasury bills. The estimated return on a proposed project may be any value, but only those above the MARR will actually be considered for funding. Strategic investment decisions, such as those discussed in Chap. 20, may have a MARR closer to the cost of capital because of less tangible reasons, but in all cases some MARR can be set for comparison with the estimated rate of return of the proposed projects.

The MARR is not a set, nonvarying value. Rather, it is altered by corporations for different types of projects. For example, a firm may use a MARR of 15% for

Figure 18.1
Relative values of cost of capital and return values for project evaluation.

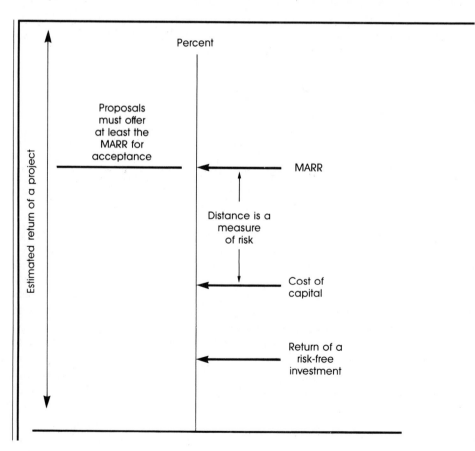

depreciable assets and a MARR of 20% for diversification investment, that is, purchasing smaller companies, land, etc.

The MARR varies from one project to another and through time because of the following:

1. *Project risk.* The more risk that is *judged* to be associated with a proposed project, the higher the MARR and, for that matter, the higher the cost of capital for the project.
2. *Sensitivity of project area.* If management is determined to diversify (or invest) in a certain area, the MARR may be lowered to encourage investment with the hope of recovering lost profit in other investment areas. This subjective reaction to investment opportunity can create much havoc with an economy study.
3. *Tax structure.* If taxes are increasing because of increased profits, recaptured depreciation and gains from retired assets, increasing local taxes, etc., the MARR will be increased. An after-tax study may assist in eliminating this reason for a fluctuating MARR.
4. *Capital-financing methods.* As capital becomes limited, the MARR is increased and management begins to look closely at the life of the project. As the demand for the limited capital exceeds supply, the MARR is further increased.
5. *Rates used by other firms.* If the rates of other firms that are used as a standard increase, a company may alter its MARR upward in response. A typical standard may be that of the firm called the *government*, even though the MARR for government projects varies and is set in a nonquantitative fashion.

Prob. 18.4

18.3 Weighted Average Cost of Capital and the Debt-Equity Mix

Most projects are funded through a combination of debt and equity financing obtained specifically for the project or taken from a common pool for the corporation. The weighted average cost of capital (WACC) is estimated using the relative fractions of debt and equity sources. If the fractions are known specifically, these should be utilized; if not known, the historical fractions for each source should be used to compute WACC.

WACC = (equity fraction)(cost of equity capital)

+ (debt fraction)(cost of debt capital)   (18.1)

If the fraction of each type of equity financing—common stock, preferred stock, and retained earnings—is detailed, Eq. (18.1) is expanded to include all four elements of financing. Example 18.1 illustrates cost-of-capital computations. The WACC value can be computed using before-tax or after-tax values for cost of capital; however, after-tax is usually the better method since debt financing has a distinct tax advantage (Sec 18.4).

The debt-equity (D-E) mix gives the percentages of debt and equity financing for a corporation. Thus, a company with a 40-60 D-E mix means that historically 40% of its capital has originated from debt sources (bonds and loans) while 60% has come from stocks and retained earnings. Since virtually all public corporations have a mixture of capital sources, the WACC is a value between the debt and equity costs

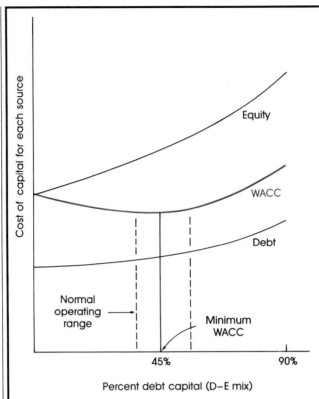

Figure 18.2   General shape of cost-of-capital curves.

of capital. Figure 18.2 shows the traditionally expected U-shape of cost-of-capital curves. This graph indicates that the WACC is a minimum at 45% debt capital. Most firms operate in a range of D-E mixes. For example, a company may find that a range of 30 to 50% debt financing on given projects is acceptable to lenders without increasing the risk and, therefore, raising the required MARR. Experience with the corporation's financial history and stability, and the proposed project helps determine the proper D-E mix for a given project.

Example 18.1

A new program in genetic engineering will require $1 million in capital. The chief financial officer has arranged for the following finances at the indicated after-tax rates:

Common stock sales	$500,000 at 13.7%
Use of retained earnings	$200,000 at 8.9%
Debt financing through bonds	$300,000 at 7.5%

Historically, this company has financed projects using a D-E mix of 40% with debt costing 7.5% and equity costing 10.0% after taxes. Compare the historical weighted-average-cost-of-capital value with that for this genetics project.

Solution By Eq. (18.1) the historical WACC is

$$WACC = 0.6(10) + 0.4(7.5) = 9.0\%$$

For the current project, the equity financing is composed of common stock (50%) and retained earnings (20%), with the remaining 30% from debt sources. The anticipated WACC is

$$\begin{aligned} WACC &= \text{stock portion} + \text{retained earnings portion} + \text{debt portion} \\ &= 0.5(13.7) + 0.2(8.9) + 0.3(7.5) \\ &= 10.88\% \end{aligned}$$

The current project is estimated to have a higher than historical cost of capital.

Estimates of after-tax or before-tax cost of capital may be made using the transformation relation

$$\text{After-tax cost} = \text{before-tax cost} \,(1 - T)$$

This relation may be used for equity and debt costs of capital separately or for the WACC rate itself.

Pro s. 18.5 to 18.9

18.4 Cost of Debt Capital

Debt financing includes borrowing via bonds, loans, and mortgages. The dividend or interest paid on debt financing can be utilized to reduce federal taxes, as discussed in Chap. 15. The CFAT (cash flow after taxes) values should be estimated and used to compute the rate-of-return value i^*, which is actually the cost of debt capital.

Example 18.2

A required amount of $500,000 will be raised by issuing 500 $1000 8%-per-year 10-year bonds. If the effective tax rate is 50% and the bonds are discounted 2% for quick sale, compute the cost of capital (*a*) before taxes and (*b*) after taxes.

Solution
(*a*) The annual dividend payment is $1000(0.08) = \$80$ and the discount sale price is $980. Find the i^* at which

$$0 = 980 - 80(P/A, i^*\%, 10) - 1000(P/F, i^*\%, 10)$$

Solution results in $i^* = 8.31\%$, which is the before-tax cost of debt capital.
(*b*) With the allowance to reduce taxes by deducting the interest on borrowed money, a tax savings of $\$80(0.5) = \$40$ per year is realized. Actual annual dividend outlay is $\$80 - \$40 = \$40$ in the equation of part (*a*). The after-tax cost of debt capital is now 4.26%.

Example 18.3

A company plans to purchase a certain asset for $20,000 with an anticipated life of 10 years. Management has decided to put $10,000 down now and borrow $10,000 at an interest rate of 6% on the unpaid balance. The repayment scheme will be $600 interest each year, with the $10,000 principal paid in year 10. What is the loan's after-tax cost of capital if the tax rate is 50%?

Solution   The cash flow for the $10,000 loan includes a tax credit of 0.5($600) = $300 of the annual $600 interest, and a payment of the $10,000 principal in year 10. The following relation is used to estimate a cost of debt capital of 3%.

$$0 = 10,000 - 300(P/A, i*\%, 10) - 10,000(P/F, i*\%, 10)$$

The 6% interest charge on the $10,000 loan is not the cost of debt capital because 6% is paid only on the borrowed funds, and the tax advantage of debt financing must be included to obtain the correct cost (i.e., 3%) of the 50% D-E mix.

Probs. 18.10 to 18.14

18.5   Cost of Equity Capital

Equity capital is derived from the following sources:

Sale of preferred stock
Sale of common stock
Use of retained earnings

The cost of each type of financing is estimated separately and entered into the WACC computation. A summary of one commonly accepted way to estimate each source's cost of capital is presented. There are several other accepted methods for estimating the cost of obtaining capital via common stock. These are not presented here.

Issuance of *preferred stock* carries with it a commitment to pay a stated dividend annually. The cost of capital is therefore the stated dividend divided by the price of the stock. However, the stock is often sold at a slight discount, so the actual proceeds from the stock, rather than the stated price, should be used as the denominator. For example, a 10% preferred stock with a value of $200, which is sold for $190 per share (5% discount), has a cost of capital equal to 10/0.95 = 10.53%.

To estimate the cost of capital for *common stock*, more is involved. The dividends paid are not a true indication of what the stock issue will actually cost the corporation in the future. Usually a valuation of the common stock is used to estimate the cost. If R_e is the equity cost of capital

$$R_e = \frac{\text{first-year dividend}}{\text{price of stock}} + \text{expected dividend growth rate}$$

$$= \frac{DV_1}{P} + g \tag{18.2}$$

The growth rate g is an estimate of the return that the shareholders expect to receive from owning stock in the company. It is the compound growth rate on dividends that the company believes is required to attract stockholders.

The *retained-earnings* cost of capital is usually set equal to the common-stock cost since it is the shareholders who will realize any returns from projects in which retained earnings are invested.

Example 18.4

A multinational corporation plans to raise capital in its United States subsidiary for a new plant in South America through the sale of $2,500,000 worth of common stocks valued at $20 each. If a $1 dividend is planned for the first year and an appreciation of 9% per year is desired for future dividends, estimate the cost of capital from this stock issue.

Solution   The values are substituted into Eq. (18.2) to compute R_e.

$$R_e = \frac{1}{20} + 0.09 = 0.14 \qquad (14\%)$$

Once the cost of capital is estimated for all planned sources, the WACC is calculated using Eq. (18.1).

Because of the fluctuations in stock prices and the higher return demanded by some corporation's stocks compared to others, there has developed over the years a valuation technique which may be utilized to estimate the cost of capital for common stocks and other investments. It is called the *capital asset pricing model (CAPM)* and is used in lieu of Eq. (18.2) for common-stock issues. The return required for a stock issue is the cost of equity capital R_e.

R_e = risk-free return + premium above risk-free return

$$= R_f + \beta(R_m - R_f) \qquad (18.3)$$

where β = volatility of company's stock relative to other stocks in the market ($\beta = 1.0$ is the norm)

R_m = return on stocks in a defined market portfolio measured by a prescribed index.

The first term in Eq. (18.3), R_f, is usually the U.S. Treasury bill return; the U.S. Treasury is about the safest place to put money in this country. The term $(R_m - R_f)$ is the premium above risk-free. The coefficient β (beta) indicates how the stock is expected to vary compared to a selected portfolio of stocks in the same general market area. If $\beta < 1.0$, the stock is less volatile, so the resulting premium can be smaller, and when $\beta > 1.0$, larger price movements are expected so the premium is increased.

Security is a word which identifies a stock, bond, or any other instrument used to develop capital. Figure 18.3 is a plot of a security market line, which is a linear fit (found by linear regression analysis) to indicate the expected return for different

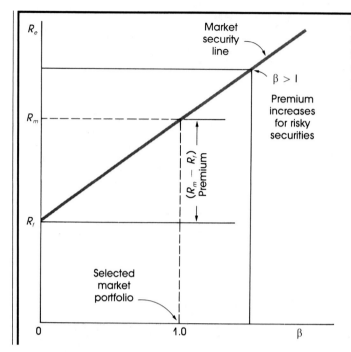

Figure 18.3 Expected return on common-stock issue (equity capital) using the capital asset pricing model.

β values. When $\beta = 0$ the risk-free return R_f is acceptable (no premium). As β increases, the premium return requirement grows. Beta values are published periodically for most stock-issuing corporations.

The use of CAPM to evaluate stock and other capital-financing sources for the cost of capital is becoming more common. Once complete, this estimated cost of equity capital can be included in the WACC computation.

Example 18.5

MegaComputers plans to develop and market intelligent systems software. A common-stock issue is a possibility if the cost of equity capital is below 15%. MegaComputers, which has an historical beta value of 1.7, uses the CAPM to determine the premium of its stock to other computer manufacturers. The security market line indicates that a 5% premium above the risk-free rate is desirable. If U.S. Treasury bills are paying 7%, estimate the cost of common-stock capital.

Solution The premium of 5% represents the term $(R_m - R_f)$ in Eq. (18.3).

$R_e = 7.0 + 1.7(5.0) = 15.5\%$

Since this cost exceeds 15%, MegaComputers should possibly use some mix of debt and equity financing for its venture into the intelligent software business.

Additional Example 18.8
Probs. 18.15 to 18.22

18.6 Setting the MARR Relative to Cost of Capital

As indicated in Fig. 18.1, the MARR is usually set above the cost of capital when the measure of risk is taken into account. There is no "hard and fast" rule to determine the MARR based upon subjectively evaluated risk. Management skills and experience are prime ingredients of the MARR determination.

It is common to separate project alternatives into risk categories and establish the MARR relative to the cost of capital for each category. For illustration purposes, Fig. 18.4 gives an example of high, medium, and low risk, plus a "no-risk or legally imposed" category which has a MARR below cost of capital, or a no-return requirement because it is legally necessary. Safety, environmental, and affirmative-action projects are often legally imposed. High-risk projects may have a MARR considerably above the cost of capital as indicated by the MARR range in Fig. 18.4.

In most large investment evaluations, there are many noneconomic factors that establish the risk for the projects. The blending of the economic and noneconomic (or intangible) factors will have an effect upon the established MARR. For example, a strategically and competitively important investment in new production technology may be in the high-risk category, but management may set the MARR low, relative to the cost of capital for the economic evaluation. Many other factors will be considered when the final go–no-go decision is made. The use of multicriteria decision making for large investments is discussed in Chap. 20.

Prob. 18.23

18.7 Effect of Debt-Equity Mix on Investment Risk

The D-E mix was introduced in Sec. 18.3. A 40% D-E mix indicates that 40% of capital is debt-financed and 60% is equity-financed. As the proportion of debt increases, the cost of capital decreases due to the tax advantages of debt capital. *The leverage offered by large debt-financing percentages tends to increase the risks taken by the firm.* Additional financing—debt and equity—gets more difficult to obtain, and the corporation can be placed in a situation where it owns only a small portion

Risk category	Example projects	Example MARR range	Figure 18.4
High	New product development International joint-venture	25–30%	Example of different MARR values by risk categories.
Medium	Capacity expansion Implementation of new, but accepted technology	18–24%	
Low	Productivity improvement	13–17%	
	Weighted average cost of capital	*12%*	
No risk or legally imposed	Cost reduction Cost-sharing projects Safety-related projects	0–11%	

of itself. Inability to obtain operating and investment capital means increased risk for the company. Thus, a balance between debt and equity financing is important to the health of a corporation. The situation presented in Example 18.6 illustrates the disadvantages of increasing D-E mixes.

Example 18.6

Three bank holding companies have the following debt and equity capital amounts and resulting D-E mixes. Assume all equity capital is in the form of common stock.

	Capital-financing source		
	Debt	Equity	
Firm	($ million)		D-E mix
First National	$10	$40	20–80%
Second State	20	20	50–50
Third City	40	10	80–20

The annual revenue is $15 million for each bank, but the interest on debts results in net incomes of $14.4, $13.4, and $10.0 million for the three, respectively. Compute the returns on common stock.

Solution Divide the net income by the stock value for each bank holding company to compute the return.

First National: Return $= \dfrac{14.4}{40} = 0.36$ or 36%

Second State: Return $= \dfrac{13.4}{20} = 0.67$ or 67%

Third City: Return $= \dfrac{10.0}{10} = 1.00$ or 100%

Of course, the return for the highly leveraged Third City is by far the largest, but only 20% of the bank is in the hands of the ownership. The return is excellent, but the risk associated with this bank is very high compared to First National where the D-E mix is 20%.

The question in Example 15.8 has now been answered. The use of large percentages of debt financing (large D-E mixes) greatly *increases the risk* taken by lenders and stock owners. Confidence in the corporation is generally poor, no matter how large the short-term return on stock.

The leverage of large D-E mixes does increase the return on equity capital, as shown in previous examples; but it can also work against the owner or investor of funds. A small percent decrease in asset value will negatively affect a highly leveraged investment much more than one with small or no leveraging. Example 18.7 illustrates this fact.

Example 18.7

Two individuals place $10,000 each in different investments. Marylynn invested $10,000 in silver mining stock and Carla leveraged the $10,000 by purchasing a $100,000 residence to be used as rental property. Compute the resulting value of the $10,000 equity capital if there is a 5% decrease in the value of the stock and residence. Neglect any dividend, income, or tax considerations.

Solution Stock value decreases by $10,000(0.05) = \$500$ and the house value decreases by $100,000(0.05) = \$5000$. The effect upon the investment is a reduction in the amount of equity funds now available, were the investment sold.

Marylynn loss: $\dfrac{500}{10,000} = 0.05$ or 5%

Carla loss: $\dfrac{5000}{10,000} = 0.50$ or 50%

The 10-1 leveraging by Carla gives her a 50% decrease in the equity while the 1-1 leveraging gives Marylynn a 1-1 or 5% reduction. Of course, the opposite is correct for a 5% increase; Carla would benefit by a 50% gain on her $10,000, while the 1-1 leverage would offer only a 5% gain. The larger leverage is much more risky. It offers a much *higher return for an increase* in the value of the investment and a much *larger loss for a decrease* in the value of the investment.

Prob. 18.24 to 18.28

Additional Example

Example 18.8

A computer repair company plans to purchase 15 delivery trucks for $150,000. Each truck will be straight-line depreciated over 10 years with SV = $1000. Debt financing for 50% of the capital or complete equity financing is possible. Equity financing requires the sale of $15 per share of stock. The first dividend is expected to be $0.50 per share and a 5% dividend growth rate is anticipated.

For debt financing, a $75,000 loan will be taken. The loan arrangements will be 10 years at 8% compounded annually, with equal end-of-year payments.

The MARR should be set at twice the cost of capital for debt and equity capital. If the CFBT is expected to be $30,000 annually and a 35% effective tax rate applies, use an after-tax analysis to determine if 100% equity or 50% debt financing is better.

Solution: 100% equity financing
Use Eq. (18.2) to estimate the equity cost of capital

$$R_e = \frac{0.5}{15} + 0.05 = 0.0833 \quad (8.3\%)$$

Because there is no tax advantage for equity financing, the MARR is $2(8.3) = 16.6\%$.

Find the after-tax return from the CFAT values. Straight-line depreciation for the 10 trucks is $13,500 annually and since there is no loan (debt), taxes and CFAT are

$$\text{Taxes} = 0.35(\text{TI}) = 0.35(30,000 - 13,500) = \$5775$$

$$\text{CFAT} = \text{CFBT} - \text{taxes} = 30,000 - 5775 = \$24,225$$

The return $i^* = 10.7\%$ is estimated from the equation

$$0 = -150,000(A/P, i^*\%, 10) + 15,000(A/F, i^*\%, 10) + 24,225$$

Since 10.7% does not exceed the MARR of 16.6%, equity financing does not meet the MARR criteria.

50% debt–50% equity financing

Find the weighted average cost of capital for both equity and debt financing. *Equity cost is 8.3% as computed above.* For the $75,000 loan, we assume a uniform reduction of principal at the rate of $7500 annually, with a payment of

$$75,000(A/P, 8\%, 10) = \$11,177$$

The annual tax advantage of the interest may be approximated as

$$(\text{Payment} - \text{principal portion})(\text{tax rate}) = \text{interest}(T)$$

$$(11,177 - 7500)(0.35) = 3677(0.35) = \$1287$$

The cost of debt capital is 5.4% from the relation

$$0 = 75,000(A/P, i^*\%, 10) - (11,177 - 1287)$$

The WACC from Eq. (18.1) is

$$\text{WACC} = 0.5(8.3) + 0.5(5.4) = 6.85\%$$

The MARR is twice the WACC or 13.7%.

To compute the return on the $75,000 of equity capital, first find the annual CFAT using Eqs. (15.5) to (15.6).

$$\text{Taxes} = \text{TI}(\text{tax rate}) = (30,000 - 13,500 - 3677)(0.35) = \$4488$$

$$\text{CFAT} = 30,000 - 4488 - 3677 - 7500 = \$14,335$$

The return for the D-E mix of 50/50 is $i^* = 15.2\%$ from

$$0 = -75,000(A/P, i^*\%, 10) + 15,000(A/F, i^*\%, 10) + 14,335$$

This is acceptable since the return exceeds the MARR of 13.7%.

Comment The use of 50% debt financing increased the expected return on equity capital by 4.5% and decreased the MARR requirement by about 3%. However, the risk is also greater due to the leverage obtained through debt financing.

Problems   **18.1**   The owner of a fast-food restaurant computed a cost of capital of 15% and has chosen a MARR $= 17\%$ per year to evaluate a new register and automated inventory system. (*a*) What return does he anticipate? Does this seem reasonable? Why? (*b*) The owner's spouse considers it necessary that all risky projects earn 8% over their cost of capital in addition to a basic return of 5%. What MARR would the spouse use in this same evaluation if this project is considered nonrisky? Risky?

18.2 State whether each of the following is in the category of debt or equity financing: (*a*) short-term note from the bank; (*b*) $5000 taken from a co-owner's savings account to pay a company bill; (*c*) a $150,000 bond issue; (*d*) an issue of preferred stock worth $55,000; (*e*) Jane borrows $50,000 from her brother at 3% interest to run her business. The brother is not a co-owner in the company.

18.3 Assume you have $5000 to invest in something. List four opportunities in which you might invest all or part of this amount. Two of the opportunities should be examples of what you consider "risky ventures" and two should be "safe ventures." After stating the four areas, explain how you personally distinguish between risky and safe investment opportunities.

18.4 Will each of the following cause the MARR to be raised or lowered? State why.
(*a*) Investment in a chain of quick-food stores is contemplated, but the president is very leery of such an undertaking.
(*b*) The Crooked Nail Construction Company built a 250-unit apartment house 3 years ago and still retains ownership. Due to the risk, when the project was undertaken a 12%-per-year return was required; however, because of the favorable outcome management feels this is safer than some other types of investments.
(*c*) Government subsidy has just been announced for manufactured products which are threatened by competition from international firms which are subsidized by their home governments.

18.5 A large company plans to purchase a small firm which has been a supplier for many years. A cost of $780,000 has been placed on the small firm. The purchasing company does not know exactly how to finance the purchase to obtain a WACC as low as that for other ventures. The WACC is presently 10%. Two schemes of financing are available. The first requires that the company invest 50% equity funds at 8% and borrow the balance at 11% per year. The second scheme requires only 25% equity funds and the balance can be borrowed at 9%. Which scheme will require the smaller earnings?

18.6 The financial committee of the company in Prob. 18.5 has decided that the WACC for the purchase must not exceed its historical average of 10% per year. What rates can be paid for debt capital in the two approaches to financing?

18.7 Mariposa, Ltd. intends to raise $100 million in new capital with 60% equity funds and 40% debt financing. The following percents and rates have been established. Compute the WACC.

Equity capital: 60% or $60 million
Common stock: 25% at 11.5%
Retained earnings: 75% at 9.0%

Debt capital: 40% or $40 million
Loans: 50% at 13.5%
Long-term bonds: 50% at 9.0%

18.8 An employee of a large corporation in which you own stock told you the company has a WACC = 12.5%. The common stocks have averaged a return of 10% over the last 5 years and the corporation states in its annual report that it uses 62% of its own funds for capitalization projects. What is your best estimate of its cost of debt capital?

18.9 A cost estimator with Sam Aerospace wants to know the debt-equity mix at which the WACC is most likely to be the lowest. She has collected the information below for large capitalization projects. (*a*) Plot the curves for debt, equity, and weighted average costs of capital. (*b*) What debt-equity mix seems to historically offer the lower WACC values?

Equity capital		Debt capital	
Percent	**Rate**	**Percent**	**Rate**
80%	14.5%	20%	5.5%
66	12.7	34	5.5
56	10.9	44	8.2
35	9.3	65	9.8
20	10.0	80	13.8

18.10 An automobile manufacturing corporation requires the infusion of $2 million in new debt capital. If 12-year 8% bonds which pay semiannually are sold to a large brokerage firm at a 4% discount, (*a*) determine the total face value of the bonds required to provide the $2 million, and (*b*) compute the after-tax debt cost of capital for $T = 50\%$.

18.11 The Barely-Making-It Swimming Suit Company has to raise $50,000 in new capital. Two methods of debt financing are available. The first is to borrow $50,000 from a bank. The company will pay an effective 8% compounded per year for 8 years to the bank. The other method will require issuing 50 $1000 10-year bonds which will pay a 6% annual dividend. If you assume that the principal on the loan is reduced uniformly for the 8 years, with the remainder going toward interest, which method of financing would you recommend after taxes are accounted for? Assume that the tax rate is 52%.

18.12 Is the answer the same for Prob. 18.11 if a before-tax analysis is made?

18.13 Purchase of new equipment for $75,000 is contemplated by the Skim Milk Dairy. The equipment will last 5 years and can be salvaged for $15,000. The company has $25,000 in available money and hopes to borrow the remainder from a bank for a 5-year period. The equipment is expected to increase cash flow before taxes by $18,000 per year. The new equipment will be straight-line-depreciated and a 50% tax rate is applicable. If the economist for the firm estimates that the taxes for this endeavor will be $1500 per year, what is the (*a*) stated interest rate paid on the loan and (*b*) effective interest rate on the loan after taxes, provided that the MARR of 10% per year is realized? Assume uniform reduction in principal over the life of the loan.

18.14 An international firm can use its own equity capital at an after-tax cost of 5.5% per year. Alternatively, it can borrow from foreign sources through the issuance of $1 million worth of 20-year bonds which pay 12% semiannually. If the tax rate is 40%, determine which source of funds is less expensive.

18.15 The common stock for HiRiz Constructors can be evaluated using a dividend method or the CAPM. Last year, the first year for dividends, stock paid $0.75 per share with an average price of $11.50 on the New York Stock Exchange. Management hopes to grow the dividend rate at 3% per year. HiRiz stock has a volatility which is higher than the norm at 1.3. If safe investments are returning 7.5% and the 3% growth rate on common stocks is also the premium above risk-free investments that HiRiz plans to pay, compare the cost of equity capital using the two estimates.

18.16 The president of a company speculates that he can use a debt-equity mix of 40/60 to finance a new $5 million venture. After-tax cost of debt capital is 9.5%, but equity capital requires issuance of preferred and common stock as well as use of retained earnings. Use the following information to determine the average cost of capital for the venture. All $ quotes are per share.

Preferred stock: $1 million of stock to sell

Face value = $30
Sales price = $27.60
Dividend rate = 8% per year

Common stock: 80,000 shares to sell

Price = $12.50
Initial dividend = $0.40
Dividend growth = 5% annually

Retained earnings: same rate as common stock

18.17 The owners of a grocery store plan to construct a laundromat next door. They will use 100% equity financing. It will cost $22,000 to build the facility, which will be straight-line-depreciated over a 15-year life using a salvage value of $7000. They have the equity funds invested at the present time and make 10% per year. If the annual CFBT is expected to be $5000 and the tax rate is 48%, is the venture expected to be profitable?

18.18 The Pure Company has always used 100% equity financing in the past. A good opportunity is now offered that will require the raising of $250,000. The owner can supply the money from personal investments, which earn an average of $8\frac{1}{2}\%$ per year. The annual CFAT is expected to be $30,000 for the next 15 years at a tax rate of 50%. Alternatively, 60% of the required amount can be borrowed at 5% compounded annually for 15 years. If it is assumed that the principal is uniformly reduced and an average annual interest is paid, use a MARR of 1.2 times WACC to determine which plan is better.

18.19 The Sno-Plow Company uses a MARR value that is 1.5 times the cost of capital. Three plans for raising $50,000 are available. These are detailed below.

Type of	Plan		
financing	1	2	3
Equity	90%	60%	20%
Debt	10	40	80

At present the before-tax cost for equity capital is 10% and for debt capital, 12%. If the project is expected to yield $10,000 for 5 years, do a before-tax analysis to determine the return for each plan and identify the plans that are acceptable.

18.20 Westfall Mattresses sees an opportunity to invest $100,000 in a new mattress line. Financing will be split between retained earnings ($25,000) and a loan with an 8% after-tax interest rate. The average annual before-tax income after all interest and loan principal payments are considered is $37,800 for the next 7 years. The effective tax rate is 50% and no depreciable assets are involved.

Westfall is a manufacturer-retailer chain which sells common stock over the counter and uses the capital asset pricing model for valuation. Recent analysis shows that Westfall has a volatility rating of only 0.85 and is paying a premium of 5% on its common-stock dividend. Nationally, the safest investment is currently paying 8% per year and Westfall uses 1.5 times its average cost of capital as an after-tax MARR. Is the $100,000 venture financially attractive?

18.21 A conscientious couple devised a plan to buy certain types of groceries now for $600 in order to save a total of 25% in the next 6 months. However, they are not sure how to finance this plan. The couple want to make a return of 50% more than the financing methods will cost over the 6-month period. The financing plans are as follows:

(a) Take $600 from a savings account now and put the monthly savings of $125 in as they are received. This account pays 6% per year compounded quarterly.

(b) Borrow $600 now from the credit union at an effective 1% per month and repay the loan at $103.54 per month for 6 months and put the difference between the payment and the amount saved each month in the 6%-per-year compounded-quarterly savings account.

(c) Use the extra $300 from this month's budget, borrow $300 at 1% per month, and repay at the rate of $51.77 per month for 6 months. Again, the difference between the payments and the savings would be saved at 6% per year compounded quarterly.

Perform a before-tax average cost-of-capital analysis to determine which financing plan is the most profitable. Assume that there is no interperiod interest paid on the savings account.

18.22 The Anxiety Attack Drug Co. has a total of 153,000 shares of common stock outstanding at a market price of $28 per share. A before-tax cost of equity capital of 24% is incurred by these shares. Stocks fund 50% of the company's undertakings. The remaining investments are financed by bonds and short-term loans. It is known that 30% of the debt capital is from $1,285,000 worth of $10,000 6%-per-year 15-year bonds, which were sold at a 2% discount. The remaining 70% of debt capital is from loans repaid at an effective 17.3% before taxes. If income taxes are 48% effectively, determine the weighted average cost of capital (a) before taxes and (b) after taxes.

18.23 It has been stated that many of the legally imposed requirements upon the manufacturing industry in safety, environment protection, disease prevention, etc. have tended to decrease the return and/or increase the cost of capital because these requirements cannot be considered in the economic evaluation of the projects. Give your opinion of this view and detail several examples of government-imposed requirements in industry that substantiate your agreement or disagreement with this statement.

18.24 Compute the debt-equity mix proposed in each of the following problems: 18.8; 18.13; 18.20; 18.21.

18.25 Why is it unhealthy for a corporation to maintain a very high D-E mix over a long period of time?

18.26 Assume all the bank holding corporations in Example 18.6 have a 50/50 D-E mix. Compute the new return values and compare them with the return values shown in the example.

18.27 The following information has been published by two companies. Total capitalization is the value of all assets for which the corporation has obtained investment capital in the past.

Corporation	Capitalization ($ million)	Percent debt capital
A	2.5	28
B	1.6	70

(a) What amount of total capitalization is from debt sources for each corporation?

(b) If a serious problem occurs in each corporation and the total capitalization of the firm is decreased by 15%, compute the new D-E mixes. Remember the amount of debt will remain constant even though the total value of assets in the firm will decrease.

18.28 Redraw the debt, equity, and WACC curves in Fig. 18.2 under the condition that a high D-E mix has been present for some time. High D-E mixes cause the debt cost to increase substantially. This makes it harder to obtain equity funds, so the cost of equity capital also increases. Explain via your graph and words the movement of the minimum WACC point under historically high D-E mixes.

Sensitivity Analysis and Decision Trees

Economic analysis utilizes estimates of future happenings to assist decision makers. Since future estimates may be incorrect to some degree, errors are likely to be present in the economic analyses. Two of the ways to account for error in estimates are discussed in this chapter.

The effect of variation in the estimates used in the economic analysis may be determined using *sensitivity analysis*. Some of the factors commonly investigated are: MARR, life, recovery period for tax purposes, all types of costs, sales, and other factors. *Usually, one factor at a time is varied and independence with other factors is assumed.* This assumption is not correct in most real-world situations, but it is necessary since the ability to account for actual dependencies is not mathematically possible.

A second way to account for variation is through probabilistic analysis, which is accomplished as accurately as possible by considering the *risk* present in alternatives. We have introduced risk in earlier chapters, but here a *decision tree* is constructed and simple expected-value computations are made to discover the most likely and best decision.

Section Objectives

After completing this chapter, you should be able to:

19.1 State the approach of *sensitivity analysis* for one or more factors and one or more projects.
19.2 Determine the sensitivity of alternative values for selected factors, given the variation range for each factor.

19.3 Select the more economical of two or more projects using *three estimates* of selected factors.

19.4 State the *expected-value* equation and compute $E(x)$, given the variable values and probabilities.

19.5 Determine the desirability of an alternative using expected-value computations, given project cash flows and their probabilities.

19.6 Construct a *decision tree* and use expected value computations to solve the tree for the most likely decision path, given information for decision alternatives, probabilities, and possible outcomes.

Study Guide

19.1 The Approach of Sensitivity Analysis

Since the workplace of economic analysis is the *future*, the estimates will likely be in error. Sensitivity analysis is a study to determine how the economic decision and measures will be altered if certain factors are varied. For example, variation in the MARR may not alter a decision when all compared proposals return more than the MARR; thus, the decision is relatively insensitive to MARR. However, if a change in the economic life is critical, the decision is sensitive to life estimates.

Usually the variations in life, annual costs, or incomes result from variations in selling price, operation at different levels of capacity, inflation, etc. For example, if 75% capacity is compared with 50% capacity for a contemplated proposal, operating costs and revenue will increase, but anticipated life will probably decrease. Usually several important factors are studied to learn how the uncertainty of estimates affect the economic study.

Plotting the sensitivity of present worth, EUAW, or rate of return versus the factor studied is illustrative. Two projects can be compared with respect to a given factor and the *breakeven point* computed. This is the variable value at which the two proposals are economically equivalent. However, the breakeven chart represents only one factor per chart. Thus, several charts must be constructed and independence of each factor assumed. In previous uses of breakeven analysis, we computed two values and connected the points with a straight line. However, if a factor generates sensitive results, several intermediate points should be used to be conscious of the sensitivity. This fact is illustrated in this chapter.

When several factors are studied, sensitivity analysis may be performed one variable at a time using manual computations. However, time may be saved by using a specially prepared computer program or spreadsheet system. The computer allows more than one basis of comparison, for example, PW and rate-of-return analysis. Additionally, the computer can rapidly plot the results. Use of computerized sensitivity analysis is further discussed in Chap. 20.

19.2 Determination of the Sensitivity of Estimates

There is a general procedure that should be followed when conducting a sensitivity analysis. The steps in this procedure are as follows:

1. Determine which factor(s) are most likely to vary from the estimated value.
2. Select the probable range and increment of variation for each factor.

3. Select an evaluation method, such as present worth, EUAW, or rate of return, to evaluate each factor's sensitivity.
4. Compute and, if desired, plot the results from the evaluation method selected in step 3.

The results of the sensitivity analysis will show the factors that should be more carefully estimated by collecting more information when possible. Example 19.1 illustrates sensitivity analysis for one project.

Example 19.1

The ACQ Company is contemplating the purchase of a new piece of automatic machinery for $80,000 with zero salvage value and an anticipated before-tax cash flow of $27,000 − 2000k per year ($k = 1, 2, \ldots, n$). The MARR for the company has varied from 10 to 25% per year for different types of investments. The economic life of similar machinery normally varies from 8 to 12 years. Use EUAW analysis to investigate the sensitivity of varying (a) the MARR using $n = 10$ and (b) the economic life, assuming a MARR of 15% per year.

Solution
(a) Allowing i to change by a 5% increment should be sufficient for sensitivity purposes. For 10%, and using the −$2000 gradient in cash flow,

$$P = -80,000 + 25,000(P/A, 10\%, 10) - 2000(P/G, 10\%, 10)$$
$$= \$27,830$$

$$\text{EUAW} = P(A/P, 10\%, 10) = \$4529$$

Similarly, other results are

i	P	EUAW
15%	$ 11,512	$ 2,294
20	−962	−229
25	−10,711	−3,000

A plot of MARR versus EUAW is shown in Fig. 19.1. The steep negative slope of EUAW indicates that the decision is quite sensitive to variations in the MARR. If it is likely that the MARR will be set at the upper end of the range, the investment will not be attractive.
(b) Using an increment of 2 years, the EUAW values at $i = 15\%$ are as follows:

n	P	EUAW
8	$ 7,221	$1,609
10	11,511	2,294
12	13,145	2,425

Figure 19.1 presents the characteristic nonlinear relation of EUAW versus n. Since EUAW is positive for all values of n, the decision to invest is not affected by the economic life.

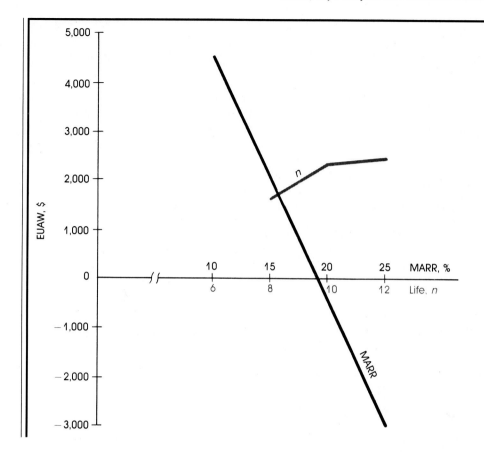

Figure 19.1 Plot of sensitivity of EUAW to MARR and life variation Example 19.1.

Comment Note that after $n = 10$, the EUAW curve seems to level out and become quite insensitive. This insensitivity to changes in cash flow in the distant future is the expected trait, because when discounted to time 0, the present-worth value gets smaller as n increases.

If two projects are compared and the sensitivity for one factor is to be determined, the actual plot may show quite nonlinear results. Take a look at the general form of the plots in Fig. 19.2. We won't be concerned with the actual computations, but the graph shows that the CFAT of each plan is a nonlinear function of hours of operation. Plan *A* is extremely sensitive in the range of 50 to 200 hours but is comparatively insensitive above 200 hours. We mentioned in Sec. 19.1 that plans which are sensitive should be studied by plotting intermediate points to reveal the nature of the variation. Since there is considerable variation with hours of operation, plan *B* would be very attractive due to its relative insensitivity, provided that plans *A* and *B* are justified. This decision will provide some assurance of a more stable CFAT.

Additional Example 19.9
Probs. 19.1 to 19.14

Figure 19.2
CFAT sensitivity
to hours of
operation for two
alternative plans.

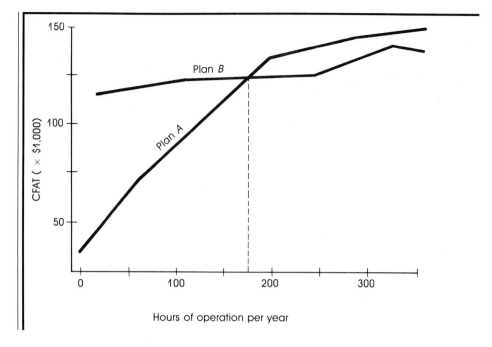

19.3 Sensitivity Analysis Using Estimates of Pertinent Factors

We can thoroughly examine the relations between two or more projects by borrowing from the field of project scheduling the concept of making three estimates for
pertinent factors: a pessimistic, a reasonable, and an optimistic estimate. This allows
us to study decision sensitivity within the most likely variation range for each factor.
The analyst should choose the decision representing the economic situation as he or
she best understands it. This involves a subjective weighing of the sensitized factors.

Example 19.2

You are an engineering economist attempting to evaluate three alternatives for which
pessimistic (P), reasonable (R), and optimistic (O) estimates are made for the life, salvage
value, and annual operating costs (Table 19.1). Determine the most economical alternative
using EUAW analysis and a before-tax MARR of 12%.

Solution   For each alternative description in Table 19.1, we compute the EUAW. For
example, using the pessimistic estimates for alternative A,

$$\text{EUAW} = 20,000(A/P, 12\%, 3) + 11,000 = \$19,327$$

Table 19.2 presents EUAW values for all situations. Figure 19.3 is a plot of EUAW
versus the three *life estimates* for each alternative. Since the EUAW calculated from the
reasonable estimates for alternative B is less than even the optimistic EUAW values for
A and C, alternative B is clearly favored.

Table 19.1 Competing alternatives with three estimates for salvage, AOC, and life

	First cost	Salvage	AOC	Life, n
Alternative A				
P	$20,000	0	$11,000	3
R	20,000	0	9,000	5
O	20,000	0	5,000	8
Alternative B				
P	$15,000	$ 500	$ 4,000	2
R	15,000	1,000	3,500	4
O	15,000	2,000	2,000	7
Alternative C				
P	$30,000	$3,000	$ 8,000	3
R	30,000	3,000	7,000	7
O	30,000	3,000	3,500	9

Table 19.2 EUAW values for three alternatives with varying factor estimates, Example 19.2

	Alternative		
Strategy	A	B	C
P	$19,327	$12,640	$19,601
R	14,548	8,229	13,276
O	9,026	5,089	8,927

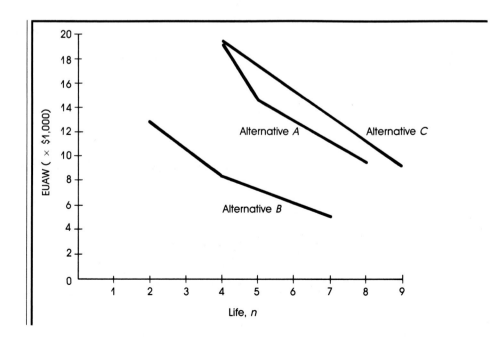

Figure 19.3 Plot of EUAW values for different life estimates.

Comment   While the alternative that should be selected in this example is obvious, this may not always be the case. For example, in Table 19.2, if the pessimistic EUAW for alternative *B* were $21,000 and the optimistic EUAW values for alternatives *A* and *C* were less than $5089, selection of *B* would not be as apparent. In this case, it would be necessary to decide which strategy (P, R, or O) would control the decision. In selecting strategy, other factors, such as the project MARR, the availability of capital, and the economic stability of the company would be taken into consideration.

Probs. 19.15 to 19.20

19.4   Economic Uncertainty and the Expected Value

The use of probability and its computations by the engineering economist are not as common as they should be; the reason for this is not that the computations are difficult to perform or understand but that realistic probability values are difficult to obtain. The economist must deal with *uncertain future* monetary and life values. Often reliance on past data, if any exist, for future cash-flow values is incorrect. However, experience and wise judgment can often be used in conjunction with the expected value to evaluate the desirability of a project. The *expected value* can be interpreted as a long-run average outcome if the project is repeated many times. However, even for a single purchase, the expected value is meaningful.

The expected value $E(X)$ may be computed using the relation below where the subscript i is omitted from X for simplicity.

$$E(X) = \sum_{i=1}^{m} XP(X) \tag{19.1}$$

where X = specific variable value
$P(X)$ = probability that a specific value of X will occur

In all probability statements, the sum of the $P(X)$ values must be 1.0; that is,

$$\sum_{i=1}^{m} P(X) = 1.0$$

If X represents the cash flows, some will be negative, such as the first cost of an asset or initial investment of a project. If the expected value is negative, the overall outcome is expected to be a cash outflow. Therefore, $E(\text{cash flow}) = \$-1500$ indicates a losing proposition.

A much more detailed explanation of probability and expected values may be found in any text on probability and statistics.

Example 19.3

You expect to be mentioned in your favorite uncle's will. You anticipate being willed $5000 with a probability of 0.5, or $50,000 with a 0.45 chance. However, there is a 0.05 chance of no inheritance. Compute your expected inheritance.

Solution   Let X be the inheritance values in dollars, and $P(X)$ represent the associated probability. The expected inheritance is

$$E(X) = 5000(0.5) + 50,000(0.45) + 0(0.05) = \$25,000$$

Note that the no-inheritance possibility is included because it makes the probability values sum to 1.0.

Probs. 19.21 to 19.23

19.5   Expected Value of Alternatives

The $E(X)$ computation and result can be utilized in several ways, and the probability values may be stated in different ways. Example 19.4 illustrates a simple $E(X)$ computation for one plan. In Example 19.5 the expected present worth for an asset is computed when different cash-flow sequences are estimated.

Example 19.4

A utility company is experiencing a difficult time in obtaining natural gas for electric generation and distribution. Monthly expenses are now averaging $7,750,000. The economist for this city-owned utility has collected average revenue figures for the past 2 years using the categories "fuel plentiful," "less than 30% other fuel purchased," and "at least 30% other fuel" (Table 19.3).

Fuels other than natural gas, purchased at a premium, have a cost which is transferable directly to the consumer. Can the utility expect to meet future expenses based on the 2 years of data?

Solution   Using the 24-month data-collection period, probabilities may be computed as follows:

Fuel	$P(X)$
Plentiful	$12/24 = 0.50$
$<30\%$	$6/24 = 0.25$
$\geqslant 30\%$	$6/24 = 0.25$

Probabilities add up to 1.0, so Eq. (19.1) results in

$$E(X) = E(\text{revenue}) = 5,270,000(0.50) + 7,850,000(0.25) + 12,130,000(0.25)$$
$$= \$7,630,000$$

Table 19.3 Fuel data for a municipal utility

Fuel situation	Months in past 2 years	Average revenue per month, X
Plentiful	12	$ 5,270,000
$<30\%$ other	6	7,850,000
$\geqslant 30\%$ other	6	12,130,000

Table 19.4 Equipment cash-flow probabilities for Example 19.5

	Economy		
Year	Receding (prob. = 0.2)	Stable (prob. = 0.6)	Expanding (prob. = 0.2)
0	$-5,000	$-5,000	$-5,000
1	+2,500	+2,000	+2,000
2	+2,000	+2,000	+3,000
3	+1,000	+2,000	+3,500

With average monthly expenses of $7,750,000, the utility will have to raise rates in the near future.

Example 19.5

The Tule Company has had experience with automatic reaming equipment. A certain piece of equipment will cost $5000 and have a life of 3 years. Estimated cash flows and probability of each are listed in Table 19.4, depending on a receding, stable, or expanding economy. Using expected-present-worth analysis, determine if the equipment should be purchased at a 15% MARR.

Solution The first step is to find the present worth of the cash flows under each type of anticipated economy and then find the $E(\text{PW})$ using Eq. (19.1). Let the subscript R represent a receding economy, S a stable economy, and E an expanding economy. The PW for each economy scenario is computed.

$$\text{PW}_R = 4344 - 5000 = \$-656$$

$$\text{PW}_S = 4566 - 5000 = -434$$

$$\text{PW}_E = 6309 - 5000 = +1309$$

The expected present worth is

$$E(\text{PW}) = \sum_{j=R,S}^{E} \text{PW}_j[P(j)]$$

$$= -656(0.2) - 434(0.6) + 1309(0.2)$$

$$= \$-310$$

Since $E(\text{PW}) < 0$, the project is not expected to be a paying proposition at a 15% required rate of return.

Additional Example 19.10
Probs. 19.24 to 19.30

19.6 Alternative Evaluation Using Decision Trees

Many alternative evaluation problems require a series of decisions where the outcome from one decision is important to the next stage of decision making. When an analyst can *clearly define each economic alternative* and wants to *explicitly account for the risk involved,* it is helpful to formulate the situation and perform the evaluation using the decision-tree approach. A decision tree includes:

Alternatives available for decision
Possible outcomes resulting from a decision
Probability that each outcome may result
Estimates of economic value (cost or revenue) for each outcome
Expected economic value of each decision
Decisions which will maximize expected revenue or minimize expected cost

The decision tree is constructed left to right and includes each possible decision and outcome. A square represents a *decision node* with the possible alternatives indicated on the *branches* from the decision node (Fig. 19.4*a*). A circle represents a *probability node* with the possible outcomes and estimated probabilities on the branches (Fig. 19.4*b*). Since outcomes always follow decisions, the treelike structure in Fig. 19.4*c* is developed as the entire situation is defined.

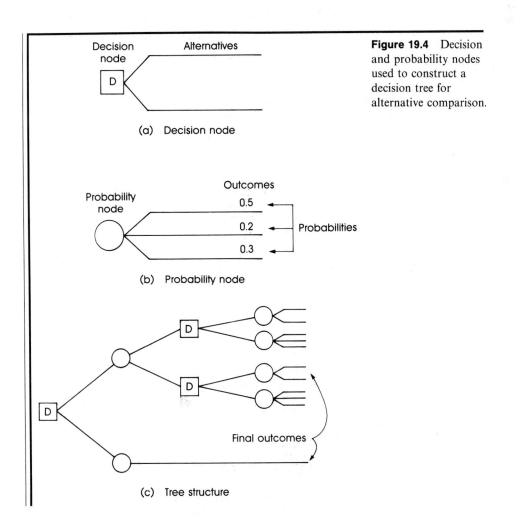

Figure 19.4 Decision and probability nodes used to construct a decision tree for alternative comparison.

Usually each branch of a decision tree has some associated economic value in cost and benefit. These cash flows are expressed in terms of PW, EUAW, or future worth (FW) values and are shown to the right of each final outcome on the tree. The cash-flow values and probability values on each outcome branch are utilized to compute the expected economic value of each decision. This process, called "solving the tree" or "fold back" is explained after Example 19.6, which illustrates the construction of a decision tree.

Example 19.6

Jerry Hill is president of a food-processing company. A large supermarket chain wants to market its own brand of low-calorie, frozen dinners which are microwave-oven prepared. The chain has offered Jerry the food-processing and packaging contract. He has decisions to make now and two years in the future. The current decision is between two alternatives: Build and own the processing and packaging facility, or lease the facility from the supermarket chain, which has agreed to convert a present processing facility into this new plant for Jerry to occupy. Outcomes are classified as good and poor depending upon market response.

The future decision alternatives are dependent upon the present lease-or-own decision. If Jerry decides to lease, good market response means that the future decision alternatives are to produce at twice, equal-to, or one-half the original volume. This will be a mutual decision between the supermarket chain and Jerry's company. A poor response will indicate a 50% level of production or complete removal from the frozen-dinner market. Outcomes for the future decisions are, again, good and poor market responses. There is no plan to anticipate future decision alternatives at this time.

As agreed by the chain, the current decision for Jerry to own the facility will allow him to set the production level 2 years hence. If market response is good, the decision alternatives are four or two times original levels. The reaction to poor response will be production at the same level or complete stoppage of production.

Construct the tree of decisions and possible outcomes for Jerry Hill and his company.

Solution   Identify the nodes and branches.

Decision now
    Identification: D1
    Alternatives: lease (L) and own (O)
Decisions 2 years hence
    Identification: D2 through D5
Choice of alternatives: Quadruple production ($4 \times$); double production ($2 \times$); level production
    ($1 \times$); one-half production ($0.5 \times$); stop production ($0 \times$)

Outcomes are market responses: Good, poor, and out-of-business

Develop the tree using the L and O branches from decision node D1 and the outcomes of good and poor. Figure 19.5 shows these branches. The alternatives on future production levels, which are different for each response, are added to the tree and followed by the market responses of good and poor. If the stop production ($0 \times$) decision is made, the only outcome is out-of-business. This completes the decision-tree structure for Jerry Hill.

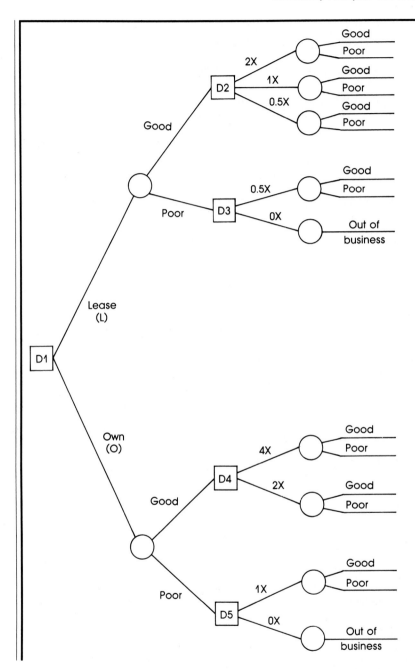

Figure 19.5 A two-stage decision tree which identifies decision alternatives and possible outcomes.

The tree in Figure 19.5 is called two-stage since there are two decision points separated by a time period of two years. The size of a tree grows very rapidly. A tree with as few as 10 decision nodes may have hundreds of final outcomes. Computer analysis to solve the tree and select an optimal decision path becomes essential as size increases.

To utilize the decision tree for alternative evaluation, the following additional information must be obtained for the branches:

1. The estimated probability that each outcome may occur. These probabilities must sum to 1.0 for each set of outcomes which result from a decision.
2. Economic information for each decision alternative and the possible outcomes, such as, initial investment and annual cash flows.

Decisions are made using the probability estimate of each outcome branch and the economic value of each outcome. Commonly the present worth is included in an expected value computation of the type in Eq. (19.1). The general procedure is:

1. Compute the present-worth value for each outcome branch taking time and the value of money into account.
2. Calculate the expected value for each decision alternative.

$$E(\text{decision}) = (\text{PW of outcome})P(\text{outcome}) \qquad (19.2)$$

where the summation is taken over all outcomes for the decision alternative.

3. For a decision, select the best expected value—minimum if only costs are involved and maximum if total payoff of costs and benefits is estimated.
4. Continue moving to the left of the tree to the root decision in order to select the best alternative.
5. Trace the best decision path through the tree.

The next two examples illustrate this procedure.

Example 19.7

A decision to market or sell a new invention has to be made. If the product is marketed, the decision to take it internationally or nationally must be made. Neglect the details of the outcome branches and assume that the decision tree and information in Fig. 19.6 are correct.
    The probabilities for each outcome and present worth of costs and benefits (payoff in $ millions) are shown. Determine which decision should be made at the node D1.

Solution Use the procedure above to determine that the alternative to sell the invention maximizes payoff.

1. Present worth of payoff is supplied in this example.
2. Compute the expected payoff at D2 and D3 using Eq. (19.2). For D2, for example, the numbers in ovals in Fig. 19.6 are:

$$E(\text{international}) = 12(0.5) + 16(0.5) = 14$$

$$E(\text{national}) = 4(0.4) - 3(0.4) - 1(0.2) = 0.2$$

Perform similar computations for D3 alternatives.

3. Select the larger expected payoff at each decision node. These are international (14) at D2 and international (4.2) at D3.
4. Calculate the expected payoff for market (6.16) and sell (9). The sell option is simple since the only outcome possible has a payoff of 9. At D1, sell yields the larger expected value.
5. Solution to the tree is to sell for a guaranteed $9,000,000.

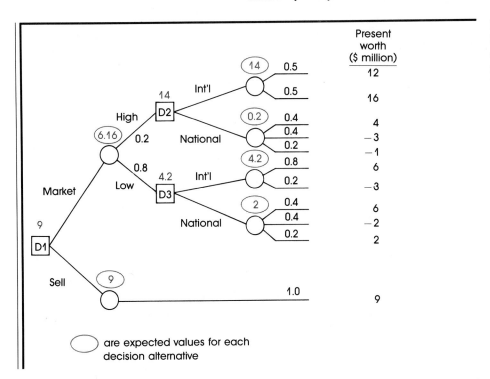

Figure 19.6
Solution of a
decision tree with
stated present-
worth values.

Example 19.8

Use the tree in Fig. 19.5 (developed in Example 19.6) to decide upon leasing or owning a building. Assume that the lease arrangement will cost $175,000 per year for 4 years. If no dinners are made in years 3 and 4 (0× decision), the lease cost is still incurred. The build-and-own decision will cost $800,000 initial investment for a building with an estimated sales value of $400,000 after 4 years. The building will not be sold immediately if a 0× decision is made for years 3 and 4.

The following revenue and probability estimates have been prepared for decision node D1.

Decision D1	Estimate	Outcome for years 1 and 2	
Lease or own		Good	Poor
	Revenue	$300,000	$150,000
	Probability	0.6	0.4

Similar data is available for D2 through D5 in Figures 19.7 through 19.10, discussed in the solution.

Use a 15% MARR and present-worth analysis to determine whether the lease-or-own option is better now, and to determine the level of production to plan for in years 3 and 4, provided the estimated information is correct.

Solution It is easier to solve such a problem in stages by constructing a small tree for each set of outcomes at a decision node. Start with D2. Figure 19.7 shows the outcomes, revenue estimates, and probabilities for each outcome. The format of Eq. (19.2) is used to compute the expected value for each outcome at D2 using PW computations. The first

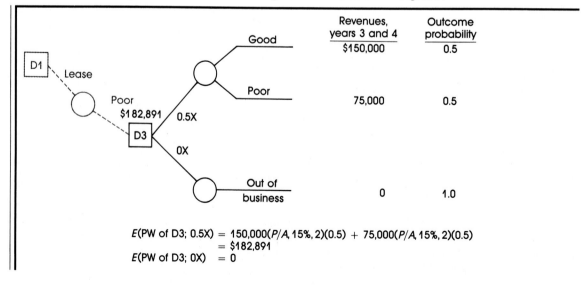

	Revenues, years 3 and 4	Outcome probability
Good	$600,000	0.8
Poor	300,000	0.2
Good	300,000	0.8
Poor	150,000	0.2
Good	150,000	0.2
Poor	75,000	0.8

Figure 19.7
Solution of D2 segment of decision tree in Fig. 19.5.

E(PW of D2; 2X) = 600,000(P/A, 15%, 2)(0.8) + 300,000(P/A, 15%, 2)(0.2)
= $877,878
E(PW of D2; 1X) = $438,939
E(PW of D2; 0.5X) = $219,470

of these computations is detailed below the tree segment in Fig. 19.7. Of the three expected values, E(PW of D2, 2×) is largest. This value is selected as the best decision at D2.

Tree segments and computation similar to that for D2 are completed for D3 through D5 in Figs. 19.8 through 19.10. The largest expected value is indicated above each decision node.

Figure 19.8   Solution of D3 segment of decision tree in Fig. 19.5.

	Revenues, years 3 and 4	Outcome probability
Good	$150,000	0.5
Poor	75,000	0.5
Out of business	0	1.0

E(PW of D3; 0.5X) = 150,000(P/A, 15%, 2)(0.5) + 75,000(P/A, 15%, 2)(0.5)
= $182,891
E(PW of D3; 0X) = 0

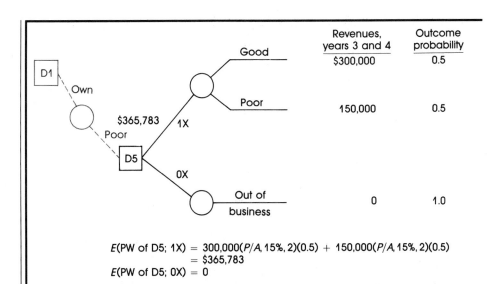

Figure 19.9
Solution of D4
segment of
decision tree in
Fig. 19.5.

Now, move back to D1. Figure 19.11 shows the tree segment with the values for each second-stage decision. The expected values for D1 options of lease and own are calculated in Fig. 19.11. Note that the lease and own costs are assumed for 4 years in each outcome and that the expected values from the previous computations for years 3 and 4 are included in the expected values for D1.

The larger value is $726,031 for the "own" alternative. Therefore, based on the current estimates, the best decision path for the company is to own the building and plan for four times the initial production level in years 3 and 4.

Figure 19.10
Solution of D5
segment of
decision tree in
Fig. 19.5.

Figure 19.11 Solution of D1 segment of decision tree in Fig. 19.5, using results of D2 through D5.

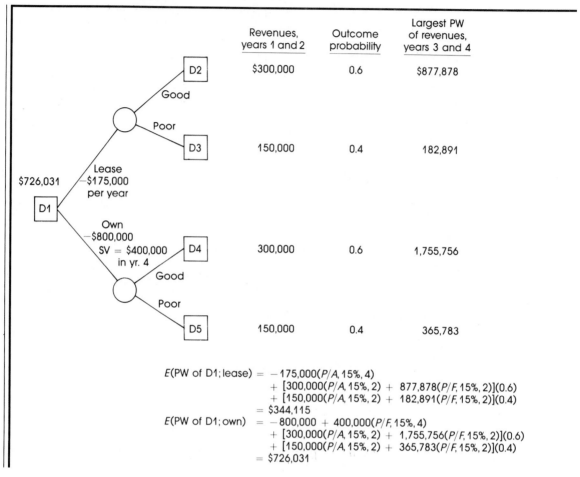

$$E(\text{PW of D1; lease}) = -175,000(P/A, 15\%, 4)$$
$$+ [300,000(P/A, 15\%, 2) + 877,878(P/F, 15\%, 2)](0.6)$$
$$+ [150,000(P/A, 15\%, 2) + 182,891(P/F, 15\%, 2)](0.4)$$
$$= \$344,115$$

$$E(\text{PW of D1; own}) = -800,000 + 400,000(P/F, 15\%, 4)$$
$$+ [300,000(P/A, 15\%, 2) + 1,755,756(P/F, 15\%, 2)](0.6)$$
$$+ [150,000(P/A, 15\%, 2) + 365,783(P/F, 15\%, 2)](0.4)$$
$$= \$726,031$$

You can see from the computations required for this relatively simple problem that a computerized "foldback" software package is needed to solve a tree of any real-world size.

Probs. 19.31 to 19.38

Additional Examples

Example 19.9

The city of Blarney has a 3-mile stretch of heavily traveled highway to resurface. The Ajax Construction Company offers two methods of resurfacing. The first is a concrete surface for a cost of $150,000 and an annual maintenance charge of $1000.

The second method is an asphalt covering with a first cost of $100,000 and a yearly service charge of $2000. However, Ajax would also request that every third year the highway be touched up at a cost of $2500 per mile.

The city uses the interest rate on revenue bonds, 6% in this case, as the MARR.
(a) Determine the breakeven number of years of the two methods. If the city expects an interstate to replace this stretch of highway in 20 years, which method should be selected?
(b) If touch-up cost increases by $500 per mile every 3 years, is the decision sensitive to this cost?

Solution
(a) Set up the EUAW equation

EUAW of concrete = EUAW of asphalt

and compute the breakeven n value.

$$150,000(A/P, 6\%, n) + 1000 = 100,000(A/P, 6\%, n) + 2000$$

$$+ 7500 \left[\sum_j (P/F, 6\%, j) \right] (A/P, 6\%, n)$$

where $j = 3, 6, 9, \ldots, n$. This may be rewritten as

$$50,000(A/P, 6\%, n) - 1000 - 7500 \left[\sum_j (P/F, 6\%, j) \right] (A/P, 6\%, n) = 0 \qquad (19.3)$$

A value of $n = 39.6$ satisfies the equation, so a life of approximately 40 years is required to break even at 6%. Since the road is required for only 20 years, the asphalt surface should be constructed.
(b) The total touch-up cost will increase by $1500 every 3 years. Equation (19.3) may be written as

$$50,000(A/P, 6\%, n) - 1000 - \sum_j \left[7500 + 1500 \left(\frac{j-3}{3} \right) (P/F, 6\%, j) \right] (A/P, 6\%, n) = 0$$

where $j = 3, 6, 9, \ldots, n$. The breakeven point is approximately 21 years, which still favors the asphalt surface. The decision is insensitive to the stated increase in the touch-up cost; however, as always, intangible factors could alter the decision.

Example 19.10

The Holden Construction Company plans to build an apartment complex close to the edge of a partially leveled hill. Support of the cliff below the complex will ensure that no damage will occur to the buildings or occupants. The amount of rainfall at any one time can cause varying amounts of damage. Table 19.5 itemizes the probability of certain rainfalls within a period of a few hours and the first cost to construct a support wall to ensure protection against the corresponding amount of water.

Table 19.5 Rainfall and support-wall cost for Example 19.10

Rainfall, inches	Probability of greater rainfall occurring	Cost of support wall to carry rainfall
2.0	0.300	$10,000
2.5	0.100	15,000
3.0	0.050	22,000
3.5	0.010	30,000
4.0	0.005	42,000

Table 19.6 EUAW for different support walls

Rainfall, inches	Support-wall cost	Annual loan cost	Expected annual damage	EUAW
2.0	$10,000	$ 973	$6,000	$6,973
2.5	15,000	1,460	2,000	3,460
3.0	22,000	2,141	1,000	3,141
3.5	30,000	2,920	200	3,120
4.0	42,000	4,088	100	4,188

The project of wall construction will be financed by a 30-year 9% loan. Data on record show that an average of $20,000 damage has often occurred when a heavy rain has fallen. Without taking the intangible (and important) fact of human life into account, determine what size support wall is most economic?

Solution   We will use EUAW analysis to find the most economic plan. For each rainfall level

EUAW = annual loan cost + expected annual damage

= cost$(A/P, 9\%, 30)$ + 20,000(probability of greater rainfall)

The $20,000 damage figure is used because this has been the previous damage on the *average*. We assume that the probabilities are a yearly value. Computations in Table 19.6 show that a $30,000 wall to protect against a 3.5-inch rainfall is most economic. A wall to protect against a 3.0-inch rain is a close second to the most economic plan.

Comment   Usually a rather hefty safety factor is automatically added when people are endangered. Pure economic analysis is not used alone as the decision maker. By building to protect to a greater degree than usually needed, the probabilities of damage are lowered— by how much we don't know, but it makes us *feel safe*.

Problems

19.1   The A-1 Salvage Company is contemplating the purchase of a new crane equipped with a magnetic pickup device to be used in moving scrap metal around the yard. The complete crane will cost $62,000 and have an 8-year life and a salvage value of $1500. Annual maintenance, fuel, and overhead costs are estimated at $0.50 per metric ton. Labor cost will be $8 per hour for regular wages and $12 for overtime. A total of 25 tons can be moved in an 8-hour period. The salvage yard has handled in the past anywhere from 10 to 30 tons of scrap per day. If the company uses a MARR of 10%, plot the sensitivity of the present worth of costs versus annual volume moved, assuming that the operator is paid for 200 days of work per year. Use a 5-metric-ton increment for the graph.

19.2   A new piece of equipment is being economically evaluated by three engineers. The first cost will be $77,000 and the life is estimated at 6 years with a salvage value of $10,000 at disposal time. The engineers disagree, however, on the annual net income to be credited to this new equipment. Engineer *A* has given an estimate of $14,000 per year. Engineer *B* states that this is too low and estimates $15,000, while *C* estimates

$18,000 per year. If the MARR is 8%, use present-worth analysis to determine if these three estimates will change the decision to buy the equipment.

19.3 Perform the same analysis in Prob. 19.2, except make it an after-tax consideration using straight-line depreciation with the salvage value considered and a 52% effective tax rate. Assume an annual operating cost of $1000 and an after-tax rate of return of 5% per year.

19.4 In Prob. 5.8 you made a present-worth comparison between building and leasing storage space. Determine the sensitivity of your decision to the following situations: (a) construction costs go up 10% and lease costs go down to $1.25 per square meter per month; (b) lease costs remain at $1.50 per square meter per month, but construction costs vary from $50 to $90 per square meter.

19.5 For the situation presented in Prob. 7.6, plot the sensitivity of the rate of return to the amount of the income gradient. Perform the computations for values of the *negative* gradient from $300 to $700 in increments of $100. If the company would like a return of at least 40% per year, would variation in this income gradient have affected the decision to buy the dump trucks?

19.6 Consider the two air-conditioning systems detailed below.

	System 1	System 2
First cost	$10,000	$17,000
Annual operating cost	200	150
Salvage value	−100	−300
New compressor and motor cost at midlife	1,750	3,000
Life, years	8	12

Use an EUAW analysis to determine the sensitivity of the economic decision to MARR values of 8, 10, and 15%.

19.7 Reread Prob. 19.6. If the MARR is 10%, plot the EUAW for each system for life values from 4 to 8 years for system 1 and 6 to 12 for system 2. Assume that the salvage values and annual operating costs are the same for each life value. Further, assume that the compressor is replaced at midlife. Plot EUAW for even-numbered years only. Which EUAW is more sensitive to a varying-life estimate?

19.8 A city couple, Joan and John Pollution, would like to buy a small section of land in the woods to be used as a weekend vacation home. Alternatively, they have thought of buying a used travel trailer and a four-wheel-drive vehicle to pull the trailer for vacations. The Pollutions have found a 5-acre tract with a cabin, well, etc., 25 miles from their home. It will cost them $30,000, but they expect to sell the acreage for $45,000 in 10 years when their children are grown. The insurance, upkeep, etc., costs are estimated at $500 per year, but this weekend site is expected to save the family $50 every day they don't go on a traveling vacation. The Pollutions estimate that even though the cabin is only 25 miles from home, they will travel 50 miles a day when at the cabin while working on it, visiting neighbors, etc. The Pollution car average 30 miles per gallon of gasoline.

    The trailer and van combination would cost $11,000 and could be sold for $2000 in 10 years. Insurance, maintenance, etc., costs will be $750 per year, but the trailer is expected to save $25 per vacation day. On a normal vacation, the Pollutions travel 300 miles each day. Mileage per gallon for the van is estimated at 60% that of the family car.

    Gas costs $1.20 per gallon and the Pollution family want a return of fun and 10% from either investment. Compute the breakeven number of days per year for the two plans, and plot the sensitivity of EUAW for each plan if the Pollutions' vacation time in the past has been from 6 to 12 days per year.

19.9 Suppose that the Pollution family of Prob. 19.8 plans to purchase the acreage with the cabin and still go on a 4-day traveling vacation in the car at a cost of $65 per day.
(a) Was the decision to buy the land still the better decision?
(b) Does the breakeven point seem to be sensitive to the types of vacation plans that combine the use of the acreage and traveling?

19.10 Plot the sensitivity of rate of return versus the life of a 5% $50,000 bond that is offered for $43,500 and has the bond interest paid quarterly. Use life values of 10, 12, 15, 18, and 20 years.

19.11 The Charley Company has been offered an investment opportunity that will require a cash outlay of $30,000 now for a cash inflow of $3500 for each year of investment. However, the company must state now the number of years it will retain the investment. If the investment is kept for 6 years, $25,000 can be gotten for the company's share, but after 10 years the resale value will be only $12,000. If money is worth 8% per year, is the decision sensitive to the retention period?

19.12 The person who did the analysis in Prob. 16.13 is concerned with the sensitivity of the breakeven point to the expected life of the project. Determine the treatment plant first cost for lives of (a) 15 years and (b) 25 years.

19.13 Determine the sensitivity of the most economic life value of Prob. 10.33 to the cost gradient. Investigate the gradient values of $60 to $140 in increments of $20 and plot the results.

19.14 Rework Prob. 19.13 at an interest rate of $i = 5\%$ and plot the results on the same graph.

19.15 Reread Prob. 6.7. The time of overhaul can vary from 2 to 4 years for the used machine and from 4 to 6 years for the new machine. Plot the EUAW values for these three estimates and determine if they will alter the decision.

19.16 If the spray method of Prob. 6.13 is used, the amount of water used can vary from an optimistic value of 60 liters to a pessimistic value of 120 liters with 80 liters being a reasonable figure. The immersion technique always takes 16 liters per ham. How will this varying use of water for the spray method affect the economic decision?

19.17 An engineer is trying to decide between two ways to pump concrete up to the top floors of a seven-story office building now under construction. Plan 1 requires the purchase of equipment costing $6000 and costing between $0.40 and $0.75 per metric ton to operate with an expected cost of $0.50 per metric ton. The asset is able to pump 100 metric tons per day. If purchased, the asset will last for 5 years, have no salvage value, and be used from 50 to 100 days per year. Plan 2 is an equipment-leasing option and is expected to cost the company $2500 per year for equipment with an optimistic cost of $1800 and a pessimistic value of $3200 per year. In addition, a $5-per-hour labor cost will be incurred for operation of the leased equipment. Plot the EUAW of each plan versus total annual operating or lease cost at $i = 12\%$. Determine which plan the engineer should select, using the reasonable estimates for a use of (a) 50 and (b) 100 days per year.

19.18 When the country's economy is expanding, the AB Investment Company is optimistic and uses a MARR of 8% on all new investments. However, in a receding economy a return of 15% on investments is required. Normally a 10% return is required. Similarly, an expanding economy causes the estimates of asset life to go down about 20% and a receding economy makes the n values increase about 10%. Plot the sensitivity of present worth versus (a) MARR and (b) life values for the two plans detailed below using the reasonable estimate for other varying factors.

	Plan *M*	Plan *Q*
Initial investment	$-100,000	$-110,000
Annual CFBT	+15,000	+19,000
Life, years	20	20

19.19 Rework Prob. 19.18 except use the data for plans *A* and *B* in Prob. 6.15.

19.20 When is it necessary to select a particular strategy of pessimistic, reasonable, or optimistic and make an economic decision on the basis of the selected strategy?

19.21 The variable *X* can take on the values $X = 5, 10, 15, 20$ with a probability of 0.30, 0.40, 0.233, 0.067, respectively. Compute the expected value of *X*.

19.22 The AOC value for a plan can take on one of two values. Your office partner told you that the high value is $2800 per year. If her computations show a probability of 0.75 for the high value and an expected AOC of $2575, what is the low AOC value used in the computation of the average AOC value?

19.23 The Lunch-a-Bunch food service company has performed an economic analysis of proposed service in a new region of the country. The three-estimate approach to sensitivity analysis has been used with each of the optimistic and pessimistic estimates expected to occur with a 15% chance. Use the EUAW values to compute the expected EUAW for the new service proposal.

	Optimistic	Reasonable	Pessimistic
EUAW	$+150,000	$+5000	$-275,000

19.24 Find the expected present worth of the following series of payments if each series is expected to be realized with the probability shown at the head of each column (assume $i = 20\%$).

	Annual cash flow		
Year	*P* = 0.5	*P* = 0.2	*P* = 0.3
0	$-5,000	$-6,000	$-4,000
1	+1,000	+500	+3,000
2	+1,000	+1,500	+1,200
3	+1,000	+2,000	-800

19.25 Compute the expected EUAW value for the cash flows of Prob. 19.24.

19.26 The officers of a resort country club are thinking of constructing an additional 18-hole golf course. Because of the northerly location of the resort, there is a 60% chance of a 120-day golf season, a 20% chance of 150 days of golfing weather, and a 20% chance of a 165-day season. The course will be used by an estimated 350 golfers each day of the 4-month season, but by only 100 per day for each extra day in the golfing season. The course will cost $375,000 to construct and will require a $25,000 rework cost after 4 years. Annual maintenance cost will be $36,000 and the green fees will be $4.25 per person. If a life of 10 years is anticipated before a major rework is required and a 12%-per-year return is desired, determine if the course should be constructed.

19.27 The owners of the Ace Roofing Company want to invest $10,000 in new equipment. A life of 6 years and a salvage value of 12% is anticipated. The annual income will depend upon the state of the housing and construction industry. The income is expected to be $2000 per year; however, a current slump in the industry is given a 50% chance of lasting 3 years and a 20% chance of continuing for 3 additional years. However, if the outlook of the depressed market does improve, either during the first or second 3-year period, the annual income of the investment is expected to be $3500. Can the company expect to make 8% per year on its investment?

19.28 The High Construction Company is building an apartment complex in the arid southwest on the top of a partially leveled hill. The road that winds around the hill to the apartment complex entrance needs retaining walls for support above and below the road surface. The probability of a rain shower greater than a given amount and the associated damage and retaining-wall construction costs are shown in the table. Determine which plan will result in the lowest annual cost over a 25-year period at an interest rate of 10% per year.

Rainfall, inches	Probability of greater rainfall occurring	Retaining-wall cost to carry rainfall	Expected annual damage for specified rainfall
1.0	0.6	$15,000	$ 1,000
2.0	0.3	16,000	1,500
2.5	0.1	18,000	2,000
3.0	0.02	21,000	5,000
3.5	0.005	28,000	9,000
4.0	0.001	35,000	14,000

19.29 Rework Prob. 19.28 using an after-tax analysis, assuming that the tax rate is 50% and the retaining-wall construction cost will be secured by an 8% compounded annually 25-year loan. Assume that the principal amount is reduced an equal amount each year with the remainder of the payment applied to interest.

19.30 A pɪ vate citizen has $5000 to invest. If he puts the money in a savings account, he is assuꞧ ed of receiving an effective 6.35% per year on the principal. If he puts the money in stocks, he has a 50/50 chance of one of the following cash-flow sequences for the next 5 years.

Year	Stock 1	Stock 2
0	$−5,000	$−5,000
1–4	+250	+600
5	+6,800	+5,400

Finally, he can invest his $5000 in improved property for 5 years with the following outcomes and probabilities.

	Cash flow		
Year	Prob. = 0.3	Prob. = 0.5	Prob. = 0.2
0	$−5,000	$−5,000	$−5,000
1	−425	0	+500
2	−425	0	+600
3	−425	0	+700
4	−425	0	+800
5	+9,500	+7,200	+5,200

Which of the three investments—savings, stocks, or property—is the best?

19.31 For the tree on the next page, determine the expected values of the two outcomes if decision D3 is made and the outcome is to be maximized. This is part of a larger tree.

19.32 A large decision tree has one outcome branch that is detailed on the bottom of the next page. If decisions D1, D2, and D3 are all options in a 1-year time period, find the decision path which maximizes outcomes. There are specific investment requirements for D1, D2, and D3 as indicated on these branches.

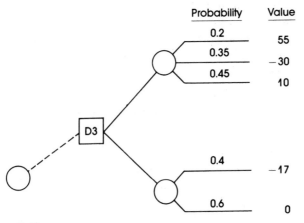

Probability	Value
0.2	55
0.35	−30
0.45	10
0.4	−17
0.6	0

Problem 19.31

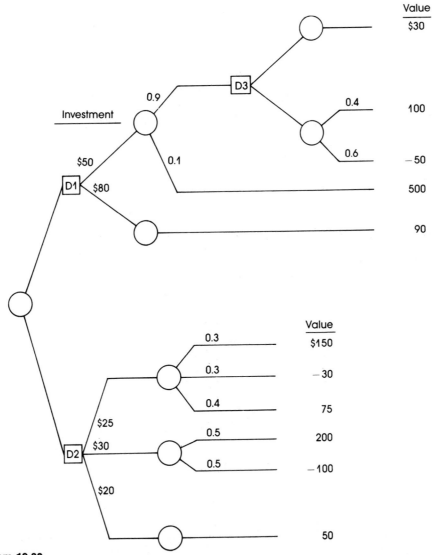

Investment

	Value
	$30
0.4	100
0.6	−50
	500
	90

	Value
0.3	$150
0.3	−30
0.4	75
0.5	200
0.5	−100
	50

Problem 19.32

19.33 The decision D4 has three possible outcomes: x, y, and z. D4 must occur in year 3 of a 6-year planning horizon to maximize the outcome. Use $i = 15\%$, the investment requirement in year 3, and the estimated revenue (cost) cash flows to determine which decision should be made in year 3. This decision is part of a larger tree.

		Decision investment, year 3	Revenues ($\times$ $1,000) Year 4	Year 5	Year 6	Outcome probability
	High	$200,000	$50	$50	$50	0.7
	Low		40	30	20	0.3
	High	75,000	30	40	50	0.45
	Low		30	30	30	0.55
	High	350,000	190	170	150	0.7
	Low		−30	−30	−30	0.3

19.34 A total of 5000 mechanical subassemblies are needed annually on a final assembly line. The subassemblies can be obtained in one of three ways: (1) *Make* them in one of three plants owned by the company; (2) *buy* them off the shelf from the one and only manufacturer, or (3) *contract* to have them made to specifications by a current vendor. The estimated annual cost for each alternative is dependent upon specific circumstances of the plant, producer, or contractor. The information below details the circumstance, a probability of occurrence, and the estimated annual cost. Construct and use a decision tree to determine the least-cost alternative to acquire the subassemblies.

Alternative	Possible outcome	Probability	Annual cost to acquire 5,000 units
(1) Make	Which plant can produce on time?		
	A	0.3	$250,000
	B	0.5	400,000
	C	0.2	350,000
(2) Buy off shelf	No control over quantity available; must obtain 5,000:		
	<5,000, pay premium	0.2	$550,000
	5,000 available	0.7	250,000
	Forced to buy too many in quantity	0.1	290,000
(3) Contract	Timely delivery	0.5	$175,000
	Late delivery; buy some off shelf	0.5	450,000

19.35 The decision tree below has been developed to determine if high- or low-volume production (D1) will maximize the value of the outcomes. Assume the investments and outcome values are coded dollar values. Do not attempt to account for the time value of money. (*a*) Perform the "foldback" using probability and expected-value computations to determine if high or low production is better. (*b*) Would the decision at node D1 be different if the initial investment for high production were increased by 10?

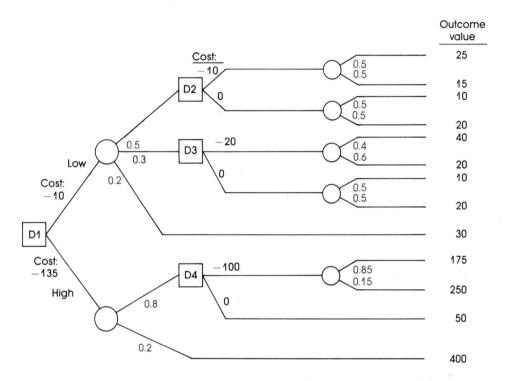

19.36 Think of your own decision making in the recent past. Select a situation which has required some thought and that has several alternatives and possible outcomes. Attempt to describe the situation in terms of a decision tree. To the degree possible, place values and probability amounts on the outcome branches. If possible, solve the tree and determine which decision path is optimal for you. Is this the decision you selected? Be objective in your appraisal of the situation.

19.37 The president of ChemTech is trying to decide whether to start a new product line or purchase a small company. It is not possible financially to do both.

To make the product for a 3-year period will require an initial investment of $250,000. The expected annual cash flows with probabilities in ( ) are: $75,000 (0.5), $90,000 (0.4), and $150,000 (0.1).

The small company will cost $450,000 now. Market surveys indicate a 55% chance of increased sales for the company, and a 45% chance of severe decreases with annual cash flow of $25,000. If decreases are experienced in the first year, the company will be sold immediately (during year 1) at a price of $200,000. Increased sales could be $100,000 the first 2 years. If this occurs, a decision to expand after 2 years at an investment of $100,000 will be considered. This expansion could generate cash flows with indicated probabilities as follows: $120,000 (0.3), $140,000 (0.3), and

$175,000 (0.4). If expansion is not chosen, the current size would be maintained with anticipated sales to continue. Assume there are no salvage values on any investments.

Use the description and a 15% per year return to:

(a) Construct a decision tree with all values and probabilities.

(b) Determine the expected values at the decision node after 2 years provided increased sales are experienced.

(c) Determine what decision should be made now to offer the greatest return possible to ChemTech?

(d) Explain in words what would happen to the expected values at each decision node if the planning horizon were extended beyond 3 years and all cash-flow values continued as forecast in the description.

19.38 Find the decision path for the tree below which will maximize the cash flows. The tree has been constructed for an inspection task where the completed item can be inspected (top branch) or placed directly into service and checked during operation.

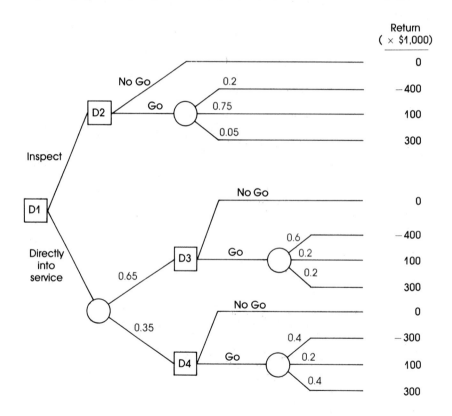

Decision Making for Large Capital Investments 20

The primary objective of this chapter is to place into proper context the economic analysis tools that you have learned thus far. These can be decision aids when the evaluation and justification of large, strategically important capital investments in new or improved technology are considered by corporate management.

Observations and criticisms about the classic engineering economy approach to justifying new technology, especially flexible manufacturing systems, concentrates upon points such as the following:

Tactical and operational, rather than strategic, in its approach to corporate investment decision making
Considers only one, rather than multiple criteria, in its computations
Requires exact estimates for factors which are difficult to estimate
Concentrates upon cost rather than benefit analysis

To place these points in perspective, this chapter discusses the difference in strategic, tactical, and operational decisions. Possible roles and uses of economic analysis techniques you have learned are examined. Some of the components used in strategic economic studies for technology advancement are defined and the use of an engineering economic perspective to evaluate them is examined.

Several of the multiple-criteria decision-making approaches used for strategic capital investments are summarized. Factors commonly present in the evaluation vary, often including the results of a PW study (which is also called a net present value, or NPV, study), or a return-on-investment study. The wise use of sensitivity analysis discussed earlier in this text offers assistance in the understanding of what costs and benefits may result from the introduction of new technology into an existing plant.

This chapter is the last in this book, but it is one of the more important in subject matter. Due to the rapidly changing scene of manufacturing and information system technology in the world, most corporate decision makers are faced with vitally important questions of how to maintain or regain nationally and internationally competitive positions for their corporations. The understanding and use of the approach and techniques presented here may be of assistance in determining the future direction of these corporations.

Section Objectives

After completing this chapter, you should be able to do the following:

20.1 State the different roles of economic analysis tools for stand-alone projects and large programs of integrated projects which require major investment and management commitment.

20.2 State how the results of an economic analysis may be interpreted differently for tactical-level and strategic-level projects.

20.3 List sample tangible and intangible benefits which may be realized from integrated systems, and explain why evaluation studies on large, long-horizon programs must consider both of these categories.

20.4 List factors which may be used to perform multiple-criteria evaluation studies, and use the *weighted evaluation method* to perform a multiple-criteria study, given the evaluation criteria and their results. Also, describe the evaluation approach taken by the *analytic hierarchy process*.

20.5 Perform a sensitivity analysis for a large investment program using percentage variation on selected economic factors, given the factors and the economic measure.

Study Guide

20.1 Role of Economic Evaluation in Capital-Investment Studies

There are three levels of management decision making in most corporate environments. These levels can be used to examine the role of economic analysis in decision making. The levels are:

Strategic—formulation of goals and policy by upper management of the corporation
Tactical—development of guidelines which set the direction for implementation of strategic decisions in the operating divisions
Operational—procedure development for programs that will carry out activities following guidelines resulting from tactical decisions.

As an illustration, a strategic corporate policy to compete with international firms in *product quality level* may result in a tactical decision by the general manager of the consumer products division to develop a *total quality system* for a specific line of products. The operational unit managers and employees of these lines may then formulate the structure and procedures of *quality improvement programs and projects*.

There will be an effect upon CAD/CAM/CAPP (computer-aided design, computer-aided manufacturing, computer-aided process planning) and other systems in the corporation as they are impacted by the strategic decision to improve product quality.

Much of the application of engineering economy has been directed to tactical- and operational-level projects through the use of PW (NPV), rate of return, EUAW, and payback analysis of stand-alone projects. In fact, the incorrect use of payback analysis (Sec. 16.1) and simplistic, incorrect uses of rate-of-return analysis are often in contradiction to the correct applications of discounted cash-flow (DCF) methods you have learned in earlier chapters.

Several studies have indicated that the basic engineering economy models of DCF, especially NPV and rate of return, are quoted and used as the standard for economic evaluation and justification studies. Often a decision to not develop and implement new technology is based on two significant factors: The proposal does *not meet the investment criteria* and *returns cannot be adequately quantified* [1]. However, if used wisely, economic analysis tools are an important part of tactical and strategic level examinations which result in 'go' decisions.

A view of economic analysis tools, suggested by Meredith and Suresh [2], which clarifies the relation between operational, tactical, and strategic levels is presented in Fig. 20.1. Manufacturing examples are given for each level. The use of DCF tools

Figure 20.1   The wide range of alternatives evaluated using operational, tactical, and strategic decision-making approaches.

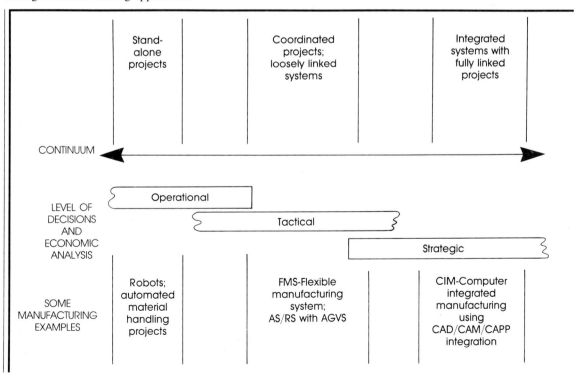

for stand-alone projects, such as robots, is common. Coordinated system projects (tactical level), such as an automated storage/retrieval system with integrated material handling using an automated guided vehicle system (AS/RS with AGVS), will most likely have advantages of intangible, noneconomic measures such as flexibility in manufacturing, shortened lead times, more-rapid customer response, reduced work-in-process inventory, and others. Ways to incorporate economic criteria with effectiveness measures for these additional, important parameters should be present in the justification study. Fully integrated programs such as CIM (computer-integrated manufacturing) must consider components across the entire continuum in the evaluation study.

Probs. 20.1 and 20.2

20.2   Distinguishing Elements in Investment Evaluation Studies

There is a large amount of material written which questions the usefulness of engineering economy when a corporation is faced with relatively large, important capital investment decisions. Often these decisions may help the company compete, or even survive, in any given market area. It is vital that corporate and technical management put economic, competitive, organizational, and other elements into perspective when evaluating projects which will have a measurable impact on the future way in which business is conducted. The approach taken in the past has been to utilize the tools of engineering economy (PW, rate of return, payback, etc.) as the primary, if not the sole, evaluation criteria. But programs which require changes in technology utilization and organization must be evaluated on the basis of elements in the operational, tactical, *and strategic* areas.

Example 20.1 presents a summary of how different a (1) stand-alone project and a (2) large, integrated program of capital investment may be viewed using specific elements defined by management. As you read this example, think of the totally different way in which the results of economic analysis tools you have learned might be interpreted in the two cases.

Example 20.1

Interactive Computing has been hit hard by international competition in the last 3 years. The prime reasons have been identified as high cost and inability to rapidly alter its microprocessor product offerings. Both reasons are caused in part by the inflexibility of Interactive's outdated assembly processes, which have been updated annually through a capital improvement program, but never studied from an integrated systems perspective.

A small task force of people in design, engineering, manufacturing, and financial management have outlined an ambitious CAAP (computed-aided assembly planning) program which interfaces with and will require extensive enhancement to current computer-based systems in design, production planning, manufacturing, quality, purchasing, and finance (cost tracking and accounting).

The task force has asked the vice president of operations to (1) define some elements which he thinks are important in the evaluation of alternatives and (2) make comments on each element for smaller, stand-alone robotics projects previously considered and for the CAAP study they have outlined.

Table 20.1 Comparison of organizational and economic elements which may be considered when tactical and strategic investment alternatives are evaluated.

| Element | Relative differences between exposure and evaluation criteria for | | Relevant reference(s) |
	Stand-alone robotics projects	Computer aided assembly planning (CAAP)	
Level of corporate exposure	One or two departments	Corporatewide	
Impact on future plans	Small; tactical level and below	Large; alter strategic plan	
Planning horizon for evaluation	2–5 years	12–20 years	[4]
Organizational level of go–no-go decision	Division level with technical management support	President level with corporate and technical management critique and support	[5]
Where major impacts will be seen	Processing methods; direct labor; work-in-process inventory	International competitiveness; market share; indirect costs; product quality	
Risk if effort fails	Low to medium risk	High risk ("you bet your company" level)	[6, 7]
Need for integration of several company functions	Some, but mostly engineering and manufacturing	Great, including management, product design, manufacturing, engineering, accounting, etc.	[5]
Effect on accounting practices	None; use same ones	Requires alteration for indirect labor allocation and auditing practices	[8]
View taken in evaluation study	Isolated replacement of old equipment with new; one-by-one replacement studies (bottom-up)	Integration of many systems involved to be well-linked (top-down)	[9]
Need to show economic success	Primarily the only factor considered	One of several factors considered, all of importance to company's future	[7, 10]
Major components considered in evaluation study	Economic, especially for direct labor reduction	Quality, cost, productivity, flexibility, impact on workforce, corporate goal attainment, and others	[2, 10]

Solution   After much deliberation the vice president summarized the elements and his comments in Table 20.1. As you read the entries, note the relative difference between the stand-alone and CAAP comments. For the strategic program the planning horizon will definitely span corporate presidents and boards of directors; the decision is made at a much higher level with virtually all functions of the corporation involved. Also, the vice president sees an impact upon the accounting standards; he comments on the top-down view for the CAAP effort; and he states that the owners are accepting a "you bet your company" environment with this program.

These differences are also present in the elements concerning economic analysis. Note in the last two elements in Table 20.1 that the stand-alone projects were justified primarily on economic terms (probably using classical engineering economy tools), but the more strategic program must consider many other factors, with economic viability still an important factor. However, it is clear from Table 20.1 that engineering economy alone cannot be used for the evaluation of the CAAP outline from the task force.

Table 20.1 includes references which discuss some of the listed elements. These articles have been reprinted in [3], edited by Meredith.

20.3 Cost and Benefit Components in Strategic Investment Studies

Cost and benefit estimates throughout this text concentrate upon small, independent projects. When several systems are integrated and have a strategic impact possibility upon the company, the components used in the study are usually more complex in nature and estimated with much less certainty. Some of the differences in cost and benefit values are discussed here.

To understand the difference in estimates used in a strategic study, compare the scope of the examples and problems you have worked in previous chapters with the following descriptions of strategic studies.

Information system to form an integral connection between marketing, engineering, manufacturing, and field service databases. Access to the contents of any database would be possible through one entry, regardless of the location of the user. Automatic interfaces of computer hardware and software would occur for all systems on the network. Estimated initial cost is $5,000,000 with an effective life of 15 years.

Computer-integrated manufacturing (CIM) system composed of eight 5-degrees-of-freedom machining centers connected by automated guided vehicles for material handling. Software systems which interface with the corporate-level design and numerical control generator system will be developed, as well as software to operate the automated machining centers in a completely personless environment (lights-out manufacturing) for one shift per day. Estimated cost is $24,000,000 with a planning horizon of 20 years.

The initial investment (P), life (n), and annual operating cost (AOC) values take on different meanings for larger studies. Independent projects and tactical-level projects have initial costs that occur in 1, 2, or 3 years (Fig. 20.2) with some planning expenditures necessary prior to implementation costs. In these cases the present value of the P estimates is computed for year zero and used in a classical economic analysis.

However, when several systems as described above are combined and evaluated as part of an integrated system, the initial cost will occur over several years of planning, implementation and enhancement (Fig. 20.2c). In fact, continued investment is usually anticipated as new features are planned.

The planning horizon or evaluation period is much longer for strategic studies. Individual project n values usually equal the economic life, anticipated product life, or recovery period for tax purposes. Many linked projects are planned to complete the development of strategic programs, thus the need for much longer evaluation periods.

The AOC term takes on the new meaning of operating, maintenance, update, direct and indirect, etc. costs as current projects are retained and new projects, all

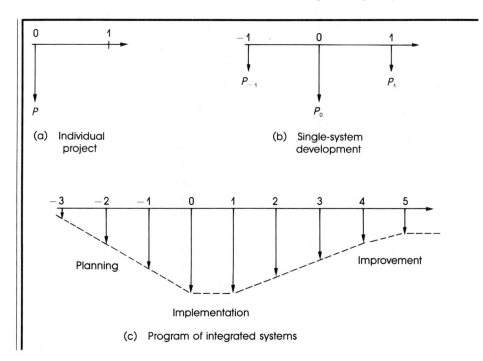

Figure 20.2
Initial-cost pro-
files for projects
ranging from
small, individual
ones to very
large, strategic
investments.

part of the overall program, are brought on-line. The AOC values, which are functionally tied to the projects in the system, will increase depending upon when new equipment/software projects are introduced. Figure 20.3 illustrates an AOC series estimate.

In our examples, the *benefit estimates* have usually been expected revenues, reduced labor and personnel costs, and AOC savings—mostly tangible benefits that

Figure 20.3   Example buildup of annual-operating-cost (AOC) profile for a large program involving phases of equipment and software systems.

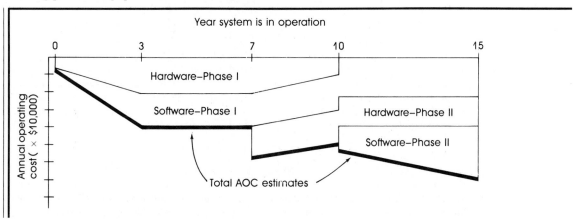

are quantifiable income and cost savings. We did not evaluate a project's less-easy-to-quantify, intangible benefits such as reduced rework, reduction in inventory, reduced design time, improved information access, and many others. For operational level projects, and usually tactical-level projects, economic evaluation is sufficient. This is not commonly so for integrated system studies involving large investments and long time horizons. The tangible benefits must be estimated as accurately as possible, and the intangible benefits must be considered in conjunction with the economic evaluation.

Operational- and tactical-level projects usually contribute to reductions which are more tangible, with some indirect cost reduction. But, strategic-level systems will most likely have contribution to tangible and intangible benefits, especially indirect cost savings, as detailed in Example 20.2. The term *indirect costs* is used generally to include costs such as:

Training programs
Public relations
Accounting and auditing
General management and administration
Contract services
Software design, development, and maintenance
Security, insurance, and workers' compensation
Taxes
Fringe benefits
Capital recovery and return requirements
Design and engineering
etc.

It is becoming vital that modern accounting systems be able to track and analyze indirect as well as direct costs for the different products, departments, services, and processing lines in a company. Improved indirect-cost tracking facilitates the quantification of intangible benefits.

Example 20.2

Jon wants to outline the evaluation study for a large, new program in medical imaging/diagnosis/surgery/treatment for the Ever-Health Hospital Corporation. The system will integrally link the imaging (x-ray, CAT scan, NMR) results to diagnosis and recommended surgery/treatment for the patient via new hardware and artificial intelligence-based software. Placement of this adventuresome system in 200 corporation-owned hospitals would cost approximately $600 million over the next 5 years, but management estimates the benefits in health service delivery to be 8 to 12 times the investment over the next 15 years. Jon needs to categorize the *tangible and intangible benefits* that might be realized by Ever-Health from this integrated system. Develop the two lists.

Solution What may be tangible benefits for one company or system may be intangible for another, depending on the nature of the system and quality of data available. When a benefit can be quantified in dollar or numerical terms, it is called tangible. Here are some of the categories of cost reduction that might be considered in this evaluation study.

Tangible benefit category	Intangible benefit category
Direct labor costs	Indirect labor costs
Indirect labor costs	Quality of health care
Energy costs	Patient processing time
Floor space requirements	New skills learned
Diagnostic costs	Flexibility of health care services
Material costs	Liability insurance costs
Surgical costs	Organizational efficiency
Patient care costs	Hospital size reduction

There are many long-term, positive results for strategic programs. In the case of this hospital corporation, some of these might be:

Increased market share of available patients
Entry into new health care markets using the systems
Development of national and international reputation
Continuing or expanding business opportunities

Of course, failure or nonacceptance of the system may jeopardize the financial viability of the corporation. Thus, a large-risk exposure is present with the suggested investment.

In each area of technology, different tangible and intangible factors are used. A good example of tangible cost-savings estimates for computer-aided manufacturing programs is presented by Sullivan [11].

Probs. 20.4 to 20.8

20.4 Multiple-Criteria Evaluation of Investment Alternatives

In addition to an economic measure of effectiveness (PW, EUAW i, n), noneconomic measures are usually involved in the final selection of available alternatives. When added factors are explicitly included in the evaluation, multiple-criteria decision making is used. It has become necessary for most large, technology-improvement decisions to include some formal method of multiple-criteria evaluation. A procedure for use of these methods is:

1. Clearly define the *alternative programs.*
2. Determine and define the *factors* to be evaluated. This should include some form of economic measure.
3. Select or develop and utilize a *multiple-criteria evaluation technique.*
4. *Choose the alternative* with the best combined result.

You can determine that these steps are no different from those used thus far. However, the several factors are combined using a technique which reflects the style and desires of the decision makers.

The *alternative programs* must be defined as specifically as possible. Each project should have cash-flow estimates, capital investment plans, and expected changes in personnel number and skill level. Required organizational changes and impact on workforce attitudes should be summarized for each alternative, because these factors are often a part of a multiple-criteria evaluation. Thus, an alternative involves much more than cash-flow estimates, tax considerations, and the like.

Table 20.2 Examples of factors used to perform multiple-criteria evaluation studies

Material-handling system [12]	Robot	Flexible manufacturing cell
Return on investment	After-tax payback	Product quality
Flexibility	Accuracy	Work-in-process inventory level
Safety	Repeatability	Annual maintenance cost
Compatibility with existing equipment	Uptime	Initial investment
Maintainability	Load capacity	Worker-skill-level requirement
	Velocity	Market share contribution

Strategic-level investment program [10]		
Factor	Definition	Decomposition of factor
Suitability	Compliance with corporate strategy	Technology position; growth; market position; workforce composition; operations management; capital source; international competition
Capability	Intrinsic ability of planned system	Design; function; reliability; availability; flexibility; human factors; technical feasibility
Performance	Anticipated level of physical performance	Throughput; quality; inventory; information rate; utilization
Productivity	Financial analysis of costs and benefits	Customer response; economic parameters (cost of capital, taxes, inflation, required return); economic measure (NPV or rate of return)

The *factors* will vary between types of industry, alternatives, and decision makers. Table 20.2 lists example factors for four cases. All include economic measures, but the flexible manufacturing cell example includes the maintenance cost and initial investment rather than a time-value-of-money measure, such as NPV. The last list for large, strategic investment programs includes high-level factors; the decomposition into more definable elements describes the criteria used in an evaluation. Each of the factors must be defined so that the people performing the evaluation, and *managers receiving the results*, can understand what they mean.

Consider the following contradictory statements and survey results about the use and justification of advanced manufacturing systems.

Most businesses use traditional return on investment or payback evaluation for decision making [1].

Payback and rate of return is the greatest barrier to the use of new manufacturing technologies [13].

About 90% of the manufacturing businesses have some form of computerized process planning or factory automation initiative [1].

Strategic and qualitative factors are equally important to economic measures [1].

Startling improvements are made when new technology is introduced into manufacturing [7].

It seems clear. The *multiple-criteria evaluation techniques*, especially those for programs which include integrated projects, must include economic factors as well as explicit evaluation of other factors historically considered too intangible to quantify. Some of the available techniques to perform multiple-criteria evaluation are summarized here. Most of these are *scoring techniques,* also called "rank and rate" techniques, the most popular of which is the weighted evaluation method.

Unweighted evaluation method: Once the evaluation factors are determined, each alternative is rated using a pre-established scale. The alternative with the largest sum of ratings is the indicated choice. The factors can include any criteria (economic, political, competitive, etc.). Sample rating scales are:

$-1, 0, +1,$ or $-2, -1, +1, +2$

0 to 1 or 0 to 100

With the 0-to-1 or 0-to-100 scale, for each factor the highest rating of 1 or 100 is assigned to one alternative and all others are rated relative to it.

Weighted evaluation method: Factors are rated as in the unweighted method, usually on a 0-to-1.0 or 0-to-100 scale. Additionally, each factor is ranked and rated for importance to the decision maker. For each alternative i, the total value V_i is then computed as

$$V_i = \sum_{j=1}^{n} w_j r_{ij} \tag{20.1}$$

where w_j = factor j weight $(j = 1, 2, \ldots, n)$
          r_{ij} = rating for alternative i on factor j

The factor weights w_j are normalized by initially assigning the score of 100 to the most important one, assigning relative scores to the other factors, then dividing each assigned weight by the total. The results are the w_j values in Eq. (20.1). This process satisfies the requirement that the sum of the factor weights be 1.0. Example 20.3 illustrates the method.

Example 20.3

The computer hardware and interactive information system used to dispatch trains has been in place for many years at Frontier Railroad. Computations and discussions have led to the definition of three distinct future alternatives and six evaluation factors of differing importance to management.

Alternative 1: Purchase new hardware and develop customized software for the company.
Alternative 2: Lease new hardware and use contract database services from an established vendor.
Alternative 3: Keep the old hardware and software and patch the software as needed in the future.

Evaluation factors

1. Payback period
2. Initial investment requirement
3. Response time
4. User interface
5. Software availability and maintainability
6. Customer service

Table 20.3 Relative ratings for three alternatives using the weighted evaluation method

Factor i	Factor importance	Relative importance (0 to 100), r_{ij}		
		Alternative 1	Alternative 2	Alternative 3
1	50	75	50	100
2	100	60	75	100
3	90	50	100	20
4	80	100	90	40
5	50	75	100	10
6	80	100	100	75
Total	450			

Use the weighted evaluation method to determine which alternative is the most desirable. Assume that the meaning of each factor has been established and that adequate numerical results are available to rate each alternative.

Solution Each factor and alternative is rated on a 0 to 100 relative scale. The factor "initial investment requirement" is considered the most important since the current system (alternative 3) is already owned. Therefore, factor 2 is assigned a score of 100. Other factors are assigned values relative to the 100 score as shown in Table 20.3. The total for all factors (450) is divided into each score to compute the weights w_j in Table 20.4.

The scores determined for the three alternatives (Table 20.3) are again relative to the best alternative for each factor using a 0 to 100 scale. For example, software availability and maintainability (factor 5) is judged best using a contract vendor (alternative 2); a score of 100 is assigned. However, the current system is very poor since it uses an outdated programming language; score assigned is 10.

TABLE 20.4 Weighted ratings for three alternatives using the weighted evaluation method

Factor i	Normalized weight w_j	Weighted ratings, V_i		
		Alternative 1	Alternative 2	Alternative 3
1	0.11	8.3	5.5	11.0
2	0.22	13.2	16.5	22.0
3	0.20	10.0	20.0	4.0
4	0.18	18.0	16.2	7.2
5	0.11	8.3	11.0	1.1
6	0.18	18.0	18.0	13.5
Totals	1.00	75.8	87.2	58.8

Factors: 1. Payback period
2. Initial investment requirement
3. Response time
4. User interface
5. Software availability and maintainability
6. Customer service

Table 20.4 presents the weighted ratings and totals using Eq. (20.1). Alternative 2 is clearly the indicated choice.

Comment   Note how the economic results you have learned can be incorporated into the multiple-criteria evaluation using this method. Of course, more or fewer economic measures can be included and their weights may vary from being highly important to virtually meaningless relative to other factors.

Analytic Hierarchy Process (AHP): Scoring techniques do not account for the inconsistency present in human judgment when several complex factors are considered and their relative importance quantified. This inconsistency may cause the chosen evaluation method to incorrectly reflect decision-maker preferences. Saaty [14] developed the AHP to remove some of the inconsistency between factors by establishing a hierarchy of decision factors and a procedure for pairwise comparison at each level of the hierarchy. Consistency checks are performed before the final results are computed for each alternative. The technique is applied to large project evaluations in areas such as financial systems, manufacturing systems, information management systems, plant construction and location decisions, etc.

Figure 20.4 shows a possible AHF hierarchy for the factors and alternatives used in Example 20.3. At the factor level a pairwise comparison results in a criterion weight with consistency considered for each factor. At the alternative level, the results

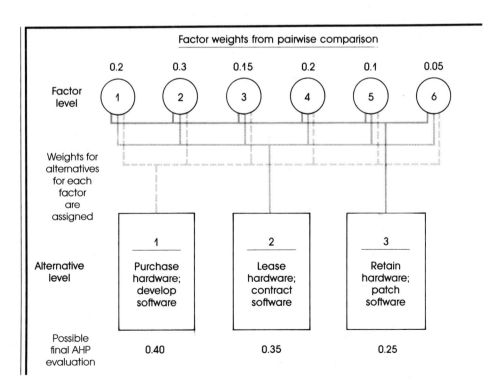

Figure 20.4
Example of the hierarchy used to evaluate multiple factors by the analytic hierarchy process (AHP).

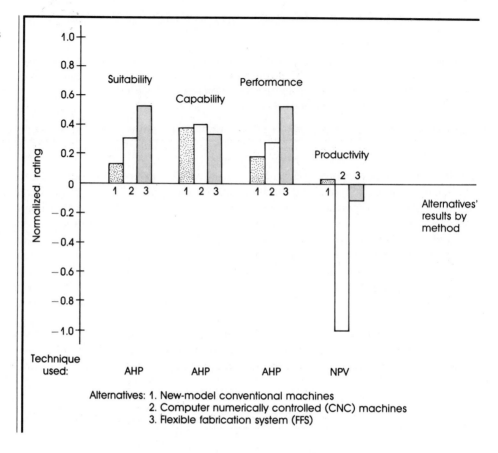

Figure 20.5 Evaluation results of three alternatives using the AHP and NPV methods.

Alternatives: 1. New-model conventional machines
2. Computer numerically controlled (CNC) machines
3. Flexible fabrication system (FFS)

of pairwise comparisons for each alternative are multiplied by factor weights and a final weighted evaluation is computed for each alternative. The largest value indicates the best alternative. See Frazelle [12] for a good example of the AHP applied to material-handling systems.

Combination of techniques such as AHP and engineering economy tools offer decision makers ways to see the differences in alternatives based upon several factors. For example, three alternatives for improved fabrication systems were evaluated using the factors listed in Table 20.2 under a strategic-level investment program. The first three factors—suitability, capability, and performance—were evaluated by the AHP technique using the decomposition terms as factors. The fourth factor, productivity, was measured using NPV (or PW) analysis. The results were normalized to range from −1 to +1 for each alternative. The results (Fig. 20.5) indicate that alternative 3 may be poor in economic terms alone, but it is excellent in terms of suitability and performance. Depending upon the relative importance of the factors, the decision maker can make a much better-informed decision using these multiple-criteria techniques than with one technique which measures a single or a few factors. See Troxler [15] for a thorough explanation of combining classical economic analysis, AHP, and other multiple-criteria techniques.

Probs. 20.9 to 20.15

20.5 Use of Sensitivity Analysis in Large Investment Decisions

The accuracy of cost and revenue estimates is usually poor for large business ventures which include strategic investment decisions like those discussed in this chapter. This is especially true during the specification and preliminary design stages. However, it is common for an economic analysis to be presented to management early in the formulation of the program and its associated projects. An expanded version of the sensitivity analysis approach presented in Secs. 19.1 to 19.3 is helpful in determining the effect of factors upon economic measures such as NPV, ROR (rate of return), or EUAW.

The amounts for selected economic factors are varied as much as 50% below and above the reasonable estimate to determine the effect upon the NPV, ROR, or EUAW measure. Some factors will affect the measure more than others. Consider the following information which uses the ROR as the economic measure.

Change, %	Annual revenue ($ thousands)	Rate of return, %	Annual labor ($ thousands)	Rate of return, %
−50	500	10.0	200	17.5
−25	750	14.0	300	17.75
0	1,000	18.0	400	18.0
+25	1,250	24.0	500	18.25
+50	1,500	30.0	600	18.5

It is clear that the return is much more sensitive to revenue than to labor costs. This indicates that when the product is in production, more attention should be paid to marketing, sales, quality, on-time delivery, and other revenue-related matters than on the strict control of labor costs.

Once all computations are complete, a composite graph of the factors and the economic measure summarizes the results. Figure 20.6 was constructed by a spreadsheet software package for a large investment program in manufacturing. The graph shows the variation in ROR for several factors with variations of ±50% on each factor. Clearly, ROR is much more sensitive to revenue than any other factor.

Recomputation of the measure, especially ROR, is too difficult for each 5 or 10% above and below the reasonable estimate for each sensitized factor. Spreadsheet software designed for such purposes makes the job much easier. Additionally, graphical output simplifies the interpretation of the numerical results. Appendix D presents an overview of a comprehensive, microprocessor-based software package that is useful in business planning, cash-flow estimation, and sensitivity analysis. The sensitivity results are generated in tabular and graphical forms for use by analysts and managers.

Probs. 20.16 to 20.20

Figure 20.6
Sensitivity of
return to several
factors as
computed and
graphed by a
spreadsheet
package. (*Adapted
with permission
from U.S.
Department of
Commerce, Office
of Productivity,
Technology, and
Innovation [16]*).

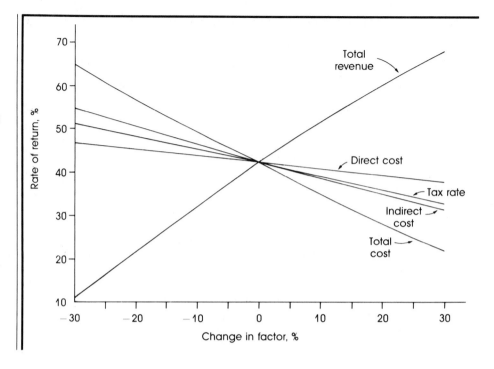

References

1. S. R. Rosenthal, "A Survey of Factory Automation in the U.S.," *Operations Management Review*, vol. 2, no. 2, 1984, pp. 3–14.
2. J. R. Meredith and N. C. Suresh, "Justification Techniques for Advanced Manufacturing Technologies," *International Journal of Production Research*, vol. 24, no. 5, 1986, pp. 1043–1058.
3. J. R. Meredith, ed., *Justifying New Manufacturing Technology*, Institute of Industrial Engineers, Atlanta, GA, 1986.
4. L. T. Blank, "The Changing Scene of Economic Analysis for the Evaluation of Manufacturing System Design and Operation," *The Engineering Economist*, vol. 30, no. 3, 1985, pp. 57–61.
5. J. F. Lardner, "The Corporate Transition to Superior Manufacturing Performance," *Manufacturing Engineering*, vol. 91, no. 2, 1983, pp. 65–67.
6. W. R. Barr, "Risk Analysis and Capital Investment Evaluation," Proceedings, Institute of Industrial Engineers Conference, 1981, pp. 418–422.
7. R. S. Kaplan, "Must CIM be Justified by Faith Alone?," *Harvard Business Review*, vol. 64, no. 2, 1986, pp. 87–95.
8. S. Hunter, "Cost Justification: The Overhead Dilemma," Proceedings, Robots 9 Conference, vol. 2, 1985, pp. 67–82.
9. W. G. Sullivan, "Models IEs Can Use to Include Strategic, Non-monetary Factors in Automation Decisions," *Industrial Engineering*, vol. 18, no. 3, 1986, pp. 42–50.
10. J. W. Troxler and L. T. Blank, "Major Attributes Used in Manufacturing System Decision Support," *Journal of Cost Management for the Manufacturing Industry*, vol. 2, no. 1, 1988, pp. 55–58.

11. W. A. Sullivan, "Computer Integrated Manufacturing Program Justification," Autofact '85 Conference and CASA/SME Technical Paper MS85–1058.
12. E. Frazelle, "Suggested Techniques Enable Multi-criteria Evaluation of Material Handling Alternatives," *Industrial Engineering*, vol. 17, no. 2, 1985, pp. 42–48.
13. R. F. Huber, "Justification: Barrier to Competitive Manufacturing," *Production*, vol. 96, no. 3, 1985, pp. 46–51.
14. T. L. Saaty, *The Analytic Hierarchy Process*, McGraw-Hill Book Company, New York, 1980.
15. J. W. Troxler, "An Economic Analysis of Flexible Automation in Batch Manufacturing with Emphasis on Fabrication," Ph.D. dissertation, Texas A&M University, May 1987.
16. J. R. Heizer, "Financial Sensitivity Analysis Software for Business Strategic Planning," U.S. Department of Commerce, Office of Productivity, Technology, and Innovation, August 1987.

Problems

20.1 Some of the economic advantages which have been historically considered intangible for evaluation of manufacturing-oriented projects are: reduced work in process, improved product quality, reduced lead time, and others. Consider the evaluation of a major project in a service industry, such as the airline industry. What are some economic intangible functions that, if quantified, might assist in justifying the project?

20.2 Based only on the descriptions below, determine if the project might have operational, tactical, or strategic implications for the company. Give a reason for your answer. For each situation, mention at least one intangible factor that, if considered, may improve the thoroughness of an economic evaluation.

(a) An estimated $2 million investment in software is planned to develop a computer interface system between the financial databases in 5 of the 10 operating divisions of the corporation. Interfaces between different-vendor computers at the sites will be necessary. An anticipated planning horizon of 8 years and a total of $350,000 in annual maintenance charges are included.

(b) A corporatewide executive education program in the benefits of improved product quality has been suggested. This first program, with an estimated initial cost of $450,000 may be followed by a program for middle management and plant supervision. A planning horizon of 20 years is to be used in this project. The mission is to assist corporate leaders in the understanding and promoting of quality mindedness. Members of the board of directors may attend selected course sessions.

(c) A new $1.5 million numerically controlled metal-removal center is being considered for a particular government contract. A 5-year life is anticipated before it is technologically obsolete. An information system connection with the currently owned machining center is planned for the project.

20.3 Fifteen years ago Perry International Farm Equipment sold 50% of its product overseas. Today, international sales account for only 5% of the business. Fifteen years ago the current president was an engineer in the firm. He remembers performing an EUAW analysis for new production equipment for tractor assembly. The three 12-year-life plans considered first costs, annual operating costs, salvage values, and depreciation in an after-tax study. Plan *C* was selected and is the one still in place today.

The president now wants to develop and evaluate plans for the use of modern production techniques in tractor assembly that may boost company revenue from international sales *and* significantly reduce their costs of tractor manufacture.

Consider yourself an economic consultant to the president. Use the organization and economic elements in Table 20.1, and any you wish to add, to suggest revenue, cost, and cost-saving elements which could be considered in this evaluation and which the president most likely did not include in the 15-year-old study.

20.4 Roadrunner Explosives, Ltd., has a substantial investment planned. The Vice President for Finance asked two people to independently estimate the first cost, life, added revenue, and costs for two plans. The information below was submitted to the vice president for analysis. Use PW analysis at 20% per year before taxes to evaluate the two plans using each person's data. Is there a difference in the results of the analyses? Why?

Category	Plan A	Plan B
Person 1		
First cost	$500,000	$375,000
Annual operating cost	75,000	80,000
Additional annual revenue	150,000	130,000
Salvage value	50,000	37,000
Expected life	5 years	5 years
Person 2		
Initial investment	$650,000	$450,000
Major upgrade, year 5	200,000	350,000
Expected life	15 years	15 years
Annual operating cost		
Years 1–5	85,000	125,000
Years 6–15	65,000	125,000
Additional annual revenue	150,000	180,000
Annual cost reduction		
Inventory carrying cost	25,000	50,000
Computer time	30,000	15,000
Scrap and rework	80,000	120,000
Salvage value	Nil	Nil

20.5 A company plans to invest approximately $1 million in a project over a 10-year period including planning and equipment acquisition as shown below. The start-up year is 1992 with planning and acquisition cash flows estimated to follow the patterns detailed below. Compute (a) the overall dollar cost of the project, (b) the separate equivalent

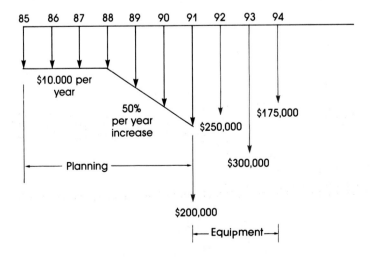

worths (PW and FW values) of the planning and the equipment expenditures at the end of year 1992, and (c) the total future-worth equivalent after all expenditures are completed. (d) Compare the answers in (a) and (b). Use $i = 18\%$ per year.

20.6 The state highway department has evaluated the "contract" versus "in-house development" options for future bridge-design projects. The contract option would have a much smaller annual cost with EUAW = $225,600 for all personnel and computer services supplied by the contractor. There is a mandatory 5-year agreement with the contractor for this service. If the current in-house design group is expanded and the necessary computer-aided design (CAD) systems are acquired, the analysis indicates that EUAW = $320,500 over a 10-year planning horizon. The EUAW analyses were performed using classical economic cost methods at MARR = 10% per year.

The department director intuitively feels that the decision to abandon the in-house design effort may have long-term negative consequences on the state's ability to design and build modern, safe, high-volume bridges for its citizens. Additionally, he feels that the economic analysis may be somewhat shortsighted in the costs (and benefits) which should be considered in the analyses. Discuss some of the benefits, cost savings, and additional costs which you feel may be included in this analysis. Categorize your factors as tangible and intangible for this situation.

20.7 Reread Prob. 16.28. What are some additional tangible and intangible benefits (and possibly some costs) which could be considered in the evaluation?

20.8 Reread Prob. 6.9. What are some tangible and intangible benefits which could be considered in the evaluation of the material-handling alternatives?

20.9 A team of three decision makers has made the following statements about the four factors to be used in a weighted evaluation method. Use them to compute the normalized factor weights for Eq. (20.1).

Factor	Comment
1. Flexibility	The most important factor
2. Safety	70% as important as uptime
3. Uptime	One-half as important as flexibility
4. Rate of return	Twice as important as safety

20.10 John has decided to use the weighted evaluation method to compare three methods of manufacturing a new watchband. The president and vice president have rated each of three factors in terms of importance to them for the alternatives, and John has placed an evaluation from 0 to 100 on each alternative for the three factors. John's ratings are as follows:

	Alternative		
Factor	1	2	3
Economic return	50	70	100
Performance of system	100	60	30
Technological competitiveness	100	40	50

Use the two sets of weights given below to evaluate the alternatives. Are the results the same for the president's and vice-president's ratings? Compare the outcomes (total value), and comment on the results for each officer of the company.

	Importance	
Factor	**President**	**Vice president**
Economic return	100	20
Performance of system	80	80
Technological competitiveness	20	100

20.11 In Example 20.3, use the *unweighted* evaluation method to choose the alternative with the largest total value. Did the factor weights change the selected alternative in Example 20.3 from the answer without the weights?

20.12 Your sister plans to purchase a new car. For three models she has evaluated the initial cost and has estimated annual costs for fuel and maintenance. She also evaluated the styling of each car for her role as a 30-year-old "new wave" engineering professional. List some additional factors (tangible and intangible) that she might use in the weighted evaluation method.

20.13 Dare Athletics has evaluated two proposals for weight lifting and exercise equipment. The present-worth analysis, at $i = 15\%$, of estimated incomes and costs resulted in

$$PW_A = \$420,500 \qquad PW_B = \$392,800$$

In addition to the economic measure, three additional factors have been given a relative importance rating of 0 to 100 by the manager and the lead trainer.

	Factor importance	
Factor	**Manager**	**Trainer**
Economic measure	100	80
Durability	35	10
Flexibility	20	100
Maintainability	20	10

Independently, you have rated on a relative scale of 0 to 1.0 each manufacturer's equipment for each factor. The economic measure factor was rated using the PW values.

Factor	**Proposal A**	**Proposal B**
Economic measure	1.00	0.90
Durability	0.35	1.00
Flexibility	1.00	0.90
Maintainability	0.25	1.00

Which proposal is selected using each of the following methods?
(a) Present-worth value.
(b) Unweighted evaluation method.
(c) Weighted evaluations of the manager.
(d) Weighted evaluations of the lead trainer.

20.14 Consider a major decision that you are currently making, or have made recently, that has economic and other factors which are important to the evaluation of two or more alternatives. Describe in writing the alternatives and list all of the factors to evaluate. Rate these factors on a relative scale of 0 to 100 so that they may be utilized in the weighted evaluation method.

20.15 In Prob. 20.10 the president and vice president are not consistent in their rating of the three factors used to evaluate watchband manufacturing alternatives. Assume you are a consultant to the company and are asked to assist in the selection of the best manufacturing method.

(a) What actions could you take to help the president and vice president in their alternative selection process using the alternative ratings in Prob. 20.10?

(b) Use the following relative ratings for each alternative and list your actions to select an alternative. Using the ratings of the president and vice president.

	Alternative		
Factor	1	2	3
Economic return	30	40	100
Performance of system	70	100	70
Technological competitiveness	100	80	90

(c) What do the different factor ratings tell you about the president and vice president?

20.16 Use the data collected by person 1 in Prob. 20.4 to graph the sensitivity of the PW value for each plan for the range -50 to $+100\%$ for each of the following factors: (a) first cost, (b) AOC, and (c) additional annual revenue.

20.17 Graph the sensitivity of the rate of return in Prob. 7.5 to a $\pm 25\%$ change in each of the factors: (a) purchase price, and (b) selling price.

20.18 Graph the sensitivity of the EUAW value for machine A in Prob. 6.6 to $\pm 50\%$ variation in the factors: (a) first cost, (b) AOC, and (c) MARR.

20.19 For process M in Prob. 6.8, graph the sensitivity of EUAW for the indicated range of each factor.

Factor	Range
(a) First cost	-30 to $+100\%$
(b) AOC	-30 to $+30\%$
(c) Annual revenue	-50 to $+25\%$

20.20 Graph the sensitivity of present worth to a $\pm 30\%$ change in the following factors for the bond purchase in Prob. 11.5: (a) purchase price, (b) dividend rate, and (c) your required nominal rate.

APPENDIXES

A   Interest Factors for Discrete Compounding, Discrete Cash-Flow
B   The Modified Accelerated Cost-Recovery System (MACRS) for Capital Recovery
C   Basics of Accounting and Cost Allocation
D   Sensitivity Analysis for Business Planning
E   Computer Applications
F   Answers to Selected Problems

APPENDIX

Interest Factors for Discrete Compounding, Discrete Cash Flow

A

Tabulated here are interest factor values when interest is compounded once each period. If interest is compounded more or less frequently, refer to Chap. 3. The computational forms of the factors are given here.

Factor	Notation	Formula
Single-payment compound amount	$(F/P, i\%, n)$	$(1 + i)^n$
Single-payment present worth	$(P/F, i\%, n)$	$\dfrac{1}{(1 + i)^n}$
Sinking fund	$(A/F, i\%, n)$	$\dfrac{i}{(1 + i)^n - 1}$
Uniform-series compound amount	$(F/A, i\%, n)$	$\dfrac{(1 + i)^n - 1}{i}$
Capital recovery	$(A/P, i\%, n)$	$\dfrac{i(1 + i)^n}{(1 + i)^n - 1}$
Uniform-series present worth	$(P/A, i\%, n)$	$\dfrac{(1 + i)^n - 1}{i(1 + i)^n}$

A few useful computational relations between factors are given below. (The $i\%$ and n are omitted from the notation when possible, simply for the sake of brevity.)

$$(P/F) = \frac{1}{(F/P)} \qquad (F/A) = \frac{1}{(A/F)} \qquad (P/A) = \frac{1}{(A/P)}$$

$$(P/A) = (F/A)(P/F) \qquad (A/P) = (A/F)(F/P)$$

$$(P/A) = \sum_{j=1}^{n} (P/F, i\%, j) \qquad F/A = \sum_{j=1}^{n} (F/P), i\%, j)$$

Tables A-31 to A-38 present factors that convert a uniform gradient (G) of $1 per period to a present worth or equivalent uniform annual series, respectively. Computational formulas are as follows:

Factor	Notation	Formula
Uniform-gradient present worth	$(P/G, i\%, n)$	$\dfrac{1}{i}\left[\dfrac{(1+i)^n - 1}{i(1+i)^n} - \dfrac{n}{(1+i)^n}\right]$
Uniform-gradient annual series	$(A/G, i\%, n)$	$\dfrac{1}{i} - \dfrac{n}{(1+i)^n - 1}$

Useful gradient relations are

$$(P/G) = (A/G)(P/A) \qquad (A/G) = (P/G)(A/P)$$

TABLE A - 1

DISCRETE CASH FLOW
0.50%   DISCRETE COMPOUND INTEREST FACTORS

	SINGLE PAYMENTS		UNIFORM SERIES PAYMENTS				
N	COMPOUND AMOUNT F/P	PRESENT WORTH P/F	SINKING FUND A/F	COMPOUND AMOUNT F/A	CAPITAL RECOVERY A/P	PRESENT WORTH P/A	N
1	1.0050	0.9950	1.00000	1.0000	1.00500	0.9950	1
2	1.0100	0.9901	0.49875	2.0050	0.50375	1.9851	2
3	1.0151	0.9851	0.33167	3.0150	0.33667	2.9702	3
4	1.0202	0.9802	0.24813	4.0301	0.25313	3.9505	4
5	1.0253	0.9754	0.19801	5.0503	0.20301	4.9259	5
6	1.0304	0.9705	0.16460	6.0755	0.16960	5.8964	6
7	1.0355	0.9657	0.14073	7.1059	0.14573	6.8621	7
8	1.0407	0.9609	0.12283	8.1414	0.12783	7.8230	8
9	1.0459	0.9561	0.10891	9.1821	0.11391	8.7791	9
10	1.0511	0.9513	0.09777	10.2280	0.10277	9.7304	10
11	1.0564	0.9466	0.08866	11.2792	0.09366	10.6770	11
12	1.0617	0.9419	0.08107	12.3356	0.08607	11.6189	12
13	1.0670	0.9372	0.07464	13.3972	0.07964	12.5562	13
14	1.0723	0.9326	0.06914	14.4642	0.07414	13.4887	14
15	1.0777	0.9279	0.06436	15.5365	0.06936	14.4166	15
16	1.0831	0.9233	0.06019	16.6142	0.06519	15.3399	16
17	1.0885	0.9187	0.05651	17.6973	0.06151	16.2586	17
18	1.0939	0.9141	0.05323	18.7858	0.05823	17.1728	18
19	1.0994	0.9096	0.05030	19.8797	0.05530	18.0824	19
20	1.1049	0.9051	0.04767	20.9791	0.05267	18.9874	20
22	1.1160	0.8961	0.04311	23.1944	0.04811	20.7841	22
24	1.1272	0.8872	0.03932	25.4320	0.04432	22.5629	24
25	1.1328	0.8828	0.03765	26.5591	0.04265	23.4456	25
26	1.1385	0.8784	0.03611	27.6919	0.04111	24.3240	26
28	1.1499	0.8697	0.03336	29.9745	0.03836	26.0677	28
30	1.1614	0.8610	0.03098	32.2800	0.03598	27.7941	30
32	1.1730	0.8525	0.02889	34.6086	0.03389	29.5033	32
34	1.1848	0.8440	0.02706	36.9606	0.03206	31.1955	34
35	1.1907	0.8398	0.02622	38.1454	0.03122	32.0354	35
36	1.1967	0.8356	0.02542	39.3361	0.03042	32.8710	36
38	1.2087	0.8274	0.02396	41.7354	0.02896	34.5299	38
40	1.2208	0.8191	0.02265	44.1588	0.02765	36.1722	40
45	1.2516	0.7990	0.01987	50.3242	0.02487	40.2072	45
50	1.2832	0.7793	0.01765	56.6452	0.02265	44.1428	50
55	1.3156	0.7601	0.01584	63.1258	0.02084	47.9814	55
60	1.3489	0.7414	0.01433	69.7700	0.01933	51.7256	60
65	1.3829	0.7231	0.01306	76.5821	0.01806	55.3775	65
70	1.4178	0.7053	0.01197	83.5661	0.01697	58.9394	70
75	1.4536	0.6879	0.01102	90.7265	0.01602	62.4136	75
80	1.4903	0.6710	0.01020	98.0677	0.01520	65.8023	80
85	1.5280	0.6545	0.00947	105.5943	0.01447	69.1075	85
90	1.5666	0.6383	0.00883	113.3109	0.01383	72.3313	90
95	1.6061	0.6226	0.00825	121.2224	0.01325	75.4757	95
100	1.6467	0.6073	0.00773	129.3337	0.01273	78.5426	100

TABLE A - 2

DISCRETE CASH FLOW
1.00%   DISCRETE COMPOUND INTEREST FACTORS

	SINGLE PAYMENTS			UNIFORM SERIES PAYMENTS				
N	COMPOUND AMOUNT F/P	PRESENT WORTH P/F	SINKING FUND A/F	COMPOUND AMOUNT F/A	CAPITAL RECOVERY A/P	PRESENT WORTH P/A	N
1	1.0100	0.9901	1.00000	1.0000	1.01000	0.9901	1
2	1.0201	0.9803	0.49751	2.0100	0.50751	1.9704	2
3	1.0303	0.9706	0.33002	3.0301	0.34002	2.9410	3
4	1.0406	0.9610	0.24628	4.0604	0.25628	3.9020	4
5	1.0510	0.9515	0.19604	5.1010	0.20604	4.8534	5
6	1.0615	0.9420	0.16255	6.1520	0.17255	5.7955	6
7	1.0721	0.9327	0.13863	7.2135	0.14863	6.7282	7
8	1.0829	0.9235	0.12069	8.2857	0.13069	7.6517	8
9	1.0937	0.9143	0.10674	9.3685	0.11674	8.5660	9
10	1.1046	0.9053	0.09558	10.4622	0.10558	9.4713	10
11	1.1157	0.8963	0.08645	11.5668	0.09645	10.3676	11
12	1.1268	0.8874	0.07885	12.6825	0.08885	11.2551	12
13	1.1381	0.8787	0.07241	13.8093	0.08241	12.1337	13
14	1.1495	0.8700	0.06690	14.9474	0.07690	13.0037	14
15	1.1610	0.8613	0.06212	16.0969	0.07212	13.8651	15
16	1.1726	0.8528	0.05794	17.2579	0.06794	14.7179	16
17	1.1843	0.8444	0.05426	18.4304	0.06426	15.5623	17
18	1.1961	0.8360	0.05098	19.6147	0.06098	16.3983	18
19	1.2081	0.8277	0.04805	20.8109	0.05805	17.2260	19
20	1.2202	0.8195	0.04542	22.0190	0.05542	18.0456	20
22	1.2447	0.8034	0.04086	24.4716	0.05086	19.6604	22
24	1.2697	0.7876	0.03707	26.9735	0.04707	21.2434	24
25	1.2824	0.7798	0.03541	28.2432	0.04541	22.0232	25
26	1.2953	0.7720	0.03387	29.5256	0.04387	22.7952	26
28	1.3213	0.7568	0.03112	32.1291	0.04112	24.3164	28
30	1.3478	0.7419	0.02875	34.7849	0.03875	25.8077	30
32	1.3749	0.7273	0.02667	37.4941	0.03667	27.2696	32
34	1.4026	0.7130	0.02484	40.2577	0.03484	28.7027	34
35	1.4166	0.7059	0.02400	41.6603	0.03400	29.4086	35
36	1.4308	0.6989	0.02321	43.0769	0.03321	30.1075	36
38	1.4595	0.6852	0.02176	45.9527	0.03176	31.4847	38
40	1.4889	0.6717	0.02046	48.8864	0.03046	32.8347	40
45	1.5648	0.6391	0.01771	56.4811	0.02771	36.0945	45
50	1.6446	0.6080	0.01551	64.4632	0.02551	39.1961	50
55	1.7285	0.5785	0.01373	72.8525	0.02373	42.1472	55
60	1.8167	0.5504	0.01224	81.6697	0.02224	44.9550	60
65	1.9094	0.5237	0.01100	90.9366	0.02100	47.6266	65
70	2.0068	0.4983	0.00993	100.6763	0.01993	50.1685	70
75	2.1091	0.4741	0.00902	110.9128	0.01902	52.5871	75
80	2.2167	0.4511	0.00822	121.6715	0.01822	54.8882	80
85	2.3298	0.4292	0.00752	132.9790	0.01752	57.0777	85
90	2.4486	0.4084	0.00690	144.8633	0.01690	59.1609	90
95	2.5735	0.3886	0.00636	157.3538	0.01636	61.1430	95
100	2.7048	0.3697	0.00587	170.4814	0.01587	63.0289	100

TABLE A - 3

DISCRETE CASH FLOW
1.50% DISCRETE COMPOUND INTEREST FACTORS

	SINGLE PAYMENTS			UNIFORM SERIES PAYMENTS			
N	COMPOUND AMOUNT F/P	PRESENT WORTH P/F	SINKING FUND A/F	COMPOUND AMOUNT F/A	CAPITAL RECOVERY A/P	PRESENT WORTH P/A	N
1	1.0150	0.9852	1.00000	1.0000	1.01500	0.9852	1
2	1.0302	0.9707	0.49628	2.0150	0.51128	1.9559	2
3	1.0457	0.9563	0.32838	3.0452	0.34338	2.9122	3
4	1.0614	0.9422	0.24444	4.0909	0.25944	3.8544	4
5	1.0773	0.9283	0.19409	5.1523	0.20909	4.7826	5
6	1.0934	0.9145	0.16053	6.2296	0.17553	5.6972	6
7	1.1098	0.9010	0.13656	7.3230	0.15156	6.5982	7
8	1.1265	0.8877	0.11858	8.4328	0.13358	7.4859	8
9	1.1434	0.8746	0.10461	9.5593	0.11961	8.3605	9
10	1.1605	0.8617	0.09343	10.7027	0.10843	9.2222	10
11	1.1779	0.8489	0.08429	11.8633	0.09929	10.0711	11
12	1.1956	0.8364	0.07668	13.0412	0.09168	10.9075	12
13	1.2136	0.8240	0.07024	14.2368	0.08524	11.7315	13
14	1.2318	0.8118	0.06472	15.4504	0.07972	12.5434	14
15	1.2502	0.7999	0.05994	16.6821	0.07494	13.3432	15
16	1.2690	0.7880	0.05577	17.9324	0.07077	14.1313	16
17	1.2880	0.7764	0.05208	19.2014	0.06708	14.9076	17
18	1.3073	0.7649	0.04881	20.4894	0.06381	15.6726	18
19	1.3270	0.7536	0.04588	21.7967	0.06088	16.4262	19
20	1.3469	0.7425	0.04325	23.1237	0.05825	17.1686	20
22	1.3876	0.7207	0.03870	25.8376	0.05370	18.6208	22
24	1.4295	0.6995	0.03492	28.6335	0.04992	20.0304	24
25	1.4509	0.6892	0.03326	30.0630	0.04826	20.7196	25
26	1.4727	0.6790	0.03173	31.5140	0.04673	21.3986	26
28	1.5172	0.6591	0.02900	34.4815	0.04400	22.7267	28
30	1.5631	0.6398	0.02664	37.5387	0.04164	24.0158	30
32	1.6103	0.6210	0.02458	40.6883	0.03958	25.2671	32
34	1.6590	0.6028	0.02276	43.9331	0.03776	26.4817	34
35	1.6839	0.5939	0.02193	45.5921	0.03693	27.0756	35
36	1.7091	0.5851	0.02115	47.2760	0.03615	27.6607	36
38	1.7608	0.5679	0.01972	50.7199	0.03472	28.8051	38
40	1.8140	0.5513	0.01843	54.2679	0.03343	29.9158	40
45	1.9542	0.5117	0.01572	63.6142	0.03072	32.5523	45
50	2.1052	0.4750	0.01357	73.6828	0.02857	34.9997	50
55	2.2679	0.4409	0.01183	84.5296	0.02683	37.2715	55
60	2.4432	0.4093	0.01039	96.2147	0.02539	39.3803	60
65	2.6320	0.3799	0.00919	108.8028	0.02419	41.3378	65
70	2.8355	0.3527	0.00817	122.3638	0.02317	43.1549	70
75	3.0546	0.3274	0.00730	136.9728	0.02230	44.8416	75
80	3.2907	0.3039	0.00655	152.7109	0.02155	46.4073	80
85	3.5450	0.2821	0.00589	169.6652	0.02089	47.8607	85
90	3.8189	0.2619	0.00532	187.9299	0.02032	49.2099	90
95	4.1141	0.2431	0.00482	207.6061	0.01982	50.4622	95
100	4.4320	0.2256	0.00437	228.8030	0.01937	51.6247	100

TABLE A - 4

DISCRETE CASH FLOW
2.00% DISCRETE COMPOUND INTEREST FACTORS

	SINGLE PAYMENTS			UNIFORM SERIES PAYMENTS			
N	COMPOUND AMOUNT F/P	PRESENT WORTH P/F	SINKING FUND A/F	COMPOUND AMOUNT F/A	CAPITAL RECOVERY A/P	PRESENT WORTH P/A	N
1	1.0200	0.9804	1.00000	1.0000	1.02000	0.9804	1
2	1.0404	0.9612	0.49505	2.0200	0.51505	1.9416	2
3	1.0612	0.9423	0.32675	3.0604	0.34675	2.8839	3
4	1.0824	0.9238	0.24262	4.1216	0.26262	3.8077	4
5	1.1041	0.9057	0.19216	5.2040	0.21216	4.7135	5
6	1.1262	0.8880	0.15853	6.3081	0.17853	5.6014	6
7	1.1487	0.8706	0.13451	7.4343	0.15451	6.4720	7
8	1.1717	0.8535	0.11651	8.5830	0.13651	7.3255	8
9	1.1951	0.8368	0.10252	9.7546	0.12252	8.1622	9
10	1.2190	0.8203	0.09133	10.9497	0.11133	8.9826	10
11	1.2434	0.8043	0.08218	12.1687	0.10218	9.7868	11
12	1.2682	0.7885	0.07456	13.4121	0.09456	10.5753	12
13	1.2936	0.7730	0.06812	14.6803	0.08812	11.3484	13
14	1.3195	0.7579	0.06260	15.9739	0.08260	12.1062	14
15	1.3459	0.7430	0.05783	17.2934	0.07783	12.8493	15
16	1.3728	0.7284	0.05365	18.6393	0.07365	13.5777	16
17	1.4002	0.7142	0.04997	20.0121	0.06997	14.2919	17
18	1.4282	0.7002	0.04670	21.4123	0.06670	14.9920	18
19	1.4568	0.6864	0.04378	22.8406	0.06378	15.6785	19
20	1.4859	0.6730	0.04116	24.2974	0.06116	16.3514	20
22	1.5460	0.6468	0.03663	27.2990	0.05663	17.6580	22
24	1.6084	0.6217	0.03287	30.4219	0.05287	18.9139	24
25	1.6406	0.6095	0.03122	32.0303	0.05122	19.5235	25
26	1.6734	0.5976	0.02970	33.6709	0.04970	20.1210	26
28	1.7410	0.5744	0.02699	37.0512	0.04699	21.2813	28
30	1.8114	0.5521	0.02465	40.5681	0.04465	22.3965	30
32	1.8845	0.5306	0.02261	44.2270	0.04261	23.4683	32
34	1.9607	0.5100	0.02082	48.0338	0.04082	24.4986	34
35	1.9999	0.5000	0.02000	49.9945	0.04000	24.9986	35
36	2.0399	0.4902	0.01923	51.9944	0.03923	25.4888	36
38	2.1223	0.4712	0.01782	56.1149	0.03782	26.4406	38
40	2.2080	0.4529	0.01656	60.4020	0.03656	27.3555	40
45	2.4379	0.4102	0.01391	71.8927	0.03391	29.4902	45
50	2.6916	0.3715	0.01182	84.5794	0.03182	31.4236	50
55	2.9717	0.3365	0.01014	98.5865	0.03014	33.1748	55
60	3.2810	0.3048	0.00877	114.0515	0.02877	34.7609	60
65	3.6225	0.2761	0.00763	131.1262	0.02763	36.1975	65
70	3.9996	0.2500	0.00667	149.9779	0.02667	37.4986	70
75	4.4158	0.2265	0.00586	170.7918	0.02586	38.6771	75
80	4.8754	0.2051	0.00516	193.7720	0.02516	39.7445	80
85	5.3829	0.1858	0.00456	219.1439	0.02456	40.7113	85
90	5.9431	0.1683	0.00405	247.1567	0.02405	41.5869	90
95	6.5617	0.1524	0.00360	278.0850	0.02360	42.3800	95
100	7.2446	0.1380	0.00320	312.2323	0.02320	43.0984	100

TABLE A - 5

DISCRETE CASH FLOW
3.00% DISCRETE COMPOUND INTEREST FACTORS

	SINGLE PAYMENTS		UNIFORM SERIES PAYMENTS				
N	COMPOUND AMOUNT F/P	PRESENT WORTH P/F	SINKING FUND A/F	COMPOUND AMOUNT F/A	CAPITAL RECOVERY A/P	PRESENT WORTH P/A	N
1	1.0300	0.9709	1.00000	1.0000	1.03000	0.9709	1
2	1.0609	0.9426	0.49261	2.0300	0.52261	1.9135	2
3	1.0927	0.9151	0.32353	3.0909	0.35353	2.8286	3
4	1.1255	0.8885	0.23903	4.1836	0.26903	3.7171	4
5	1.1593	0.8626	0.18835	5.3091	0.21835	4.5797	5
6	1.1941	0.8375	0.15460	6.4684	0.18460	5.4172	6
7	1.2299	0.8131	0.13051	7.6625	0.16051	6.2303	7
8	1.2668	0.7894	0.11246	8.8923	0.14246	7.0197	8
9	1.3048	0.7664	0.09843	10.1591	0.12843	7.7861	9
10	1.3439	0.7441	0.08723	11.4639	0.11723	8.5302	10
11	1.3842	0.7224	0.07808	12.8078	0.10808	9.2526	11
12	1.4258	0.7014	0.07046	14.1920	0.10046	9.9540	12
13	1.4685	0.6810	0.06403	15.6178	0.09403	10.6350	13
14	1.5126	0.6611	0.05853	17.0863	0.08853	11.2961	14
15	1.5580	0.6419	0.05377	18.5989	0.08377	11.9379	15
16	1.6047	0.6232	0.04961	20.1569	0.07961	12.5611	16
17	1.6528	0.6050	0.04595	21.7616	0.07595	13.1661	17
18	1.7024	0.5874	0.04271	23.4144	0.07271	13.7535	18
19	1.7535	0.5703	0.03981	25.1169	0.06981	14.3238	19
20	1.8061	0.5537	0.03722	26.8704	0.06722	14.8775	20
22	1.9161	0.5219	0.03275	30.5368	0.06275	15.9369	22
24	2.0328	0.4919	0.02905	34.4265	0.05905	16.9355	24
25	2.0938	0.4776	0.02743	36.4593	0.05743	17.4131	25
26	2.1566	0.4637	0.02594	38.5530	0.05594	17.8768	26
28	2.2879	0.4371	0.02329	42.9309	0.05329	18.7641	28
30	2.4273	0.4120	0.02102	47.5754	0.05102	19.6004	30
32	2.5751	0.3883	0.01905	52.5028	0.04905	20.3888	32
34	2.7319	0.3660	0.01732	57.7302	0.04732	21.1318	34
35	2.8139	0.3554	0.01654	60.4621	0.04654	21.4872	35
36	2.8983	0.3450	0.01580	63.2759	0.04580	21.8323	36
38	3.0748	0.3252	0.01446	69.1594	0.04446	22.4925	38
40	3.2620	0.3066	0.01326	75.4013	0.04326	23.1148	40
45	3.7816	0.2644	0.01079	92.7199	0.04079	24.5187	45
50	4.3839	0.2281	0.00887	112.7969	0.03887	25.7298	50
55	5.0821	0.1968	0.00735	136.0716	0.03735	26.7744	55
60	5.8916	0.1697	0.00613	163.0534	0.03613	27.6756	60
65	6.8300	0.1464	0.00515	194.3328	0.03515	28.4529	65
70	7.9178	0.1263	0.00434	230.5941	0.03434	29.1234	70
75	9.1789	0.1089	0.00367	272.6309	0.03367	29.7018	75
80	10.6409	0.0940	0.00311	321.3630	0.03311	30.2008	80
85	12.3357	0.0811	0.00265	377.8570	0.03265	30.6312	85
90	14.3005	0.0699	0.00226	443.3489	0.03226	31.0024	90
95	16.5782	0.0603	0.00193	519.2720	0.03193	31.3227	95
100	19.2186	0.0520	0.00165	607.2877	0.03165	31.5989	100

TABLE A - 6

DISCRETE CASH FLOW
4.00% DISCRETE COMPOUND INTEREST FACTORS

	SINGLE PAYMENTS			UNIFORM SERIES PAYMENTS			
N	COMPOUND AMOUNT F/P	PRESENT WORTH P/F	SINKING FUND A/F	COMPOUND AMOUNT F/A	CAPITAL RECOVERY A/P	PRESENT WORTH P/A	N
1	1.0400	0.9615	1.00000	1.000	1.04000	0.9615	1
2	1.0816	0.9246	0.49020	2.040	0.53020	1.8861	2
3	1.1249	0.8890	0.32035	3.122	0.36035	2.7751	3
4	1.1699	0.8548	0.23549	4.246	0.27549	3.6299	4
5	1.2167	0.8219	0.18463	5.416	0.22463	4.4518	5
6	1.2653	0.7903	0.15076	6.633	0.19076	5.2421	6
7	1.3159	0.7599	0.12661	7.898	0.16661	6.0021	7
8	1.3686	0.7307	0.10853	9.214	0.14853	6.7327	8
9	1.4233	0.7026	0.09449	10.583	0.13449	7.4353	9
10	1.4802	0.6756	0.08329	12.006	0.12329	8.1109	10
11	1.5395	0.6496	0.07415	13.486	0.11415	8.7605	11
12	1.6010	0.6246	0.06655	15.026	0.10655	9.3851	12
13	1.6651	0.6006	0.06014	16.627	0.10014	9.9856	13
14	1.7317	0.5775	0.05467	18.292	0.09467	10.5631	14
15	1.8009	0.5553	0.04994	20.024	0.08994	11.1184	15
16	1.8730	0.5339	0.04582	21.825	0.08582	11.6523	16
17	1.9479	0.5134	0.04220	23.698	0.08220	12.1657	17
18	2.0258	0.4936	0.03899	25.645	0.07899	12.6593	18
19	2.1068	0.4746	0.03614	27.671	0.07614	13.1339	19
20	2.1911	0.4564	0.03358	29.778	0.07358	13.5903	20
22	2.3699	0.4220	0.02920	34.248	0.06920	14.4511	22
24	2.5633	0.3901	0.02559	39.083	0.06559	15.2470	24
25	2.6658	0.3751	0.02401	41.646	0.06401	15.6221	25
26	2.7725	0.3607	0.02257	44.312	0.06257	15.9828	26
28	2.9987	0.3335	0.02001	49.968	0.06001	16.6631	28
30	3.2434	0.3083	0.01783	56.085	0.05783	17.2920	30
32	3.5081	0.2851	0.01595	62.701	0.05595	17.8736	32
34	3.7943	0.2636	0.01431	69.858	0.05431	18.4112	34
35	3.9461	0.2534	0.01358	73.652	0.05358	18.6646	35
36	4.1039	0.2437	0.01289	77.598	0.05289	18.9083	36
38	4.4388	0.2253	0.01163	85.970	0.05163	19.3679	38
40	4.8010	0.2083	0.01052	95.026	0.05052	19.7928	40
45	5.8412	0.1712	0.00826	121.029	0.04826	20.7200	45
50	7.1067	0.1407	0.00655	152.667	0.04655	21.4822	50
55	8.6464	0.1157	0.00523	191.159	0.04523	22.1086	55
60	10.5196	0.0951	0.00420	237.991	0.04420	22.6235	60
65	12.7987	0.0781	0.00339	294.968	0.04339	23.0467	65
70	15.5716	0.0642	0.00275	364.290	0.04275	23.3945	70
75	18.9453	0.0528	0.00223	448.631	0.04223	23.6804	75
80	23.0498	0.0434	0.00181	551.245	0.04181	23.9154	80
85	28.0436	0.0357	0.00148	676.090	0.04148	24.1085	85
90	34.1193	0.0293	0.00121	827.983	0.04121	24.2673	90
95	41.5114	0.0241	0.00099	1012.785	0.04099	24.3978	95
100	50.5049	0.0198	0.00081	1237.624	0.04081	24.5050	100

TABLE A - 7

DISCRETE CASH FLOW
5.00%   DISCRETE COMPOUND INTEREST FACTORS

	SINGLE PAYMENTS			UNIFORM SERIES PAYMENTS			
N	COMPOUND AMOUNT F/P	PRESENT WORTH P/F	SINKING FUND A/F	COMPOUND AMOUNT F/A	CAPITAL RECOVERY A/P	PRESENT WORTH P/A	N
1	1.0500	0.9524	1.00000	1.000	1.05000	0.9524	1
2	1.1025	0.9070	0.48780	2.050	0.53780	1.8594	2
3	1.1576	0.8638	0.31721	3.152	0.36721	2.7232	3
4	1.2155	0.8227	0.23201	4.310	0.28201	3.5460	4
5	1.2763	0.7835	0.18097	5.526	0.23097	4.3295	5
6	1.3401	0.7462	0.14702	6.802	0.19702	5.0757	6
7	1.4071	0.7107	0.12282	8.142	0.17282	5.7864	7
8	1.4775	0.6768	0.10472	9.549	0.15472	6.4632	8
9	1.5513	0.6446	0.09069	11.027	0.14069	7.1078	9
10	1.6289	0.6139	0.07950	12.578	0.12950	7.7217	10
11	1.7103	0.5847	0.07039	14.207	0.12039	8.3064	11
12	1.7959	0.5568	0.06283	15.917	0.11283	8.8633	12
13	1.8856	0.5303	0.05646	17.713	0.10646	9.3936	13
14	1.9799	0.5051	0.05102	19.599	0.10102	9.8986	14
15	2.0789	0.4810	0.04634	21.579	0.09634	10.3797	15
16	2.1329	0.4581	0.04227	23.657	0.09227	10.8378	16
17	2.2920	0.4363	0.03870	25.840	0.08870	11.2741	17
18	2.4066	0.4155	0.03555	28.132	0.08555	11.6896	18
19	2.5270	0.3957	0.03275	30.539	0.08275	12.0853	19
20	2.6533	0.3769	0.03024	33.066	0.08024	12.4622	20
22	2.9253	0.3418	0.02597	38.505	0.07597	13.1630	22
24	3.2251	0.3101	0.02247	44.502	0.07247	13.7986	24
25	3.3864	0.2953	0.02095	47.727	0.07095	14.0939	25
26	3.5557	0.2812	0.01956	51.113	0.06956	14.3752	26
28	3.9201	0.2551	0.01712	58.403	0.06712	14.8981	28
30	4.3219	0.2314	0.01505	66.439	0.06505	15.3725	30
32	4.7649	0.2099	0.01328	75.299	0.06328	15.8027	32
34	5.2533	0.1904	0.01176	85.067	0.06176	16.1929	34
35	5.5160	0.1813	0.01107	90.320	0.06107	16.3742	35
36	5.7918	0.1727	0.01043	95.836	0.06043	16.5469	36
38	6.3855	0.1566	0.00928	107.710	0.05928	16.8679	38
40	7.0400	0.1420	0.00828	120.800	0.05828	17.1591	40
45	8.9850	0.1113	0.00626	159.700	0.05626	17.7741	45
50	11.4674	0.0872	0.00478	209.348	0.05478	18.2559	50
55	14.6356	0.0683	0.00367	272.713	0.05367	18.6335	55
60	18.6792	0.0535	0.00283	353.584	0.05283	18.9293	60
65	23.8399	0.0419	0.00219	456.798	0.05219	19.1611	65
70	30.4264	0.0329	0.00170	588.529	0.05170	19.3427	70
75	38.8327	0.0258	0.00132	756.654	0.05132	19.4850	75
80	49.5614	0.0202	0.00103	971.229	0.05103	19.5965	80
85	63.2544	0.0158	0.00080	1245.087	0.05080	19.6838	85
90	80.7304	0.0124	0.00063	1594.607	0.05063	19.7523	90
95	103.035	0.0097	0.00049	2040.694	0.05049	19.8059	95
100	131.501	0.0076	0.00038	2610.025	0.05038	19.8479	100

TABLE A - 8

DISCRETE CASH FLOW
6.00% DISCRETE COMPOUND INTEREST FACTORS

	SINGLE PAYMENTS		UNIFORM SERIES PAYMENTS				
N	COMPOUND AMOUNT F/P	PRESENT WORTH P/F	SINKING FUND A/F	COMPOUND AMOUNT F/A	CAPITAL RECOVERY A/P	PRESENT WORTH P/A	N
1	1.0600	0.9434	1.00000	1.000	1.06000	0.9434	1
2	1.1236	0.8900	0.48544	2.060	0.54544	1.8334	2
3	1.1910	0.8396	0.31411	3.184	0.37411	2.6730	3
4	1.2625	0.7921	0.22859	4.375	0.28859	3.4651	4
5	1.3382	0.7473	0.17740	5.637	0.23740	4.2124	5
6	1.4185	0.7050	0.14336	6.975	0.20336	4.9173	6
7	1.5036	0.6651	0.11914	8.394	0.17914	5.5824	7
8	1.5938	0.6274	0.10104	9.897	0.16104	6.2098	8
9	1.6895	0.5919	0.08702	11.491	0.14702	6.8017	9
10	1.7908	0.5584	0.07587	13.181	0.13587	7.3601	10
11	1.8983	0.5268	0.06679	14.972	0.12679	7.8869	11
12	2.0122	0.4970	0.05928	16.870	0.11928	8.3838	12
13	2.1329	0.4688	0.05296	18.882	0.11296	8.8527	13
14	2.2609	0.4423	0.04758	21.015	0.10758	9.2950	14
15	2.3966	0.4173	0.04296	23.276	0.10296	9.7122	15
16	2.5404	0.3936	0.03895	25.673	0.09895	10.1059	16
17	2.6928	0.3714	0.03544	28.213	0.09544	10.4773	17
18	2.8543	0.3503	0.03236	30.906	0.09236	10.8276	18
19	3.0256	0.3305	0.02962	33.760	0.08962	11.1581	19
20	3.2071	0.3118	0.02718	36.786	0.08718	11.4699	20
22	3.6035	0.2775	0.02305	43.392	0.08305	12.0416	22
24	4.0489	0.2470	0.01968	50.816	0.07968	12.5504	24
25	4.2919	0.2330	0.01823	54.865	0.07823	12.7834	25
26	4.5494	0.2198	0.01690	59.156	0.07690	13.0032	26
28	5.1117	0.1956	0.01459	68.528	0.07459	13.4062	28
30	5.7435	0.1741	0.01265	79.058	0.07265	13.7648	30
32	6.4534	0.1550	0.01100	90.890	0.07100	14.0840	32
34	7.2510	0.1379	0.00960	104.184	0.06960	14.3681	34
35	7.6861	0.1301	0.00897	111.435	0.06897	14.4982	35
36	8.1473	0.1227	0.00839	119.121	0.06839	14.6210	36
38	9.1543	0.1092	0.00736	135.904	0.06736	14.8460	38
40	10.2857	0.0972	0.00646	154.762	0.06646	15.0463	40
45	13.7646	0.0727	0.00470	212.744	0.06470	15.4558	45
50	18.4202	0.0543	0.00344	290.336	0.06344	15.7619	50
55	24.6503	0.0406	0.00254	394.172	0.06254	15.9905	55
60	32.9877	0.0303	0.00189	533.128	0.06188	16.1614	60
65	44.1450	0.0227	0.00139	719.083	0.06139	16.2891	65
70	59.0759	0.0169	0.00103	967.932	0.06103	16.3845	70
75	79.0569	0.0126	0.00077	1300.949	0.06077	16.4558	75
80	105.796	0.0095	0.00057	1746.600	0.06057	16.5091	80
85	141.579	0.0071	0.00043	2342.982	0.06043	16.5489	85
90	189.465	0.0053	0.00032	3141.075	0.06032	16.5787	90
95	253.546	0.0039	0.00024	4209.104	0.06024	16.6009	95
100	339.302	0.0029	0.00018	5638.368	0.06018	16.6175	100

TABLE A - 9

DISCRETE CASH FLOW
7.00% DISCRETE COMPOUND INTEREST FACTORS

| | SINGLE PAYMENTS | | | UNIFORM SERIES PAYMENTS | | | |
| | COMPOUND AMOUNT | PRESENT WORTH | SINKING FUND | COMPOUND AMOUNT | CAPITAL RECOVERY | PRESENT WORTH | |
N	F/P	P/F	A/F	F/A	A/P	P/A	N
1	1.0700	0.9346	1.00000	1.000	1.07000	0.9346	1
2	1.1449	0.8734	0.48309	2.070	0.55309	1.8080	2
3	1.2250	0.8163	0.31105	3.215	0.38105	2.6243	3
4	1.3108	0.7629	0.22523	4.440	0.29523	3.3872	4
5	1.4026	0.7130	0.17389	5.751	0.24389	4.1002	5
6	1.5007	0.6663	0.13980	7.153	0.20980	4.7665	6
7	1.6058	0.6227	0.11555	8.654	0.18555	5.3893	7
8	1.7182	0.5820	0.09747	10.260	0.16747	5.9713	8
9	1.8385	0.5439	0.08349	11.978	0.15349	6.5152	9
10	1.9672	0.5083	0.07238	13.816	0.14238	7.0236	10
11	2.1049	0.4751	0.06336	15.784	0.13336	7.4987	11
12	2.2522	0.4440	0.05590	17.889	0.12590	7.9427	12
13	2.4098	0.4150	0.04965	20.141	0.11965	8.3577	13
14	2.5785	0.3878	0.04434	22.550	0.11434	8.7455	14
15	2.7590	0.3624	0.03979	25.129	0.10979	9.1079	15
16	2.9522	0.3387	0.03586	27.888	0.10586	9.4466	16
17	3.1589	0.3166	0.03243	30.840	0.10243	9.7632	17
18	3.3799	0.2959	0.02941	33.999	0.09941	10.0591	18
19	3.6165	0.2765	0.02675	37.379	0.09675	10.3356	19
20	3.8697	0.2584	0.02439	40.995	0.09439	10.5940	20
22	4.4304	0.2257	0.02041	49.006	0.09041	11.0612	22
24	5.0724	0.1971	0.01719	58.177	0.08719	11.4693	24
25	5.4274	0.1842	0.01581	63.249	0.08581	11.6536	25
26	5.8074	0.1722	0.01456	68.676	0.08456	11.8258	26
28	6.6488	0.1504	0.01239	80.698	0.08239	12.1371	28
30	7.6123	0.1314	0.01059	94.461	0.08059	12.4090	30
32	8.7153	0.1147	0.00907	110.218	0.07907	12.6466	32
34	9.9781	0.1002	0.00780	128.259	0.07780	12.8540	34
35	10.6766	0.0937	0.00723	138.237	0.07723	12.9477	35
36	11.4239	0.0875	0.00672	148.913	0.07672	13.0352	36
38	13.0793	0.0765	0.00580	172.561	0.07580	13.1935	38
40	14.9745	0.0668	0.00501	199.635	0.07501	13.3317	40
45	21.0025	0.0476	0.00350	285.749	0.07350	13.6055	45
50	29.4570	0.0339	0.00246	406.529	0.07246	13.8007	50
55	41.3150	0.0242	0.00174	575.929	0.07174	13.9399	55
60	57.9464	0.0173	0.00123	813.520	0.07123	14.0392	60
65	81.2729	0.0123	0.00087	1146.755	0.07087	14.1099	65
70	113.989	0.0088	0.00062	1614.134	0.07062	14.1604	70
75	159.876	0.0063	0.00044	2269.657	0.07044	14.1964	75
80	224.234	0.0045	0.00031	3189.063	0.07031	14.2220	80
85	314.500	0.0032	0.00022	4478.576	0.07022	14.2403	85
90	441.103	0.0023	0.00016	6287.185	0.07016	14.2533	90
95	618.670	0.0016	0.00011	8823.854	0.07011	14.2626	95
100	867.716	0.0012	0.00008	12381.662	0.07008	14.2693	100

TABLE A - 10

DISCRETE CASH FLOW
8.00% DISCRETE COMPOUND INTEREST FACTORS

	SINGLE PAYMENTS		UNIFORM SERIES PAYMENTS				
	COMPOUND AMOUNT	PRESENT WORTH	SINKING FUND	COMPOUND AMOUNT	CAPITAL RECOVERY	PRESENT WORTH	
N	F/P	P/F	A/F	F/A	A/P	P/A	N
1	1.0800	0.9259	1.00000	1.000	1.08000	0.9259	1
2	1.1664	0.8573	0.48077	2.080	0.56077	1.7833	2
3	1.2597	0.7938	0.30803	3.246	0.38803	2.5771	3
4	1.3605	0.7350	0.22192	4.506	0.30192	3.3121	4
5	1.4693	0.6806	0.17046	5.867	0.25046	3.9927	5
6	1.5869	0.6302	0.13632	7.336	0.21632	4.6229	6
7	1.7138	0.5835	0.11207	8.923	0.19207	5.2064	7
8	1.8509	0.5403	0.09401	10.637	0.17401	5.7466	8
9	1.9990	0.5002	0.08008	12.488	0.16008	6.2469	9
10	2.1589	0.4632	0.06903	14.487	0.14903	6.7101	10
11	2.3316	0.4289	0.06008	16.645	0.14008	7.1390	11
12	2.5182	0.3971	0.05270	18.977	0.13270	7.5361	12
13	2.7196	0.3677	0.04652	21.495	0.12652	7.9038	13
14	2.9372	0.3405	0.04130	24.215	0.12130	8.2442	14
15	3.1722	0.3152	0.03683	27.152	0.11683	8.5595	15
16	3.4259	0.2919	0.03298	30.324	0.11298	8.8514	16
17	3.7000	0.2703	0.02963	33.750	0.10963	9.1216	17
18	3.9960	0.2502	0.02670	37.450	0.10670	9.3719	18
19	4.3157	0.2317	0.02413	41.446	0.10413	9.6036	19
20	4.6610	0.2145	0.02185	45.762	0.10185	9.8181	20
22	5.4365	0.1839	0.01803	55.457	0.09803	10.2007	22
24	6.3412	0.1577	0.01498	66.765	0.09498	10.5288	24
25	6.8485	0.1460	0.01368	73.106	0.09368	10.6748	25
26	7.3964	0.1352	0.01251	79.954	0.09251	10.8100	26
28	8.6271	0.1159	0.01049	95.339	0.09049	11.0511	28
30	10.0627	0.0994	0.00883	113.283	0.08883	11.2578	30
32	11.7371	0.0852	0.00745	134.214	0.08745	11.4350	32
34	13.6901	0.0730	0.00630	158.627	0.08630	11.5869	34
35	14.7853	0.0676	0.00580	172.317	0.08580	11.6546	35
36	15.9682	0.0626	0.00534	187.102	0.08534	11.7172	36
38	18.6253	0.0537	0.00454	220.316	0.08454	11.8289	38
40	21.7245	0.0460	0.00386	259.057	0.08386	11.9246	40
45	31.9204	0.0313	0.00259	386.506	0.08259	12.1084	45
50	46.9016	0.0213	0.00174	573.770	0.08174	12.2335	50
55	68.9139	0.0145	0.00118	848.923	0.08118	12.3186	55
60	101.257	0.0099	0.00080	1253.213	0.08080	12.3766	60
65	148.780	0.0067	0.00054	1847.248	0.08054	12.4160	65
70	218.606	0.0046	0.00037	2720.080	0.08037	12.4428	70
75	321.205	0.0031	0.00025	4002.557	0.08025	12.4611	75
80	471.955	0.0021	0.00017	5886.935	0.08017	12.4735	80
85	693.456	0.0014	0.00012	8655.706	0.08012	12.4820	85
90	1018.915	0.0010	0.00008	12723.939	0.08008	12.4877	90
95	1497.121	0.0007	0.00005	18701.507	0.08005	12.4917	95
100	2199.761	0.0005	0.00004	27484.516	0.08004	12.4943	100

TABLE A - 11

DISCRETE CASH FLOW
9.00% DISCRETE COMPOUND INTEREST FACTORS

	SINGLE PAYMENTS			UNIFORM SERIES PAYMENTS			
N	COMPOUND AMOUNT F/P	PRESENT WORTH P/F	SINKING FUND A/F	COMPOUND AMOUNT F/A	CAPITAL RECOVERY A/P	PRESENT WORTH P/A	N
1	1.0900	0.9174	1.00000	1.000	1.09000	0.9174	1
2	1.1881	0.8417	0.47847	2.090	0.56847	1.7591	2
3	1.2950	0.7722	0.30505	3.278	0.39505	2.5313	3
4	1.4116	0.7084	0.21867	4.573	0.30867	3.2397	4
5	1.5386	0.6499	0.16709	5.985	0.25709	3.8897	5
6	1.6771	0.5963	0.13292	7.523	0.22292	4.4859	6
7	1.8280	0.5470	0.10869	9.200	0.19869	5.0330	7
8	1.9926	0.5019	0.09067	11.028	0.18067	5.5348	8
9	2.1719	0.4604	0.07680	13.021	0.16680	5.9952	9
10	2.3674	0.4224	0.06582	15.193	0.15582	6.4177	10
11	2.5804	0.3875	0.05695	17.560	0.14695	6.8052	11
12	2.8127	0.3555	0.04965	20.141	0.13965	7.1607	12
13	3.0658	0.3262	0.04357	22.953	0.13357	7.4869	13
14	3.3417	0.2992	0.03843	26.019	0.12843	7.7862	14
15	3.6425	0.2745	0.03406	29.361	0.12406	8.0607	15
16	3.9703	0.2519	0.03030	33.003	0.12030	8.3126	16
17	4.3276	0.2311	0.02705	36.974	0.11705	8.5436	17
18	4.7171	0.2120	0.02421	41.301	0.11421	8.7556	18
19	5.1417	0.1945	0.02173	46.018	0.11173	8.9501	19
20	5.6044	0.1784	0.01955	51.160	0.10955	9.1285	20
22	6.6586	0.1502	0.01590	62.873	0.10590	9.4424	22
24	7.9111	0.1264	0.01302	76.790	0.10302	9.7066	24
25	8.6231	0.1160	0.01181	84.701	0.10181	9.8226	25
26	9.3992	0.1064	0.01072	93.324	0.10072	9.9290	26
28	11.1671	0.0895	0.00885	112.968	0.09885	10.1161	28
30	13.2677	0.0754	0.00734	136.308	0.09734	10.2737	30
32	15.7633	0.0634	0.00610	164.037	0.09610	10.4062	32
34	18.7284	0.0534	0.00508	196.982	0.09508	10.5178	34
35	20.4140	0.0490	0.00464	215.711	0.09464	10.5668	35
36	22.2512	0.0449	0.00424	236.125	0.09424	10.6118	36
38	26.4367	0.0378	0.00354	282.630	0.09354	10.6908	38
40	31.4094	0.0318	0.00296	337.882	0.09296	10.7574	40
45	48.3273	0.0207	0.00190	525.859	0.09190	10.8812	45
50	74.3575	0.0134	0.00123	815.084	0.09123	10.9617	50
55	114.408	0.0087	0.00079	1260.092	0.09079	11.0140	55
60	176.031	0.0057	0.00051	1944.792	0.09051	11.0480	60
65	270.846	0.0037	0.00033	2993.288	0.09033	11.0701	65
70	416.730	0.0024	0.00022	4619.223	0.09022	11.0844	70
75	641.191	0.0016	0.00014	7113.232	0.09014	11.0938	75
80	986.552	0.0010	0.00009	10950.574	0.09009	11.0998	80
85	1517.932	0.0007	0.00006	16854.800	0.09006	11.1038	85
90	2335.527	0.0004	0.00004	25939.184	0.09004	11.1064	90
95	3593.497	0.0003	0.00003	39916.635	0.09003	11.1080	95
100	5529.041	0.0002	0.00002	61422.675	0.09002	11.1091	100

TABLE A - 12

DISCRETE CASH FLOW
10.00% DISCRETE COMPOUND INTEREST FACTORS

| | SINGLE PAYMENTS | | | UNIFORM SERIES PAYMENTS | | | |
| | COMPOUND AMOUNT F/P | PRESENT WORTH P/F | SINKING FUND A/F | COMPOUND AMOUNT F/A | CAPITAL RECOVERY A/P | PRESENT WORTH P/A | |
N							N
1	1.1000	0.9091	1.00000	1.000	1.10000	0.9091	1
2	1.2100	0.8264	0.47619	2.100	0.57619	1.7355	2
3	1.3310	0.7513	0.30211	3.310	0.40211	2.4869	3
4	1.4641	0.6830	0.21547	4.641	0.31547	3.1699	4
5	1.6105	0.6209	0.16380	6.105	0.26380	3.7908	5
6	1.7716	0.5645	0.12961	7.716	0.22961	4.3553	6
7	1.9487	0.5132	0.10541	9.487	0.20541	4.8684	7
8	2.1436	0.4665	0.08744	11.436	0.18744	5.3349	8
9	2.3579	0.4241	0.07364	13.579	0.17364	5.7590	9
10	2.5937	0.3855	0.06275	15.937	0.16275	6.1446	10
11	2.8531	0.3505	0.05396	18.531	0.15396	6.4951	11
12	3.1384	0.3186	0.04676	21.384	0.14676	6.8137	12
13	3.4523	0.2897	0.04078	24.523	0.14078	7.1034	13
14	3.7975	0.2633	0.03575	27.975	0.13575	7.3667	14
15	4.1772	0.2394	0.03147	31.772	0.13147	7.6061	15
16	4.5950	0.2176	0.02782	35.950	0.12782	7.8237	16
17	5.0545	0.1978	0.02466	40.545	0.12466	8.0216	17
18	5.5599	0.1799	0.02193	45.599	0.12193	8.2014	18
19	6.1159	0.1635	0.01955	51.159	0.11955	8.3649	19
20	6.7275	0.1486	0.01746	57.275	0.11746	8.5136	20
22	8.1403	0.1228	0.01401	71.403	0.11401	8.7715	22
24	9.8497	0.1015	0.01130	88.497	0.11130	8.9847	24
25	10.8347	0.0923	0.01017	98.347	0.11017	9.0770	25
26	11.9182	0.0839	0.00916	109.182	0.10916	9.1609	26
28	14.4210	0.0693	0.00745	134.210	0.10745	9.3066	28
30	17.4494	0.0573	0.00608	164.494	0.10608	9.4269	30
32	21.1138	0.0474	0.00497	201.138	0.10497	9.5264	32
34	25.5477	0.0391	0.00407	245.477	0.10407	9.6086	34
35	28.1024	0.0356	0.00369	271.024	0.10369	9.6442	35
36	30.9127	0.0323	0.00334	299.127	0.10334	9.6765	36
38	37.4043	0.0267	0.00275	364.043	0.10275	9.7327	38
40	45.2593	0.0221	0.00226	442.593	0.10226	9.7791	40
45	72.8905	0.0137	0.00139	718.905	0.10139	9.8628	45
50	117.391	0.0085	0.00086	1163.909	0.10086	9.9148	50
55	189.059	0.0053	0.00053	1880.591	0.10053	9.9471	55
60	304.482	0.0033	0.00033	3034.816	0.10033	9.9672	60
65	490.371	0.0020	0.00020	4893.707	0.10020	9.9796	65
70	789.747	0.0013	0.00013	7887.470	0.10013	9.9873	70
75	1271.895	0.0008	0.00008	12708.954	0.10008	9.9921	75
80	2048.400	0.0005	0.00005	20474.002	0.10005	9.9951	80
85	3298.969	0.0003	0.00003	32979.690	0.10003	9.9970	85
90	5313.023	0.0002	0.00002	53120.226	0.10002	9.9981	90
95	8556.676	0.0001	0.00001	85556.760	0.10001	9.9988	95

TABLE A - 13

DISCRETE CASH FLOW
11.00%   DISCRETE COMPOUND INTEREST FACTORS

| | SINGLE PAYMENTS | | UNIFORM SERIES PAYMENT | | | | |
| | COMPOUND AMOUNT | PRESENT WORTH | SINKING FUND | COMPOUND AMOUNT | CAPITAL RECOVERY | PRESENT WORTH | |
N	F/P	P/F	A/F	F/A	A/P	P/A	N
1	1.1100	0.9009	1.00000	1.00	1.11000	0.9009	1
2	1.2321	0.8116	0.47393	2.11	0.58393	1.7125	2
3	1.3676	0.7312	0.29921	3.34	0.40921	2.4437	3
4	1.5181	0.6587	0.21233	4.71	0.32233	3.1024	4
5	1.6851	0.5935	0.16057	6.23	0.27057	3.6959	5
6	1.8704	0.5346	0.12638	7.91	0.23638	4.2305	6
7	2.0762	0.4817	0.10222	9.78	0.21222	4.7122	7
8	2.3045	0.4339	0.08432	11.86	0.19432	5.1461	8
9	2.5580	0.3909	0.07060	14.16	0.18060	5.5370	9
10	2.8394	0.3522	0.05980	16.72	0.16980	5.8892	10
11	3.1518	0.3173	0.05112	19.56	0.16112	6.2065	11
12	3.4985	0.2858	0.04403	22.71	0.15403	6.4924	12
13	3.8833	0.2575	0.03815	26.21	0.14815	6.7499	13
14	4.3104	0.2320	0.03323	30.09	0.14323	6.9819	14
15	4.7846	0.2090	0.02907	34.41	0.13907	7.1909	15
16	5.3109	0.1883	0.02552	39.19	0.13552	7.3792	16
17	5.8951	0.1696	0.02247	44.50	0.13247	7.5488	17
18	6.5436	0.1528	0.01984	50.40	0.12984	7.7016	18
19	7.2633	0.1377	0.01756	56.94	0.12756	7.8393	19
20	8.0623	0.1240	0.01558	64.20	0.12558	7.9633	20
22	9.9336	0.1007	0.01231	81.21	0.12231	8.1757	22
24	12.2392	0.0817	0.00979	102.17	0.11979	8.3481	24
25	13.5855	0.0736	0.00874	114.41	0.11874	8.4217	25
26	15.0799	0.0663	0.00781	128.00	0.11781	8.4881	26
28	18.5799	0.0538	0.00626	159.82	0.11626	8.6016	28
30	22.8923	0.0437	0.00502	199.02	0.11502	8.6938	30
32	28.2056	0.0355	0.00404	247.32	0.11404	8.7686	32
34	34.7521	0.0288	0.00326	306.84	0.11326	8.8293	34
35	38.5749	0.0259	0.00293	341.59	0.11293	8.8552	35
36	42.8181	0.0234	0.00263	380.16	0.11263	8.8786	36
38	52.7562	0.0190	0.00213	470.51	0.11213	8.9186	38
40	65.0009	0.0154	0.00172	581.83	0.11172	8.9511	40
45	109.530	0.0091	0.00101	986.64	0.11101	9.0079	45
50	184.565	0.0054	0.00060	1668.77	0.11060	9.0417	50

TABLE A - 14

DISCRETE CASH FLOW
12.00% DISCRETE COMPOUND INTEREST FACTORS

	SINGLE PAYMENTS			UNIFORM SERIES PAYMENTS			
N	COMPOUND AMOUNT F/P	PRESENT WORTH P/F	SINKING FUND A/F	COMPOUND AMOUNT F/A	CAPITAL RECOVERY A/P	PRESENT WORTH P/A	N
1	1.1200	0.8929	1.00000	1.000	1.12000	0.8929	1
2	1.2544	0.7972	0.47170	2.120	0.59170	1.6901	2
3	1.4049	0.7118	0.29635	3.374	0.41635	2.4018	3
4	1.5735	0.6355	0.20923	4.779	0.32923	3.0373	4
5	1.7623	0.5674	0.15741	6.353	0.27741	3.6048	5
6	1.9738	0.5066	0.12323	8.115	0.24323	4.1114	6
7	2.2107	0.4523	0.09912	10.089	0.21912	4.5638	7
8	2.4760	0.4039	0.08130	12.300	0.20130	4.9676	8
9	2.7731	0.3606	0.06768	14.776	0.18768	5.3282	9
10	3.1058	0.3220	0.05698	17.549	0.17698	5.6502	10
11	3.4785	0.2875	0.04842	20.655	0.16842	5.9377	11
12	3.8960	0.2567	0.04144	24.133	0.16144	6.1944	12
13	4.3635	0.2292	0.03568	28.029	0.15568	6.4235	13
14	4.8871	0.2046	0.03087	32.393	0.15087	6.6282	14
15	5.4736	0.1827	0.02682	37.280	0.14682	6.8109	15
16	6.1304	0.1631	0.02339	42.753	0.14339	6.9740	16
17	6.8660	0.1456	0.02046	48.884	0.14046	7.1196	17
18	7.6900	0.1300	0.01794	55.750	0.13794	7.2497	18
19	8.6128	0.1161	0.01576	63.440	0.13576	7.3658	19
20	9.6463	0.1037	0.01388	72.052	0.13388	7.4694	20
22	12.1003	0.0826	0.01081	92.503	0.13081	7.6446	22
24	15.1786	0.0659	0.00846	118.155	0.12846	7.7843	24
25	17.0001	0.0588	0.00750	133.334	0.12750	7.8431	25
26	19.0401	0.0525	0.00665	150.334	0.12665	7.8957	26
28	23.8839	0.0419	0.00524	190.699	0.12524	7.9844	28
30	29.9599	0.0334	0.00414	241.333	0.12414	8.0552	30
32	37.5817	0.0266	0.00328	304.848	0.12328	8.1116	32
34	47.1425	0.0212	0.00260	384.521	0.12260	8.1566	34
35	52.7996	0.0189	0.00232	431.663	0.12232	8.1755	35
36	59.1356	0.0169	0.00206	484.463	0.12206	8.1924	36
38	74.1797	0.0135	0.00164	609.831	0.12164	8.2210	38
40	93.0510	0.0107	0.00130	767.091	0.12130	8.2438	40
45	163.988	0.0061	0.00074	1358.230	0.12074	8.2825	45
50	289.002	0.0035	0.00042	2400.018	0.12042	8.3045	50

TABLE A - 15

DISCRETE CASH FLOW

13.00%   DISCRETE COMPOUND INTEREST FACTORS

	SINGLE PAYMENTS		UNIFORM SERIES PAYMENT				
N	COMPOUND AMOUNT F/P	PRESENT WORTH P/F	SINKING FUND A/F	COMPOUND AMOUNT F/A	CAPITAL RECOVERY A/P	PRESENT WORTH P/A	N
1	1.1300	0.8850	1.00000	1.00	1.13000	0.8850	1
2	1.2769	0.7831	0.46948	2.13	0.59948	1.6681	2
3	1.4429	0.6931	0.29352	3.41	0.42352	2.3612	3
4	1.6305	0.6133	0.20619	4.85	0.33619	2.9745	4
5	1.8424	0.5428	0.15431	6.48	0.28431	3.5172	5
6	2.0820	0.4803	0.12015	8.32	0.25015	3.9975	6
7	2.3526	0.4251	0.09611	10.40	0.22611	4.4226	7
8	2.6584	0.3762	0.07839	12.76	0.20839	4.7988	8
9	3.0040	0.3329	0.06487	15.42	0.19487	5.1317	9
10	3.3946	0.2946	0.05429	18.42	0.18429	5.4262	10
11	3.8359	0.2607	0.04584	21.81	0.17584	5.6869	11
12	4.3345	0.2307	0.03899	25.65	0.16899	5.9176	12
13	4.8980	0.2042	0.03335	29.98	0.16335	6.1218	13
14	5.5348	0.1807	0.02867	34.88	0.15867	6.3025	14
15	6.2543	0.1599	0.02474	40.42	0.15474	6.4624	15
16	7.0673	0.1415	0.02143	46.67	0.15143	6.6039	16
17	7.9861	0.1252	0.01861	53.74	0.14861	6.7291	17
18	9.0243	0.1108	0.01620	61.73	0.14620	6.8399	18
19	10.1974	0.0981	0.01413	70.75	0.14413	6.9380	19
20	11.5231	0.0868	0.01235	80.95	0.14235	7.0248	20
22	14.7138	0.0680	0.00948	105.49	0.13948	7.1695	22
24	18.7881	0.0532	0.00731	136.83	0.13731	7.2829	24
25	21.2305	0.0471	0.00643	155.62	0.13643	7.3300	25
26	23.9905	0.0417	0.00565	176.85	0.13565	7.3717	26
28	30.6335	0.0326	0.00439	227.95	0.13439	7.4412	28
30	39.1159	0.0256	0.00341	293.20	0.13341	7.4957	30
32	49.9471	0.0200	0.00266	376.52	0.13266	7.5383	32
34	63.7774	0.0157	0.00207	482.90	0.13207	7.5717	34
35	72.0685	0.0139	0.00183	546.68	0.13183	7.5856	35
36	81.4374	0.0123	0.00162	618.75	0.13162	7.5979	36
38	103.987	0.0096	0.00126	792.21	0.13126	7.6183	38
40	132.782	0.0075	0.00099	1013.70	0.13099	7.6344	40
45	244.641	0.0041	0.00053	1874.16	0.13053	7.6609	45
50	450.736	0.0022	0.00029	3459.51	0.13029	7.6752	50

TABLE A - 16

DISCRETE CASH FLOW

14.00% DISCRETE COMPOUND INTEREST FACTORS

| | SINGLE PAYMENTS | | UNIFORM SERIES PAYMENT | | | | |
| | COMPOUND AMOUNT F/P | PRESENT WORTH P/F | SINKING FUND A/F | COMPOUND AMOUNT F/A | CAPITAL RECOVERY A/P | PRESENT WORTH P/A | |
N							N
1	1.1400	0.8772	1.00000	1.00	1.14000	0.8772	1
2	1.2996	0.7695	0.46729	2.14	0.60729	1.6467	2
3	1.4815	0.6750	0.29073	3.44	0.43073	2.3216	3
4	1.6890	0.5921	0.20320	4.92	0.34320	2.9137	4
5	1.9254	0.5194	0.15128	6.61	0.29128	3.4331	5
6	2.1950	0.4556	0.11716	8.54	0.25716	3.8887	6
7	2.5023	0.3996	0.09319	10.73	0.23319	4.2883	7
8	2.8526	0.3506	0.07557	13.23	0.21557	4.6389	8
9	3.2519	0.3075	0.06217	16.09	0.20217	4.9464	9
10	3.7072	0.2697	0.05171	19.34	0.19171	5.2161	10
11	4.2262	0.2366	0.04339	23.04	0.18339	5.4527	11
12	4.8179	0.2076	0.03667	27.27	0.17667	5.6603	12
13	5.4924	0.1821	0.03116	32.09	0.17116	5.8424	13
14	6.2613	0.1597	0.02661	37.58	0.16661	6.0021	14
15	7.1379	0.1401	0.02281	43.84	0.16281	6.1422	15
16	8.1372	0.1229	0.01962	50.98	0.15962	6.2651	16
17	9.2765	0.1078	0.01692	59.12	0.15692	6.3729	17
18	10.5752	0.0946	0.01462	68.39	0.15462	6.4674	18
19	12.0557	0.0829	0.01266	78.97	0.15266	6.5504	19
20	13.7435	0.0728	0.01099	91.02	0.15099	6.6231	20
22	17.8610	0.0560	0.00830	120.44	0.14830	6.7429	22
24	23.2122	0.0431	0.00630	158.66	0.14630	6.8351	24
25	26.4619	0.0378	0.00550	181.87	0.14550	6.8729	25
26	30.1666	0.0331	0.00480	208.33	0.14480	6.9061	26
28	39.2045	0.0255	0.00366	272.89	0.14366	6.9607	28
30	50.9502	0.0196	0.00280	356.79	0.14280	7.0027	30
32	66.2148	0.0151	0.00215	465.82	0.14215	7.0350	32
34	86.0528	0.0116	0.00165	607.52	0.14165	7.0599	34
35	98.1002	0.0102	0.00144	693.57	0.14144	7.0700	35
36	111.834	0.0089	0.00126	791.67	0.14126	7.0790	36
38	145.340	0.0069	0.00097	1031.00	0.14097	7.0937	38
40	188.884	0.0053	0.00075	1342.03	0.14075	7.1050	40
45	363.679	0.0027	0.00039	2590.56	0.14039	7.1232	45
50	700.233	0.0014	0.00020	4994.52	0.14020	7.1327	50

TABLE A - 17

DISCRETE CASH FLOW
15.00% DISCRETE COMPOUND INTEREST FACTORS

	SINGLE PAYMENTS		UNIFORM SERIES PAYMENTS				
N	COMPOUND AMOUNT F/P	PRESENT WORTH P/F	SINKING FUND A/F	COMPOUND AMOUNT F/A	CAPITAL RECOVERY A/P	PRESENT WORTH P/A	N
1	1.1500	0.8696	1.00000	1.000	1.15000	0.8696	1
2	1.3225	0.7561	0.46512	2.150	0.61512	1.6257	2
3	1.5209	0.6575	0.28798	3.472	0.43798	2.2832	3
4	1.7490	0.5718	0.20027	4.993	0.35027	2.8550	4
5	2.0114	0.4972	0.14832	6.742	0.29832	3.3522	5
6	2.3131	0.4323	0.11424	8.754	0.26424	3.7845	6
7	2.6600	0.3759	0.09036	11.067	0.24036	4.1604	7
8	3.0590	0.3269	0.07285	13.727	0.22285	4.4873	8
9	3.5179	0.2843	0.05957	16.786	0.20957	4.7716	9
10	4.0456	0.2472	0.04925	20.304	0.19925	5.0188	10
11	4.6524	0.2149	0.04107	24.349	0.19107	5.2337	11
12	5.3503	0.1869	0.03448	29.002	0.18448	5.4206	12
13	6.1528	0.1625	0.02911	34.352	0.17911	5.5831	13
14	7.0757	0.1413	0.02469	40.505	0.17469	5.7245	14
15	8.1371	0.1229	0.02102	47.580	0.17102	5.8474	15
16	9.3576	0.1069	0.01795	55.717	0.16795	5.9542	16
17	10.7613	0.0929	0.01537	65.075	0.16537	6.0472	17
18	12.3755	0.0808	0.01319	75.836	0.16319	6.1280	18
19	14.2318	0.0703	0.01134	88.212	0.16134	6.1982	19
20	16.3665	0.0611	0.00976	102.444	0.15976	6.2593	20
22	21.6447	0.0462	0.00727	137.632	0.15727	6.3587	22
24	29.6252	0.0349	0.00543	184.168	0.15543	6.4338	24
25	32.9190	0.0304	0.00470	212.793	0.15470	6.4641	25
26	37.8569	0.0264	0.00407	245.712	0.15407	6.4906	26
28	50.0656	0.0200	0.00306	327.104	0.15306	6.5335	28
30	66.2118	0.0151	0.00230	434.745	0.15230	6.5660	30
32	87.5651	0.0114	0.00173	577.100	0.15173	6.5905	32
34	115.805	0.0086	0.00131	765.365	0.15131	6.6091	34
35	133.176	0.0075	0.00113	881.170	0.15113	6.6166	35
36	153.152	0.0065	0.00099	1014.346	0.15099	6.6231	36
38	202.543	0.0049	0.00074	1343.622	0.15074	6.6338	38
40	267.864	0.0037	0.00056	1779.090	0.15056	6.6418	40
45	538.769	0.0019	0.00028	3585.128	0.15028	6.6543	45
50	1083.657	0.0009	0.00014	7217.716	0.15014	6.6605	50

TABLE A - 18

DISCRETE CASH FLOW

16.00% DISCRETE COMPOUND INTEREST FACTORS

	SINGLE PAYMENTS		UNIFORM SERIES PAYMENT				
N	COMPOUND AMOUNT F/P	PRESENT WORTH P/F	SINKING FUND A/F	COMPOUND AMOUNT F/A	CAPITAL RECOVERY A/P	PRESENT WORTH P/A	N
1	1.1600	0.8621	1.00000	1.00	1.16000	0.8621	1
2	1.3456	0.7432	0.46296	2.16	0.62296	1.6052	2
3	1.5609	0.6407	0.28526	3.51	0.44526	2.2459	3
4	1.8106	0.5523	0.19738	5.07	0.35738	2.7982	4
5	2.1003	0.4761	0.14541	6.88	0.30541	3.2743	5
6	2.4364	0.4104	0.11139	8.98	0.27139	3.6847	6
7	2.8262	0.3538	0.08761	11.41	0.24761	4.0386	7
8	3.2784	0.3050	0.07022	14.24	0.23022	4.3436	8
9	3.8030	0.2630	0.05708	17.52	0.21708	4.6065	9
10	4.4114	0.2267	0.04690	21.32	0.20690	4.8332	10
11	5.1173	0.1954	0.03886	25.73	0.19886	5.0286	11
12	5.9360	0.1685	0.03241	30.85	0.19241	5.1971	12
13	6.8858	0.1452	0.02718	36.79	0.18718	5.3423	13
14	7.9875	0.1252	0.02290	43.67	0.18290	5.4675	14
15	9.2655	0.1079	0.01936	51.66	0.17936	5.5755	15
16	10.7480	0.0930	0.01641	60.93	0.17641	5.6685	16
17	12.4677	0.0802	0.01395	71.67	0.17395	5.7487	17
18	14.4625	0.0691	0.01188	84.14	0.17188	5.8178	18
19	16.7765	0.0596	0.01014	98.60	0.17014	5.8775	19
20	19.4608	0.0514	0.00867	115.38	0.16867	5.9288	20
22	26.1864	0.0382	0.00635	157.41	0.16635	6.0113	22
24	35.2364	0.0284	0.00467	213.98	0.16467	6.0726	24
25	40.8742	0.0245	0.00401	249.21	0.16401	6.0971	25
26	47.4141	0.0211	0.00345	290.09	0.16345	6.1182	26
28	63.8004	0.0157	0.00255	392.50	0.16255	6.1520	28
30	85.8499	0.0116	0.00189	530.31	0.16189	6.1772	30
32	115.520	0.0087	0.00140	715.75	0.16140	6.1959	32
34	155.443	0.0064	0.00104	965.27	0.16104	6.2098	34
35	180.314	0.0055	0.00089	1120.71	0.16089	6.2153	35
36	209.164	0.0048	0.00077	1301.03	0.16077	6.2201	36
38	281.452	0.0036	0.00057	1752.82	0.16057	6.2278	38
40	378.721	0.0026	0.00042	2360.76	0.16042	6.2335	40
45	795.444	0.0013	0.00020	4965.27	0.16020	6.2421	45
50	1670.704	0.0006	0.00010	10435.65	0.16010	6.2463	50

TABLE A - 19

DISCRETE CASH FLOW

17.00% DISCRETE COMPOUND INTEREST FACTORS

	SINGLE PAYMENTS		UNIFORM SERIES PAYMENT				
N	COMPOUND AMOUNT F/P	PRESENT WORTH P/F	SINKING FUND A/F	COMPOUND AMOUNT F/A	CAPITAL RECOVERY A/P	PRESENT WORTH P/A	N
1	1.1700	0.8547	1.00000	1.00	1.17000	0.8547	1
2	1.3689	0.7305	0.46083	2.17	0.63083	1.5852	2
3	1.6016	0.6244	0.28257	3.54	0.45257	2.2096	3
4	1.8739	0.5337	0.19453	5.14	0.36453	2.7432	4
5	2.1924	0.4561	0.14256	7.01	0.31256	3.1993	5
6	2.5652	0.3898	0.10861	9.21	0.27861	3.5892	6
7	3.0012	0.3332	0.08495	11.77	0.25495	3.9224	7
8	3.5115	0.2848	0.06769	14.77	0.23769	4.2072	8
9	4.1084	0.2434	0.05469	18.28	0.22469	4.4506	9
10	4.8068	0.2080	0.04466	22.39	0.21466	4.6586	10
11	5.6240	0.1778	0.03676	27.20	0.20676	4.8364	11
12	6.5801	0.1520	0.03047	32.82	0.20047	4.9884	12
13	7.6987	0.1299	0.02538	39.40	0.19538	5.1183	13
14	9.0075	0.1110	0.02123	47.10	0.19123	5.2293	14
15	10.5387	0.0949	0.01782	56.11	0.18782	5.3242	15
16	12.3303	0.0811	0.01500	66.65	0.18500	5.4053	16
17	14.4265	0.0693	0.01266	78.98	0.18266	5.4746	17
18	16.8790	0.0592	0.01071	93.41	0.18071	5.5339	18
19	19.7484	0.0506	0.00907	110.28	0.17907	5.5845	19
20	23.1056	0.0433	0.00769	130.03	0.17769	5.6278	20
22	31.6293	0.0316	0.00555	180.17	0.17555	5.6964	22
24	43.2973	0.0231	0.00402	248.81	0.17402	5.7465	24
25	50.6578	0.0197	0.00342	292.10	0.17342	5.7662	25
26	59.2697	0.0169	0.00292	342.76	0.17292	5.7831	26
28	81.1342	0.0123	0.00212	471.38	0.17212	5.8099	28
30	111.065	0.0090	0.00154	647.44	0.17154	5.8294	30
32	152.036	0.0066	0.00113	888.45	0.17113	5.8437	32
34	208.123	0.0048	0.00082	1218.37	0.17082	5.8541	34
35	243.503	0.0041	0.00070	1426.49	0.17070	5.8582	35
36	284.899	0.0035	0.00060	1669.99	0.17060	5.8617	36
38	389.998	0.0026	0.00044	2288.23	0.17044	5.8673	38
40	533.869	0.0019	0.00032	3134.52	0.17032	5.8713	40
45	1170.479	0.0009	0.00015	6879.29	0.17015	5.8773	45
50	2566.215	0.0004	0.00007	15089.50	0.17007	5.8801	50

TABLE A - 20

DISCRETE CASH FLOW
18.00% DISCRETE COMPOUND INTEREST FACTORS

	SINGLE PAYMENTS			UNIFORM SERIES PAYMENTS			
N	COMPOUND AMOUNT F/P	PRESENT WORTH P/F	SINKING FUND A/F	COMPOUND AMOUNT F/A	CAPITAL RECOVERY A/P	PRESENT WORTH P/A	N
1	1.1800	0.8475	1.00000	1.000	1.18000	0.8475	1
2	1.3924	0.7182	0.45872	2.180	0.63872	1.5656	2
3	1.6430	0.6086	0.27992	3.572	0.45992	2.1743	3
4	1.9388	0.5158	0.19174	5.215	0.37174	2.6901	4
5	2.2878	0.4371	0.13978	7.154	0.31978	3.1272	5
6	2.6996	0.3704	0.10591	9.442	0.28591	3.4976	6
7	3.1855	0.3139	0.08236	12.142	0.26236	3.8115	7
8	3.7589	0.2660	0.06524	15.327	0.24524	4.0776	8
9	4.4355	0.2255	0.05239	19.086	0.23239	4.3030	9
10	5.2338	0.1911	0.04251	23.521	0.22251	4.4941	10
11	6.1759	0.1619	0.03478	28.755	0.21478	4.6560	11
12	7.2876	0.1372	0.02863	34.931	0.20863	4.7932	12
13	8.5994	0.1163	0.02369	42.219	0.20369	4.9095	13
14	10.1472	0.0985	0.01968	50.818	0.19968	5.0081	14
15	11.9737	0.0835	0.01640	60.965	0.19640	5.0916	15
16	14.1290	0.0708	0.01371	72.939	0.19371	5.1624	16
17	16.6722	0.0600	0.01149	87.068	0.19149	5.2223	17
18	19.6733	0.0508	0.00964	103.740	0.18964	5.2732	18
19	23.2144	0.0431	0.00810	123.414	0.18810	5.3162	19
20	27.3930	0.0365	0.00682	146.628	0.18682	5.3527	20
22	38.1421	0.0262	0.00485	206.345	0.18485	5.4099	22
24	53.1090	0.0188	0.00345	289.494	0.18345	5.4509	24
25	62.6686	0.0160	0.00292	342.603	0.18292	5.4669	25
26	73.9490	0.0135	0.00247	405.272	0.18247	5.4804	26
28	102.9666	0.0097	0.00177	566.481	0.18177	5.5016	28
30	143.3706	0.0070	0.00126	790.948	0.18126	5.5168	30
32	199.6293	0.0050	0.00091	1103.496	0.18091	5.5277	32
34	277.9638	0.0036	0.00065	1538.688	0.18065	5.5356	34
35	327.9973	0.0030	0.00055	1816.652	0.18055	5.5386	35
36	387.0368	0.0026	0.00047	2144.649	0.18047	5.5412	36
38	538.9100	0.0019	0.00033	2988.389	0.18033	5.5452	38
40	750.3783	0.0013	0.00024	4163.213	0.18024	5.5482	40
45	1716.684	0.0006	0.18010	9531.577	0.18010	5.5523	45
50	3927.357	0.0003	0.18005	21813.094	0.18005	5.5541	50

TABLE A - 21

DISCRETE CASH FLOW

19.00% DISCRETE COMPOUND INTEREST FACTORS

	SINGLE PAYMENTS		UNIFORM SERIES PAYMENT				
N	COMPOUND AMOUNT F/P	PRESENT WORTH P/F	SINKING FUND A/F	COMPOUND AMOUNT F/A	CAPITAL RECOVERY A/P	PRESENT WORTH P/A	N
1	1.1900	0.8403	1.00000	1.00	1.19000	0.8403	1
2	1.4161	0.7062	0.45662	2.19	0.64662	1.5465	2
3	1.6852	0.5934	0.27731	3.61	0.46731	2.1399	3
4	2.0053	0.4987	0.18899	5.29	0.37899	2.6386	4
5	2.3864	0.4190	0.13705	7.30	0.32705	3.0576	5
6	2.8398	0.3521	0.10327	9.68	0.29327	3.4098	6
7	3.3793	0.2959	0.07985	12.52	0.26985	3.7057	7
8	4.0214	0.2487	0.06289	15.90	0.25289	3.9544	8
9	4.7854	0.2090	0.05019	19.92	0.24019	4.1633	9
10	5.6947	0.1756	0.04047	24.71	0.23047	4.3389	10
11	6.7767	0.1476	0.03289	30.40	0.22289	4.4865	11
12	8.0642	0.1240	0.02690	37.18	0.21690	4.6105	12
13	9.5964	0.1042	0.02210	45.24	0.21210	4.7147	13
14	11.4198	0.0876	0.01823	54.84	0.20823	4.8023	14
15	13.5895	0.0736	0.01509	66.26	0.20509	4.8759	15
16	16.1715	0.0618	0.01252	79.85	0.20252	4.9377	16
17	19.2441	0.0520	0.01041	96.02	0.20041	4.9897	17
18	22.9005	0.0437	0.00868	115.27	0.19868	5.0333	18
19	27.2516	0.0367	0.00724	138.17	0.19724	5.0700	19
20	32.4294	0.0308	0.00605	165.42	0.19605	5.1009	20
22	45.9233	0.0218	0.00423	236.44	0.19423	5.1486	22
24	65.0320	0.0154	0.00297	337.01	0.19297	5.1822	24
25	77.3881	0.0129	0.00249	402.04	0.19249	5.1951	25
26	92.0918	0.0109	0.00209	479.43	0.19209	5.2060	26
28	130.4112	0.0077	0.00147	681.11	0.19147	5.2228	28
30	184.6753	0.0054	0.00103	966.71	0.19103	5.2347	30
32	261.5187	0.0038	0.00073	1371.15	0.19073	5.2430	32
34	370.3366	0.0027	0.00051	1943.88	0.19051	5.2489	34
35	440.7006	0.0023	0.00043	2314.21	0.19043	5.2512	35
36	524.4337	0.0019	0.00036	2754.91	0.19036	5.2531	36
38	742.6506	0.0013	0.00026	3903.42	0.19026	5.2561	38
40	1051.668	0.0010	0.19018	5529.83	0.19018	5.2582	40
45	2509.651	0.0004	0.19008	13203.42	0.19008	5.2611	45
50	5988.914	0.0002	0.19003	31515.34	0.19003	5.2623	50

TABLE A - 22

DISCRETE CASH FLOW
20.00% DISCRETE COMPOUND INTEREST FACTORS

	SINGLE PAYMENTS			UNIFORM SERIES PAYMENTS			
N	COMPOUND AMOUNT F/P	PRESENT WORTH P/F	SINKING FUND A/F	COMPOUND AMOUNT F/A	CAPITAL RECOVERY A/P	PRESENT WORTH P/A	N
1	1.2000	0.8333	1.00000	1.000	1.20000	0.8333	1
2	1.4400	0.6944	0.45455	2.200	0.65455	1.5278	2
3	1.7280	0.5787	0.27473	3.640	0.47473	2.1065	3
4	2.0736	0.4823	0.18629	5.368	0.38629	2.5887	4
5	2.4883	0.4019	0.13438	7.442	0.33438	2.9906	5
6	2.9860	0.3349	0.10071	9.930	0.30071	3.3255	6
7	3.5832	0.2791	0.07742	12.916	0.27742	3.6046	7
8	4.2998	0.2326	0.06061	16.499	0.26061	3.8372	8
9	5.1598	0.1938	0.04808	20.799	0.24808	4.0310	9
10	6.1917	0.1615	0.03852	25.959	0.23852	4.1925	10
11	7.4301	0.1346	0.03110	32.150	0.23110	4.3271	11
12	8.9161	0.1122	0.02526	39.581	0.22526	4.4392	12
13	10.6993	0.0935	0.02062	48.497	0.22062	4.5327	13
14	12.8392	0.0779	0.01689	59.196	0.21689	4.6106	14
15	15.4070	0.0649	0.01388	72.035	0.21388	4.6755	15
16	18.4884	0.0541	0.01144	87.442	0.21144	4.7296	16
17	22.1861	0.0451	0.00944	105.931	0.20944	4.7746	17
18	26.6233	0.0376	0.00781	128.117	0.20781	4.8122	18
19	31.9480	0.0313	0.00646	154.740	0.20646	4.8435	19
20	38.3376	0.0261	0.00536	186.688	0.20536	4.8696	20
22	55.2061	0.0181	0.00369	271.031	0.20369	4.9094	22
24	79.4968	0.0126	0.00255	392.484	0.20255	4.9371	24
25	95.3962	0.0105	0.00212	471.981	0.20212	4.9476	25
26	114.4755	0.0087	0.00176	567.377	0.20176	4.9563	26
28	164.8447	0.0061	0.00122	819.223	0.20122	4.9697	28
30	237.3763	0.0042	0.00085	1181.882	0.20085	4.9789	30
32	341.8219	0.0029	0.00059	1704.109	0.20059	4.9854	32
34	492.2235	0.0020	0.00041	2456.118	0.20041	4.9898	34
35	590.6682	0.0017	0.00034	2948.341	0.20034	4.9915	35
36	708.8019	0.0014	0.00028	3539.009	0.20028	4.9929	36
38	1020.675	0.0010	0.20020	5098.373	0.20020	4.9951	38
40	1469.772	0.0007	0.20014	7343.858	0.20014	4.9966	40
45	3657.262	0.0003	0.20005	18281.310	0.20005	4.9986	45
50	9100.438	0.0001	0.20002	45497.191	0.20002	4.9995	50

TABLE A - 23

DISCRETE CASH FLOW

22.00% DISCRETE COMPOUND INTEREST FACTORS

	SINGLE PAYMENTS		UNIFORM SERIES PAYMENT				
N	COMPOUND AMOUNT F/P	PRESENT WORTH P/F	SINKING FUND A/F	COMPOUND AMOUNT F/A	CAPITAL RECOVERY A/P	PRESENT WORTH P/A	N
1	1.2200	0.8197	1.00000	1.00	1.22000	0.8197	1
2	1.4884	0.6719	0.45045	2.22	0.67045	1.4915	2
3	1.8158	0.5507	0.26966	3.71	0.48966	2.0422	3
4	2.2153	0.4514	0.18102	5.52	0.40102	2.4936	4
5	2.7027	0.3700	0.12921	7.74	0.34921	2.8636	5
6	3.2973	0.3033	0.09576	10.44	0.31576	3.1669	6
7	4.0227	0.2486	0.07278	13.74	0.29278	3.4155	7
8	4.9077	0.2038	0.05630	17.76	0.27630	3.6193	8
9	5.9874	0.1670	0.04411	22.67	0.26411	3.7863	9
10	7.3046	0.1369	0.03489	28.66	0.25489	3.9232	10
11	8.9117	0.1122	0.02781	35.96	0.24781	4.0354	11
12	10.8722	0.0920	0.02228	44.87	0.24228	4.1274	12
13	13.2641	0.0754	0.01794	55.75	0.23794	4.2028	13
14	16.1822	0.0618	0.01449	69.01	0.23449	4.2646	14
15	19.7423	0.0507	0.01174	85.19	0.23174	4.3152	15
16	24.0856	0.0415	0.00953	104.93	0.22953	4.3567	16
17	29.3844	0.0340	0.00775	129.02	0.22775	4.3908	17
18	35.8490	0.0279	0.00631	158.40	0.22631	4.4187	18
19	43.7358	0.0229	0.00515	194.25	0.22515	4.4415	19
20	53.3576	0.0187	0.00420	237.99	0.22420	4.4603	20
22	79.4175	0.0126	0.00281	356.44	0.22281	4.4882	22
24	118.2050	0.0085	0.00188	532.75	0.22188	4.5070	24
25	144.2101	0.0069	0.00154	650.96	0.22154	4.5139	25
26	175.9364	0.0057	0.00126	795.17	0.22126	4.5196	26
28	261.8637	0.0038	0.00084	1185.74	0.22084	4.5281	28
30	389.7579	0.0026	0.00057	1767.08	0.22057	4.5338	30
32	580.1156	0.0017	0.00038	2632.34	0.22038	4.5376	32
34	863.4441	0.0012	0.00026	3920.20	0.22026	4.5402	34
35	1053.402	0.0009	0.00021	4783.64	0.22021	4.5411	35
36	1285.150	0.0008	0.00017	5837.05	0.22017	4.5419	36
38	1912.818	0.0005	0.00012	8690.08	0.22012	4.5431	38
40	2847.038	0.0004	0.00008	12936.54	0.22008	4.5439	40
45	7694.712	0.0001	0.00003	34971.42	0.22003	4.5449	45
50	20796.56	0.0000	0.00001	94525.28	0.22001	4.5452	50

TABLE A - 24

DISCRETE CASH FLOW

24.00% DISCRETE COMPOUND INTEREST FACTORS

	SINGLE PAYMENTS		UNIFORM SERIES PAYMENT				
N	COMPOUND AMOUNT F/P	PRESENT WORTH P/F	SINKING FUND A/F	COMPOUND AMOUNT F/A	CAPITAL RECOVERY A/P	PRESENT WORTH P/A	N
1	1.2400	0.8065	1.00000	1.00	1.24000	0.8065	1
2	1.5376	0.6504	0.44643	2.24	0.68643	1.4568	2
3	1.9066	0.5245	0.26472	3.78	0.50472	1.9813	3
4	2.3642	0.4230	0.17593	5.68	0.41593	2.4043	4
5	2.9316	0.3411	0.12425	8.05	0.36425	2.7454	5
6	3.6352	0.2751	0.09107	10.98	0.33107	3.0205	6
7	4.5077	0.2218	0.06842	14.62	0.30842	3.2423	7
8	5.5895	0.1789	0.05229	19.12	0.29229	3.4212	8
9	6.9310	0.1443	0.04047	24.71	0.28047	3.5655	9
10	8.5944	0.1164	0.03160	31.64	0.27160	3.6819	10
11	10.6571	0.0938	0.02485	40.24	0.26485	3.7757	11
12	13.2148	0.0757	0.01965	50.89	0.25965	3.8514	12
13	16.3863	0.0610	0.01560	64.11	0.25560	3.9124	13
14	20.3191	0.0492	0.01242	80.50	0.25242	3.9616	14
15	25.1956	0.0397	0.00992	100.82	0.24992	4.0013	15
16	31.2426	0.0320	0.00794	126.01	0.24794	4.0333	16
17	38.7408	0.0258	0.00636	157.25	0.24636	4.0591	17
18	48.0386	0.0208	0.00510	195.99	0.24510	4.0799	18
19	59.5679	0.0168	0.00410	244.03	0.24410	4.0967	19
20	73.8641	0.0135	0.00329	303.60	0.24329	4.1103	20
22	113.5735	0.0088	0.00213	469.06	0.24213	4.1300	22
24	174.6306	0.0057	0.00138	723.46	0.24138	4.1428	24
25	216.5420	0.0046	0.00111	898.09	0.24111	4.1474	25
26	268.5121	0.0037	0.00090	1114.63	0.24090	4.1511	26
28	412.8642	0.0024	0.00058	1716.10	0.24058	4.1566	28
30	634.8199	0.0016	0.00038	2640.92	0.24038	4.1601	30
32	976.0991	0.0010	0.00025	4062.91	0.24025	4.1624	32
34	1500.850	0.0007	0.00016	6249.38	0.24016	4.1639	34
35	1861.054	0.0005	0.00013	7750.23	0.24013	4.1644	35
36	2307.707	0.0004	0.00010	9611.28	0.24010	4.1649	36
38	3548.330	0.0003	0.00007	14780.54	0.24007	4.1655	38
40	5455.913	0.0002	0.00004	22728.80	0.24004	4.1659	40
45	15994.69	0.0001	0.00002	66640.38	0.24002	4.1664	45
50	46890.43	0.0000	0.00001	195372.64	0.24001	4.1666	50

TABLE A - 25

DISCRETE CASH FLOW
25.00%   DISCRETE COMPOUND INTEREST FACTORS

	SINGLE PAYMENTS		UNIFORM SERIES PAYMENTS				
N	COMPOUND AMOUNT F/P	PRESENT WORTH P/F	SINKING FUND A/F	COMPOUND AMOUNT F/A	CAPITAL RECOVERY A/P	PRESENT WORTH P/A	N
1	1.2500	0.8000	1.00000	1.000	1.25000	0.8000	1
2	1.5625	0.6400	0.44444	2.250	0.69444	1.4400	2
3	1.9531	0.5120	0.26230	3.813	0.51230	1.9520	3
4	2.4414	0.4096	0.17344	5.766	0.42344	2.3616	4
5	3.0518	0.3277	0.12185	8.207	0.37185	2.6893	5
6	3.8147	0.2621	0.08882	11.259	0.33882	2.9514	6
7	4.7684	0.2097	0.06634	15.073	0.31634	3.1611	7
8	5.9605	0.1678	0.05040	19.842	0.30040	3.3289	8
9	7.4506	0.1342	0.03876	25.802	0.28876	3.4631	9
10	9.3132	0.1074	0.03007	33.253	0.28007	3.5705	10
11	11.6415	0.0859	0.02349	42.566	0.27349	3.6564	11
12	14.5519	0.0687	0.01845	54.208	0.26845	3.7251	12
13	18.1899	0.0550	0.01454	68.760	0.26454	3.7801	13
14	22.7374	0.0440	0.01150	86.949	0.26150	3.8241	14
15	28.4217	0.0352	0.00912	109.687	0.25912	3.8593	15
16	35.5271	0.0281	0.00724	138.109	0.25724	3.8874	16
17	44.4089	0.0225	0.00576	173.636	0.25576	3.9099	17
18	55.5112	0.0180	0.00459	218.045	0.25459	3.9279	18
19	69.3889	0.0144	0.00366	273.556	0.25366	3.9424	19
20	86.7362	0.0115	0.00292	342.945	0.25292	3.9539	20
22	135.5253	0.0074	0.00186	538.101	0.25186	3.9705	22
24	211.7582	0.0047	0.00119	843.033	0.25119	3.9811	24
25	264.6978	0.0038	0.00095	1054.791	0.25095	3.9849	25
26	330.8722	0.0030	0.00076	1319.489	0.25076	3.9879	26
28	516.9879	0.0019	0.00048	2063.952	0.25048	3.9923	28
30	807.7936	0.0012	0.00031	3227.174	0.25031	3.9950	30
32	1262.177	0.0008	0.00020	5044.710	0.25020	3.9968	32
34	1972.152	0.0005	0.00013	7884.609	0.25013	3.9980	34
35	2465.190	0.0004	0.00010	9856.761	0.25010	3.9984	35
36	3081.488	0.0003	0.00008	12321.952	0.25008	3.9987	36
38	4814.825	0.0002	0.00005	19255.299	0.25005	3.9992	38
40	7523.164	0.0001	0.00003	30088.655	0.25003	3.9995	40
45	22958.87	0.0000	0.00001	91831.496	0.25001	3.9998	45

TABLE A - 26

DISCRETE CASH FLOW
30.00%   DISCRETE COMPOUND INTEREST FACTORS

	SINGLE PAYMENTS		UNIFORM SERIES PAYMENTS		
N	COMPOUND AMOUNT F/P	PRESENT WORTH P/F	SINKING FUND A/F	COMPOUND AMOUNT F/A	CAPITAL RECOVERY A/P	PRESENT WORTH P/A	N
1	1.3000	0.7692	1.00000	1.000	1.30000	0.7692	1
2	1.6900	0.5917	0.43478	2.300	0.73478	1.3609	2
3	2.1970	0.4552	0.25063	3.990	0.55063	1.8161	3
4	2.8561	0.3501	0.16163	6.187	0.46163	2.1662	4
5	3.7129	0.2693	0.11058	9.043	0.41058	2.4356	5
6	4.8268	0.2072	0.07839	12.756	0.37839	2.6427	6
7	6.2749	0.1594	0.05687	17.583	0.35687	2.8021	7
8	8.1573	0.1226	0.04192	23.858	0.34192	2.9247	8
9	10.6045	0.0943	0.03124	32.015	0.33124	3.0190	9
10	13.7858	0.0725	0.02346	42.619	0.32346	3.0915	10
11	17.9216	0.0558	0.01773	56.405	0.31773	3.1473	11
12	23.2981	0.0429	0.01345	74.327	0.31345	3.1903	12
13	30.2875	0.0330	0.01024	97.625	0.31024	3.2233	13
14	39.3738	0.0254	0.00782	127.913	0.30782	3.2487	14
15	51.1859	0.0195	0.00598	167.286	0.30598	3.2682	15
16	66.5417	0.0150	0.00458	218.472	0.30458	3.2832	16
17	86.5042	0.0116	0.00351	285.014	0.30351	3.2948	17
18	112.4554	0.0089	0.00269	371.518	0.30269	3.3037	18
19	146.1920	0.0068	0.00207	483.973	0.30207	3.3105	19
20	190.0496	0.0053	0.00159	630.165	0.30159	3.3158	20
22	321.1839	0.0031	0.00094	1067.280	0.30094	3.3230	22
24	542.8008	0.0018	0.00055	1806.003	0.30055	3.3272	24
25	705.6410	0.0014	0.00043	2348.803	0.30043	3.3286	25
26	917.3333	0.0011	0.00033	3054.444	0.30033	3.3297	26
28	1550.293	0.0006	0.00019	5164.311	0.30019	3.3312	28
30	2619.996	0.0004	0.00011	8729.985	0.30011	3.3321	30
32	4427.793	0.0002	0.00007	14755.975	0.30007	3.3326	32
34	7482.970	0.0001	0.00004	24939.899	0.30004	3.3329	34
35	9727.860	0.0001	0.00003	32422.868	0.30003	3.3330	35

TABLE A - 27

DISCRETE CASH FLOW
35.00% DISCRETE COMPOUND INTEREST FACTORS

	SINGLE PAYMENTS		UNIFORM SERIES PAYMENTS				
N	COMPOUND AMOUNT F/P	PRESENT WORTH P/F	SINKING FUND A/F	COMPOUND AMOUNT F/A	CAPITAL RECOVERY A/P	PRESENT WORTH P/A	N
1	1.3500	0.7407	1.00000	1.000	1.35000	0.7407	1
2	1.8225	0.5487	0.42553	2.350	0.77553	1.2894	2
3	2.4604	0.4064	0.23966	4.172	0.58966	1.6959	3
4	3.3215	0.3011	0.15076	6.633	0.50076	1.9969	4
5	4.4840	0.2230	0.10046	9.954	0.45046	2.2200	5
6	6.0534	0.1652	0.06926	14.438	0.41926	2.3852	6
7	8.1722	0.1224	0.04880	20.492	0.39880	2.5075	7
8	11.0324	0.0906	0.03489	28.664	0.38489	2.5982	8
9	14.8937	0.0671	0.02519	39.696	0.37519	2.6653	9
10	20.1066	0.0497	0.01832	54.590	0.36832	2.7150	10
11	27.1439	0.0368	0.01339	74.697	0.36339	2.7519	11
12	36.6442	0.0273	0.00982	101.841	0.35982	2.7792	12
13	49.4697	0.0202	0.00722	138.485	0.35722	2.7994	13
14	66.7841	0.0150	0.00532	187.954	0.35532	2.8144	14
15	90.1585	0.0111	0.00393	254.738	0.35393	2.8255	15
16	121.7139	0.0082	0.00290	344.897	0.35290	2.8337	16
17	164.3138	0.0061	0.00214	466.611	0.35214	2.8398	17
18	221.8236	0.0045	0.00158	630.925	0.35158	2.8443	18
19	299.4619	0.0033	0.00117	852.748	0.35117	2.8476	19
20	404.2736	0.0025	0.00087	1152.210	0.35087	2.8501	20
22	736.7886	0.0014	0.00048	2102.253	0.35048	2.8533	22
24	1342.797	0.0007	0.00026	3833.706	0.35026	2.8550	24
25	1812.776	0.0006	0.00019	5176.504	0.35019	2.8556	25
26	2447.248	0.0004	0.00014	6989.280	0.35014	2.8560	26
28	4460.109	0.0002	0.00008	12740.313	0.35008	2.8565	28
30	8128.550	0.0001	0.00004	23221.570	0.35004	2.8568	30
32	14814.28	0.0001	0.00002	42323.661	0.35002	2.8569	32
34	26999.03	0.0000	0.00001	77137.223	0.35001	2.8570	34
35	36448.69	0.0000	0.00001	104136.25	0.35001	2.8571	35

TABLE A - 28

DISCRETE CASH FLOW
40.00% DISCRETE COMPOUND INTEREST FACTORS

	SINGLE PAYMENTS		UNIFORM SERIES PAYMENTS				
N	COMPOUND AMOUNT F/P	PRESENT WORTH P/F	SINKING FUND A/F	COMPOUND AMOUNT F/A	CAPITAL RECOVERY A/P	PRESENT WORTH P/A	N
1	1.4000	0.7143	1.00000	1.000	1.40000	0.7143	1
2	1.9600	0.5102	0.41667	2.400	0.81667	1.2245	2
3	2.7440	0.3644	0.22936	4.360	0.62936	1.5889	3
4	3.8416	0.2603	0.14077	7.104	0.54077	1.8492	4
5	5.3782	0.1859	0.09136	10.946	0.49136	2.0352	5
6	7.5295	0.1328	0.06126	16.324	0.46126	2.1680	6
7	10.5414	0.0949	0.04192	23.853	0.44192	2.2628	7
8	14.7579	0.0678	0.02907	34.395	0.42907	2.3306	8
9	20.6610	0.0484	0.02034	49.153	0.42034	2.3790	9
10	28.9255	0.0346	0.01432	69.814	0.41432	2.4136	10
11	40.4957	0.0247	0.01013	98.739	0.41013	2.4383	11
12	56.6939	0.0176	0.00718	139.235	0.40718	2.4559	12
13	79.3715	0.0126	0.00510	195.929	0.40510	2.4685	13
14	111.1201	0.0090	0.00363	275.300	0.40363	2.4775	14
15	155.5681	0.0064	0.00259	386.420	0.40259	2.4839	15
16	217.7953	0.0046	0.00185	541.988	0.40185	2.4885	16
17	304.9135	0.0033	0.00132	759.784	0.40132	2.4918	17
18	426.8789	0.0023	0.00094	1064.697	0.40094	2.4941	18
19	597.6304	0.0017	0.00067	1491.576	0.40067	2.4958	19
20	836.6826	0.0012	0.00048	2089.206	0.40048	2.4970	20
22	1639.898	0.0006	0.00024	4097.245	0.40024	2.4985	22
24	3214.200	0.0003	0.00012	8032.999	0.40012	2.4992	24
25	4499.880	0.0002	0.00009	11247.199	0.40009	2.4994	25
26	6299.831	0.0002	0.00006	15747.079	0.40006	2.4996	26
28	12347.67	0.0001	0.00003	30866.674	0.40003	2.4998	28
30	24201.43	0.0000	0.00002	60501.081	0.40002	2.4999	30
32	47434.81	0.0000	0.00001	118584.52	0.40001	2.4999	32
34	92972.22	0.0000	0.00000	232428.06	0.40000	2.5000	34
35	130161.1	0.0000	0.00000	325400.28	0.40000	2.5000	35

TABLE A - 29

DISCRETE CASH FLOW
45.00% DISCRETE COMPOUND INTEREST FACTORS

	SINGLE PAYMENTS			UNIFORM SERIES PAYMENTS			
N	COMPOUND AMOUNT F/P	PRESENT WORTH P/F	SINKING FUND A/F	COMPOUND AMOUNT F/A	CAPITAL RECOVERY A/P	PRESENT WORTH P/A	N
1	1.4500	0.6897	1.00000	1.000	1.45000	0.6897	1
2	2.1025	0.4756	0.40816	2.450	0.85816	1.1653	2
3	3.0486	0.3280	0.21966	4.552	0.66966	1.4933	3
4	4.4205	0.2262	0.13156	7.601	0.58156	1.7195	4
5	6.4097	0.1560	0.08318	12.022	0.53318	1.8755	5
6	9.2941	0.1076	0.05426	18.431	0.50426	1.9831	6
7	13.4765	0.0742	0.03607	27.725	0.48607	2.0573	7
8	19.5409	0.0512	0.02427	41.202	0.47427	2.1085	8
9	28.3343	0.0353	0.01646	60.743	0.46646	2.1438	9
10	41.0347	0.0243	0.01123	89.077	0.46123	2.1681	10
11	59.5728	0.0168	0.00768	130.162	0.45768	2.1849	11
12	86.3806	0.0116	0.00527	189.735	0.45527	2.1965	12
13	125.2518	0.0080	0.00362	276.115	0.45362	2.2045	13
14	181.6151	0.0055	0.00249	401.367	0.45249	2.2100	14
15	263.3419	0.0038	0.00172	582.982	0.45172	2.2138	15
16	381.8458	0.0026	0.00118	846.324	0.45118	2.2164	16
17	553.6764	0.0018	0.00081	1228.170	0.45081	2.2182	17
18	802.8308	0.0012	0.00056	1781.846	0.45056	2.2195	18
19	1164.105	0.0009	0.00039	2584.677	0.45039	2.2203	19
20	1637.952	0.0006	0.00027	3749.782	0.45027	2.2209	20
22	3548.919	0.0003	0.00013	7884.264	0.45013	2.2216	22
24	7461.602	0.0001	0.00006	16579.115	0.45006	2.2219	24
25	10319.32	0.0001	0.00004	24040.716	0.45004	2.2220	25
26	15689.02	0.0001	0.00003	34860.038	0.45003	2.2221	26
28	32984.06	0.0000	0.00001	73295.681	0.45001	2.2222	28
30	69349.99	0.0000	0.00001	154106.62	0.45001	2.2222	30
32	145806.2	0.0000	0.00000	324011.62	0.45000	2.2222	32
34	306557.6	0.0000	0.00000	681236.87	0.45000	2.2222	34
35	444508.5	0.0000	0.00000	987794.46	0.45000	2.2222	35

TABLE A - 30

DISCRETE CASH FLOW
50.00% DISCRETE COMPOUND INTEREST FACTORS

	SINGLE PAYMENTS		UNIFORM SERIES PAYMENTS				
N	COMPOUND AMOUNT F/P	PRESENT WORTH P/F	SINKING FUND A/F	COMPOUND AMOUNT F/A	CAPITAL RECOVERY A/P	PRESENT WORTH P/A	N
1	1.5000	0.6667	1.00000	1.000	1.50000	0.6667	1
2	2.2500	0.4444	0.40000	2.500	0.90000	1.1111	2
3	3.3750	0.2963	0.21053	4.750	0.71053	1.4074	3
4	5.0625	0.1975	0.12308	8.125	0.62308	1.6049	4
5	7.5938	0.1317	0.07583	13.188	0.57583	1.7366	5
6	11.3906	0.0878	0.04812	20.781	0.54812	1.8244	6
7	17.0859	0.0585	0.03108	32.172	0.53108	1.8829	7
8	25.6289	0.0390	0.02030	49.258	0.52030	1.9220	8
9	38.4434	0.0260	0.01335	74.887	0.51335	1.9480	9
10	57.6650	0.0173	0.00882	113.330	0.50882	1.9653	10
11	86.4976	0.0116	0.00585	170.995	0.50585	1.9769	11
12	129.7463	0.0077	0.00388	257.493	0.50388	1.9846	12
13	194.6195	0.0051	0.00258	387.239	0.50258	1.9897	13
14	291.9293	0.0034	0.00172	581.859	0.50172	1.9931	14
15	437.8939	0.0023	0.00114	873.788	0.50114	1.9954	15
16	656.8408	0.0015	0.00076	1311.682	0.50076	1.9970	16
17	985.2613	0.0010	0.00051	1968.523	0.50051	1.9980	17
18	1477.892	0.0007	0.00034	2953.784	0.50034	1.9986	18
19	2216.838	0.0005	0.00023	4431.676	0.50023	1.9991	19
20	3325.257	0.0003	0.00015	6648.513	0.50015	1.9994	20
22	7481.828	0.0001	0.00007	14961.655	0.50007	1.9997	22
24	16834.11	0.0001	0.00003	33666.224	0.50003	1.9999	24
25	25251.17	0.0000	0.00002	50500.337	0.50002	1.9999	25
26	37876.75	0.0000	0.00001	75751.505	0.50001	1.9999	26
28	85222.69	0.0000	0.00001	170443.39	0.50001	2.0000	28
30	191751.1	0.0000	0.00000	383500.12	0.50000	2.0000	30
32	431439.9	0.0000	0.00000	862877.77	0.50000	2.0000	32
34	970739.7	0.0000	0.00000	1941477.5	0.50000	2.0000	34

TABLE A - 31

PRESENT WORTH GRADIENT FACTORS (P/G)
DISCRETE CASH FLOW, DISCRETE COMPOUNDING

N	1%	2%	3%	4%	5%	6%	N
2	0.980	0.961	0.943	0.925	0.907	0.890	2
3	2.921	2.846	2.773	2.703	2.635	2.569	3
4	5.804	5.617	5.438	5.267	5.103	4.946	4
5	9.610	9.240	8.889	8.555	8.237	7.935	5
6	14.321	13.680	13.076	12.506	11.968	11.459	6
7	19.917	18.903	17.955	17.066	16.232	15.450	7
8	26.381	24.878	23.481	22.181	20.970	19.842	8
9	33.696	31.572	29.612	27.801	26.127	24.577	9
10	41.843	38.955	36.309	33.881	31.652	29.602	10
11	50.807	46.998	43.533	40.377	37.499	34.870	11
12	60.569	55.671	51.248	47.248	43.624	40.337	12
13	71.113	64.948	59.420	54.455	49.988	45.963	13
14	82.422	74.800	68.014	61.962	56.554	51.713	14
15	94.481	85.202	77.000	69.735	63.288	57.555	15
16	107.273	96.129	86.348	77.744	70.160	63.459	16
17	120.783	107.555	96.028	85.958	77.140	69.401	17
18	134.996	119.458	106.014	94.350	84.204	75.357	18
19	149.895	131.814	116.279	102.893	91.328	81.306	19
20	165.466	144.600	126.799	111.565	98.488	87.230	20
21	181.695	157.796	137.550	120.341	105.667	93.114	21
22	198.566	171.379	148.509	129.202	112.846	98.941	22
23	216.066	185.331	159.657	138.128	120.009	104.701	23
24	234.180	199.630	170.971	147.101	127.140	110.381	24
25	252.894	214.259	182.434	156.104	134.228	115.973	25
26	272.196	229.199	194.026	165.121	141.259	121.468	26
27	292.070	244.431	205.731	174.138	148.223	126.860	27
28	312.505	259.939	217.532	183.142	155.110	132.142	28
29	333.486	275.706	229.414	192.121	161.913	137.310	29
30	355.002	291.716	241.361	201.062	168.623	142.359	30
31	377.039	307.954	253.361	209.956	175.233	147.286	31
32	399.586	324.403	265.399	218.792	181.739	152.090	32
33	422.629	341.051	277.464	227.563	188.135	156.768	33
34	446.157	357.882	289.544	236.261	194.417	161.319	34
35	470.158	374.883	301.627	244.877	200.581	165.743	35
36	494.621	392.040	313.703	253.405	206.624	170.039	36
37	519.533	409.342	325.762	261.840	212.543	174.207	37
38	544.834	426.776	337.796	270.175	218.338	178.249	38
39	570.662	444.330	349.794	278.407	224.005	182.165	39
40	596.856	461.993	361.750	286.530	229.545	185.957	40
42	650.451	497.601	385.502	302.437	240.239	193.173	42
44	705.585	533.517	408.997	317.870	250.417	199.913	44
46	762.176	569.662	432.186	332.810	260.084	206.194	46
48	820.146	605.966	455.025	347.245	269.247	212.035	48
50	879.418	642.361	477.480	361.164	277.915	217.457	50

TABLE A - 32

PRESENT WORTH GRADIENT FACTORS (P/G)
DISCRETE CASH FLOW, DISCRETE COMPOUNDING

N	7%	8%	9%	10%	11%	12%	N
2	0.873	0.857	0.842	0.826	0.812	0.797	2
3	2.506	2.445	2.386	2.329	2.274	2.221	3
4	4.795	4.650	4.511	4.378	4.250	4.127	4
5	7.647	7.372	7.111	6.862	6.624	6.397	5
6	10.978	10.523	10.092	9.684	9.297	8.930	6
7	14.715	14.024	13.375	12.763	12.187	11.644	7
8	18.789	17.806	16.888	16.029	15.225	14.471	8
9	23.140	21.808	20.571	19.421	18.352	17.356	9
10	27.716	25.977	24.373	22.891	21.522	20.254	10
11	32.466	30.266	28.248	26.396	24.695	23.129	11
12	37.351	34.634	32.159	29.901	27.839	25.952	12
13	42.330	39.046	36.073	33.377	30.929	28.702	13
14	47.372	43.472	39.963	36.800	33.945	31.362	14
15	52.446	47.886	43.807	40.152	36.871	33.920	15
16	57.527	52.264	47.585	43.416	39.695	36.367	16
17	62.592	56.588	51.282	46.582	42.409	38.697	17
18	67.622	60.843	54.886	49.640	45.007	40.908	18
19	72.599	65.013	58.387	52.583	47.486	42.998	19
20	77.509	69.090	61.777	55.407	49.842	44.968	20
21	82.339	73.063	65.051	58.110	52.077	46.819	21
22	87.079	76.926	68.205	60.689	54.191	48.554	22
23	91.720	80.673	71.236	63.146	56.186	50.178	23
24	96.255	84.300	74.143	65.481	58.066	51.693	24
25	100.676	87.804	76.926	67.696	59.832	53.105	25
26	104.981	91.184	79.586	69.794	61.490	54.418	26
27	109.166	94.439	82.124	71.777	63.043	55.637	27
28	113.226	97.569	84.542	73.650	64.497	56.767	28
29	117.162	100.574	86.842	75.415	65.854	57.814	29
30	120.972	103.456	89.028	77.077	67.121	58.782	30
31	124.655	106.216	91.102	78.640	68.302	59.676	31
32	128.212	108.857	93.069	80.108	69.401	60.501	32
33	131.643	111.382	94.931	81.486	70.423	61.261	33
34	134.951	113.792	96.693	82.777	71.372	61.961	34
35	138.135	116.092	98.359	83.987	72.254	62.605	35
36	141.199	118.284	99.932	85.119	73.071	63.197	36
37	144.144	120.371	101.416	86.178	73.829	63.741	37
38	146.973	122.358	102.816	87.167	74.530	64.239	38
39	149.588	124.247	104.135	88.091	75.179	64.697	39
40	152.293	126.042	105.376	88.953	75.779	65.116	40
42	157.181	129.365	107.643	90.505	76.845	65.851	42
44	161.661	132.355	109.646	91.851	77.753	66.466	44
46	165.758	135.038	111.410	93.016	78.525	66.979	46
48	169.498	137.443	112.962	94.022	79.180	67.407	48
50	172.905	139.593	114.325	94.889	79.734	67.762	50

TABLE A - 33

PRESENT WORTH GRADIENT FACTORS (P/G)
DISCRETE CASH FLOW, DISCRETE COMPOUNDING

N	13%	14%	15%	18%	20%	22%	N
2	0.783	0.769	0.756	0.718	0.694	0.672	2
3	2.169	2.119	2.071	1.935	1.852	1.773	3
4	4.009	3.896	3.786	3.483	3.299	3.127	4
5	6.180	5.973	5.775	5.231	4.906	4.607	5
6	8.582	8.251	7.937	7.083	6.581	6.124	6
7	11.132	10.649	10.192	8.967	8.255	7.615	7
8	13.765	13.103	12.481	10.829	9.883	9.042	8
9	16.428	15.563	14.755	12.633	11.434	10.378	9
10	19.080	17.991	16.979	14.352	12.887	11.610	10
11	21.687	20.357	19.129	15.972	14.233	12.732	11
12	24.224	22.640	21.185	17.481	15.467	13.744	12
13	26.574	24.825	23.135	18.877	16.588	14.649	13
14	29.023	26.901	24.972	20.158	17.601	15.452	14
15	31.262	28.862	26.693	21.327	18.509	16.161	15
16	33.384	30.706	28.296	22.389	19.321	16.784	16
17	35.388	32.430	29.783	23.348	20.042	17.328	17
18	37.271	34.038	31.156	24.212	20.680	17.803	18
19	39.037	35.531	32.421	24.988	21.244	18.214	19
20	40.685	36.914	33.582	25.681	21.739	18.570	20
21	42.221	38.190	34.645	26.300	22.174	18.877	21
22	43.649	39.366	35.615	26.851	22.555	19.142	22
23	44.972	40.446	36.499	27.339	22.887	19.369	23
24	46.196	41.437	37.302	27.772	23.176	19.563	24
25	47.326	42.344	38.031	28.155	23.428	19.730	25
26	48.369	43.173	38.692	28.494	23.646	19.872	26
27	49.328	43.929	39.289	28.791	23.835	19.993	27
28	50.209	44.618	39.828	29.054	23.999	20.096	28
29	51.018	45.244	40.315	29.284	24.141	20.184	29
30	51.759	45.813	40.753	29.486	24.263	20.258	30
31	52.438	46.330	41.147	29.664	24.368	20.321	31
32	53.059	46.798	41.501	29.819	24.459	20.375	32
33	53.626	47.222	41.818	29.955	24.537	20.420	33
34	54.143	47.605	42.103	30.074	24.604	20.458	34
35	54.615	47.952	42.359	30.177	24.661	20.491	35
36	55.045	48.265	42.587	30.268	24.711	20.518	36
37	55.436	48.547	42.792	30.347	24.753	20.541	37
38	55.792	48.802	42.974	30.415	24.789	20.560	38
39	56.115	49.031	43.137	30.475	24.820	20.576	39
40	56.409	49.238	43.283	30.527	24.847	20.590	40
42	56.917	49.590	43.529	30.611	24.889	20.611	42
44	57.335	49.875	43.723	30.675	24.920	20.626	44
46	57.578	50.105	43.878	30.723	24.942	20.637	46
48	57.958	50.289	44.000	30.759	24.958	20.644	48
50	58.187	50.438	44.096	30.786	24.970	20.649	50

TABLE A - 34

PRESENT WORTH GRADIENT FACTORS (P/G)
DISCRETE CASH FLOW, DISCRETE COMPOUNDING

N	25%	30%	35%	40%	45%	50%	N
2	0.640	0.592	0.549	0.510	0.476	0.444	2
3	1.664	1.502	1.362	1.239	1.132	1.037	3
4	2.893	2.552	2.265	2.020	1.810	1.630	4
5	4.204	3.630	3.157	2.764	2.434	2.156	5
6	5.514	4.666	3.983	3.428	2.972	2.595	6
7	6.773	5.622	4.717	3.997	3.418	2.947	7
8	7.947	6.480	5.352	4.471	3.776	3.220	8
9	9.021	7.234	5.889	4.858	4.058	3.428	9
10	9.987	7.887	6.336	5.170	4.277	3.584	10
11	10.846	8.445	6.705	5.417	4.445	3.699	11
12	11.602	8.917	7.005	5.611	4.572	3.784	12
13	12.262	9.314	7.247	5.762	4.668	3.846	13
14	12.833	9.644	7.442	5.879	4.740	3.890	14
15	13.326	9.917	7.597	5.969	4.793	3.922	15
16	13.748	10.143	7.721	6.038	4.832	3.945	16
17	14.108	10.328	7.818	6.090	4.861	3.961	17
18	14.415	10.479	7.895	6.130	4.882	3.973	18
19	14.674	10.602	7.955	6.160	4.898	3.981	19
20	14.893	10.702	8.002	6.183	4.909	3.987	20
21	15.078	10.783	8.038	6.200	4.917	3.991	21
22	15.233	10.848	8.067	6.213	4.923	3.994	22
23	15.362	10.901	8.089	6.222	4.927	3.996	23
24	15.471	10.943	8.106	6.229	4.930	3.997	24
25	15.562	10.977	8.119	6.235	4.933	3.998	25
26	15.637	11.005	8.130	6.239	4.934	3.999	26
27	15.700	11.026	8.137	6.242	4.935	3.999	27
28	15.752	11.044	8.143	6.244	4.936	3.999	28
29	15.796	11.058	8.148	6.245	4.937	4.000	29
30	15.832	11.069	8.152	6.247	4.937	4.000	30
31	15.861	11.078	8.154	6.248	4.938	4.000	31
32	15.886	11.085	8.157	6.248	4.938	4.000	32
33	15.906	11.090	8.158	6.249	4.938	4.000	33
34	15.923	11.094	8.159	6.249	4.938	4.000	34
35	15.937	11.098	8.160	6.249	4.938	4.000	35
36	15.948	11.101	8.161	6.249	4.938	4.000	36
37	15.957	11.103	8.162	6.250	4.938	4.000	37
38	15.965	11.105	8.162	6.250	4.938	4.000	38
39	15.971	11.106	8.162	6.250	4.938	4.000	39
40	15.977	11.107	8.163	6.250	4.938	4.000	40
42	15.984	11.109	8.163	6.250	4.938	4.000	42
44	15.990	11.110	8.163	6.250	4.938	4.000	44
46	15.993	11.110	8.163	6.250	4.938	4.000	46
48	15.995	11.111	8.163	6.250	4.938	4.000	48
50	15.997	11.111	8.163	6.250	4.938	4.000	50

TABLE A - 35

ANNUAL COST GRADIENT FACTORS(A/G)
DISCRETE CASH FLOW, DISCRETE COMPOUNDING

N	1/2%	1%	1.5%	2%	3%	4%	5%	N
2	0.499	0.498	0.496	0.495	0.493	0.490	0.488	2
3	0.997	0.993	0.990	0.987	0.980	0.974	0.967	3
4	1.494	1.488	1.481	1.475	1.463	1.451	1.439	4
5	1.990	1.980	1.970	1.960	1.941	1.922	1.903	5
6	2.485	2.471	2.457	2.442	2.414	2.386	2.358	6
7	2.980	2.960	2.940	2.921	2.882	2.843	2.805	7
8	3.474	3.448	3.422	3.396	3.345	3.294	3.245	8
9	3.967	3.934	3.901	3.868	3.803	3.739	3.676	9
10	4.459	4.418	4.377	4.337	4.256	4.177	4.099	10
11	4.950	4.901	4.851	4.802	4.705	4.609	4.514	11
12	5.441	5.381	5.323	5.264	5.148	5.034	4.922	12
13	5.930	5.861	5.792	5.723	5.587	5.453	5.322	13
14	6.419	6.338	6.258	6.179	6.021	5.866	5.713	14
15	6.907	6.814	6.722	6.631	6.450	6.272	6.097	15
16	7.394	7.289	7.184	7.080	6.874	6.672	6.474	16
17	7.880	7.761	7.643	7.526	7.294	7.066	6.842	17
18	8.366	8.232	8.100	7.968	7.708	7.453	7.203	18
19	8.850	8.702	8.554	8.407	8.118	7.834	7.557	19
20	9.334	9.169	9.006	8.843	8.523	8.209	7.903	20
22	10.299	10.100	9.902	9.705	9.319	8.941	8.573	22
24	11.261	11.024	10.788	10.555	10.095	9.648	9.214	24
25	11.741	11.483	11.228	10.974	10.477	9.993	9.524	25
26	12.220	11.941	11.665	11.391	10.853	10.331	9.827	26
28	13.175	12.852	12.531	12.214	11.593	10.991	10.411	28
30	14.126	13.756	13.388	13.025	12.314	11.627	10.969	30
32	15.075	14.653	14.236	13.823	13.017	12.241	11.501	32
34	16.020	15.544	15.073	14.608	13.702	12.832	12.006	34
35	16.492	15.987	15.488	14.996	14.037	13.120	12.250	35
36	16.962	16.428	15.901	15.381	14.369	13.402	12.487	36
38	17.901	17.306	16.719	16.141	15.018	13.950	12.944	38
40	18.836	18.178	17.528	16.889	15.650	14.477	13.377	40
45	21.159	20.327	19.507	18.703	17.156	15.705	14.364	45
50	23.462	22.436	21.428	20.442	18.558	16.812	15.223	50
55	25.745	24.505	23.289	22.106	19.860	17.807	15.966	55
60	28.006	26.533	25.093	23.696	21.067	18.697	16.606	60
65	30.247	28.522	26.839	25.215	22.184	19.491	17.154	65
70	32.468	30.470	28.529	26.663	23.215	20.196	17.621	70
75	34.668	32.379	30.163	28.043	24.163	20.821	18.018	75
80	36.847	34.249	31.742	29.357	25.035	21.372	18.353	80
85	39.006	36.080	33.268	30.606	25.835	21.857	18.635	85
90	41.145	37.872	34.740	31.793	26.567	22.283	18.871	90
95	43.263	39.626	36.160	32.919	27.235	22.655	19.069	95
100	45.361	41.343	37.530	33.986	27.844	22.980	19.234	100

TABLE A - 36

ANNUAL COST GRADIENT FACTORS(A/G)
DISCRETE CASH FLOW, DISCRETE COMPOUNDING

N	6%	7%	8%	9%	10%	11%	12%	N
2	0.485	0.483	0.481	0.478	0.476	0.474	0.472	2
3	0.961	0.955	0.949	0.943	0.937	0.931	0.925	3
4	1.427	1.416	1.404	1.393	1.381	1.370	1.359	4
5	1.884	1.865	1.846	1.828	1.810	1.792	1.775	5
6	2.330	2.303	2.276	2.250	2.224	2.198	2.172	6
7	2.768	2.730	2.694	2.657	2.622	2.586	2.551	7
8	3.195	3.147	3.099	3.051	3.004	2.958	2.913	8
9	3.613	3.552	3.491	3.431	3.372	3.314	3.257	9
10	4.022	3.946	3.871	3.798	3.725	3.654	3.585	10
11	4.421	4.330	4.240	4.151	4.064	3.979	3.895	11
12	4.811	4.703	4.596	4.491	4.388	4.288	4.190	12
13	5.192	5.065	4.940	4.818	4.699	4.582	4.468	13
14	5.564	5.417	5.273	5.133	4.996	4.862	4.732	14
15	5.926	5.758	5.594	5.435	5.279	5.127	4.980	15
16	6.279	6.090	5.905	5.724	5.549	5.379	5.215	16
17	6.624	6.411	6.204	6.002	5.807	5.618	5.435	17
18	6.960	6.722	6.492	6.269	6.053	5.844	5.643	18
19	7.287	7.024	6.770	6.524	6.286	6.057	5.838	19
20	7.605	7.316	7.037	6.767	6.508	6.259	6.020	20
22	8.217	7.872	7.541	7.223	6.919	6.628	6.351	22
24	8.795	8.392	8.007	7.638	7.288	6.956	6.641	24
25	9.072	8.639	8.225	7.832	7.458	7.104	6.771	25
26	9.341	8.877	8.435	8.016	7.619	7.244	6.892	26
28	9.857	9.329	8.829	8.357	7.914	7.498	7.110	28
30	10.342	9.749	9.190	8.666	8.176	7.721	7.297	30
32	10.799	10.138	9.520	8.944	8.409	7.915	7.459	32
34	11.228	10.499	9.821	9.193	8.615	8.084	7.596	34
35	11.432	10.669	9.961	9.308	8.709	8.159	7.658	35
36	11.630	10.832	10.095	9.417	8.796	8.230	7.714	36
38	12.007	11.140	10.344	9.617	8.956	8.357	7.814	38
40	12.359	11.423	10.570	9.796	9.096	8.466	7.899	40
45	13.141	12.036	11.045	10.160	9.374	8.676	8.057	45
50	13.796	12.529	11.411	10.430	9.570	8.819	8.160	50
55	14.341	12.921	11.690	10.626	9.708	8.913	8.225	55
60	14.791	13.232	11.902	10.768	9.802	8.976	8.266	60
65	15.160	13.476	12.060	10.870	9.867	9.017	8.292	65
70	15.461	13.666	12.178	10.943	9.911	9.044	8.308	70
75	15.706	13.814	12.266	10.994	9.941	9.061	8.318	75
80	15.903	13.927	12.330	11.030	9.961	9.072	8.324	80
85	16.062	14.015	12.377	11.055	9.974	9.079	8.328	85
90	16.189	14.081	12.412	11.073	9.983	9.083	8.330	90
95	16.290	14.132	12.437	11.085	9.989	9.086	8.331	95
100	16.371	14.170	12.455	11.093	9.993	9.088	8.332	100

TABLE A - 37

ANNUAL COST GRADIENT FACTORS(A/G)
DISCRETE CASH FLOW, DISCRETE COMPOUNDING

N	13%	14%	15%	16%	17%	18%	19%	N
2	0.469	0.467	0.465	0.463	0.461	0.459	0.457	2
3	0.919	0.913	0.907	0.901	0.896	0.890	0.885	3
4	1.348	1.337	1.326	1.316	1.305	1.295	1.284	4
5	1.757	1.740	1.723	1.706	1.689	1.673	1.657	5
6	2.147	2.122	2.097	2.073	2.049	2.025	2.002	6
7	2.517	2.483	2.450	2.417	2.385	2.353	2.321	7
8	2.869	2.825	2.781	2.739	2.697	2.656	2.615	8
9	3.201	3.146	3.092	3.039	2.987	2.936	2.886	9
10	3.516	3.449	3.383	3.319	3.255	3.194	3.133	10
11	3.813	3.733	3.655	3.578	3.503	3.430	3.359	11
12	4.094	4.000	3.908	3.819	3.732	3.647	3.564	12
13	4.357	4.249	4.144	4.041	3.942	3.845	3.751	13
14	4.605	4.482	4.362	4.246	4.134	4.025	3.920	14
15	4.837	4.699	4.565	4.435	4.310	4.189	4.072	15
16	5.055	4.901	4.752	4.609	4.470	4.337	4.209	16
17	5.259	5.089	4.925	4.768	4.616	4.471	4.331	17
18	5.449	5.263	5.084	4.913	4.749	4.592	4.441	18
19	5.627	5.424	5.231	5.046	4.869	4.700	4.539	19
20	5.792	5.573	5.365	5.167	4.978	4.798	4.627	20
22	6.088	5.838	5.601	5.377	5.164	4.963	4.773	22
24	6.343	6.062	5.798	5.549	5.315	5.095	4.888	24
25	6.457	6.161	5.883	5.623	5.379	5.150	4.936	25
26	6.561	6.251	5.961	5.690	5.436	5.199	4.978	26
28	6.747	6.410	6.096	5.804	5.533	5.281	5.047	28
30	6.905	6.542	6.207	5.896	5.610	5.345	5.100	30
32	7.039	6.652	6.297	5.971	5.670	5.394	5.140	32
34	7.151	6.743	6.371	6.030	5.718	5.433	5.171	34
35	7.200	6.782	6.402	6.055	5.738	5.449	5.184	35
36	7.245	6.818	6.430	6.077	5.756	5.462	5.194	36
38	7.323	6.880	6.478	6.115	5.785	5.485	5.212	38
40	7.389	6.930	6.517	6.144	5.807	5.502	5.225	40
45	7.508	7.019	6.583	6.193	5.844	5.529	5.245	45
50	7.581	7.071	6.620	6.220	5.863	5.543	5.255	50
55	7.626	7.102	6.641	6.234	5.873	5.549	5.259	55
60	7.653	7.120	6.653	6.242	5.877	5.553	5.261	60
65	7.669	7.130	6.659	6.246	5.880	5.554	5.262	65
70	7.679	7.136	6.663	6.248	5.881	5.555	5.263	70
75	7.684	7.139	6.665	6.249	5.882	5.555	5.263	75
80	7.688	7.141	6.666	6.249	5.882	5.555	5.263	80
85	7.690	7.142	6.666	6.250	5.882	5.555	5.263	85
90	7.691	7.142	6.666	6.250	5.882	5.556	5.263	90
95	7.691	7.142	6.667	6.250	5.882	5.556	5.263	95
100	7.692	7.143	6.667	6.250	5.882	5.556	5.263	100

TABLE A - 38

ANNUAL CCST GRADIENT FACTORS (A/G)
DISCRETE CASH FLOW, DISCRETE COMPOUNDING

N	20%	25%	30%	35%	40%	45%	50%	N
2	0.455	0.444	0.435	0.426	0.417	0.408	0.400	2
3	0.879	0.852	0.827	0.803	0.780	0.758	0.737	3
4	1.274	1.225	1.178	1.134	1.092	1.053	1.015	4
5	1.641	1.563	1.490	1.422	1.358	1.298	1.242	5
6	1.979	1.868	1.765	1.670	1.581	1.499	1.423	6
7	2.290	2.142	2.006	1.881	1.766	1.661	1.565	7
8	2.576	2.387	2.216	2.060	1.919	1.791	1.675	8
9	2.836	2.605	2.396	2.209	2.042	1.893	1.760	9
10	3.074	2.797	2.551	2.334	2.142	1.973	1.824	10
11	3.289	2.966	2.683	2.436	2.221	2.034	1.871	11
12	3.484	3.115	2.795	2.520	2.285	2.082	1.907	12
13	3.660	3.244	2.889	2.589	2.334	2.118	1.933	13
14	3.817	3.356	2.969	2.644	2.373	2.145	1.952	14
15	3.959	3.453	3.034	2.689	2.403	2.165	1.966	15
16	4.085	3.537	3.089	2.725	2.426	2.180	1.976	16
17	4.198	3.608	3.135	2.753	2.444	2.191	1.983	17
18	4.298	3.670	3.172	2.776	2.458	2.200	1.988	18
19	4.386	3.722	3.202	2.793	2.468	2.206	1.991	19
20	4.464	3.767	3.228	2.808	2.476	2.210	1.994	20
22	4.594	3.836	3.265	2.827	2.487	2.216	1.997	22
24	4.694	3.886	3.289	2.839	2.493	2.219	1.999	24
25	4.735	3.905	3.298	2.843	2.494	2.220	1.999	25
26	4.771	3.921	3.305	2.847	2.496	2.221	1.999	26
28	4.829	3.946	3.315	2.851	2.498	2.221	2.000	28
30	4.873	3.963	3.322	2.853	2.499	2.222	2.000	30
32	4.906	3.975	3.326	2.855	2.499	2.222	2.000	32
34	4.931	3.983	3.329	2.856	2.500	2.222	2.000	34
35	4.941	3.986	3.330	2.856	2.500	2.222	2.000	35
36	4.949	3.988	3.330	2.856	2.500	2.222	2.000	36
38	4.963	3.992	3.332	2.857	2.500	2.222	2.000	38
40	4.973	3.995	3.332	2.857	2.500	2.222	2.000	40
45	4.988	3.998	3.333	2.857	2.500	2.222	2.000	45
50	4.995	3.999	3.333	2.857	2.500	2.222	2.000	50
55	4.998	4.000	3.333	2.857	2.500	2.222	2.000	55
60	4.999	4.000	3.333	2.857	2.500	2.222	2.000	60
65	5.000	4.000	3.333	2.857	2.500	2.222	2.000	65
70	5.000	4.000	3.333	2.857	2.500	2.222	2.000	70
75	5.000	4.000	3.333	2.857	2.500	2.222	2.000	75
80	5.000	4.000	3.333	2.857	2.500	2.222	2.000	80
85	5.000	4.000	3.333	2.857	2.500	2.222	2.000	85
90	5.000	4.000	3.333	2.857	2.500	2.222	2.000	90
95	5.000	4.000	3.333	2.857	2.500	2.222	2.000	95
100	5.000	4.000	3.333	2.857	2.500	2.222	2.000	100

The Modified Accelerated Cost Recovery System (MACRS) for Capital Recovery

B

This appendix presents a brief description of the accelerated cost recovery system (ACRS), which was the depreciation method in the United States from 1981 through 1986. The modified ACRS method was introduced in 1987 through the 1986 Tax Reform Act. Both systems provide statutory depreciation rates for all personal and real property while taking advantage of accelerated depreciation methods. Depreciation is computed using the relation

$$D_t = d_t B \qquad (B.1)$$

The unadjusted basis is completely removed; that is, the assumption $SV = 0$ is inherent to both systems.

ACRS and MACRS did simplify annual depreciation computations, but they removed much of the flexibility that companies previously had in selecting their own methods. Additionally, since any economic analysis utilizes future estimates as its basis, it may be performed more rapidly and about as accurately using the classic straight-line method.

Since alterations to the current methods of capital recovery are inevitable in the United States, current methods of tax-related depreciation computations may be different by the time you read this material.

Section Objectives

After completing this appendix, you should be able to do the following:

B.1 Compute the annual depreciation charge and present worth of depreciation using the ACRS method, given the depreciation rates, unadjusted basis, and recovery period.

B.2 Compute the annual depreciation charge and present worth of depreciation using the MACRS method, given the depreciation rates, recovery period, and applicable conventions.

B.3 Derive the MACRS depreciation rate for any year during the recovery period, given the recovery period and guidelines for the MACRS method.

Study Guide

B.1 Accelerated Cost Recovery System (ACRS)

The ACRS recovery rates were standardized to 3, 5, and 10 years for personal property using ADR life values similar to those of Table 13.1. The ACRS rates shown in Table B.1 have embedded DB-to-SL switching. They may start with DDB or 150% DB and switch to SL when the SL method offers faster write-off. Switching rules are not adhered to strictly for all n values and the half-year convention was imposed, thus disallowing one-half the first year depreciation for tax purposes. The alternative to ACRS was the straight-line method with longer recovery periods and the imposed half-year convention. This made SL an unattractive alternative compared to ACRS.

Example B.1

Compute the present worth of depreciation P_D using ACRS for a \$10,000 8-year asset at $i = 15\%$. Compare the result with those summarized in Example 13.4. A 5-year recovery period is used for ACRS in this case.

Table B.1 ACRS personal property recovery rates (percent) applied to unadjusted basis B effective 1981 to 1986

Recovery period, n	Recovery rate, d_t (percent) Year, t									
	1	2	3	4	5	6	7	8	9	10
3	25	38	37							
5	15	22	21	21	21					
10	8	14	12	10	10	10	9	9	9	9

Solution When the annual depreciation values and their present worth are computed using the 5-year ACRS rates in Table B.1, the result is $P_D = \$6593.47$. This is less than any of the values in Example 13.4, even straight-line values; thus, ACRS provides a less accelerated write-off.

If a true 8-year life is used for depreciation purposes, which would be the case for straight-line, the annual depreciation is \$1250 and $P_D = \$5609.13$. Now, ACRS offers an improvement in accelerated depreciation.

Probs. B.1 to B.2

B.2 Modified ACRS Depreciation Method

The modified accelerated cost recovery system (MACRS) is based upon the same philosophy as ACRS with some notable convention and numerical adjustments to recovery periods, and a number of alterations which are of more interest to the depreciation accountant than the investment analyst. Accelerated and switching depreciation techniques discussed in Chap. 13 are inherent parts of MACRS. To utilize this depreciation method the following information must be determined for the personal or real property evaluated: recovery period; applicable convention(s); depreciation method with switching allowed.

Recovery period. The ADR system is used to determine the MACRS recovery period. Table 13.1, which reflects these periods for MACRS, is used to determine n.

Applicable conventions. Two conventions apply to personal property for all n values from 3 to 20. The *half-year convention* assumes all property is placed in service at the midpoint of the year; therefore, only 50% of the applicable depreciation charge for the first year is allowed for tax purposes. This has the effect of removing some of the accelerated depreciation advantage in the first year and requiring recovery to extend to $t = n + 1$. Since this convention materially affects depreciation rates, it should be utilized in economic studies.

The *midquarter convention* must be considered in lieu of the half-year convention when actual depreciation charges are made for tax computation purposes. If more than 40% of the entire bases of property is placed in service during the final quarter of the tax year, the midquarter convention must be applied to all personal property bases for the installation year. The following recovery rates apply to the first year of a 5-year recovery period. The DDB rate for the first year, $2/n = 0.40$, is reduced by the midquarter convention multiplier.

Quarter property is placed in service	Percent of DDB depreciation allowed for first year	First year recovery rate for 5-year asset
1	87.5%	$0.875(40\%) = 35\%$
2	62.5%	$0.625(40\%) = 25\%$
3	37.5%	$0.375(40\%) = 15\%$
4	12.5%	$0.125(40\%) = 5\%$

Different rates are then used throughout the recovery period to write off the entire basis B in $n + 1$ years. Because the impact of the midquarter convention upon economic evaluation results is difficult to predict and probably small, it is not included in most economic studies nor in problems of this text. Details of its use can be found in most tax publications.

Depreciation method. The method for all personal property is the DB-to-SL model, and for real property the SL method is used to compute annual recovery rates. The half-year convention is imposed and any remaining basis is depreciated in the year $n + 1$. The value SV $= 0$ is assumed for all MACRS schedules. Since different DB percentages are used for different n values, the following summary and formulas apply in determining D_t and BV_t values. The symbols D_D and D_S are used to identify DDB and SL depreciation amounts, respectively. The year subscript t has been omitted for simplicity, but in all cases $t = 1, 2, \ldots, n$.

For $n = 3, 5, 7,$ and 10: Use DDB with the half-year convention switching to SL in the year t when $D_S > D_D$. Use the switching rules of Sec. 13.5 but add 0.5 year to the life when computing D_S to account for the half-year convention. For DDB, the uniform depreciation rate d is

$$d = \begin{cases} \dfrac{1}{n} & t = 1 \\[2mm] \dfrac{2}{n} & t = 2, 3, \ldots \end{cases} \tag{B.2}$$

Annual depreciation values for each year t applied to the adjusted basis are

$$D_D = (d)BV_{t-1} \tag{B.3}$$

$$D_S = \frac{BV_{t-1}}{n - t + 1.5} \tag{B.4}$$

When the switch to SL takes place, which is usually in the last 1 to 3 years of the recovery period, any remaining basis is charged off in year $n + 1$ to reach a book value of zero. This is usually 50% of the applicable SL amount after the switch has occurred.

For $n = 15$ and 20: Use 150% DB with the half-year convention and the switch to SL when $D_S \geqslant D_D$. Until SL is more advantageous, the annual DB depreciation is computed as $D_D = (d)BV_{t-1}$ where

$$d = \begin{cases} \dfrac{0.75}{n} & t = 1 \\[2mm] \dfrac{1.50}{n} & t = 2, 3, \ldots \end{cases} \tag{B.5}$$

For $n = 27.5$ and 31.5 (real property): Straight-line depreciation is used for all real property with only one-half the annual charge allowed in years $t = 1$ and $n + 1$. These recovery periods are not used in the examples and problems of this text.

Example B.2

A truck with a 5-year recovery has been purchased for $10,000. (a) Use MACRS to obtain the annual depreciation and book value for this asset. (b) Compute the present worth of depreciation at $i = 15\%$ per year.

Solution
(a) With $n = 5$ and the half-year convention for MACRS, use the DDB-to-SL method. Equations (B.2) through (B.4) are used in conjunction with the switching procedure of Sec. 13.5 to obtain the results in Table B.2. The switch to SL occurs in year 4 when both depreciation values are equal. Using Eqs. (B.3) and (B.4) for $t = 4$:

$$D_D = 0.4(2880) = \$1152$$

$$D_S = \frac{2880}{5 - 4 + 1.5} = \$1152$$

The recovery amount of $576 in year 6 is the result of the half-year convention. The remaining book value is removed at the equivalent of one-half the SL rate, that is, $d_6 = 0.5(1152) = \$576$. The amount for $t = 5$, $D_D = \$691.20$ is shown for illustration purposes only; it need not be computed since the SL values will always exceed DB values once the switch is made.
(b) Equation (13.11) and the selected D_t values in Table B.2 are used to compute

$$P_D = \sum_{t=1}^{6} D_t(P/F, 15\%, t) = \$6901.61$$

This number, as expected, is about the same as $P_D = \$6901.20$ computed in Example 13.4 using the tabulated rates in Table 13.4.

Comment Since the data for this example and that for Examples 13.3, 13.4, and B.1 are the same, the P_D values can be compared to determine which method actually offers the most accelerated write-off of $10,000 over a 5-year period. Figure B.1 presents a comparison for five methods with P_D values rounded to the nearest dollar. The switching method is the best performer. ACRS is less accelerated than MACRS because 150% DB-to-SL switching is inherent in the 5-year ACRS, whereas the 200% DB (that is, DDB) is used in the MACRS method. The half-year convention is included in both methods.

Table B.2 Computation of depreciation using MACRS for $n = 5$ and $B = \$10,000$

Years, t	DDB d	DDB D_D	SL D_S	Selected D_t	BV_t
0	—	—	—	—	$10,000
1	0.2	$2,000.00	$1,818.18	$2,000	8,000
2	0.4	3,200.00	1,777.78	3,200	4,800
3	0.4	1,920.00	1,371.43	1,920	2,880
4	0.4	1,152.00	1,152.00	1,152	1,728
5	0.4	691.20	1,152.00	1,152	576
6	—	—	576.00	576	0
				$10,000	

Figure B.1
Comparison of
the present worth
of depreciation
values for different
methods of capital
recovery.

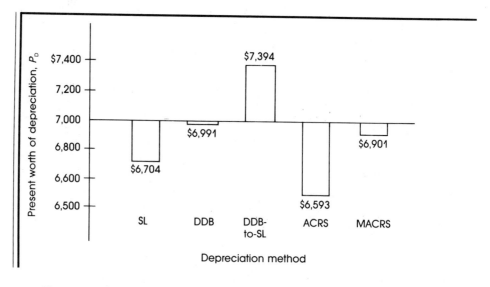

However, until tax laws are altered once again, MACRS is the only method that is legal for most depreciable assets.

Probs. B.3 to B.6

B.3 Derivation of the MACRS Depreciation Rates

For computation purposes, it is easier to use the recovery rates in Table 13.4 and Eq. (B.1) to determine annual depreciation directly from the unadjusted basis B. The logic behind the MACRS rates is described here for those interested in their derivation. (This rationale is the same for the ACRS rates, but the numbers are slightly different.)

The annual MACRS recovery rates may be derived using the applicable uniform percentage rate d for the DB method. The subscripts D and S have been inserted with the year t for the DDB and SL methods, respectively. For the first year

$$d_{D,1} = \frac{1}{n} \tag{B.6}$$

For years $t = 2, 3, \ldots, n$,

$$d_{D,t} = d\left(1 - \sum_{i=1}^{t-1} d_i\right) \tag{B.7}$$

$$d_{S,t} = \frac{1 - \sum_{i=1}^{t-1} d_i}{n - t + 1.5} \tag{B.8}$$

Also, for the year $n + 1$, the rate is

$$d_{S,n+1} = \frac{1}{2} d_{S,n} \tag{B.9}$$

The results of Eqs. (B.7) and (B.8) are compared for each t value to determine which is larger and when the switch to SL depreciation should occur. The MACRS rates in Table 3.4 (rounded) present the resulting d_t values.

Example B.3

Verify the MACRS rates in Table 13.4 for a 3-year recovery period. The depreciation rates in percent are 33.3, 44.5, 14.8, and 7.4 for years 1, 2, 3, and 4, respectively.

Solution The uniform rate for DDB when $n = 3$ is $d = \frac{2}{3} = 0.6667$. Using the half-year convention in year 1 and Eqs. (B.6) through (B.9), the results are:

d_1: $d_{D,1} = 0.5d = 0.5(0.6667) = 0.3333$
d_2: Cumulative depreciation rate is 0.3333

$$d_{D,2} = 0.6667(1 - 0.3333) = 0.4445 \qquad \text{(larger } d_2)$$

$$d_{S,2} = \frac{1 - 0.3333}{3 - 2 + 1.5} = 0.2667$$

d_3: Cumulative depreciation rate is 0.7778

$$d_{D,3} = 0.6667(1 - 0.7778) = 0.1481$$

$$d_{S,3} = \frac{1 - 0.7778}{3 - 3 + 1.5} = 0.1481$$

Both d_3 values are the same; switch to straight line.
d_4: This rate is 50% of the last SL rate

$$d_{S,4} = 0.5(d_{S,3}) = 0.5(0.1481) = 0.0741$$

The selected rates for the 4 years are 0.3333, 0.4445, 0.1481, and 0.0741 which are easily rounded to the values in Table 13.4.

Comment The DDB and SL rates for year 3 are equal because the 0.5 year is added to the denominator for $d_{S,3}$. The multipliers are the same for both methods.

As mentioned in Sec. 13.6, the only alternative to MACRS for most personal depreciable property is the SL method with the half-year convention and a prescribed recovery period. This alternative is often used by companies which are young and do not need the tax benefit of accelerated depreciation during the first few years of operation and capital-asset ownership.

Probs. B.7, B.8

Further Information

Standard Federal Tax Reports, *Depreciation Guide: Featuring ACRS and Modified ACRS (MACRS)*, Commerce Clearing House, Inc., Chicago, 1987.

Problems

B.1 Examine the current depreciation tax laws for capital recovery in your country and compare them with the classical depreciation methods of SL, DB, and SYD. Also compare them with the ACRS and MACRS methods. In your comparison, concentrate upon the differences in annual recovery rates for the methods.

B.2 Derive the 3-year recovery period rates for ACRS given in Table B.1. Start with the 150% DB method in year 1 and switch to SL when it offers a larger recovery rate. Do not forget the half-year convention imposed by ACRS in the first year.

B.3 Using a single graph, compare the book value under ACRS and MACRS depreciation for a 3-year recovery period and the following asset data: $B = \$50,000$ and an expected $SV = \$5000$.

B.4 Give an example in which the MACRS midquarter convention must be imposed. What is the net effect of this convention upon the amount of depreciation allowed (*a*) in the first year of the recovery period? (*b*) Over the life of the asset? (*c*) Does this convention tend to increase or decrease the present worth of depreciation? Why?

B.5 An automated drafting system was purchased 3 years ago at a cost of $30,000. A 5-year recovery period and MACRS depreciation have been used to write off the basis. The system is to be prematurely replaced with a trade-in value of $5000. What is the difference between the book value and the trade-in value?

B.6 Use the computations in Eqs. (B.2) through (B.4) to determine the MACRS annual depreciation for the following asset data: $B = \$50,000$, ADR life $= 13$ years.

B.7 Verify the MACRS rates in Table 13.4 for a 5-year recovery period by switching from DDB to SL rates when the straight-line method offers faster write-off.

B.8 The 3-year MACRS recovery rates are 33.3%, 44.5%, 14.8%, and 7.4% respectively. (*a*) What are the corresponding SL rates for the MACRS alternative with half-year convention? (*b*) Compare the present worth of depreciation for these two sets of rates if $B = \$80,000$ and $i = 10\%$ per year.

Basics of Accounting and Cost Allocation

C

The overall objective of this appendix is to give you an idea about the basics of financial statements and cost-accounting records, the information they contain, and how that information may be used in an economic study.

By no means is this material thorough enough to permit the engineering economist who has mastered it to practice accounting. However, some of the basic approaches of cost accounting discussed here will assist in gathering data for the study to be performed.

Section Objectives

After completing this appendix, you should be able to do the following:

C.1 For a *balance sheet*, name and explain the major accounting categories and state the basic equation used in its preparation. Given the financial data, prepare a balance sheet.

C.2 For an *income statement* and a *cost-of-goods-sold statement*, name and explain the major categories and state the basic equations used in their preparation. Given the financial data, prepare each of these statements.

C.3 Compute and give a meaning for each of the following *business ratios*, given a balance sheet, income statement, and cost-of-goods-sold statement: current ratio, acid-test ratio, equity ratio, income ratio, and inventory-turnover ratio.

C.4 Compute *overhead rates*, given the allocation basis, total overhead expenses, and estimated activity level.

C.5 Compute the *actual cost of overhead* and the *variance*, given the allocation basis, overhead rate, and observed activity level.

Study Guide

C.1   The Balance Sheet

At the end of each fiscal year, a company publishes a *balance sheet*. A sample balance sheet for the Wimble Corporation is presented in Table C.1. This is a yearly presentation of the state of the firm at a particular time—for example, December, 19XX; however, a balance sheet is also usually prepared quarterly and monthly. Note that three main categories are used.

Assets. A summary of all resources owned by or owed to the firm. Two classes of assets are distinguishable. *Current assets* represent the short-lived working capital of the company (cash, accounts receivable, etc.), which can be converted to cash in approximately 1 year. Long-lived assets are referred to as *fixed*. Conver-

Table C.1 Sample balance sheet

Wimble Corporation
Balance Sheet
December 31, 19XX

Assets				Liabilities		
Current						
Cash	$ 10,500			Accounts payable	$ 19,700	
Accounts receivable	18,700			Dividends payable	7,000	
Interest accrued receivable	500			Long-term notes payable	16,000	
Inventories	52,000			Bonds payable	20,000	
Total current assets		$ 81,700		Total liabilities		$ 62,700
Fixed				**Net worth**		
Land		$ 25,000		Common stock	$275,000	
Building and equipment	$438,000			Preferred stock	100,000	
Less:						
Depreciation allowance	82,000	356,000		Retained earnings	25,000	
Total fixed assets			381,000	Total net worth		400,000
Total assets			$462,700	Total liabilities and net worth		$462,700

sion of these holdings (land, equipment, etc.) to cash in a short period of time would require a major company reorientation.

Liabilities. A summary of financial obligations of the company.

Net worth (equity). A summary of the worth of ownership, including outstanding stock issues and earnings retained for expansion.

The balance sheet is constructed using the basic equation

Assets = liabilities + net worth

Note in Table C.1 that each major category is subdivided into particular titles. For example, accounts receivable represents all money owed to the firm by its customers. Further utilization of some of these subcategories is made when business ratios are discussed.

Prob. C.1

C.2 The Income Statement and the Cost-of-Goods-Sold Statement

A second important financial statement utilized to present the relations between the major categories in the balance sheet is the *income statement* (Table C.2). The income statement shows this relation by summarizing the profit or loss situation of the firm for a preceding, stated amount of time. The major categories of an income statement are:

Revenues. All sales and interest revenue that the company has received in the past accounting period.

Expenses. A summary of all expenses for the accounting period. Some expense values (for example, income taxes, and cost of goods sold) are itemized in other statements.

Table C.2 Sample income statement

<table>
<tr><td colspan="3">**Wimble Corporation**
Income Statement
Year Ended December 31, 19XX</td></tr>
<tr><td>Revenues</td><td></td><td></td></tr>
<tr><td>Sales</td><td>$505,000</td><td></td></tr>
<tr><td>Interest revenue</td><td>3,500</td><td></td></tr>
<tr><td>Total revenues</td><td></td><td>$508,500</td></tr>
<tr><td>Expenses</td><td></td><td></td></tr>
<tr><td>Cost of goods sold (Table C.3)</td><td>$290,000</td><td></td></tr>
<tr><td>Selling</td><td>28,000</td><td></td></tr>
<tr><td>Administrative</td><td>35,000</td><td></td></tr>
<tr><td>Other</td><td>12,000</td><td></td></tr>
<tr><td>Total expenses</td><td></td><td>365,000</td></tr>
<tr><td>Income before taxes</td><td></td><td>143,500</td></tr>
<tr><td>Taxes for year</td><td></td><td>71,750</td></tr>
<tr><td>Net profit for year</td><td></td><td>$ 71,750</td></tr>
</table>

Table C.3 Sample cost-of-goods-sold statement

Wimble Corporation
Statement of Cost of Goods Sold
Year Ended December 31, 19XX

Materials

Inventory, January 1, 19XX	$ 54,000	
Purchases during year	174,500	
Total	$228,500	
Less: Inventory December 31, 19XX	50,000	
Cost of materials		$178,500
Direct labor		110,000
Prime cost		288,500
Factory expense		7,000
Factory cost		295,500
Less: Increase in finished goods inventory during year		5,500
Cost of goods sold Table C.2)		$290,000

The income statement is published at the same time as the balance sheet. The income statement uses the equation

Revenues − expenses = profit (or loss)

The *cost of goods sold* is an important accounting term. It represents the net cost of producing the product marketed by the firm. A statement of the cost of goods sold, such as that shown in Table C.3, is useful in determining exactly how much it costs to make a particular product over a stated time period, usually a year. Note that the total of the cost-of-goods-sold statement is entered as an expense item on the income statement. This total is determined using the relations

Direct materials + direct labor = prime cost

Prime cost + factory expense = cost of goods sold   (C.1)

The item "factory expense" includes all overhead charged to produce a product. Overhead accumulation systems are discussed later in this appendix. Cost of goods sold is also called *factory cost*.

Probs. C.2 and C.3

C.3 Business Ratios

Accountants frequently utilize ratio analysis to evaluate the state of a company over time and in relation to industry norms. Because the economic analyst must communicate with the accountant, the analyst should have an idea of the ratios commonly employed. Information for these ratios is extracted from the balance sheet and income statement. Tables C.1 and C.2 are used for the numerical examples of this section.

Current ratio. This ratio, utilized to analyze the company's working-capital condition, is defined as

$$\text{Current ratio} = \frac{\text{current assets}}{\text{current liabilities}}$$

Current liabilities include all short-term debts, such as accounts and dividends payable. Note that only balance-sheet data are utilized in the current ratio; thus, no association with revenues or expenses is made. For the balance sheet of Table C.1 current liabilities amount to $19,700 + $7000 = $26,700 and

$$\text{Current ratio} = \frac{81,700}{26,700} = 3.06$$

Since current liabilities are those debts payable in the next year, the current-ratio value of 3.06 means that the current assets would "cover" short-term debts approximately three times. Current-ratio values of 2 to 3 are common. For comparison purposes, it is necessary to compute the current ratio, and all other ratios, for several companies in the same industry. Industrywide median ratio values are periodically published by firms such as Dun and Bradstreet in *Key Business Ratios*. You should realize that the current ratio assumes that the working capital invested in inventory can be converted to cash quite rapidly. Often, however, a better idea of a company's *immediate* financial position can be obtained by using the acid-test ratio.

Acid-test ratio (quick ratio). This ratio, defined by the relation

$$\text{Acid-test ratio} = \frac{\text{quick assets}}{\text{current liabilities}} = \frac{\text{current assets} - \text{inventories}}{\text{current liabilities}}$$

is meaningful for the emergency situation when the firm must cover short-term debts using its readily convertible assets. For the Wimble Corporation,

$$\text{Acid-test ratio} = \frac{81,700 - 52,000}{26,700} = 1.11$$

Comparison of this and the current ratio shows that approximately two times the current debts of the company are invested in inventories. However, an acid-test ratio of approximately 1.0 is generally regarded as a strong *current* position, regardless of the amount of assets in inventories.

Equity ratio. This ratio has historically been a measure of financial strength since it is defined as

$$\text{Equity ratio} = \frac{\text{stockholder's equity}}{\text{total assets}}$$

For the Wimble Corporation,

$$\text{Equity ratio} = \frac{400,000}{462,700} = 0.865$$

that is, 86.5% of Wimble is stockholder-owned. The $25,000 retained earnings is also called *equity* since it is actually owned by the stockholders, not the corporation. An equity ratio in the range 0.80 to 1.0 usually indicates sound financial condition, with little fear of forced reorganization because of unpaid liabilities. However, a company with virtually no debts, that is, one with a very high equity ratio, may not have a promising future, because of its inexperience in dealing with short- and long-term debt financing. The debt-equity (D-E) mix is another measure of financial strength (Chap. 18).

Income ratio. This often quoted, and often misused, ratio is defined as

$$\text{Income ratio} = \frac{\text{net profit}}{\text{total revenue}} \ (100\%)$$

Net profit is the after-tax value from the income statement. For the Wimble Corporation,

$$\text{Income ratio} = \frac{71,750}{508,500} \ (100\%) = 14.1\%$$

Such a high percentage as this for the Wimble Corporation may, at first glance, look highly favorable. However, often a large income ratio reflects an inability to earn a high rate of return on resources owned and employed by the firm. In recent years, corporations have pointed to small income ratios, say 2.5 to 4.0%, as indications of sagging economic conditions. In truth, for a relatively large-volume, high-turnover business, an income ratio of 3% is quite healthy. Of course, a steadily decreasing ratio indicates rising company expenses, which absorb the potentially higher net profit after taxes.

Inventory turnover ratio. This ratio indicates the number of times the average inventory value passes through the operations of the company. If turnover of inventory to net sales is desired, the formula is

$$\text{Net sales to inventory} = \frac{\text{net sales}}{\text{average inventory}}$$

where average inventory is the figure recorded in the balance sheet. For the Wimble Corporation this ratio is

$$\text{Net sales to inventory} = \frac{505,000}{52,000} = 9.71$$

which means that the average value of the inventory has been sold 9.71 times during the year. Values of this ratio vary greatly from one industry to another.

If inventory turnover is related to cost of goods sold, the ratio to use is

$$\text{Cost of goods sold to inventory} = \frac{\text{cost of goods sold}}{\text{average inventory}}$$

where average inventory is computed as the average of the beginning and ending inventory values in the statement of cost of goods sold. This ratio is commonly used as a measure of the inventory turnover rate in manufacturing companies. It varies with industries, but management likes to see it remain relatively constant as business increases. For Wimble, using the values in Table C.3

$$\text{Cost of goods sold to inventory} = \frac{290{,}000}{1/2(54{,}000 + 50{,}000)} = 5.58$$

There are naturally many other ratios a company's personnel can use in various circumstances; however, the ones presented here are commonly used by both accountants and economic analysts.

Example C.1

The median value for three ratios of nationally surveyed companies are presented here. Compare the Wimble Corporation ratios with these norms and comment on differences and similarities.

Ratio	House furnishings (SIC 2392)[†]	Manufactured ice (SIC 2097)	Soaps and detergents (SIC 2841)
Current	1.89	1.37	2.71
Net sales to inventory	7.7	37.3	8.9
Income	2.85%	6.71%	3.37%

[†] SIC—Standard Industrial Code

Solution   The corresponding ratio values for Wimble computed above are

Current ratio $= 3.06$

Net sales to inventory $= 9.71$

Income ratio $= 14.1\%$

If Wimble makes furniture or ice, its current ratio is higher than expected, since it can cover current liabilities over three times compared with the averages of 1.89 and 1.37 found in the survey. If Wimble makes ice, its inventory turnover is too low compared with the 37.3 times published for other companies.

If Wimble makes soaps and detergents, it is probably better off using these few ratios as criteria, since the current and inventory turnover are comparable and the income ratio is much higher than that of other companies in the same SIC classification.

Probs. C.4 and C.5

C.4   Allocation of Factory Expense

All costs incurred in the production of an item are accounted for by the cost-accounting system. It can be generally stated that the statement of cost of goods sold (Table C.3) is the end product of this system. Cost accounting accumulates the material, direct labor, and factory-expense costs by using *cost centers*. For example, a department or a machine may be a cost center. All costs incurred in the department or in the utilization of the machine are thus collected under one cost-center title, such as machine 300. Since direct materials and labor are assignable to a cost center, the accountant has only to keep track of these costs. Of course, this is no easy chore

Table C.4 Factory-expense allocation bases

Overhead category	Allocation basis
Taxes	Space occupied
Heat, light	Space, usage, number of outlets
Power	Space, direct labor hours, direct labor cost, machine hours
Receiving, purchasing	Cost of materials, number of orders, number of items
Personnel, machine shop	Direct labor hours, direct labor cost
Building maintenance	Space occupied, direct labor cost
Software development	Cycle time, throughput
Computers in manufacturing	Number of transactions, processing time

and often the cost of the cost-accounting system prohibits collection of direct costing data in the detail that the accountant or the economist might desire.

By far the most burdensome chore of cost accounting is the allocation of factory expense, or overhead. The costs associated with property taxes, service and maintenance departments, personnel, supervision, utilities, etc., must be allocated to the using cost center. Detailed collection of these data is cost prohibitive and often impossible; thus, allocation schemes are utilized to distribute the expenses on a reasonable basis. A description of some of the bases is given in Table C.4. The most common bases used are direct labor cost and direct labor hours.

Most allocation is accomplished utilizing a predetermined factory expense rate, computed using the relation

$$\text{Overhead rate} = \frac{\text{estimated overhead costs}}{\text{estimated basis level}} \qquad (C.2)$$

As automation has increased, the number of direct labor hours expended to manufacture a product or complete a service has decreased substantially. Where once as much as 35 to 50% of the final cost of a product was represented in labor, now in a large percentage of industry, the labor component is 5 to 15%. The use of direct labor hours as a basis for overhead allocation is no longer appropriate. This has led to the development of *indirect cost-tracking systems* which are able to directly charge much of the overhead that was historically allocated. Alternatively, different bases have become useful. As shown in Table C.4, categories such as computer software used in manufacturing support and control functions—a heavy user of indirect labor—can be allocated using bases such as throughput (production rate) and cycle time (production time per unit). Further discussion of indirect costs is included in Chap. 20, especially Sec. 20.3.

The estimated overhead cost is the amount allocated to the particular cost center. For example, if a company has two producing departments, the overhead allocated to *each department* is used in Eq. (C.2) when determining the departmental overhead rate. Example C.2 illustrates the use of Eq. (C.2) when the cost center is a machine.

Example C.2

You are attempting to compute overhead rates for the production of glass bookends. The following information is obtained from last year's budget for the three machines used to produce this product.

Cost source	Allocation basis	Estimated activity level
Machine 1	Labor cost	$10,000
Machine 2	Labor hours	2,000 hours
Machine 3	Material cost	$12,000

Determine overhead rates for each machine if the estimated factory expense budget for producing the bookends is $5000 per machine.

Solution   Applying Eq. (C.2) for each machine, rates for the year are

$$\text{Machine 1 rate} = \frac{\text{overhead budget}}{\text{direct labor cost}} = \frac{5000}{10,000}$$

$$= \$0.50 \text{ per dollar labor}$$

$$\text{Machine 2 rate} = \frac{\text{overhead budget}}{\text{direct labor hours}} = \frac{5000}{2000}$$

$$= \$2.50 \text{ per labor hour}$$

$$\text{Machine 3 rate} = \frac{\text{overhead budget}}{\text{material cost}} = \frac{5000}{12,000}$$

$$= \$0.42 \text{ per material dollar}$$

Comment   Once the product has been manufactured and actual direct labor costs, hours, and material costs are computed, each dollar of direct labor cost spent on machine 1 implies that $0.50 in factory expense will be added to the cost of the product. Similar expense values will be added for machines 2 and 3.

The amount of overhead allocated to an asset is a component of the annual operating cost (AOC). Other elements of AOC are directly assignable expenses such as insurance, direct labor cost, and some indirect costs.

Additional Example C.4
Probs. C.6 to C.9

C.5   Factory-Expense Computation and Variance

Once the overhead rates are computed and data collected, it is possible to actually determine the cost of production. This cost may be to operate a department or a machine or to manufacture a product, depending upon the manner in which the cost centers are set up. A *cost center* is simply a term used for the specific overhead account; for example, department *A* may be one cost center.

Assuming that the estimated overhead budget was correct, we know that total allocated overhead should equal estimated overhead. However, since some error in budgeting always exists, there will be some over- or underallocation of overhead, termed *variance*. Experience in overhead estimation assists in reducing the variance

Table C.5 Actual data used for overhead allocation

Cost source	Machine number	Actual cost	Actual hours
Material	1	$ 3,800	
	3	19,550	
Labor	1	2,500	650
	2	3,200	750
	3	2,800	720

at the end of the accounting period. Example C.3 illustrates overhead allocation and variance computation.

Example C.3

Since you collected the data and determined overhead rates in Example C.2, you are now ready to compute the actual cost of production. Perform the computations using the basis data in Table C.5. Also, calculate the variance for factory expense.

Solution To determine actual production costs, use the factory cost relation given in Eq. (C.1). That is,

Factory cost = materials + labor + factory expense

To determine factory expense, which is the overhead expense, the rates from Example C.2 are utilized as follows:

Machine 1 overhead = (labor cost)(rate) = 2500(0.50)

$\qquad$ = $1250

Machine 2 overhead = (labor hours)(rate) = 750(2.50)

$\qquad$ = $1875

Machine 3 overhead = (material cost)(rate) = 19,550(0.42)

$\qquad$ = $8211

Total factory overhead expense = $11,336

Thus, factory cost is the sum of material, labor, and overhead charges, or a total of $43,186 for the year.

$\qquad$ The variance for factory expense in this example is 3($5000) − $11,336 = $3664 under the budget. The $15,000 budget for the three machines represents a 32.3% allocation over actual overhead. This analysis should allow a more realistic overhead budget for the production in the future years.

$\qquad$ Actually, once estimates of overhead costs are determined, it is possible to make an economic analysis of the present operation versus a proposed or anticipated operation. Such a study is explained in the Additional Examples.

Additional Example C.5
Probs. C.10 to C.13

Table C.6 Overhead allocation for 1 year for Example C.4

Allocation	Stereo processing	Dinette processing	Finishing
Overhead dollars	$145,000	$145,000	$10,000
Direct labor hours	20,000	10,000	6,000

Additional Examples

Example C.4

The ABC Company produces many products, two of which are stereo cabinets and Formica-topped dinette sets. A total of $33,000 is allocated to overhead for next year. Management wants to determine overhead rates on the basis of direct labor hours for the two processing and one finishing departments (a) individually and (b) using an overall (blanket) basis. Table C.6 presents departmental overhead allocation and processing time. Develop the rates for management.

Solution
(a) The rate for each department is computed using Eq. (C.2):

$$\text{Stereo processing} = \frac{145,000}{20,000} = \$7.25 \text{ per hour}$$

$$\text{Dinette processing} = \frac{145,000}{10,000} = \$14.50 \text{ per hour}$$

$$\text{Finishing} = \frac{10,000}{6000} = \$1.67 \text{ per hour}$$

(b) A blanket rate is found by computing

$$\text{Overhead rate} = \frac{\text{total overhead allocation}}{\text{total labor hours}}$$

$$= \frac{300,000}{36,000}$$

$$= \$8.33 \text{ per hour}$$

Comment   Of course, an overall rate is easier to compute and use; however, it cannot account for differences in the nature of the work of departments "blanketed" under the rate.

Example C.5

For several years a certain company has purchased the motor and frame assembly of its major product line at an annual cost of $1.5 million. The thought now is to make the components in-house, using existing departmental facilities. For the three departments involved the overhead rates, estimated material, labor, and hours are quoted in Table C.7. The allocated hours column is the time necessary to produce the motor and frame only.

To produce the products, equipment must be purchased. The machinery has a first cost of $2 million, a salvage value of $50,000, and a life of 10 years. Perform an economic analysis of the suggestion to make the components, assuming a 15% per year return is required.

Table C.7 Production cost estimates for Example C.5

Department	Basis, hours	Overhead Rate per hour	Allocated hours	Material cost	Direct labor cost
A	Labor	$10	25,000	$200,000	$200,000
B	Machine	5	25,000	50,000	200,000
C	Labor	15	10,000	50,000	100,000
				$300,000	$500,000

Solution For making the components in-house, the annual operating costs (AOC) are composed of labor, material, and overhead costs. Using the data of Table C.7, we find that overhead is

Department A: 25,000(10) =	$250,000
Department B: 25,000(5) =	125,000
Department C: 10,000(15) =	150,000
	$525,000

Thus,

$$AOC = 500,000 + 300,000 + 525,000 = \$1,325,000$$

The EUAW is

$$EUAW_{make} = P(A/P, i\%, n) - SV(A/F, i\%, n) + AOC$$
$$= 2,000,000(A/P, 15\%, 10) - 50,000(A/F, 15\%, 10) + 1,325,000$$
$$= \$1,721,037$$

Currently,

$$EUAW_{buy} = \$1,500,000$$

Therefore, it is cheaper to continue to buy the motor and frames assembly.

Problems The following financial data have been gathered for the month of July 19XX for the Stop-Gap Company. You will use this information in Probs. C.1 to C.5.

Present situation, July 31, 19XX

Account	Balance
Accounts payable	$ 35,000
Accounts receivable	29,000
Bonds payable (20 year)	110,000
Buildings (net value)	605,000
Cash on hand	17,000
Dividends payable	8,000
Inventory value (all inventories)	31,000
Land value	450,000
Long-term mortgage payable	450,000
Retained earnings	154,000
Stock value outstanding	375,000

Transactions for July, 19XX

Category		Amount
Direct labor		$ 50,000
Expenses		
Insurance	$ 20,000	
Selling	62,000	
Rent and lease	40,000	
Salaries	110,000	
Other	62,000	
Total		294,000
Income taxes		20,000
Increase in finished goods inventory		25,000
Materials inventory, July 1, 19XX		46,000
Materials inventory, July 31, 19XX		25,000
Materials purchases		20,000
Overhead charges		75,000
Revenue from sales		500,000

C.1 Using the data summary in the tables, (a) construct a balance sheet for the Stop-Gap Company as of July 31, 19XX and (b) determine the value of each term in the basic equation of the balance sheet.

C.2 (a) Prepare the cost-of-goods-sold statement for July 19XX. (b) What was the net change in materials inventory value during the month?

C.3 Use the account summary above to develop (a) an income statement for July 19XX and (b) the basic equation of the income statement. (c) What percent of revenue is reported as after-tax income?

C.4 (a) Compute the value of each accounting ratio that uses only balance-sheet information from the statement you constructed in Prob. C.1. (b) What percent of the company's current debt is "tied up" in inventory?

C.5 (a) Compute the turnover of inventory (based on net sales) and the income ratio for the Stop-Gap Company. (b) What percent of each sales dollar can the company rely upon as profit?

C.6 A company has a processing department with 25 machines. Because of the nature and use of three of these machines, each is considered a separate account for factory overhead. The remaining 22 machines are grouped under one account number, M104. Machine hours are used as an overhead allocation basis for all machines. A total of $25,000 in overhead is allocated to the department for next year. Using the data below, determine (a) the overhead rate for each account and (b) the blanket rate for the department.

Account number	Overhead allocated	Estimated machine hours
M101	$ 5,000	600
M102	5,000	200
M103	5,000	800
M104	10,000	1,600

C.7 A machining department's overhead is allocated by accounting. The department manager has just obtained records of allocation rates and actual charges for the prior months and estimates for this month (May) and next month. However, the basis of allocation is not indicated and accounting has no record of the basis used. You, an engineer with the company, are asked to investigate the allocation for each month. You are given the following information by the department manager as it comes from accounting.

		Overhead	
Month	Rate	Allocated	Charged
February	$1.40	$2,800	$2,600
March	1.33	3,400	3,800
April	1.37	3,500	3,500
May	1.03	3,600	
June	0.92	6,000	

You collect the following from departmental and accounting records.

	Direct labor		Material	Space, square
Month	Hours	Costs	costs	feet
February	640	$2,560	$5,400	2,000
March	640	2,560	4,600	2,000
April	640	2,560	5,700	3,500
May	640	2,720	6,300	3,500
June	800	3,320	6,500	3,500

(*a*) Determine the actual allocation basis used each month and (*b*) comment on the decreasing overhead rate published each month by accounting.

C.8 An electronics manufacturing company has five separate departments. Departmental charges for a certain month are given below. Also detailed are the space allocations, direct labor hours, and direct labor costs for each producing department.

	Overhead	Space, square	Direct labor	
Department	costs	feet	Hours	Costs
Preparation	$20,000	10,000	480	$1,680
Subassemblies	15,000	18,000	1,000	3,250
Final assembly	10,000	6,000	600	2,460
Quality control	5,000	1,200		
Engineering	9,000	2,000		

Determine the departmental overhead rates to be used in making a redistribution of the overhead cost for quality control and engineering ($14,000) to the other three departments on the basis of (*a*) space, (*b*) direct labor hours, and (*c*) direct labor costs.

C.9 Compute the actual overhead allocation and department variance for Prob. C.6 using (*a*) the individual account rates and (*b*) the blanket rate. The actual hours credited to each account are as follows: M101, 700 hours; M102, 350 hours; M103, 650 hours; M104, 1300 hours.

C.10 Use Eq. (C.2) and the different bases listed in Table C.4 to explain why a decrease in direct labor hours (and an increase in indirect labor hours) due to automation on a production line may require the use of new bases to allocate overhead costs.

C.11 Overhead allocation rates and the allocation bases for the six producing departments of the E-Z-Duz-It Calculator Manufacturers are listed below. (*a*) Use the reported data to distribute overhead to the departments. (*b*) Compute the variance of the allocation if a total of $750,000 had been allocated.

Department	Allocation basis[†]	Rate	Direct labor hours	Direct labor cost	Machine hours
1	DLH	$2.50	5,000	$20,000	3,500
2	MH	0.95	5,000	35,000	25,000
3	DLH	1.25	10,500	44,100	5,000
4	DLC	5.75	12,000	84,000	40,000
5	DLC	3.45	10,200	54,700	10,200
6	DLH	1.00	29,000	89,000	60,500

[†] DLH = direct labor hours; MH = machine hours; DLC = direct labor costs.

C.12 Perform the overhead allocation of Prob. C.8 using the rates you determined. Use a basis of direct labor hours for the preparation and subassembly departments and direct labor cost for final assembly.

C.13 (*a*) The electronics firm in Prob. C.8 presently makes all of the components required by the preparation department. The firm is considering the possibility of purchasing, rather than making, these components. The firm has a quote of $67,500 a month from an outside contractor to make the items. If the costs for the month stated in Prob. C.8 are representative and if $41,000 worth of materials is charged to preparation, make an economic comparison of the present versus proposed situation. Assume that the preparation department's share of the quality control and engineering costs is a total of $3230 per month.

(*b*) Another alternative for the company is to purchase new equipment in the preparation department. The machinery will cost $375,000 and have a 5-year life, no salvage value, and a monthly operating cost of $475. This purchase is expected to reduce the quality control and engineering costs by $2000 and $3000, respectively, and reduce direct labor hours to 200 and direct labor cost to $850 for the preparation department. The redistribution of the overhead costs in quality control and engineering to the three production departments is on the basis of direct labor hours. If other costs remain the same, compare the present cost of making the components with the cost of the equipment purchase alternative. A return of a nominal 12% per year compounded monthly is required on all capital investments.

APPENDIX D

Sensitivity Analysis for Business Planning

John R. Heizer
U.S. Department of Commerce

D.1 Introduction

Sensitivity analysis is a technique which examines the response of a system to changes in its various parameters (see Chap. 19). In this appendix we discuss business systems represented by cash-flow models that are associated with strategic business plans. The economic analysis techniques presented and used in Chaps. 1 to 20 are utilized throughout this appendix.

Strategic business plans are long-range plans whose time frame is long enough to capture the benefits of specific capital investments. As an example, if a business wants to invest $10 million in building a factory that takes 3 years to build, a strategic business plan for this operation should include enough time to capture the productive benefits of this new plant. The plan should cover at least 10 years from the time the capital is raised. The productive benefits would appear in the cash-flow model in terms of the rate of return on the invested capital. The sensitivity of the rate of return to the various factors in the plan can be examined by using sensitivity analysis. Thus, if a 5% increase in the direct labor rate of the production operation will produce a 3% decrease in the rate of return, and if a 5% increase in the cost of materials will produce a 6% decrease in the rate of return, then the rate of return is more sensitive to the cost of materials than to the labor rate.

Through sensitivity analysis, the planner can identify those factors that have a major influence on the rate of return and measure the degree of sensitivity of each of these factors. Thus, sensitivity analysis can be used as a tool for both business plan development and business plan analysis.

To perform sensitivity analysis it is first necessary to obtain or develop a cash-flow model of the plan. A cash-flow model for a strategic business plan will incorporate a performance measure such as the rate of return on invested capital. This performance measure must be an explicit function of the *parameters* to be examined. Examples of parameters are: raw material, labor cost, production rate, inventory cost, and so on. The analysis is usually performed using a computerized spreadsheet system. In the cash-flow spreadsheet each parameter of interest must occupy a line in the spreadsheet.

With the advent of microcomputers and electronic spreadsheets, it is much easier for a planner to build a strategic business plan and to create a cash-flow model. Then, using the microcomputer, it is easy to explore the implications of the plan through asking "what if" questions and getting immediate answers.

D.2 Sensitivity Analysis Concept

Sensitivity analysis can identify assumptions that are especially critical in a business plan and thus alert the planner to the possibility that these critical assumptions may need more study and evaluation. Sensitivity analysis can assess the risk of the various cost and market assumptions in terms of a specified performance criteria. As an example, if there is a degree of uncertainty in sales estimates, sensitivity analysis can indicate what impact a worst-case sales assumption would have on the rate of return, and alert the planner to possible problems. If the worst-case market assumption generates a serious problem, then it will be necessary to develop a more-detailed assessment of the market assumptions. Thus, sensitivity analysis will help build an analytic base for the business plan that can be used by the planner to sell the plan to top-level management and/or investors.

Sensitivity analysis can also be used to evaluate the impact of new technologies on a business, such as the conversion of a manufacturing plant to a flexible manufacturing facility. In the case of evaluating new technologies, the engineering parameters associated with the technology must be converted to financial parameters which are used in the cash-flow model as the basis for sensitivity analysis.

With the model defined on an electronic spreadsheet, a sensitivity analysis can be performed quickly for a wide variety of parameters and the results graphed in a sensitivity diagram (See Fig. D.1) Each curve on the graph represents the sensitivity of the performance measure, for example, rate of return on invested capital in terms of one parameter. The degree of sensitivity is indicated in the slope of the curve, especially near the base case, which is the intersection of all of the sensitivity curves.

A planner or analyst must do the following before performing a sensitivity analysis:

1. Specify the performance measure of the project and develop functional relationships between this measure and those financial parameters that are to be examined. For a strategic business plan, rate of return on invested capital (ROI), rate of return on investor's equity, or cumulative cash flow can be considered as performance measures.
2. Develop a cash-flow model of the business project in enough detail to make the analysis meaningful. Thus, for a strategic business plan, the time frame must be long enough to capture the benefits of the proposed capital expenditures.
3. Make explicit all of the assumptions that will influence the analysis. This can include assumptions such as the cost of capital, availability of materials and labor, production levels, and so on. These are assumptions which will affect the cash-flow estimates for each of the parameters represented in the plan.

D.3 Department of Commerce Sensitivity Analysis Software Package

The sensitivity analysis program developed by the Office of Productivity, Technology, and Innovation (OPTI), U.S. Department of Commerce, was designed to be used as a strategic business planning tool. It will model a business operation, calculate the rate of return on invested capital (ROI) and determine the sensitivity of this rate of return to various marketing and cost factors. In the cash-flow model, the ROI is calculated using the Lotus 1-2-3 Symphony internal rate-of-return (IRR) function.

This software package is composed of two major programs, the Model Generator and the Analysis Program. The model generator generates the sensitivity analysis model based on the user specifications, such as the number of time periods, number of product categories, number of direct labor categories, etc. It can take into account a difference in production volume and sales volume during one or more time periods and thus calculate inventory costs. It can handle up to three categories of capital investment in which each category can have a different depreciation life.

The analysis program uses the model produced by the model generator for specific sensitivity analysis cases.

These two programs are written in the Symphony Command language on a Symphony worksheet and are menu driven.

D.4 Cash-Flow Model Relations

The equations built into the cash-flow model are summarized below. The word "sum" is used to indicate that the sum (Σ) is taken over all the categories for each parameter.

Total revenue:
Total revenue = sum (product sales)(sales price)

Direct costs:
Labor cost   = sum (unit labor cost)(production volume)
Material cost = sum (unit material cost)(production volume)
Energy cost   = sum (unit energy cost)(production volume)

Units in inventory:
Units = production volume − sales volume

Inventory costs (time based):
Inventory cost = (units in inventory)(unit inventory cost)

Depreciation:
$$\text{Depreciation} = \frac{\text{capital investment}}{\text{recovery period}}$$

Once the specific capital amount for a given time period has been fully depreciated over the appropriate number of time periods, the depreciation amount for the rest of the time periods for this capital amount is zero.

Total indirect costs:
Total indirect costs = sum (indirect costs) + depreciation

Gross profit:
Gross profit = revenue − total costs

Net operating profit:
Net operating profit = gross profit − depreciation

Net income before taxes:
Net income before taxes = net operating profit − other income adjustments

Income tax:
Income tax = (net income before taxes)(effective tax rate)

Adjusted income tax:
Adjusted income tax = income tax − applicable tax credits

Net income after taxes:
Net income after taxes = net income before taxes − income tax

Annual cash flow:
Annual cash flow = net income after taxes + depreciation − capital invested

The capital invested is the annual amount, if any, invested in the business project.
The internal rate of return is that interest rate which is required to make the net present value (NPV) of the annual cash flows equal to zero, that is, the i value at which
$$0 = \text{sum (annual cash flow)(present-worth factor)}$$

$$= \sum_{t=1}^{n} (CF_t)(P/F, i\%, t)$$

This is the same procedure presented in Chap. 7, Eq. (7.1), where the relation $0 = -P_D + P_R$ is used to determine the rate of return of a project. P_D is the present worth of disbursements and P_R is the present worth of the receipts.

Table D.1 Sensitivity analysis input table

PROJECT TITLE: EXAMPLE 2

TIME PERIOD	YEAR 1	YEAR 2	YEAR 3	YEAR 4	YEAR 5	YEAR 6	YEAR 7
CAPITAL INVESTMENT							
Capital 1	$1,000,000	$1,000,000	$500,000				
Capital 2	$5,000,000						
PRODUCTION VOLUME							
Product 1		100	500	1,000	900	800	500
Product 2			300	1,000	2,000	3,100	4,000
SALES VOLUME							
Product 1		100	500	900	950	800	500
Product 2			250	800	1,900	3,200	3,500
SALES PRICE							
Product 1		$4,000.00	$4,000.00	$3,000.00	$2,800.00	$2,000.00	$2,000.00
Product 2			$6,000.00	$5,250.00	$5,000.00	$4,000.00	$4,000.00
DIRECT COSTS							
PRODUCT 1							
Labor 1		$100.00	$100.00	$100.00	$100.00	$100.00	$100.00
Labor 2		$200.00	$200.00	$200.00	$200.00	$200.00	$200.00

PRODUCT 2					
Labor 1	$300.00	$300.00	$300.00	$300.00	$300.00
Labor 2	$400.00	$400.00	$400.00	$400.00	$400.00
PRODUCT 1					
Material 1	$500.00	$500.00	$500.00	$500.00	$500.00
Material 2	$600.00	$600.00	$600.00	$600.00	$600.00
PRODUCT 2					
Material 1	$700.00	$700.00	$700.00	$700.00	$700.00
Material 2	$800.00	$800.00	$800.00	$800.00	$800.00
PRODUCT 1					
Energy 1	$50.00	$50.00	$50.00	$50.00	$50.00
PRODUCT 2					
Energy 1	$70.00	$70.00	$70.00	$70.00	$70.00
INVENTORY COST					
Product 1	$300.00	$300.00	$300.00	$300.00	$300.00
Product 2	$400.00	$400.00	$400.00	$400.00	$400.00
INDIRECT COSTS					
Cost 1	$250,000	$250,000	$250,000	$250,000	$250,000
OTHER INCOME DEDUCTIONS					
TAX CREDITS					
INCOME TAX RATE(%)	30.00%				

Table D.2 Cash-flow table for sensitivity analysis

PROJECT TITLE: EXAMPLE 2

TIME PERIOD	YEAR 1	YEAR 2	YEAR 3	YEAR 4	YEAR 5	YEAR 6	YEAR 7
PRODUCT SALES							
Product 1	$0	$400,000	$2,000,000	$2,700,000	$2,660,000	$1,600,000	$1,000,000
Product 2	$0	$0	$1,500,000	$4,200,000	$9,500,000	$12,800,000	$14,000,000
TOTAL REVENUE	$0	$400,000	$3,500,000	$6,900,000	$12,160,000	$14,400,000	$15,000,000
CAPITAL INVESTMENT							
Capital 1	$1,000,000	$1,000,000	$500,000	$0	$0	$0	$0
Capital 2	$5,000,000	$0	$0	$0	$0	$0	$0
TOTAL CAPITAL INVESTMENT	$6,000,000	$1,000,000	$500,000	$0	$0	$0	$0
COSTS							
DIRECT COSTS							
PRODUCT 1							
Labor 1	$0	$10,000	$50,000	$100,000	$90,000	$80,000	$50,000
Labor 2	$0	$20,000	$100,000	$200,000	$180,000	$160,000	$100,000
TOTAL LABOR COST	$0	$30,000	$150,000	$300,000	$270,000	$240,000	$150,000
Material 1	$0	$50,000	$250,000	$500,000	$450,000	$400,000	$250,000
Material 2	$0	$60,000	$300,000	$600,000	$540,000	$480,000	$300,000
TOTAL MATERIAL COST	$0	$110,000	$550,000	$1,100,000	$990,000	$880,000	$550,000
Energy 1	$0	$5,000	$25,000	$50,000	$45,000	$40,000	$25,000
TOTAL ENERGY COST	$0	$5,000	$25,000	$50,000	$45,000	$40,000	$25,000
TOTAL PRODUCT COST	$0	$145,000	$725,000	$1,450,000	$1,305,000	$1,160,000	$725,000

PRODUCT 2							
Labor 1................:	$0	$0	$90,000	$300,000	$600,000	$930,000	$1,200,000
Labor 2................:	$0	$0	$120,000	$400,000	$800,000	$1,240,000	$1,600,000
TOTAL LABOR COST......:	$0	$0	$210,000	$700,000	$1,400,000	$2,170,000	$2,800,000
Material 1.............:	$0	$0	$210,000	$700,000	$1,400,000	$2,170,000	$2,800,000
Material 2.............:	$0	$0	$240,000	$800,000	$1,600,000	$2,480,000	$3,200,000
TOTAL MATERIAL COST...:	$0	$0	$450,000	$1,500,000	$3,000,000	$4,650,000	$6,000,000
Energy 1...............:	$0	$0	$21,000	$70,000	$140,000	$217,000	$280,000
TOTAL ENERGY COST.....:	$0	$0	$21,000	$70,000	$140,000	$217,000	$280,000
TOTAL PRODUCT COST....:	$0	$0	$681,000	$2,270,000	$4,540,000	$7,037,000	$9,080,000
TOTAL PRODUCT LABOR COST.......:	$0	$30,000	$360,000	$1,000,000	$1,670,000	$2,410,000	$2,950,000
TOTAL PRODUCT MATERIAL COST....:	$0	$110,000	$1,000,000	$2,600,000	$3,990,000	$5,530,000	$6,550,000
TOTAL PRODUCT ENERGY COST......:	$0	$5,000	$46,000	$120,000	$185,000	$257,000	$305,000
TOTAL DIRECT COSTS.............:	$0	$145,000	$1,406,000	$3,720,000	$5,845,000	$8,197,000	$9,805,000
UNITS IN INVENTORY							
Product 1..............:	0	0	0	100	50	50	50
Product 2..............:	0	0	50	-250	350	250	750
INVENTORY COSTS							
Product 1..............:	$0	$0	$0	$30,000	$15,000	$15,000	$15,000
Product 2..............:	$0	$0	$20,000	$100,000	$140,000	$100,000	$300,000
TOTAL INVENTORY COST...:	$0	$0	$20,000	$130,000	$155,000	$115,000	$315,000

(continued)

Table D.2 (continued)

TIME PERIOD	YEAR 1	YEAR 2	YEAR 3	YEAR 4	YEAR 5	YEAR 6	YEAR 7
INDIRECT COSTS (EX. DEPRECIATION)							
Cost 1	$0	$250,000	$250,000	$250,000	$250,000	$250,000	$250,000
TOTAL INDIRECT COSTS	$0	$250,000	$250,000	$250,000	$250,000	$250,000	$250,000
TOTAL PRODUCT SALES	$0	$400,000	$3,500,000	$6,900,000	$12,160,000	$14,400,000	$15,000,000
COST OF GOODS PRODUCED	$0	$145,000	$1,406,000	$3,720,000	$5,845,000	$8,197,000	$9,805,000
INVENTORY COSTS	$0	$0	$20,000	$130,000	$155,000	$115,000	$315,000
GROSS PROFIT	$0	$255,000	$2,074,000	$3,050,000	$6,160,000	$6,088,000	$4,880,000
DEPRECIATION	$1,033,333	$1,233,333	$1,333,333	$1,333,333	$1,333,333	$1,333,333	$300,000
INDIRECT COSTS	$0	$250,000	$250,000	$250,000	$250,000	$250,000	$250,000
NET OPERATING PROFIT	($1,033,333)	($1,228,333)	$490,667	$1,466,667	$4,576,667	$4,504,667	$4,330,000
OTHER INCOME ADJUSTMENTS	$0	$0	$0	$0	$0	$0	$0
NET INCOME BEFORE TAXES	($1,033,333)	($1,228,333)	$490,667	$1,466,667	$4,576,667	$4,504,667	$4,330,000
INCOME TAX	$0	$0	$147,200	$440,000	$1,373,000	$1,351,400	$1,299,000
INCOME TAX CREDITS	$0	$0	$0	$0	$0	$0	$0
ADJUSTED INCOME TAX	$0	$0	$147,200	$440,000	$1,373,000	$1,351,400	$1,299,000
NET INCOME AFTER TAXES	($1,033,333)	($1,228,333)	$343,467	$1,026,667	$3,203,667	$3,153,267	$3,031,000
ANNUAL CASH FLOW	($6,000,000)	($995,000)	$1,176,800	$2,360,000	$4,537,000	$4,486,600	$3,331,000
CUMULATIVE CASH FLOW	($6,000,000)	($6,995,000)	($5,818,200)	($3,458,200)	$1,078,800	$5,565,400	$8,896,400
INTERNAL RATE OF RETURN	21.97%						

D.5   Sample Input and Output of Sensitivity Analysis Software

A relatively small cash-flow model is described here using the sensitivity analysis software. Characteristics of the sample are as follows:

Product types:	2
Capital categories:	2
Material types:	2 per product
Energy type:	1
Indirect cost category:	1
Tax rate:	30%

Table D.3 Detailed sensitivity table by parameter

PROJECT TITLE: EXAMPLE 2

SALES PRICE		SALES VOLUME		LABOR COSTS		MATERIAL COSTS	
%CHG	IRR	%CHG	IRR	%CHG	IRR	%CHG	IRR
-50%	ERR	-50%	1.65%	-50%	33.01%	-50%	33.01%
-40%	ERR	-40%	6.72%	-40%	31.01%	-40%	31.01%
-30%	-10.48%	-30%	11.15%	-30%	28.93%	-30%	28.93%
-20%	4.75%	-20%	15.13%	-20%	26.73%	-20%	26.73%
-10%	14.42%	-10%	18.67%	-10%	24.42%	-10%	24.42%
0%	21.97%	0%	21.97%	0%	21.97%	0%	21.97%
10%	28.38%	10%	25.07%	10%	19.35%	10%	19.35%
20%	34.03%	20%	27.99%	20%	16.54%	20%	16.54%
30%	39.13%	30%	30.77%	30%	13.48%	30%	13.48%
40%	43.80%	40%	33.42%	40%	10.11%	40%	10.11%
50%	48.14%	50%	35.96%	50%	6.33%	50%	6.33%

INVENTORY COSTS		TAX RATE		INDIRECT COSTS		CAPITAL	
%CHG	IRR	%CHG	IRR	%CHG	IRR	%CHG	IRR
-50%	22.43%	-50%	25.91%	-50%	23.35%	-50%	44.47%
-40%	22.33%	-40%	25.15%	-40%	23.07%	-40%	37.94%
-30%	22.24%	-30%	24.38%	-30%	22.80%	-30%	32.78%
-20%	22.15%	-20%	23.59%	-20%	22.52%	-20%	28.55%
-10%	22.06%	-10%	22.79%	-10%	22.24%	-10%	25.01%
0%	21.97%	0%	21.97%	0%	21.97%	0%	21.97%
10%	21.88%	10%	21.13%	10%	21.69%	10%	19.33%
20%	21.78%	20%	20.27%	20%	21.42%	20%	17.00%
30%	21.69%	30%	19.38%	30%	21.14%	30%	14.93%
40%	21.60%	40%	18.48%	40%	20.87%	40%	13.03%
50%	21.51%	50%	17.55%	50%	20.59%	50%	11.27%

The complete input table for a 7-year horizon is shown in Table D.1. The planner or analyst has obtained data to estimate each value in this input table. The parameters are all the categories on the left of Table D.1.

The software uses the input to generate the following output.

Cash-flow table by year: Table D.2
Sensitivity analysis table: Table D.3
Sensitivity analysis graph: Figure D.1

The user is given the choice of performance measure and parameters to be included in the sensitivity analysis. Figure D.1 shows ROI versus 6 different parameters. In this case, ROI is most sensitive to sales price and least sensitive to indirect cost.

Figure D.1 Sensitivity analysis graph by parameter.

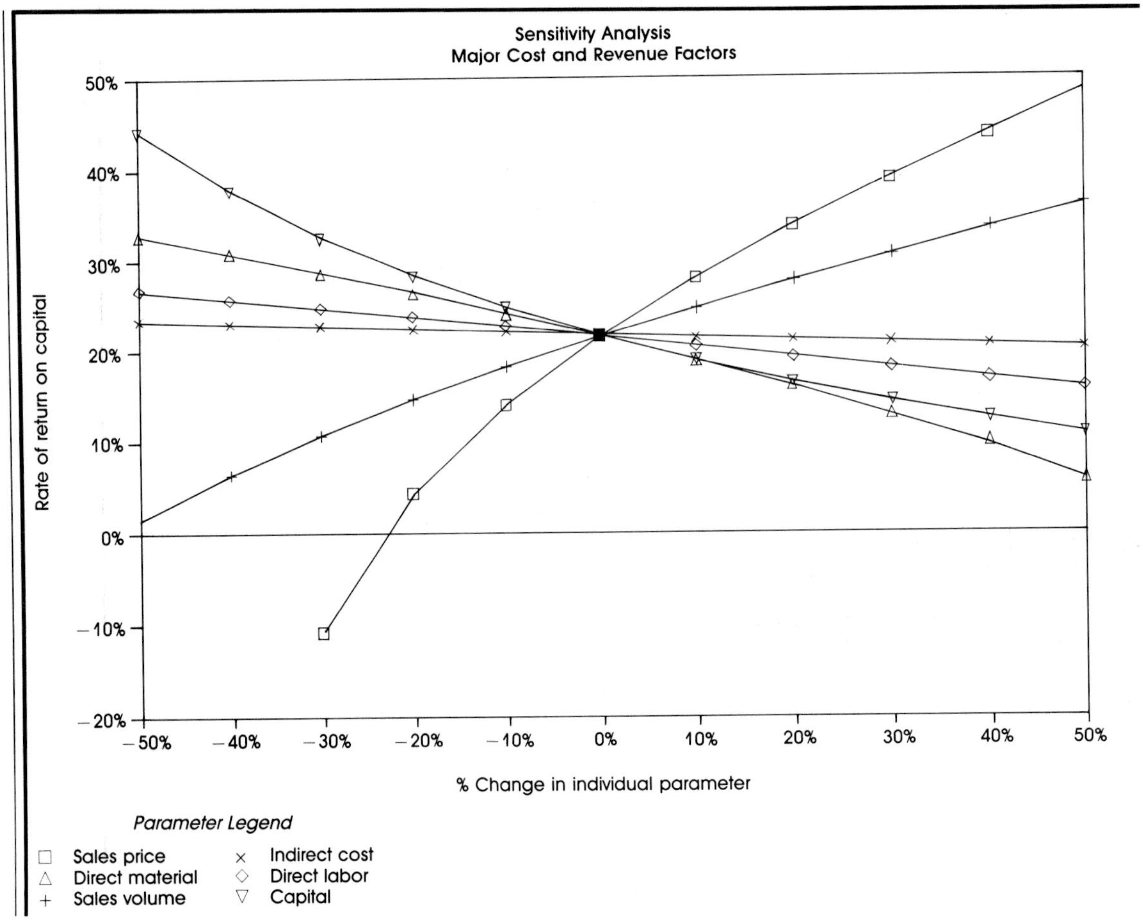
Sensitivity Analysis
Major Cost and Revenue Factors
50%
40%
30%
20%
10%
0%
−10%
−20%
Rate of return on capital
−50%
−40%
−30%
−20%
−10%
0%
10%
20%
30%
40%
50%
% Change in individual parameter
Parameter Legend
□ Sales price
△ Direct material
+ Sales volume
× Indirect cost
◇ Direct labor
▽ Capital

D.6   Availability of Sensitivity Analysis Software

The business planning/sensitivity analysis software is available for IBM compatible computers with at least 512K RAM. A users' manual is supplied and the example in the previous section is included as a sample on the diskettes. For further information contact:

Mr. John R. Heizer
Financial Economist
Office of Productivity, Technology and Innovation
US Department of Commerce
Washington, D.C. 20230
(202) 377-8879

E APPENDIX
Computer Applications

This appendix gives an overview of the computer programs available from the publisher which assist in the solution of problems in economic decision analysis. The programs may be used in conjunction with this text in a laboratory or classroom setting or as separate packages which can help in the performance of engineering economy computations and alternative comparison. Contents of the package include programs such as the following:

1. Fundamental economic analysis computations (FEAC)
2. Return-on-investment determination system (ROIDS)
3. Minimum cost life determination (MINCL)
4. Basic Economic Analysis Tutorial (BEAT)

The programs are interactive and written in BASIC for easy implementation and alteration. They are usable on all IBM-compatible personal computers and microcomputers. No expertise in the techniques is necessary to operate the programs, but a working familiarity with the material in this text will assist the user in better utilizing and interpreting program output.

Code for the programs may be altered to increase the capabilities of any package. The code, a brief users guide for each system, and all software is available from the publisher free of charge. As new capabilities are incorporated into the packages, new releases will be made available to instructors.

FEAC—Fundamental Economic Analysis Computations

Many of the fundamental computations necessary to perform engineering economic analyses and alternative comparison are included in this package. It is possible to utilize the software in close conjunction with the level 1, 2, and 3 chapters of this text as you learn the techniques of economic analysis and find it necessary to perform basic computations.

Some of the features of the FEAC package are: present worth, equivalent uniform annual worth, future worth, nominal and effective interest rates, nonconventional cash flows, and inflation. For alternative evaluation, this package will compute the PW, EUAW, or FW values for each alternative, with the final decision-making left to the user.

ROIDS—Return on Investment Determination System

This program will find the rate-of-return value i^* given any cash-flow sequence for a single project (Chap. 7) or the incremental cash-flow sequence for two projects (Chap. 8). Both fixed amount and fixed percent gradients are allowed in the cash-flow sequence. The program will find all real i^* values between -12.5 and 500% to an accuracy of 0.1%. If there is no i^* found in this range, the user is so informed.

If the conditions for multiple rate-of-return values are present in the cash-flow sequence, all real roots within the specified range are determined. As discussed in Sec. 7.5, these rates each assume that any net positive cash flow in a particular year is reinvested at the rate of return which balances the rate-of-return equation. If a specific reinvestment rate applies (Sec. 7.5), the user simply inputs this figure and ROIDS proceeds by converting the cash-flow sequence to a conventional sequence and finds the correct i^* value to balance the equation.

It is possible to see the effect of inflation on the rate of return using ROIDS. The user can input an inflation factor for each year of the sequence—for example, 10% per year. When inflation is accounted for, the user is given the i^* value with inflation considered. Once i^* is computed, ROIDS allows the user to input a new cash-flow sequence, or the user may edit the current sequence to solve for i^* under slightly different conditions. This edit capability makes ROIDS a useful tool for sensitivity analysis of specific factors (Chap. 19 and Chap. 20, Sec. 20.5).

MINCL—Minimum Cost Life Determination

This program will calculate the EUAW value of costs for an asset, assuming it is retained for 1 year, 2 years, and so on until the EUAW value reaches a minimum and begins to increase thereafter (Chap. 10, Sec. 10.5). The year value n and the minimum

EUAW are presented for use in an economic analysis. Additionally, the EUAW for each year value considered is shown.

The user inputs the initial cost, salvage value, and annual cost figure for each year of possible retention. The salvage value and annual cost values may be input in any of the following forms: constant for all years; with a positive or negative gradient (uniform or percentage); or different for each year.

If the minimum EUAW life cannot be found in the range of years input, the user is given the opportunity to input salvage values and annual costs for additional years, which will extend the possible retention period.

BEAT—Basic Economic Analysis Tutorial

This program provides a formula-manipulation and graphical plotter capability for use by students learning engineering economic computations. Present, future or annual worth relations are entered using standard factor notation. Entry errors are indicated on the screen via a tutorial troubleshooting feature. The equation is solved once it is completely entered.

BEAT can be utilized for sensitivity analysis because of its equation-plotting function. The equation results are displayed and plotted as a function of one of the variables (P, A, n, i) and the range may be varied to plot for several values. This program was developed at the University of Puerto Rico at Mayaguez in the Department of Industrial Engineering.

APPENDIX
Answers to Selected Problems*

Chapter 1

1.1 See Sec 1.1.
1.3 See Sec 1.1.
1.5 6 months.
1.7 $1666.67.
1.9 See Sec 1.3.
1.11 There is some risk associated with investments other than those considered safe. Higher risk requires higher potential return.
1.13 (a) 27.41; (b) 4.57% of P.
1.15 $1280.
1.17 $756.50.
1.19 Invest for 4 years.

* This appendix includes the final answers to odd-numbered and some even-numbered problems for all chapters in the text.

1.21 5.36 years.

1.23 (*a*) 12.5 years; (*b*) 9 years.

1.25 (*a*) 8.33%; (*b*) 6%.

1.27 $P = \$1000$ every 2 years, $F = ?$, $i = 10\%$, $n = 9$ years.

1.29 $P = \$1400$, $F = \$4200$, $i = 6\%$, $n = ?$

1.31

Year	Payment	Income	Net cash flow
0	0	$2,000	$+2,000
1–4	0	0	0
5	$F = ?$	0	$-F = ?$

1.33

Year	0, 2, 4	1, 3	5, 7, 9, 11, 12, 13	6, 8, 10
Net cash flow	-500	0	$+300$	-200

1.35 $P = ?$ in year 0, $F = \$+3000$ in year 5, $i = 8\%$ per year

1.37

Year	Deposit	Withdrawal	Net cash flow
0–1	$0	$ 0	$ 0
2	$P_1 = -?$	0	$P_1 = -?$
3	0	0	0
4	$P_2 = -?$	0	$P_2 = -?$
4–8	0	100	100
9	0	500	500

1.39 $P = -?$, $F = \$+580$ in year 8.

1.41 $P = +?$ in year 0, $F_1 = \$-1200$ in year 5, $F_2 = \$-2200$ in year 8, $i = 10\%$ per year.

1.43 $P = \$-10,000$ in year 0, $F = +?$ in year 10, $n = 10$, $i = 12\%$

1.45

Year, k	Deposit	Withdrawal	Net cash flow
0	$500	$0	$-500
1–6	500 + 50(k)	0	$-500 - 50(k)$
6	0	$F = +?$	$F = +?$

Chapter 2

2.1 SSPWF: Move F to year $n - 1$ to get a factor of $1/(1 + i)^{n-1}$.
USPWF: Move all A values back 1 year to get a P/A factor in year 0 of $(1 + i)^n - 1/[i(1 + i)^{n-1}]$.
USCAF: Move all A values back 1 year to get an F/A factor in year n of $[(1 + i)/i][(1 + i)^n - 1]$.

2.3

Year	1	2	3	4	5	6	7
Cash flow, $	60	60	60	100	110	120	130

2.5 3.484.

2.7 (*a*) 7.6954, 5513.3834, 0.5133, 0.00001; (*b*) 7.6910, 5162.5537, 0.5129, 0.00001.

2.9 (*a*) 0.7509, 10.3126, 44.374, 0.00488; (*b*) 0.7504, 10.2903, 43.6445, 0.00472.

2.11 $F = \$13,636$.

2.13 $P = \$2369.68$.

2.15 $A = \$1073.34$.

2.17 $P = \$8233.77$.

2.19 $P = \$3331.50$.
2.21 $F = \$153,220.02$.
2.23 $P = \$5956.63$.
2.25 $P = \$1357.16$.
2.27 $P = \$4105.60$.
2.29 $A = \$5239.25$.
2.31 $F = \$164,494$.
2.33 $A = \$990$.
2.35 $\$10,497.23$.
2.37 $P_1 = \$2500$, $P_2 = \$2380.50$.
2.39 P later $= \$22,569.20$, buy later.
2.41 (a) $A = \$5710$; (b) $P = \$20,582.80$.
2.43 $P = \$12,019.14$.
2.45 $P = \$17,545.72$.
2.47 $\$3147.42$.
2.49 $P = \$2755.65$.
2.51 $G = \$-6391.97$; reduction: year 2 is $\$33,608.03$, year 3 is $\$27,216.06$, year 4 is $\$20,824.09$.
2.53 $P_E = \$48,435.60$, $P = \$65,132$.
2.55 $P = \$3425.10$, $D = \$239.76$.
2.57 $n = 7.8$ years.
2.59 $i = 3.62\%$.
2.61 $i = 12.05\%$.
2.63 $i = 6.87\%$.
2.65 $i = 8.88\%$.
2.67 $n = 20.95$ years.
2.69 $n = 6.88$ or 7 years.
2.71 $n = 7.11$ years.
2.73 $n = 17.7$ years.

Chapter 3

3.1 (a) 13% per year; (b) 4% per year.
3.3 The time period between payments or receipts for a uniform cash-flow series.
3.5 (a) Month; (b) quarter; (c) year; (d) month; (e) week.
3.7 Percent per compounding period.
3.9 $r = 8\%$, $i = 8.16\%$ per year.
3.11 16.99%.
3.13 1.47% per quarter.
3.15 Yes, because effective rates per year are about the same.
3.17 (a) 12.75% per year; (b) 1% per month; (c) 1.005% per month.
3.19 $A = \$2672.91$.
3.21 $A = \$937.44$.
3.23 (a) $i = 8.15\%$; (b) $r = 7.84\%$.
3.25 $P = \$273,842$
3.27 (a) $r = 6.97\%$; (b) $i = 7.22\%$.
3.29 $r = 18.44\%$, $i = 20.08\%$ per year.
3.31 $F = 17,252.90$.
3.33 $P = \$5906.15$.
3.35 $F = \$80,353.90$.

3.37 $r = 0.656\%$.
3.39 $P = \$3531.30$.
3.41 $F = \$2253.68$.
3.45 (a) $F = \$6801.30$; (b) $F = \$6802.30$.
3.47 $F = \$338.64$.
3.49 $A = \$199.02$ per month.
3.51 $n = 140$.

Chapter 4

4.1 (a) $A = \$-600$ for years 4 to 10, $P = ?$ in year 3, $F = ?$ in year 10; (b) $A_1 = \$-2000$ for months 1 to 6 and $A_2 = \$+1500$ for months 3 to 6, $P_1 = ?$ in month 0 and $F_1 = ?$ in month 6, $P_2 = ?$ in month 2 and $F_2 = ?$ in month 6, (c)

Month	Deposit	Withdrawal
December	$P_1 = ?$	$\$\ 0$
January	$A_1 = \$20$	0
February	$A_1 = 20$	$P_4 = ?$
March	$A_1 = 20$	$A_4 = 10$
April	$F_1 = ?, A_2 = 20, P_2 = ?$	$A_4 = 10$
May	$A_2 = 75$	$F_4 = ?, A_4 = 10, P_5 = ?$
June	$A_2 = 75$	$A_5 = 25$
July	$A_2 = 75$	$A_5 = 25$
August	$F_2 = ?, A_2 = 75, P_3 = ?$	$F_5 = ?, A_5 = 25, P_6 = ?$
September	$A_3 = 25$	$A_6 = 10$
October	$A_3 = 25$	$A_6 = 10$
November	$F_3 = ?, A_3 = 25$	$F_6 = ?, A_6 = 10$
December	0	250

4.3 F in year $16 = \$9895.25$, $A = \$113.20$.
4.5 $F = \$27,418$.
4.7 $P = \$4456$.
4.9 $P = \$452.92$.
4.11 $A = \$172$.
4.13 $F = \$28,880$.
4.15 $A = \$1014.93$.
4.17 $x = \$1361.62$.
4.19 $F = \$63,719$.
4.21 $x = \$91,373$.
4.23 $P = \$25,297$.
4.25 (a) $P = \$25,959$; (b) $F = \$120,575$ at effective $i = 16.6\%$.
4.27 $A = \$2555.33$.
4.29 $A = \$3339$.
4.31 (a) $A = \$615.20$, (b) $P = \$3601.01$.
4.33 $A = \$684.11$.
4.35 (a) $P = \$2820$ (P for gradient is in year 1); (b) $A = \$436.32$.
4.37 P in period -1 is $\$7862.40$ and $A = \$560.06$ at effective semiannual i of 3.53%.
4.39 $n = 14$, last payment is $\$240$.
4.41 $G = \$877.52$.
4.43 $P = \$24,394$.
4.45 $x = \$818.70$.
4.47 $F = \$5142.61$.
4.49 $P = \$918.83$.

4.51 $P = \$94,520.$
4.53 $P = \$111,582.$
4.55 $P = \$1847.10.$
4.57 $P = \$9821.38.$
4.59 $P = \$949.97.$
4.61 $P = \$4501.60, F = \$14,758.$

Chapter 5

5.1 $PW_A = \$30,985, PW_B = \$28,865;$ select B.
5.3 $PW_{buy} = \$53,529, PW_{rent} = \$43,367;$ select rent.
5.5 $PW_{purchase} = \$9573.78, PW_{lease} = \$13,238.10;$ select purchase.
5.7 $PW_{none} = \$61,679, PW_{auto} = \$24,416;$ select automatic system.
5.9 $PW_G = \$190,344, PW_H = \$217,928;$ select G.
5.11 $PW_{manual} = \$9981.23, PW_{auto} = \$6287.02;$ select automatically cleaned screen.
5.13 $PW_{buy} = \$100,560, PW_{lease} = \$132,465;$ purchase clamshell.
5.15 $PW_{high} = \$16,195, PW_{low} = \$43,186;$ select high-pressure system.
5.17 $PW_{radial} = \$1909.81, PW_{recap} = \$1970.06;$ purchase radial.
5.19 $PW_{regular} = \$66,969, PW_{auto} = \$88,475;$ select regular.
5.21 $PW_{disposable} = \$14,330, PW_{reusable} = \$28,203;$ use disposable utensils.
5.23 $PW_{YEA} = \$17,529, PW_{TEAM} = \$16,016.$
5.25 Expense is $\$375,790$, income is $\$462,441$; total capitalized cost is $\$86,652$ (net profit).
5.27 Capitalized cost $= \$120,212.$
5.29 Capitalized cost $= \$42,417.$
5.31 Capitalized cost $= \$772,532.$
5.33 EUAW WHY $= \$27,857.02$, capitalized cost WHY $= \$185,713$, EUAW NOT $= \$29,562$, capitalized cost NOT $= \$197,084$; select WHY.
5.35 EUAW $X = \$72,920.20$, capitalized cost $X = \$503,245$, capitalized cost $Z = \$437,841$; select Z.
5.37 Capitalized cost of PAY LATER $= \$284,560$, capitalized cost KNOT NOW $= \$60,407.$

Chapter 6

6.1 EUAW is cost for infinite number of renewals, and PW is cost for one life cycle.
6.3 $A = \$7167.34.$
6.5 (a) $A = \$2801.92$; (b) $A = \$7167.34.$
6.7 $EUAW_{new} = \$16,094.45, EUAW_{used} = \$15,542.38;$ select used.
6.9 $EUAW_1 = \$11,857.44, EUAW_2 = \$15,218.91;$ select proposal 1.
6.11 $EUAW_{R-11} = \$93.86, EUAW_{R-19} = \$96.40;$ use $R-11$.
6.13 $EUAW_{spray} = \$128,000, EUAW_{immersion} = \$26,098.50;$ use immersion.
6.15 $EUAW_A = \$2005.04, EUAW_B = \$5614.75;$ select plan A.
6.17 $EUAW_{spray} = \$163,109, EUAW_{truck} = \$175,956;$ select spray.
6.19 n diesel $= 9$, n gasoline $= 6$, $EUAW_{diesel} = \$1186.45$, $EUAW_{gasoline} = \$1369.07$; purchase diesel trucks.
6.21 $EUAW_{metal} = \$38,992.82, EUAW_{concrete} = \$26,127.59;$ select concrete.
6.23 $EUAW_A = \$10,900, EUAW_B = \$12,545;$ select plan A.
6.25 $EUAW_{PAY} = \$11,184, EUAW_{KNOT} = \$12,074;$ select alternative PAY LATER.
6.27 $EUAW_X = \$13,902, EUAW_Y = \$16,141;$ select alternative X.

6.29 Effective $i = 18.81\%$, EUAW = $8088.73.
6.31 EUAW = $1,483,779.
6.33 P in year 14 is $108,333.33, A = $3344.25.
6.35 F in year 12 = $19,930.15, A = $2989.52.
6.37 F in year 11 = $23,225.42, A = $3019.30.
6.39 $EUAW_{U.R.}$ = $2,527,930, $EUAW_{O.K.}$ = $7,382,000; select U.R.
6.41 $EUAW_{MAX}$ = $89,301, $EUAW_{MIN}$ = $109,000; select MAX.

Chapter 7

7.1 See Sec. 7.1.
7.3 $i^* = 6.26\%$.
7.5 $i^* = 1.91\%$.
7.7 $i^* = 10.78\%$.
7.9 $i^* = 3.63\%$.
7.11 $i^* = 6.22\%$ per month nominal, or 74.64% per year.
7.13 $i^* = 10.06\%$.
7.15 $i_1^* = 39.9\%$, $i_2^* = 53.77\%$.
7.17 (a) $i_1^* = 25\%$, $i_2^* = 400\%$.
7.19 (a) $i^* = 400\%$; (b) $i_1^* = 21.55\%$, $i_2^* = 278.45\%$; (c) $i' = 46.15\%$; (d) $i' = 0\%$.

7.21 (a)

c, %	15	25	35	45	50
i', %	0	32.0	59.26	82.76	93.33

7.23 By trial and error, $i = 1.2\%$ per month, or 15.39% per year.

Chapter 8

8.1

Year	0	1–4	5	6–9	10
S − R	$−4,000	−500	6,500	−500	500

8.3

Year	0	1–3	4	5–7	8	9–11	12
Y − X	$−9,000	1,800	−700	1,800	−700	1,800	3,300

8.5 See Sec. 8.2.
8.7 $i_{A-B}^* = 58.2\%$; select A.
8.9 $i_{D-G}^* = 4.87\%$; select grass.
8.11 $i^* = 11.85\%$; select X.
8.13 $i_{A-M}^* = 37.85\%$, no difference; select A.
8.15 (a) $i^* = 11.6\%$; select gravel; (b) $i^* = 6.71\%$; select gravel.
8.17 See Sec. 8.5.
8.19 Select method 3, incremental $i^* > 20\%$.
8.21 (a) Select machine 4; incremental $i^* = 38.45\%$; (b) $P = \$ − 104,929$ at 18%.
8.23 (a) Select 10-square-meter truck, incremental $i^* = 20.54\%$; (b) EUAW = $1720 at 18%.
8.25 (a) Select pressing 2, slicing 2, weighing 2, and wrapping 1; (b) investment = $45,000, AOC = $37,000.
8.27 Incremental rate of return = 27.4%; select B.
8.29 Incremental return of $10,000 = 5\%$; select (f).
8.31 (a) Select A, B, and C; (b) select A; (c) select D.

Chapter 9

9.1 Disbenefits are losses to the people and should therefore be considered in the numerator.

9.3 B/C = 0.29 by EUAW; do not construct dam.

9.5 EUAW disbenefits = $24,072, B = $513,452, G = $18,017.

9.7 EUAW = $173,560, B/C = 0.92; no.

9.9 (a) B = user cost savings = $576,000 per year, cost is extra EUAW = $2,117,320 for transmountain, B/C = 0.27; select long route; (b) B − C = $−1,541,320; select long route.

9.11 Resurface EUAW = $296,850, new EUAW = $918,603; annual user costs: resurface $720,000, new $600,000, B/C = 0.99; select resurface.

9.13 (a) Lined EUAW = $337,920, not lined EUAW = $122,905, B/C = 0.56; do not line; (b) B − C = $−95,015; do not line.

9.15 Only the best alternative should be selected.

9.17 Do nothing versus 1 B/C = 1.67—select 1; method 2 versus method 1 B/C = 3.76—eliminate method 1; method 3 versus method 2 B/C = 1.67—eliminate method 2; method 4 versus method 3 B/C = 0.63—eliminate method 4; method 5 versus method 3 B/C = 0.87—eliminate method 5, select method 3.

9.19 Rank alternatives: DN, 2, 3, 6, 1, 5, 4, 7; DN versus 2 B/C = 1.39, eliminate DN; 2 versus 3 B/C = 3.73, eliminate 2; 3 versus 6 B/C = 63.55, eliminate 3; 6 versus 1 B/C = −15.11, eliminate 1; 6 versus 5 B/C < 1.0 eliminate 5; 6 versus 4 B/C < 1.0, eliminate 4; 6 versus 7 B/C < 1.0; select proposal 6.

9.21 200 mm versus 240 mm: benefits same but costs lower for 240-mm pipe; therefore, eliminate 200 mm; 240 mm versus 300 mm B/C = 0.07—eliminate 300-mm pipe; select 240-mm pipe.

9.23 Modified B/C_x = 0.002, modified B/C_y = 0.004; select do nothing.

9.25 Incremental B/C = 0.45; use armor coat.

Chapter 10

10.3 (a) P = $18,000, n = 8, SV = $1000, AOC = $150; (b) sunk cost = $−8000.

10.5 (a) $2925; (b) $5300.

10.7 Select old with EUAW = $79,731.

10.9 Select new with EUAW = $2646.

10.11 Select conveyor plus old mover with EUAW = $23,955.50.

10.13 Select plan III with EUAW = $6348.

10.15 Negotiate down to $23,195.76.

10.17 Select old with EUAW = $64,736.

10.19 Select plan I with EUAW ≈ $5275.27.

10.21 RV = $11,527.

10.23 (a) RV = $54,622; (b) 5.84% increase.

10.24 RV = $−26,620 (pay less than RV for disposal).

10.25 Purchase challenger now.

10.27 (a) Correct decision to keep defender; (b) keep defender both years.

10.29 Keep old car all 3 years.

10.31 Retain 8 or 9 years at EUAW = $2148.

10.33 n^* = 12.6 years.

10.34 Retain for a total of 14 years with EUAW = $1879.

10.35 (a) 4 years; (b) EUAW relatively insensitive for n = 4, 5, and 6 years.

Chapter 11

11.1 See Sec. 11.1.

11.3 Three months (quarterly), $I = \$150$.

11.5 $P = \$10,680$.

11.7 $P = \$6611$.

11.9 Semiannual $i = 6.77\%$, $P = \$10,372$ (interpolation).

11.11 Conventional bonds, $i = 3.5\%$ per quarter, $I = \$300$ per quarter, $P = \$8804$; put bonds, $i = 3.125\%$ per quarter, $P = \$9702$; difference $= \$898$ per \$100,000 bond.

11.13 Semiannual $i = 5.06\%$, $P = \$2,461,776$.

11.15 $I = \$400$ per 6 months, sales price $= \$13,810$.

11.17 Original $i = 8.38\%$, $F = \$1380$.

11.19 $n = 9.32$ semiannual periods, or 4.7 years.

11.21 $P = \$5,680,164$, face value of convertible bond $= \$14,668,273$; for continuous compounding face value $= \$14,775,482$.

11.23 $P = \$9,403,450$, loss \$96,550.

11.25 $P = \$1008.85$.

11.27 Nominal $i = 12.1\%$ per year compounded semiannually.

11.29 Nominal $i = 5.8\%$ per year compounded semiannually.

11.31 $b = 10.78\%$ per year.

11.33 $b = 11.8\%$ per year payable semiannually.

11.35 Nominal $i = 14.82\%$ per year.

11.37 Total semiannual expense $= \$265,000$, nominal $i = 11.88\%$ per year.

11.39 (a) Nominal $i = 12.46\%$ per year; (b) call bonds.

Chapter 12

12.1 $i_f = 26.5\%$, $P = \$5612.87$ (from equation).

12.3 (a) $P = \$2287.71$; (b) $P = \$1575.22$.

12.5 $i = 2.515\%$ per month, $P = \$5,509.37$.

12.7 (a) $P = \$11,142.68$; (b) $= \$11,142.54$.

12.9 $P_A = \$35,328$, $P_C = \$29,110$; select machine C.

12.11 $P_A = \$215,302$ ($i = 15\%$), $P_B = 214,489$ ($i_f = 26.5\%$); purchase from B.

12.13 $P_E = \$6018.66$; $P = \$15,018.66$.

12.15 $F = \$84,387$.

12.17 $F = \$426,275.30$.

12.19 $F = \$58,007$.

12.21 $F = \$48,470.59$.

12.23 $A = \$10,461$.

12.25 F in year 16 $= \$422,535$ ($i_f = 23.2\%$), $A = \$9883$ ($i = 12\%$).

12.27 $A = \$30,585.07$.

12.29 $C = \$27,290$.

12.31 Cost $= \$73,055$.

12.33 Divide by 1.257; index in 1983 $= 252.1$.

12.35 Index in 1996 $= 1428.4$.

12.37 $C_2 = \$238,365$.

12.39 $C_T = \$233,600$.

12.41 Cost in 1970 $= \$22,595$; cost in 1986 $= \$59,419$.

Chapter 13

13.1 $B = \$55,000$, $n = 10$, $SV = \$5200$, actual life $= 5$, market value $= \$8700$, book value $=$ $\$13,750$.

13.3 (a) $B = \$350,000$; (b) $SV = \$27,500$; (c) $D_t = \$10,750$; (d) $BV_{20} = \$135,000$.

13.9 For DDB depreciation;

t	D_t	BV_t	d_t
1	$3,000.00	$9,000.00	0.2500
2	2,250.00	6,750.00	0.1875
3	1,687.50	5,062.50	0.1406
4	1,265.63	3,796.88	0.1055
5	949.22	2,847.66	0.0791
6	711.91	2,135.74	0.0593
7	135.74	2,000.00	0.0113
8	0	2,000.00	0

13.11 (a) $BV_{13} = \$216,000$; (b) $BV_{13} = \$130,499.22$.

13.12 (a)

t	D_t DDB	D_t DB	BV_t DDB	BV_t DB
2	$8,099	$6,719	$64,790	$67,897
7	4,494	4,192	35,954	42,358
12	2,494	2,615	19,952	26,426
18	0	1,485	15,000	15,000

13.13 (a)

t	D_t	BV_t
1	$21,875.00	$153,125.00
12	5,035.43	35,248.02

(b) SV for DB is larger.

13.15 (a) SL: $BV_t = 25,000 - t(2500)$ for $t = 1$ to 10.
(b) DDB uses $d = 0.7143$.

t	0	1	2	3	4	5	6	7
BV_t	$25,000	17,857	12,755	9,111	6,508	4,648	3,320	2,372

13.17 $P_D = \$64,211$.

13.18 (a) No; (b) $d < 0.04425$.

13.19 (a) Switch in year 7 with $P_D = \$5897$; (b) Switch in year 5 with $P_D = \$5499$.

13.21 (b) SL yields $P_D = \$12,577$ and MACRS yields $P_D = \$14,105.55$.

13.23 (a) $P_D = \$13,802$; (b) $P_D = \$11,125$ for DDB.

13.25 $45,000.

13.27 Selected answers are as follows:

t	D_t	BV_t	d_t
1	$2,222.22	$9,777.78	8/36
4	1,388.89	4,777.78	5/36
8	277.78	2,000.00	1/36

13.29 (a) $4714; (b) 1/7; (c) $1457.

13.31 Expense = $10,000; D_1 = $21,160, ..., D_6 = $6136.

13.33 (a)

t	1	2	3	4	5
Depletion	$70,000	91,000	104,400	129,600	120,960

(b) 14.74%

13.34

t	D_t	BV_t
1	$5,000	$77,000
2	6,250	70,750
3	4,375	66,375

Chapter 14

14.1 (a) Recaptured depreciation; (b) taxable income; (c) operating loss; (d) taxable income; (e) capital loss; (f) taxable income.

14.3 Dough: no difference; Broke: −9.93%.

14.4 (a) 32.3%; (b) 39%; (c) 33.2%; (d) $24,400.

14.5 $636,550.

14.7 $780 increase.

14.8 $29,952.

14.9 (a) LTL = $1000; (b) STL = $333 (c) STG = $800; (d) RD = $300; (e) STL = $1750; taxes = $6015.

14.11 LTG = $5000; RD = $2000; asset 2 has RD = $9000; LTG = $1000.

14.13 CFBT = $1,905,797.

14.15 DDB: P_{tax} = $43,541; MACRS: P_{tax} = $45,388.

14.16 MACRS is $3556 lower in P_{tax} value.

14.17 SYD is $743 lower in P_{tax} value.

14.19 P_{tax} = $12,094.

14.21 P_{TS} = $2326.

14.23 P_{tax} = $1999 for n = 3 years is lower.

14.25 MACRS with n = 5 years with P_{tax} = $44,499.

14.27 57.9%.

Chapter 15

15.1

Year	0	1–7	8
Cash flow, $	−350,000	17,460	385,940

15.3

Year	0	1–2	3	4	5
CFAT, $	−16,000	13,700	32,100	4,500	−25,500

15.5

Year	0	1–5	6
CFAT, $	−10,000	3,560	4,199

15.7

Year	0	1–5	6
CFAT, $	−5,000	2,187	5,675

15.9 Select used ($EUAW_U$ = $10,031 is slightly smaller).

15.11 (a) Select Round with PW_R = $−2043; (b) $EUAW_R$ = $−300.

15.13 Select B with $PW_B = \$9841$.
15.15 $i^* = 5.54\%$.
15.17 (a) $i^* = 42.1\%$; (b) 28%.
15.19 (a) $i^* = 9.75\%$; (b) B, B, A, A.
15.21 0.12% is breakeven i value.
15.22 (a) Plan A: 19.15%, plan B: 24.26%; (b) 33.52%.
15.23 (a) $B = \$42,666$; (b) $SV = \$50,000$; (c) $D = \$13,438$.
15.25 None.
15.26 (a) Defender, $EUAW_D = \$1362$; (b) defender, $EUAW_D = \$866$.
15.27 Plan III with $EUAW = \$3087$.
15.29 (a) Defender, $EUAW_D = \$7794$; (b) more than B by $\$26,758$.
15.31 Retain defender 2 years.
15.33 $RR = \$14,300$
15.35 Challenger with $EUAW_C = \$4230$.

Chapter 16

16.1 (a) 1,200,000 units; (b) $\$300,000$.
16.3 (a) $r = \$53$ per unit; (b) profit $= \$60,000$.
16.5 (b) $Q_p = 1571$ units, $p = \$17,284$.
16.7 2486 hours per year.
16.9 94 days per year.
16.11 (a) 14.52 years; (b) $\$94,450$.
16.12 (a) 333 samples per year; (b) 202 samples per year; (c) select present condition.
16.13 $\$288,557$.
16.15 $\$850.79$.
16.17 (a) 25.9 days per year; (b) $\$34,744$ for 52 weeks per year.
16.19 $n = 5.66$ years.
16.21 (a) $n = 15.55$ years; (b) no.
16.23 $n = 3.9$ years.
16.25 (a) Purchase $n = 5.9$ months, lease $n = 8.65$, 17.01, or 25.06 months (multiple answers); (b) select purchase with $EUAW = \$155.50$ per month.
16.27 Select B, level 1, with $EUAW = \$481,054$.

Chapter 17

17.1 Select A and C with $P = \$23,749$.
17.3 Select 1, 2, and 4 with $P = \$2240$.
17.5 Select 3 with $P = \$4261$.
17.7 Select C and D.
17.9 Select 3 with incremental $i^* > 15\%$.
17.13 (a) $X_1 = X_3 = 1$, $X_2 = X_4 = 0$; (b) $X_1 = X_3 = X_4 = 1$, $X_2 = 0$.

Chapter 18

18.1 (a) 2%; (b) 20%, 28%.
18.5 Scheme 2 with $WACC = 8.75\%$.
18.7 10.275%.

18.9 (*b*) D-E mix range of 60–40% to 65–35%.
18.10 (*a*) $2,083,333; (b) $i = 4.48\%$.
18.11 Recommend bonds with $i = 2.9\%$.
18.13 (*a*) $i = 9.4\%$; (b) i = 5.4%.
18.15 Dividend: 9.52%; CAPM: 11.4%.
18.17 $i = 12.2\%$, yes.
18.18 100% equity: $i = 8.5\%$, no; D-E = 40–60%: $i = 15.9\%$, yes.
18.19 Plan 1: $i = 3.6\%$, no; Plan 2: $i = 19.9\%$, yes; Plan 3: $i = 100\%$, yes.
18.21 Use plan *a* with $i = 3.02\%$ on a 6-month basis.
18.22 (*a*) 18.99%; (*b*) 9.87%.
18.26 In bank name order: 57.6%, 67%, 40%.
18.27 *A*: D-E = 33–67%; *B*: 82.4–17.6%.

Chapter 19

19.1

Tons	10	15	20	25	30
PW, $	−134,922	−137,589	−140,257	−142,924	−166,078

19.3 Buy using estimate of *C* with $PW_C = \$6428$.

19.5

G, $	300	400	500	600	700
i*, %	42.0	41.5	41.0	40.4	39.8

19.7

Life, *n*	4	6	8	10	12
EUAW₁	$3,832	2,811	2,307		
EUAW₂		$4,610	3,746	3,239	2,907

19.8 Breakeven is 3.35 days per year.
19.9 (*a*) Yes; (*b*) breakeven is now 9.4 days per year, which is sensitive.
19.11 Decision is sensitive; PW = $−956 for $n = 10$ years; PW = $1935 for $n = 6$ years.

19.13

G, $	60	80	100	120	140
n*	16.3	14.1	12.6	11.5	10.7

19.15 Same decision as 6.7; used is less.

n	2	3	4
EUAW_used, $	−15,542	−15,251	−15,035

19.17

AOC or lease cost, $/year	EUAW			
	50 days per year		100 days per year	
	Plan 1	Plan 2	Plan 1	Plan 2
1,800	. . .	$3,800	. . .	$5,800
2,000	$3,664			
2,500	4,164	4,500	. . .	6,500
3,200	. . .	5,200	. . .	7,200
3,750	5,414			
4,000	. . .	. . .	$5,664	
5,000	. . .	. . .	6,664	
7,500	. . .	. . .	9,164	

(*a*) Plan 1; (*b*) plan 2.

19.19

Interest rate or life	Plan *A*	Plan *B*
(a) Pessimistic (15%)	$-13,317	$-37,293
Reasonable (10%)	-14,867	-38,600
Optimistic (8%)	-15,916	-39,472
(b) Pessimistic (44, 22)	-14,908	-38,519
Reasonable (40, 20)	-14,867	-38,600
Optimistic (32, 10)	-14,716	-38,762

19.21 $E(X) = 10.335$.

19.23 $E(\text{EUAW}) = \$-15,250$.

19.25 $E(\text{EUAW}) = \$-1169$.

19.27 $E(\text{PW}) = \$4389$, yes.

19.29 Build for 2.5 inches with EUAW = $1403.

19.30 Stocks with $E(\text{return}) = 10.7\%$

19.31 Top: $E(\text{value}) = 5.0$; bottom: $E(\text{value}) = -6.8$.

19.33 Select *y* with $E(\text{PW of D4}, y) = \$28,160$.

19.35 (*a*) High production at $14; (*b*) yes.

19.37 (*b*) No expansion: $E(\text{PW}) = \$86,960$; expand $E(\text{PW}) = \$28,700$; (*c*) select produce with $E(\text{PW})\ \$-47,937$.

Chapter 20

20.4 Yes; same plan, *B*, selected, but person 2 indicates more than 20% return over longer horizon.

20.5 (*a*) $1,036,250; (*b*) planning FW = $196,907; equipment PW = $865,935; (*c*) FW = $1,479,901; (*d*) approximately the same.

20.7 Tangible benefits: scrap, labor, material reduction; intangible benefits: improved delivery schedule and customer image.

20.9 0.39, 0.14, 0.20, 0.27.

20.11 Alternative 2; same choice.

20.13 (*a*) *A*; (*b*) *B*; (*c*) *B*; (*d*) *A*.

20.15 (*b*) President: 3; vice president: 2; (*c*) their perspectives are quite different.

20.17

	ROR (%)	
Variation	(a) Purchase	(b) Selling
-25%	10.55%	<0%
0	1.91	1.91
25%	<0	11.09

20.19

	EUAW$_M$		
Variation	(a) Purchase	(b) AOC	(c) Revenue
-50%	—	—	$-12,876
-30	$+11,964	$+11,124	—
0	+ 6,624	+ 6,624	+ 6,624
25	—	—	+16,374
30	—	+ 2,124	—
50	- 2,276	—	—
100	-11,177	—	—

Appendix B

B.3 Year	1	2	3	4
ACRS	$37,500	18,500	0	0
MACRS	$33,350	11,100	3,700	0

B.5 $3,640 loss.

B.7 Switch to SL in year 4, where DDB and SL rates are 0.115.

Appendix C

C.1 (a) Current assets = $77,000, fixed assets = $1,055,000, liabilities = $603,000, equity = $529,000; (b) $1,132,000 = 603,000 + 529,000.

C.3 (a) Net profit = $45,000; (c) 9%.

C.5 Inventory turnover = 16.13; (a) income ratio = 9%; (b) 9%.

C.7 (a) Space (February), DLC (March, April), space (May), material (June).

C.9 (a) $5831 (M101), $8750 (M102), $4063 (M103), $8125 (M104); $1769 under; (b) $23,430; $1570 over.

C.11 (a) $12,500 (1), $23,750 (2), $29,000 (6); (b) $90.

C.13 (a) Select make at cost of $65,910; (b) continue to make (new equipment cost is $71,669 per month).

Bibliography

Books

AMERICAN TELEPHONE AND TELEGRAPH: *Engineering Economy*, 3d ed., McGraw-Hill, New York, 1980.

BARISH, N. N., AND S. KAPLAN: *Economic Analysis for Engineering and Management Decision Making*, 2d ed., McGraw-Hill, New York, 1978.

BUSSEY, L. E.: *The Economic Analysis of Industrial Projects*, Prentice-Hall, Englewood Cliffs, NJ, 1978.

CANADA, J. R., AND J. A., WHITE: *Capital Investment Decision Analysis for Management and Engineering*, Prentice-Hall, Englewood Cliffs, NJ, 1980.

COLLIER, C. A., AND W. B. LEDBETTER: *Engineering Cost Analysis*, Harper & Row, New York, 1982.

DeGARMO, E. P., J. R. CANADA, AND W. G. SULLIVAN: *Engineering Economy*, 7th ed., Macmillan, New York, 1984.

FABRYCKY, W. J., and G. J. THUESEN: *Economic Decision Analysis*, 2d ed., Prentice-Hall, Englewood Cliffs, NJ, 1980.

FLEISCHER, G.: *Engineering Economy: Capital Allocation Theory*, Wadsworth Publishing, Boston, 1984.

GRANT, E. L., W. G. IRESON, AND R. S. LEVENWORTH: *Principles of Engineering Economy*, 7th ed., John Wiley and Sons, New York, 1982.

KASNER, E.: *Essentials of Engineering Economics*, McGraw-Hill, New York, 1979.

KLEINFELD, I. H.: *Engineering and Managerial Economics*, Holt, Rinehart and Winston, New York, 1986.

NEWNAN, D.G.: *Engineering Economic Analysis*, 3d ed., Engineering Press, San Jose, CA, 1988.

OSTWALD, P. F.: *Cost Estimating*, Prentice-Hall, Englewood Cliffs, NJ, 1984.

REISMAN, A.: *Managerial and Engineering Economics*, Allyn and Bacon, Boston, 1971.

RIGGS, J. L., AND T. M. WEST: *Engineering Economics*, 3d ed., McGraw-Hill, New York, 1986.

SMITH, G. W.: *Engineering Economy: Analysis of Capital Expenditures*, 4th ed., Iowa State University Press, Ames, IA, 1987.

SPRAQUE, J. C., AND J. D. WHITTAKER: *Economic Analysis for Engineers and Managers*, Prentice-Hall, Englewood Cliffs, NJ, 1986.

STEVENS, G. T., JR.: *The Economic Analysis of Capital Investments for Managers and Engineers*, Reston Publishing, Reston, VA, 1983.

STEWART, R. D., AND R. M. WYSKIDA: *Cost Estimator's Reference Manual*, John Wiley and Sons, New York, 1987.

TAYLOR, G. A.: *Managerial and Engineering Economy: Economic Decision-Making*, 3d ed., Van Nostrand, New York, 1980.

THUESEN, H. G., AND W. J. FABRYCKY: *Engineering Economy*, 7th ed., Prentice-Hall, Englewood Cliffs, NJ, 1988.

WHITE, J. A., M. H. AGEE, AND K. E. CASE: *Principles of Engineering Economic Analysis*, 3d ed., John Wiley and Sons, New York, 1989.

Selected Periodicals and Annual Publications

American Journal of Agricultural Economics, American Agricultural Economics Association, Lexington, KY, 5 issues per year.

Engineering Economy Abstracts, Industrial Engineering Department, Iowa State University, Ames, IA.

Harvard Business Review, Harvard University Press, Boston, 6 issues per year.

Journal of Finance, American Finance Association, New York, 5 issues per year.

Journal of Financial and Quantitative Analysis, University of Washington, Seattle, WA, 5 issues per year.

Lasser, J. K., *Your Income Tax*, Prentice-Hall, New York.

Public Utilities Fortnightly, Public Utilities Reports, Washington, DC, fortnightly.

Tax Information on Corporations, U.S. Internal Revenue Service Publication 542, Government Printing Office, Washington, DC.

The Engineering Economist, Institute of Industrial Engineers, Norcross, GA, 4 issues per year.

U.S. Master Tax Guide, Commerce Clearing House, Chicago.

Index

Index

A/F factor, 28–30
A/G factor, 34–36
A/P factor, 28–30
Accelerated Cost Recovery System (ACRS), 280, 432
Accelerated write-off, 271, 274–284
Accounting, 477–488
  ratios (factors) used in, 480–483
  statements used in, 478–480
Acid-test ratio, 481
After-tax analysis, 308–320
  rate of return, 313–316
After-tax replacement analysis, 316–318
Alternatives:
  comparison: by benefit/cost ratio, 202–205
    by capitalized-cost method, 122–124
    by decision trees, 386–394
    by equivalent-uniform-annual-worth method, 139–142
    by payback-period method, 338–341
    by present-worth method, 116–120
    by rate-of-return method, 172–188
  definition, 4
  independent (nonmutually exclusive), 173
  multiple (see Multiple alternatives)
  mutually exclusive, 173, 182–188
  strategic, 406–413
  unweighted evaluation, 415
  weighted evaluation, 415
Amortization (see Depreciation)
Analytic hierarchy process, 417–419
Annual worth, equivalent uniform (EUAW), 134–142
  advantages, 135
  after-tax analysis using, 312
  alternative evaluation by, 139
  capital-recovery-plus-interest method, 138
  capitalized cost and, 120–124
  incremental rate-of-return analysis, 181–182
  inflation considered, 254–255
  salvage present-worth method, 137
  salvage sinking-fund method, 136
Asset Depreciation Range (ADR), 276, 471
Assets:
  book value, 270, 272, 274
  depreciation of, 269–286
  evaluation, 367–369
  replacement studies and, 212–222, 316–318
  sunk cost for, 214

Balance sheet:
  basic equation for, 479

Balanced sheet (*continued*)
  categories on, 479
  ratios from, 480–483
Basis, unadjusted, 271
Before-tax rate of return, 302
Benefit/cost ratio:
  alternative evaluation by, 202–205
  calculation of, 198–200
  classifications for, 198
  incremental analysis of, 204–205
  interpretation of, 202
  modified benefit/cost, 198–200
  for three or more alternatives, 204
Benefits, 198–200
  and cost difference, 199–200
Bonds:
  classification and types of, 234–235
  interest computation for, 235–236
  present worth of, 236–239
  rate of return on, 239–240
  Standard and Poor's rating of, 234
  treatment of, in debt financing, 365
Book value:
  by declining-balance method, 274
  definition of, 270
  by double-declining-balance method, 274
  by MACRS, 282, 472
  by straight-line method, 272
  by sum-of-year-digits method, 283
Borrowed money (*see* Debt capital; Equity capital)
Breakeven analysis, 330–338
  general description of, 330, 335
  one variable, 330–335
  in rate-of-return comparisons, 179–180
  three or more alternatives, 338
  two alternatives, 335–338

Capital asset pricing model (CAPM), 367–369
Capital, cost of (*see* Cost of capital)
Capital budgeting:
  general problem of, 351
  math programming formulation, 356–358
  present-worth method, 351–355
  rate-of-return method, 355
Capital expensing, 285
Capital financing:
  cost of capital (*see* Cost of capital)
  debt, 361, 363–366
  equity, 361, 366–368

Capital gains and losses:
  short-term and long-term, 297
  taxes for, 296–298
Capital management (*see* Capital budgeting)
Capital-recovery factor (A/P), 28–30
Capital recovery for assets (*see* Depreciation)
Capital-recovery-plus-interest method, 138
Capitalized cost:
  alternative evaluation by, 120–124
  definition, 120
Carry-back and carry-forward income tax
    provisions, 298
Cash flow:
  after taxes, tabulation of, 298, 309–311
  continuous (*see* Continuous compounding of
      interest)
  conventional sequence of, 160
  discounted, 115
  net, definition of, 14, 174–176
  tabulation of, 174
  (*See also* Cash-flow diagrams)
Cash-flow diagrams, 14–16
  partitioned, 43–44, 95
Cash-flow models, 494
Cash-flow replacement analysis, 218–220
Cash-flow sequence:
  conventional, 160
  nonconventional, 160
Challenger:
  in multiple-alternative evaluation, 182–188
  in replacement analysis, 214
Class life, 271, 276
Collateral bonds, 234
Common multiple of lives, 117
Common stock, 366
Composite rate of return, 163
Compound-amount factors:
  single-payment (F/P), 25
  summary of, 29–30, 429
  uniform-series (F/A), 28
Compound interest, 9–11
Compounding:
  annual, 10, 60
  doubling time, 11
Compounding period, 60
  longer than payment period, 71–74
  shorter than payment period, 67–71
Computer integrated manufacturing, 407–410
Computer programs, 504–506
Consultant's viewpoint, 214

Continuous compounding of interest:
   compared to discrete cash flow, 74–75
   effective interest rate for, 66–67
Conventional cash-flow sequence, 160
Conventional replacement analysis, 218–220
Convertible bonds, 234
Cost of capital:
   for debt financing, 365
   definition of, 363
   effect of debt-equity mix on, 369–371
   effect of income taxes on, 365
   for equity financing, 366
   weighted average, 363–365
Cost-capacity equations, 258–259
Cost-of-goods-sold statement, 480
Cost allocation (*see* Factory expense)
Cost center, 483
Cost estimation:
   cost-capacity method, 258
   cost indexes and, 258–259
   factor method, 259
Cost indexes, 255–260
Costs:
   capitalized (*see* Capitalized cost)
   EUAW (*see* Annual worth, equivalent uniform)
   factory overhead (*see* Factory expense)
   fixed, 330
   as function of life, 222, 410–412
   incremental (*see* Incremental costs)
   indirect, 412
   life-cycle, 341–342
   sunk, in replacement studies, 214
   variable, 330
Current assets, 478
Current ratio, 481

Debenture bonds, 234
Debt capital, 361, 363–366
Debit-equity mix, 363–365, 369–371
Decision trees, 386–394
Declining-balance depreciation, 274–275
Decreasing gradients, 94–96
Defender:
   in multiple-alternative evaluation, 182–188
   in replacement analysis, 214
Depletion, 286–287
Depletion factor, 286
Depreciable property, disposition of, 276

Depreciation, 269–286
   ACRS, 280, 470
   Asset Depreciation Range (ADR) system, 276
   class life, 271, 276
   declining-balance, 274–275
   definition of, 270
   double-declining-balance, 274–275
   effect of time on, 270–272
   half-year convention, 281, 433
   income taxes and, 299–302
   MACRS, 280–282, 471–475
   mid-quarter convention, 471
   modified straight-line, 282
   rate of, 271
   recaptured (*see* Recaptured depreciation)
   revenue requirements and, 319
   straight-line, 271–273
   sum-of-year-digits, 283–284
   switching between methods, 277–279, 471–473
   tax effects of, 299–302
   types of property, 276
   unit-of-production method of, 287
Depreciation rate, 271, 474
Disbenefits, 198–200
Discount rate, 115
Discounted-cash-flow method, 115, 407
Discrete cash flows:
   continuous compounding, 66–67
   discrete compounding, 62–66
Do-nothing alternative, 183
Double-declining-balance method, 274–275

Effective income tax rate, 295
Effective interest rate, 60–62
   computation of, 62–66
   for continuous compounding, 66–67
   equation for, 63
   interpretation of, 61
End-of-period convention, 14
Equal-service alternatives, 116
Equipment trust bonds, 234
Equity capital, 361, 366–368
Equity ratio, 481
Equivalence, 7–9
Equivalent uniform annual series (*see* Annual worth,
   equivalent uniform)
Escalating series, 36–37, 45, 91–94
EUAW (*see* Annual worth, equivalent uniform)
Evaluation criterion, 4, 407

Expected value:
  computation of, 384
  decision trees and, 388–394
  in economy studies, 385–386
Expenses (*see* Costs)
Extra investment, rate of return on, 176–188

F/A factor, 28
F/P factor, 25
Face value of bonds, 235
Factor cost estimation, 259–260
Factor depletion, 286
Factors:
  continuous compound interest, 66–67
  discrete compound interest, 24–37
  intangible, 5
  multiple, use of, 80–96
  present-worth (*see* Present-worth factors)
  standard notation for, 29–30, 429
  (*See also* Interest factors)
Factory expense:
  allocation of, 483
  overhead rates, 484
  variance, 485–486
Financing:
  debt, 361, 363–366
  equity, 361, 366–368
First cost, 271
Fixed assets, 478
Fixed costs, 330
Future worth:
  computation of, 25, 28, 39
  definition, 12, 25–26
  inflation considered, 251–254
  placement of, 82

Gains and losses (*see* Capital gains and losses)
General obligation bonds, 234
Gradients:
  base amount, 31
  conventional, 31
  decreasing, 94–96
  definition, 30
  derivation of factors for, 30–37
  equivalent uniform annual series for, 225
  escalating series, 36–37, 91–94
  n value, 42, 88
  present worth of, 42–45, 88–91

  shifted, 88–91
  use of tables, 35
Gross income, 293

Half-year convention, 281, 471

Income ratio, 482
Income statement:
  basic equation for, 480
  categories of, 479
  ratios from, 482
Income taxes:
  capital gains and losses and, 296–298
  cash flow after, 298, 309–311
  corporate, 294–296
  definitions of, 269, 293–294
  depreciation and, 299–302
    recaptured, 294
  effective rates of, 295
  investment tax credit, 285
  present worth of, 299
  rate of return considering, 313–316
  replacement studies and, 316–318
  revenue requirements and, 318–320
  state and local, 295
Incremental benefit/cost analysis for
  multiple alternatives, 203–204
Incremental costs:
  B/C analysis and, 203–204
  definition of, 173
  rate-of-return analysis, 176–188
Incremental effective income tax rate, 295
Incremental rate of return:
  for multiple alternatives, 182–188
  for two alternatives, 173–174, 176–182
Independent alternatives, 173, 351
Indirect costs, 260, 412
Inflated before-tax rate of return, 302
Inflated interest rate, 248
Inflation:
  annual worth with, 254–255
  definition of, 247
  future worth with, 251–254
  present worth with, 247–250
  uniform series calculations, 254–255
Intangible factors, 5, 410–413
Integer linear programming, 356–358
Interest:
  compound, 9–11

Interest (*continued*)
  computation of, for bonds, 235
  continuous compounding of (*see* Continuous
    compounding of interest)
  definition, 10
  rate of (*see* Interest rates)
Interest factors:
  for continuous compounding (*see* Continuous
    compounding of interest)
  derivation of, 24–49
  for discrete compounding, 429
  symbols for, 29–30
Interest period, 6, 60
Interest rates:
  definition, 5–11
  determination of unknown, 37, 156–160
  effective (*see* Effective interest rate)
  estimation for trial-and-error solutions, 157
  inflated, 248
  nominal, 60–62
  real, 251
Interest tables:
  discrete compounding, 429–468
  interpolation in, 37
Internal rate of return, 162
Interperiod interest, 71–74
Interpolation, 37–39
Inventory turnover ratio, 482
Investment tax credit, 285

Junk bonds, 235

Large investments, 405–420
Least-cost life, 222–223
Least-common multiple, 117
Leverage, 369–371
Life:
  minimum cost, 222–223
  remaining (*see* Replacement studies)
Life cycle, xviii, 135, 341
Life-cycle costing (LCC), 341–342
Lives:
  common multiple of, 117
  comparing alternatives: with equal, 116
    with perpetual, 120–122, 135
    with unequal, 117–120, 135

MACRS (Modified Accelerated Cost Recovery
    System):
  depreciation, 280–282, 471–475

  depreciation rate derived, 474–475
  straight-line alternative, 282
Market value, 214, 270
Math programming in capital budgeting, 356–358
Mid-quarter convention, 471
Minimum attractive rate of return (MARR):
  comparison to other returns, 362
  use of: for capital budgeting, 351–356
    for mutually exclusive alternatives, 182–188
  variations in, 362
Minimum cost life of asset, 222–223
Modified benefit/cost ratio, 198–200
Mortgage bonds, 234
Multiple alternatives:
  benefit/cost analysis for, 202–205
  comparison of: using incremental benefit/cost
    ratio, 204–205
    using incremental rate of return, 176–188
  independent, 173, 351
  mutually exclusive, 182–188
Multiple-criteria evaluation, 413–419
Multiple rate-of-return computation, 160–166
Municipal bonds, 234
Mutually exclusive alternatives, evaluation of,
    182–188
Mutually exclusive bundles in capital budgeting,
    351–352

Net cash flow, 174–176
Net present value, 115, 405, 419
Net worth, 479
Nominal interest rate, 60–62
Nonmutually exclusive (independent) alternatives,
    173, 351–355
Nonconventional cash-flow sequence, 160

One-additional-year replacement studies, 220–222
Operating loss, 294, 298
Optimistic estimate, 382
Overhead rates, factory, 484–485

P/A factor, 28
P/F factor, 26
P/G factor, 34–36
Payback analysis:
  calculations for, 338–340
  limitations of, 120, 340
  payback period, 119–120, 338–339
Payment period, 62

Payout period (*see* Payback analysis)
Percentage depletion, 287
Perpetual investment (*see* Capitalized cost)
Personal property, 276
Pessimistic estimates, 382
Planning horizon (*see* Study period)
Portfolio selection (*see* Capital budgeting)
Preferred stock, 366
Present worth:
    after-tax analysis using, 312
    alternative evaluation by, 116–120
    bond, 236–239
    in capital budgeting, 351–355
    computation of: for equal lives, 116
        for unequal lives, 117
    of escalating series, 45, 91
    of income taxes, 299–302
    incremental-rate-of-return analysis, 177–180
    inflation considered, 247–250
    placement of, 81–83
Present-worth factors (P/F, P/A, P/G):
    gradient, 30–37, 88–96
    single-payment, 26
    uniform-series, 28
    use of tables for, 29
Probability:
    decision tree nodes, 387
    expected value and, 384
    use of, in economy studies, 385–394
Profit-and-loss statement (*see* Income statement)
Project net investment, 163–166

Rate of depreciation:
    declining-balance, 274
    double-declining-balance, 274
    MACRS, 281, 474
    straight-line, 273
    sum-of-year-digits, 283
Rate of return:
    after-tax, 313–316
    on bond investments, 239–240
    composite, 162–166
    by EUAW, 159–160
    incremental analysis and, 173
    on debt capital, 365
    definition of, 154
    on equity capital, 366
    on extra investment, 173, 176–188
    guessing for trial-error solutions, 157
    inflated, before taxes, 302

    internal, 162–166
    minimum attractive (*see* Minimum attractive
        rate of return)
    multiple values, 160–166
    by present worth, 155–159
    reinvestment rate, 163
    use of, in capital budgeting, 355
Real interest rate, 251
Real property, 276
Reasonable estimate, 382
Recaptured depreciation:
    versus capital gain, 294
    definition of, 294
    taxes for, 294
Recovery period, 271, 301–302, 433
Recovery rate in depreciation, 271
Reinvestment rate, 163–166
Replacement, reasons for, 212
Replacement analysis:
    after-tax, 316–318
    before-tax, 212–222
    cash-flow approach, 218–220
    conventional approach, 218–220
    equal-lived assets, 215
    gains, losses, and recaptured depreciation in, 316
    one-additional-year, 220–222
    and replacement value, 219
    sunk costs in, 214
    unequal-lived assets, 216
    use of: market value in, 214
        study periods in, 215–218
Replacement value, 219
Retained earnings, 478
Revenue bonds, 234
Revenue requirements, 318–320
Risk, 362, 384–394
Rule of 72, 11

Salvage present-worth method, 137
Salvage sinking-fund method, 136
Salvage value, 271
Section 179 property capital expense deduction,
    285
Sensitivity analysis:
    approach, 379, 493
    of large investments, 419–420
    of one factor, 379–381
    of several factors, 382, 492–503
    using three estimates, 382–384
Shifted gradients, 88–91

Simple interest, 11
Single-payment factors (P/F, F/P):
    derivation of, 25–27
    summary of, 29
Sinking-fund factor (A/F), 28–30
Standard factor notation, 29
Stock volatility, 367
Stocks:
    common, 366
    in equity financing, 366
    preferred, 366
Straight-line depreciation, 271–273
Strategic decisions, 406–419
Stripped bonds, 236
Study period:
    capital budgeting, 353
    capitalized cost studies, 120
    EUAW evaluation, 135–136
    PW evaluation, 117–119
    replacement analysis, 215–218
    (see also Planning horizon)
Sum-of-year-digits depreciation, 283–284
Sunk cost, 214
Switching between depreciation methods, 277–279,
    471–473
Symbols in engineering economy, 12

Tangible versus intangible benefits, 5, 410–413
Tax credit, investment, 285

Taxable income, 293, 309
Taxes (see Income taxes)
Time placement of dollars, 12, 14
Time value of money, 5

Unadjusted basis, 271
Uncertainty in estimates, 379, 384
Uniform gradient (see Gradients)
Uniform-series factors, 27–29
Unit-of-production depreciation, 287
Unknown interest rate, 46, 153–160
Unknown years (life), 48
Unrecovered balance, 154
Unweighted evaluation method, 415

Variable costs, 330
Variance in overhead allocation, 485–486
Variations in economic decisions, 479

Weighted average cost of capital, 363–365
Weighted evaluation method, 415

Zero-coupon bond, 236